Functions of Bounded Variation

Theory · Methods · Applications

Department of Mathematics
Julius-Maximilians-University
Würzburg, Germany

Functions of Bounded Variation
Theory · Methods · Applications

Dissertation submitted by

Simon Reinwand

in Partial Fulfillment of the Requirements for the Degree of
Doctor of Natural Sciences
(Doctor rerum naturalium)
in Mathematics

March 25, 2021

Supervisor: Prof. Dr. Jürgen Appell (University of Würzburg, Germany)
Co-Supervisor: Prof. Dr. Daria Bugajewska (University of Poznań, Poland)
Co-Referee: Prof. Dr. Gianluca Vinti (University of Perugia, Italy)

Bibliografische Information der Deutschen Nationalbibliothek
Die Deutsche Nationalbibliothek verzeichnet diese Publikation in der Deutschen Nationalbibliografie; detaillierte bibliografische Daten sind im Internet über http://dnb.d-nb.de abrufbar.
1. Aufl. - Göttingen: Cuvillier, 2021
Zugl.: Würzburg, Univ., Diss., 2021

Nonnenstieg 8, 37075 Göttingen
Telefon: 0551-54724-0
Telefax: 0551-54724-21
www.cuvillier.de

1. Auflage, 2021
Gedruckt auf umweltfreundlichem, säurefreiem Papier aus nachhaltiger Forstwirtschaft.

ISBN 978-3-7369-7403-6
eISBN 978-3-7369-6403-7

"Ich leb' allein in meinem Himmel, in meinem Lieben, in meinem Lied."

Friedrich Rückert

Preface

The doctoral thesis at hand is the result of my research during the time as a doctoral student at the Department of Mathematics of the University of Würzburg. Of course, during a long research project like this, there are always some ups and downs, and one is lucky to have some people around who help to overcome the downs and share the ups. Therefore, I would like to take this opportunity to thank several people who have supported me to write this thesis, either directly or indirectly.

First of all, I would like to express my utmost gratitude to my supervisor Jürgen Appell for his outstanding support and guidance through the last five years both in mathematical and in personal matters. He provided me with a very interesting and challenging research project, shared his time generously and often ignited new ideas by fruitful discussions. I could substantially benefit from his great experience in the fields of functional analysis as well as operator and fixed point theory which guided me carefully on the path that eventually led to this thesis. Several joint publications are witnesses of an inspiring collaboration which clearly had a big impact on the material presented here.
Moreover, I would like to thank Jürgen for his mentoring and motivating manners which were always embossed with infectious and immortal humor, honest and impartial critique, patience and endurance, respect and acceptance on both a scientific and a personal level as well as with an admirable skill to always find the right words within the right moment.

But I thank Jürgen also for his engagement that extends far beyond plain discussions. First, through his immense network of scientific relationships to mathematicians all around the globe I got to know and collaborate with a couple of prominent researchers in my field of work like, for example, Umberto Mosco and the research group around Dariusz Bugajewski and Daria Bugajewska. Especially Daria also deserves my deepest appreciation for agreeing to co-referee this dissertation as well as for welcoming me several times in Poznań with open arms full of the greatest hospitality and support. Second, Jürgen also assisted me in getting a scholarship of the German Academic Scholarship Foundation (Studienstiftung) which provided not only a financial backup but also a huge range of intangible values that eventually ended into the offering of being a one-time member of the commission to select new scholarship holders. Third, our joint participation in the "Research in Pairs" program together with two of our Polish colleagues and friends was definitely an unforgettable experience from which most of the mathematical ideas in this thesis and some joint publications arose. In particular, I would like to thank the Mathematical Research Institute in Oberwolfach, where the four of us spent a couple of weeks and enjoyed the unique spirit, excellent working facilities, and overwhelming hospitality.

Apart from my supervisor there are some more persons to whom I owe my gratitude since this entire project could hardly have been realized without their support. First, I would like to thank Jens and Richard for their continuous trust giving me every semester the opportunity to teach and work with students. Furthermore, I thank my friends Carolin for many amusing discussions, shopping experiences and horror movie nights, Jochen for always helping me out whenever professional obligations demanded too much as well as Stefan for reliably having an open ear in both dark and bright moments.

A very special thanks goes to my closest friend and partner Helmut to whom I feel deeply indebted for caring through the lows and for sharing all the highs, for being fun and childish as well as kind and serious when needed.

I thank my aunt Gitti who - in total contrast to me - never had a single bit of doubt as to the quality of my work. It is a matter of particular concern to me to also thank my aunt Marliese who since my childhood had been a motherlike and nearby anchor, but, unfortunately, had to leave this world way too early.

Finally, I would like to express my deepest gratitude to my beloved parents for their generous financial support during my studies, for their everlasting unquestioning love and trust and for their unrestrained emotional backup.

Würzburg, March 25, 2021. Simon Reinwand

Contents

Introduction

Functions of bounded variation of a single variable, or short BV-functions, were first introduced in 1881 by Camille Jordan [74]. He extended a result about the pointwise convergence of Fourier series of periodic and piecewise monotone functions proven around 50 years earlier by Johann Peter Gustav Lejeune Dirichlet [52] who gave the first rigorous proof of a conjecture on the representability of functions by means of trigonometric series originally raised in 1808 by Jean Baptiste Joseph Fourier [62]. Jordan proved that the Fourier series of any 2π-periodic function $x : \mathbb{R} \to \mathbb{R}$ of bounded variation converges at each point to the arithmetic mean of the right and left sided limits of x; in particular, if x is continuous, then its Fourier series converges even uniformly to x. This is nowadays known as the Dirichlet-Jordan-Theorem. In the same paper, Jordan also proved that any function of bounded variation may be written as a difference of two monotonically increasing functions. In this sense, the class BV of all real-valued functions of bounded variation defined on the real interval $[0,1]$ is the linear hull of the set of monotone functions on that interval which do not form a linear space on their own.

The class BV has also been extended in many interesting directions. For instance, in the early 1920s, Norbert Wiener made the first noteworthy extension to Jordan's bounded variation concept by introducing the space WBV_2 of functions of bounded quadratic variation [154]. He proved that the Dirichlet-Jordan-Theorem still holds for functions of this type. In 1937, Laurence Chisholm Young showed that this theorem could be further extended to higher exponents and introduced the class WBV_p of functions of bounded p-variation for arbitrary $p \geq 1$ [159]. Together with Eric Russel Love, Young gave a comprehensive study of these functions [92] and finally went on to generalize Wiener's ideas by replacing the exponentiation by p by a composition with a suitable convex and increasing "gauge function" $\varphi : [0, \infty) \to [0, \infty)$ [160]. By doing so, he was hoping to extend the Dirichlet-Jordan-Theorem beyond the result he already proved for functions in WBV_p. In 1940, Raphaël Salem found a condition on φ ensuring that the Dirichlet-Jordan-Theorem holds for functions in the resulting more general space YBV_φ [139]. Moreover, in 1972, Albert Baernstein showed that among the YBV_φ-spaces, Salem's result concerning Fourier series is the best possible [19].

Also in 1972, Daniel Waterman extended the class of BV-functions in another direction by weighting the summands in Jordan's definition not by a composition but by a multiplication with a decreasing sequence Λ of positive numbers instead [151]. The resulting class ΛBV of such functions is of particular interest if one takes Λ to be the harmonic sequence $\Lambda = (1/n)_{n\in\mathbb{N}}$; in this case, ΛBV is denoted by HBV, and functions in this spaces are called "of bounded harmonic variation". Waterman showed not only that the Dirichlet-Jordan-Theorem about Fourier series holds for functions in HBV, he also pointed out that his result is best possible among all ΛBV-spaces. Moreover, he showed that for any Young function φ satisfying Salem's condition, the inclusion $YBV_\varphi \subseteq HBV$ holds. Consequently, among all the generalizations of the Dirichlet-Jordan-Theorem mentioned here, Waterman's version is the strongest.

Another notion of "bounded variation" has been introduced by Frigyes Riesz in 1910 [135, 136]. His type of variation seems very natural from a functional analytic point of view. In fact, an important result states that for fixed $p \in (1,\infty)$ a function $x : [0,1] \to \mathbb{R}$ is of bounded p-variation in the sense of Riesz if and only if x is absolutely continuous and its derivative x' belongs to the Lebesgue space L_p. In this case, we write $x \in RBV_p$, and the Riesz variation of x may be calculated explicitly by an integral over x'. Clearly, such a formula cannot be true for $p = 1$, because functions in $RBV_1 = BV$ are in general not continuous, let alone absolutely continuous.

Remarkably, any function $x \in RBV_p$ belongs to the Sobolev space $W^{1,p}$, and any function in $W^{1,p}$ in turn agrees almost everywhere with a function in RBV_p [56]. This means that RBV_p consists precisely of the continuous representatives of $W^{1,p}$. In this sense Riesz introduced Sobolev spaces, at least in the scalar case, around 25 years prior to Sobolev.

A very comprehensive overview about properties of functions of bounded variation and their various generalizations may be found in the monograph [6].

Besides the development of the theory of Fourier series, BV-type functions have been extensively studied also in other fields of mathematics, for instance, in geometric measure theory, calculus of variations, and mathematical physics. Renato Caccioppoli and Ennio de Giorgi used them to define measures of nonsmooth boundaries of sets [34, 35, 48]. Olga Arsenievna Oleinik introduced her view of generalized solutions for nonlinear partial differential equations as functions from the space BV [125], and was able to construct a generalized solution of bounded variation of a first order partial differential equation [126]. A few years later, Edward D. Conway and Joel A. Smoller applied BV-functions to the study of a single nonlinear hyperbolic partial differential equation of first order [44], proving that the solution of the Cauchy problem for such equations is a function of bounded variation, provided the initial value belongs to the same class.

But functions of bounded variation turn out to be useful even when it comes to questions from the very foundations of analysis. For instance, it is clear that the sum of two functions with primitive again has a primitive, but this is wrong when "sum" is replaced by "product". This raises the question what the multipliers of the set Δ of functions with primitive are, that is, how the functions $g : [0,1] \to \mathbb{R}$ look like such that xg belongs to Δ whenever x belongs to Δ. A discussion of these and more general questions will be the starting point of this thesis: We will discuss some natural "habitats" of functions of bounded variation and how they are related to other function classes.

This thesis is organized in seven chapters. The first chapter will be introductory in which we collect basic definitions, notations and function classes that we use the most. To be a little more precise we introduce in Section 1.1 the class C of continuous functions, the class B of bounded functions as well as the class D of Darboux functions (that is, functions with the intermediate value property) and discuss their relation to BV and to each other. For instance, the inclusions $BV \cap D \subsetneq C \subsetneq D \cap B$ hold, but none of these inclusions may be inverted. We also consider Lebesgue measurable and integrable functions, regular functions, absolutely and Lipschitz continuous functions and summarize how these classes are related to the class BV.

Section 1.2 is then devoted to functions of generalized bounded variation. We formally introduce the Wiener spaces WBV_p, the Young spaces YBV_φ, the Waterman spaces ΛBV and the Riesz spaces RBV_p. Equipped with a suitable norm building upon the corresponding type of variation, all these spaces become Banach spaces. Since functions which are zero everywhere except on a countable set become very important throughout this thesis, a major part of Section 1.2 is reserved for this kind of functions and how they behave in the various BV-type spaces. At the end of Section 1.2 we quickly discuss Helly's Selection Principle which provides a certain type of compactness in BV-spaces: Accordingly, every sequence in one of the BV-spaces that is bounded in its norm possesses a pointwise convergent subsequence.

The class Δ of derivatives to which we will give our main attention in Chapter 2 is situated between the classes C and D. From Lebesgue's and Riemann's integration theory it is well known that there are functions with primitive which are neither Lebesgue nor Riemann integrable. Consequently, in order to characterize the functions in Δ we need to pass in Section 2.1 to another notion of integration which will be functions that are integrable in the sense of Kurzweil and Henstock (KH-integrable). Every derivative is KH-integrable automatically and fulfills the Fundamental Theorem of Calculus. We also discuss another stronger form of integrability which enshrines both being KH-integrable and having a primitive. We then move on to other attempts that have been made in order to find integral free characterizations of the functions in Δ. However, it turns out that even if these attempts pretend to be integral free, they are in fact not. Nowadays, it is still not clear whether functions in Δ can be characterized without any kind of integration process; most mathematicians believe that this is impossible.

Moving on to more algebraic questions we discuss what happens when derivatives are multiplied or composed; we will do this in the Sections 2.2 and 2.3, where Section 2.2 is the largest part of this chapter. Therein we slowly approach a full discussion of the set Δ/Δ of multipliers of the class Δ as described above which simultaneously serves as a bridge to the class BV. Indeed, being continuous and of bounded variation is sufficient but not necessary to be a member of Δ/Δ, while there are functions in Δ/Δ that are bounded but neither continuous nor of bounded variation. In fact, functions in Δ/Δ turn out to be those functions that have a primitive and are in a certain sense of "local" bounded variation [59, 111].

Besides multipliers of the class Δ we also consider multipliers in other function spaces X and Y of real-valued functions on $[0,1]$. We denote by

$$Y/X := \left\{g : [0,1] \to \mathbb{R} \mid xg \in Y \text{ for all } x \in X\right\}$$

the multiplier set of Y over X. While we identify multiplier sets for some classical function spaces only in case $X = Y$ in Section 2.2 we pass in Section 3.1 to other combinations, where we also allow $X \neq Y$. Some of these combinations are easy to find. For instance, it is straightforward to show that $BV/BV = BV$, and that $D/B = C/B = C/BV$ contains only the zero function $\mathbb{0}$. However, other combinations are very difficult to find or even unknown, especially when $Y = D$. Here, the three classes D/C, D/Δ and D/D will be of particular importance for us. Some authors claim without proof that the class D/D is easily deduced from the following result due to Radaković [133]: If a function g has the property that $x + g$ is a Darboux function whenever x is a Darboux function, then g is constant. We show in Section 3.1 that D/D may indeed be deduced from Radaković's result, but this deduction is by far not so easy, especially when g has zeros. Moreover, since we do not know how the classes D/C and D/Δ look like, we give only partial results and show how they are related to other multiplier classes and function spaces.

Section 3.2 is then dedicated to multipliers of spaces of functions of generalized variation. Conveniently, the results are quite similar for all such spaces. Since all BV-type spaces considered in this thesis are algebras, we have $X/X = X$ whenever X is one of these spaces. On the other hand, if X and Y are two Wiener spaces, then $Y/X = Y$ for $X \subseteq Y$. If $X \not\subseteq Y$, then Y/X contains only functions from Y with countable support. The same is true if X and Y are two Young spaces or two Waterman spaces. We will also see that for two Riesz spaces X and Y the condition $X \not\subseteq Y$ yields the strong degeneracy $Y/X = \{\mathbb{0}\}$.

Especially for applications it is quite handy that many differential equations may be solved by rewriting them into integral equations. Those can then often be handled with fixed point theory, even in the space BV and its various generalizations. In order to use classical fixed point theorems like those named after Stefan Banach, Juliusz Schauder, Gabriele Darbo or Mark Alexandrovich Krasnoselskii, one has to check several sometimes complicated conditions on the linear and nonlinear operators involved. This has been done many times in the BV-type spaces mentioned above; we refer the reader to the work of the Polish mathematicians Daria Bugajewska, Dariusz Bugajewski and their colleagues [25, 26, 27, 29, 30, 31, 32, 33, 46].

However, many analytic and set theoretic properties of such operators are either extremely complicated to characterize or just unknown. While linear operators such as multiplication, substitution or integral operators are mostly relatively easy to handle, nonlinear operators like composition or superposition operators behave sometimes in a rather strange way. The aim of the Chapters 4 and 5 of this thesis is to extend the theory concerning properties of these operators in the various BV-spaces.

Here, we consider the following three linear operators in Chapter 4 on two function spaces X and Y of real-valued functions on $[0,1]$. The multiplication operator

$$M_g : X \to Y,\ M_g x(t) = x(t)g(t)$$

for a generating function $g : [0,1] \to \mathbb{R}$ in Section 4.1, the substitution operator

$$S_g : X \to Y,\ S_g x(t) = x\big(g(t)\big)$$

for a generating function $g : [0,1] \to [0,1]$ in Section 4.2, and the integral operator

$$I_g : X \to Y,\ I_g x(t) = \int_0^1 g(t,s)x(s)\,\mathrm{d}s$$

for a generating function $g : [0,1] \times [0,1] \to \mathbb{R}$ in Section 4.3. For all three operators we are particularly interested in analytic properties like acting conditions for various BV-spaces X and Y, as well as continuity (which is for linear operators equivalent to boundedness) and compactness. Especially for the multiplication operator the results of Chapter 3 will be useful: Indeed, a multiplication operator $M_g : X \to Y$ is well-defined if and only if its generator g belongs to the multiplier space Y/X. In particular, recalling the sample results from above, the operator M_g maps BV into itself if and only if $g \in BV$. Moreover, regarding compactness the operator $M_g : BV \to BV$ is compact if and only if the support of g is countable, while $S_g : BV \to BV$ is compact if and only if g has finite range. We show these and similar results for other BV-type spaces in the Sections 4.1 and 4.2 for the multiplication and substitution operator, respectively. But we also give some remarks on set theoretic properties like injectivity, surjectivity and bijectivity. For instance, $M_g : BV \to BV$ is injective, if and only if g has no zeros, while $S_g : BV \to BV$ is injective if and only if g is surjective. Thus, mapping properties of M_g may often be described in terms of the support of g, while mapping properties of S_g can often be characterized in terms of the image of g.
Especially for integral equations a comprehensive investigation of the integral operator I_g is of particular importance for us. Therefore, Section 4.3 is by far the largest section of Chapter 4. Our main concern is analytic properties, and from the aforementioned cited papers of the Polish mathematicians Bugajewska, Bugajewski and colleagues many sometimes quite technical conditions are known guaranteeing that the integral operator maps a BV-space into itself and is bounded or compact. For instance, if $g(t,\cdot) \in L_1$ for any $t \in [0,1]$ and the variation of the function $g(\cdot,s)$ is almost everywhere bounded with respect to s by some L_1-function, then I_g maps BV into itself and is bounded and compact. Similar results are known for a few other BV-spaces. We generalize the known results in two directions: The first is that we give a unified approach to tackle all BV-spaces at once. The second is that we also consider the operator I_g from L_∞ into a BV-space X and give conditions under which such operators are well-defined, bounded and compact. This will be one of our main ingredients in the investigation of integral equations.

In Chapter 5 we discuss mapping properties of the following two nonlinear operators on two function spaces X and Y of real-valued functions on $[0,1]$. The (autonomous) composition operator

$$C_g : X \to Y,\ C_g x(t) = g\big(x(t)\big)$$

for a generating function $g : \mathbb{R} \to \mathbb{R}$ in Section 5.1, and the (nonautonomous) superposition operator

$$N_g : X \to Y,\ N_g x(t) = g\big(t, x(t)\big)$$

for a generating function $g : [0,1] \times \mathbb{R} \to \mathbb{R}$ in Section 5.2. As for the composition operator C_g it is well known that C_g maps BV into itself if and only if g is locally Lipschitz continuous [75]. Similar results are also known for the other BV-spaces. We then give some remarks about injectivity and surjectivity in $X = Y = BV$ and other BV-spaces. Here, $C_g : BV \to BV$ is injective if and only if g is injective. However, surjectivity is not so easy to describe. We give a sufficient condition which states that $C_g : BV \to BV$ is surjective if the slope of g is at suitable points in a certain sense bounded away from zero; unfortunately, we were not able to decide whether this condition is also necessary, but we give some indication why we think that it is. We then move on to different types of continuity. In summary, one can say that the more regular g is, the more "continuous" C_g is in BV and other spaces. For instance, C_g is uniformly continuous on bounded sets if and only if g is continuously differentiable, locally Lipschitz continuous if and only if g is continuously differentiable with locally Lipschitz continuous derivative, globally uniformly continuous if and only if g is affine, and compact if and only if g is constant. Similar results hold also in other BV-spaces, where the Riesz spaces have to be treated separately. We prove all these results using a unified approach. Surprisingly, the question of whether C_g is automatically pointwise continuous in BV if g meets the acting condition has an interesting history. The first proof given in [118] is very long and complicated, the second was given only recently in [96]. We give a third proof, but for this purpose we develop some new theory in Chapter 6 and therefore present the proof there also. Nonetheless, all proofs cannot be generalized to other BV-spaces, at least to the best of our knowledge.

Section 5.2 is dedicated to the superposition operator, and we only focus on analytic properties. Although both operators C_g and N_g are defined by an outer composition, the additional dependence of t allows N_g to behave rather chaotic and complicated compared to C_g. Again, many conditions guaranteeing analytic properties are known, but the behavior of the operator N_g even in the space BV is by far not fully understood. For instance, there is no (useful) criterion for the pointwise continuity of N_g in BV. Again, we provide a unified approach to handle all BV-spaces at once. The aim of Section 5.2 is to discuss the weird properties of N_g and reveal disparities between N_g and C_g. For instance, in contrast to C_g too weak kinds of regularity of g seem not directly connected to any kind of regularity of N_g. It is possible to find a discontinuous function $g : [0,1] \times \mathbb{R} \to \mathbb{R}$ that generates a constant and therefore utmost regular operator N_g, while it is also possible to construct a globally Lipschitz continuous generator g that induces a discontinuous operator $N_g : BV \to BV$; we give a general technique on how to construct such examples. Also, in contrast to C_g, there are compact operators $N_g : BV \to BV$ generated by nonconstant functions g. Our main result, however, is

Theorem 5.2.31. It provides for the first time a sufficient condition on g guaranteeing that N_g maps any of our BV-spaces into itself and is locally Lipschitz continuous. We show that our condition also covers the corresponding results for multiplication and composition operators which can be seen as special superposition operators. Theorem 5.2.31 will also serve as one of the main ingredients in the theory of integral equations in Chapter 7.

As mentioned before, the aim of Chapter 6 is to provide a new proof for the fact that C_g is continuous in BV if g is locally Lipschitz continuous. In order to do that we approximate C_g by other composition operator C_{g_n} for sufficiently smooth generators $g_n : \mathbb{R} \to \mathbb{R}$, where $n \in \mathbb{N}$. This approximation has to be done in such a way that the continuity of each C_{g_n} carries over to C_g. Therefore, we investigate in Section 6.1 on the abstract level of metric spaces the following four types of convergence: Quasi uniform, semi uniform, continuously uniform and locally uniform convergence. All of these are able to transmit continuity to the limit function. Historically, quasi uniformly convergence was introduced by Cesare Arzelà [14, 15], who answered the question what on top of pointwise convergence has to be assumed in order to guarantee that the limit function of a sequence of continuous functions is again continuous. Moreover, we give criteria on such sequences and their underlying spaces under which convergent subsequences can be extracted and recall that several types of convergence can even be used to characterize compactness of the domains the functions under consideration live in. Eventually, we compare all five types of convergence (pointwise convergence included) with each other.

In Section 6.2 we then pass to the proof of the fact that C_g is continuous in BV provided that it is well-defined. For this we first develop some theory and introduce the restricted variation, another more general type of variation measuring the variation of that part of a function that falls into a given set. The main result in this section is Theorem 6.2.7. It states that a sequence (C_{g_n}) converges in BV locally semi uniformly to a given composition operator C_g if and only if the corresponding generators g_n converge in BV to g and locally have a uniformly bounded Lipschitz constant. The continuity of C_g is then a simple consequence.

As for applications Chapter 7 will probably be the most relevant. Here, we consider Hammerstein and Volterra integral equations, where the latter are only special cases of the former. A starting point of our considerations in Section 7.1 is the Hammerstein integral equation

$$x(t) = h(t) + \lambda \int_0^1 k(t,s) g\big(x(s)\big)\,\mathrm{d}s$$

and some slight modifications that have already been studied in some BV-spaces, where h, k and g are given and x is unknown. Building on our results presented in the Chapters 4 and 5 we investigate the much more general equation

$$x(t) = h\big(t, x(t)\big) + \lambda f\big(t, x(t)\big) \int_0^1 k(t,s) g\big(s, x(s)\big)\,\mathrm{d}s$$

for given data h, f, k and g and prove existence and sometimes also uniqueness for solutions in BV-spaces. Again, we use a unified approach in order to handle all our

BV-spaces simultaneously. To get uniqueness of solutions we mostly use the fixed point theorem of Banach and Caccioppoli which requires strong conditions on the data involved. We also use other fixed point theorems that require less restrictive conditions on the data for the price that they guarantee only existence of solutions. Especially for boundary and initial value problems we investigate the Hammerstein integral equation

$$x(t) = Ax(t) + \lambda \int_0^1 k(t,s)g\big(x(s)\big)\,\mathrm{d}s,$$

where A is a linear operator from one BV-space into itself and provide some existence results building on Schauder's fixed point theorem. In the short Section 7.2 we reformulate all results about Hammerstein integral equation to the corresponding Volterra integral equations, where the upper limit of integration is replaced by the variable t. The final Section 7.3 is then dedicated to boundary and initial value problems. In [27] the boundary value problem

$$x''(t) = -\lambda g\big(t, x(t)\big)$$

subject to the nonclassical boundary conditions

$$x(0) = A_0 x, \quad x(1) = A_1 x$$

are solved, where A_0 and A_1 are linear functionals on BV. Two results are presented in this paper each of which giving conditions under which the boundary value problem has a solution. We generalize the ideas, simplify the conditions and summarize everything in one stronger result that is even able to handle cases that have not been covered yet. We also give some remarks on how the theory may be applied to other similar boundary value problems. We end the section with initial value problems

$$x''(t) = -\lambda g\big(t, x(t)\big)$$

subject to the nonclassical initial conditions

$$x(0) = A_0 x, \quad x'(0) = A_1 x.$$

we present very similar results and conditions guaranteeing the existence of solutions.

Throughout this thesis we give a lot of estimates, proofs and results, and many of them are quite technical. Therefore, it is of particular concern to us to illustrate most results by special cases, remarks, comparisons and summaries to make the presentation as clear as possible. This will be done by a total of 14 figures, 20 tables and 166 examples and counterexamples.

Chapter 1

Preliminaries

1.1 Classical Classes of Functions

In this section we recall the definitions of classical function spaces which we use the most. We repeat several mostly well-known results the proofs of which (unless stated otherwise) may be found in the monograph [6] and in the recent paper [10], respectively. Before we start, let us make some comments on notations. We primarily consider sets X of real-valued functions defined on the real interval $[0,1]$. Sometimes, the functions are defined on a compact interval $[a,b] = \{(1-t)a + tb \mid t \in [0,1]\}$ with $a, b \in \mathbb{R}$ (we allow $a > b$ for technical reasons). We then write $X[a,b]$ instead of X and mean the set of functions $x \circ \psi$, where $x \in X$ and $\psi : [a,b] \to [0,1]$, $t \mapsto (t-a)/(b-a)$. The symbols $\mathbb{0}$ and $\mathbb{1}$ always denote the functions that are 0 respectively 1 everywhere on their domain of definition. Finally, for a set $A \subseteq \mathbb{R}$ we denote by

$$\chi_A(t) := \begin{cases} 1 & \text{for } t \in A, \\ 0 & \text{for } t \in \mathbb{R}\backslash A \end{cases}$$

its *characteristic function* that satisfies $\chi_\emptyset = \mathbb{0}$ and $\chi_\mathbb{R} = \mathbb{1}$.

To construct examples and counterexamples, we will frequently make use of the following *oscillation function* $\varphi_{\alpha,\beta,n} : [0,1] \to \mathbb{R}$ which we define for $\alpha, \beta \in \mathbb{R}$ and $n \in \mathbb{N}$ by the formula

$$\varphi_{\alpha,\beta,n}(t) = \begin{cases} t^\alpha \sin^n \dfrac{1}{t} & \text{for } 0 < t \leq 1, \\ \beta & \text{for } t = 0. \end{cases} \tag{1.1.1}$$

In Figure 1.1.1 are some pictures showing the behavior of $\varphi_{\alpha,\beta,n}$ for different parameter values. While x_1, x_2 and x_4 are discontinuous and x_3 is continuous, the functions x_1, x_2 and x_3 are bounded, but x_4 is not.

Generally speaking, the function $\varphi_{\alpha,\beta,n}$ is bounded if and only if $\alpha \geq 0$ and continuous at $t = 0$ and hence on all of $[0,1]$ if and only if $\alpha > 0$ and $\beta = 0$. Consequently, denoting by C the space of real-valued continuous functions on $[0,1]$ and by B the

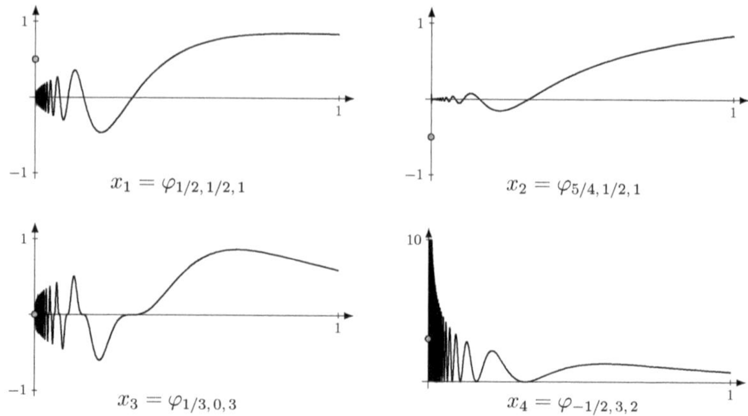

Figure 1.1.1: Some examples for $\varphi_{\alpha,\beta,n}$.

space of real-valued bounded functions on $[0,1]$ the function $\varphi_{0,0,1}$ shows that the well-known inclusion

$$C \subseteq B \tag{1.1.2}$$

is strict. Equipped with the supremum norm

$$\|x\|_\infty := \sup_{0\le t\le 1} |x(t)| \tag{1.1.3}$$

both B and C become Banach spaces and we even have the continuous embedding $C \hookrightarrow B$. The spaces $C(\mathbb{R})$ and $B(\mathbb{R})$ denote the spaces of real-valued continuous respectively bounded functions on all of $\mathbb{R}$ equipped with the norm

$$\|x\|_\infty := \sup_{t\in\mathbb{R}} |x(t)|.$$

For functions in $C[a,b]$ and $B[a,b]$, respectively, we write $\|x\|_{[a,b]} = \sup_{a\le t\le b} |x(t)|$ for the supremum norm.

By the Intermediate Value Theorem any continuous function $x \in C$ is a Darboux function, that is, $x(I)$ is an interval whenever $I \subseteq [0,1]$ is an interval. Consequently, if D denotes the set of Darboux functions defined on $[0,1]$, then we have the inclusion

$$C \subseteq D. \tag{1.1.4}$$

The following example shows that this inclusion is strict as well.

Example 1.1.1. The function $\varphi_{-1,\beta,1}$, which for $\beta \in \mathbb{R}$ according to (1.1.1) is given by

$$\varphi_{-1,\beta,1}(t) = \begin{cases} \dfrac{1}{t}\sin\dfrac{1}{t} & \text{for } 0 < t \le 1, \\ \beta & \text{for } t = 0, \end{cases} \tag{1.1.5}$$

belongs to D for any β and to C or B for no β. ◇

While B and C are linear spaces, the space D is not. Although D is stable under multiplication with real scalars, it is not closed under addition.

Example 1.1.2. The functions $x := \varphi_{0,1,1}$ and $y := -\varphi_{0,0,1}$ which are given by

$$x(t) = \begin{cases} \sin\frac{1}{t} & \text{for } 0 < t \le 1, \\ 1 & \text{for } t = 0 \end{cases} \qquad \text{and} \qquad y(t) = \begin{cases} -\sin\frac{1}{t} & \text{for } 0 < t \le 1, \\ 0 & \text{for } t = 0, \end{cases}$$

are both Darboux functions. But their sum which is the (bounded) characteristic function $x + y = \chi_{\{0\}}$ is clearly not a Darboux function. ◇

The preceding two Examples 1.1.1 and 1.1.2 show that there is no inclusion between the spaces D and B.

Generally speaking, the functions $\varphi_{\alpha,\beta,n}$ belong to D if and only if one of the following five cases is satisfied. Either $\alpha > 0$ and $\beta = 0$ and n is arbitrary, or $\alpha = 0$ and $\beta \in [0,1]$ and n is even, or $\alpha = 0$ and $\beta \in [-1,1]$ and n is odd, or $\alpha < 0$ and $\beta \ge 0$ and n is even, or $\alpha < 0$ and β is arbitrary and n is odd. For instance, the two functions x_3 and x_4 in Figure 1.1.1 are Darboux functions, whereas x_1 and x_2 are not; we will prove this for the general functions (1.1.1) in Proposition 1.1.12 below.

The space C is also contained in L_p for any $p \in [1,\infty]$, where L_p denotes the space of all (equivalence classes of) measurable[1] functions $x : [0,1] \to \mathbb{R}$ such that for $1 \le p < \infty$ the function $|x|^p$ is Lebesgue integrable and for $p = \infty$ the function x is essentially bounded. Here, two functions are considered to be equivalent if they agree everywhere on their domain of definition except on a set of Lebesgue measure zero. The spaces $L_p(I)$ of functions defined on an arbitrary measurable set $I \subseteq \mathbb{R}$ of finite positive measure are strictly decreasing with respect to p, i.e.

$$L_q(I) \subsetneq L_p(I) \quad \text{for } 1 \le p < q \le \infty.$$

This is no longer true if I has infinite measure.

Example 1.1.3. The function $f(t) = 1/t$ for $t \in I = [1,\infty)$ does not belong to $L_1(I)$, because

$$\int_1^\infty f(t)\,\mathrm{d}t = \lim_{b\to\infty} \log(b) = \infty.$$

However, $f \in L_q(I)$ for any $q > 1$, as

$$\int_1^\infty f(t)^q\,\mathrm{d}t = \lim_{b\to\infty} \frac{1-b^{1-q}}{q-1} = \frac{1}{q-1} < \infty.$$

Consequently, $L_q(I) \not\subseteq L_1(I)$ in this case. ◇

It is well known that the L_p-spaces when endowed with the norms

$$\|x\|_{L_p} = \left(\int_I |x(t)|^p\,\mathrm{d}t\right)^{1/p} \quad \text{for } 1 \le p < \infty, \tag{1.1.6}$$

$$\|x\|_{L_\infty} = \operatorname*{ess\,sup}_{t\in I} |x(t)|, \tag{1.1.7}$$

[1]Whenever we talk about measurable functions or sets, we mean measurable with respect to the Lebesgue measure.

are Banach spaces. However, only $L_\infty(I)$ is an algebra, because among all Lebesgue spaces this is the only one which is closed under multiplication. The Hölder inequality estimates the product of two functions from Lebesgue spaces; we will use this in more detail in Section 3.2 on $I = [0,1]$.

Although functions in either of the spaces B and L_∞ are called "bounded" respectively "essentially bounded", there is no inclusion between these two spaces.

Example 1.1.4. The characteristic function χ_A for A being a nonmeasurable[2] set $A \subseteq [0,1]$ is bounded, but does not belong to L_∞ as it is not measurable. The function

$$x(t) = \begin{cases} \dfrac{\chi_{\mathbb{Q}}(t)}{t} & \text{for } 0 < t \leq 1, \\ 0 & \text{for } t = 0, \end{cases}$$

is measurable and essentially bounded, since $\mathbb{Q} \cap [0,1]$ has measure zero. But it is not bounded, since $x(1/n) = n$ for all $n \in \mathbb{N}$. ♢

The function in the previous example is a function with countable *support.* Such functions will be of great importance in the sequel, in particular in the Chapters 3 and 4. For a function $x : [0,1] \to \mathbb{R}$ we write

$$\operatorname{supp}(x) := \Big\{t \in [0,1] \mid x(t) \neq 0\Big\} \tag{1.1.8}$$

for its support. Note that in contrast to the standard definition we do not take the closure here. More generally, for $\delta > 0$ we also define

$$\operatorname{supp}_\delta(x) := \Big\{t \in [0,1] \mid |x(t)| \geq \delta\Big\}. \tag{1.1.9}$$

Then $\operatorname{supp}_\delta(x)$ is decreasing with respect to δ in the sense that $\operatorname{supp}_\delta(x) \subseteq \operatorname{supp}_\eta(x)$ whenever $\delta \geq \eta$, and it is related to $\operatorname{supp}(x)$ via

$$\operatorname{supp}(x) = \bigcup_{\delta>0} \operatorname{supp}_\delta(x) = \bigcup_{n\in\mathbb{N}} \operatorname{supp}_{1/n}(x). \tag{1.1.10}$$

In particular, if $\operatorname{supp}(x)$ is uncountable, then $\operatorname{supp}_\delta(x)$ is also uncountable for some $\delta > 0$. Conversely, if $\operatorname{supp}_\delta(x)$ is countable for all $\delta > 0$, then also $\operatorname{supp}(x)$ is countable as it is then a countable union of countable sets; this fact also has the following consequence which we will use later on in Section 4.1.

Lemma 1.1.5. *Let $x : [0,1] \to \mathbb{R}$ have uncountable support. Then there is some $m > 0$ and a strictly monotone sequence (t_j) in $(0,1)$ such that $|x(t_j)| \geq m$ for all $j \in \mathbb{N}$.*

Proof. If $\operatorname{supp}(x)$ is uncountable, then because of (1.1.10) there is some $n \in \mathbb{N}$ such that $\operatorname{supp}_{1/n}(x)$ is uncountable, as well. For $m := 1/n$ we therefore find infinitely many pairwise distinct numbers $s_j \in \operatorname{supp}_m(x)$ for which $|x(s_j)| \geq m$ holds for all $j \in \mathbb{N}$. But

[2] Giuseppe Vitali was the first who proved in 1905 in [148] the existence of such sets, but he used the axiom of choice. Without it an example of Vitali's type is in fact not possible, and this was proved later in 1970 by Solovay [143].

then, since (s_j) is a bounded sequence, by a classical argument we can extract from it a monotone subsequence (t_j). Since the s_j are pairwise distinct, so are the t_j which makes them form a strictly monotone sequence. ■

Finally, by $\mathcal{S}_c$ we denote the family of functions on $[0,1]$ with countable support, and by S_f the family of functions on $[0,1]$ with finite support. Both families are related by the strict inclusion $\mathcal{S}_f \subsetneq \mathcal{S}_c$.

Example 1.1.6. The function $x := \chi_A$ for some set $A \subseteq [0,1]$ belongs to $\mathcal{S}_c$ if and only if A is countable, and to $\mathcal{S}_f$ if and only if A is finite. Moreover, $\operatorname{supp}_\delta(x) = A$ for $0 < \delta \leq 1$ and $\operatorname{supp}_\delta(x) = \emptyset$ for $\delta > 1$. ♢

We now turn to the most important space of this thesis, the space BV of functions of *bounded variation* which was introduced in 1881 in [74] by Camille Jordan.

Definition 1.1.7. For a function $x : [0,1] \to \mathbb{R}$ we call the possibly infinite number

$$\operatorname{Var}(x) = \sup_P \sum_{j=1}^{n} |x(t_{j-1}) - x(t_j)| \tag{1.1.11}$$

the *(Jordan) variation of x on* $[0,1]$, where the supremum is taken over all finite partitions $P : 0 = t_0 < \ldots < t_n = 1$ of $[0,1]$. If $\operatorname{Var}(x) < \infty$, we say that x has bounded variation and write $x \in BV$.
For functions $x : [a,b] \to \mathbb{R}$ with $a < b$ we write $\operatorname{Var}(x,[a,b])$ instead of $\operatorname{Var}(x)$.

Clearly, every monotone function $x : [0,1] \to \mathbb{R}$ belongs to BV with $\operatorname{Var}(x) = |x(0) - x(1)|$. Conversely, $\operatorname{Var}(x) = |x(0) - x(1)|$ implies that x is monotone. However, BV-functions can be quite chaotic. The following auxiliary result deals with functions that have countable support, that is, with functions in $\mathcal{S}_c$.

Proposition 1.1.8. *For $x \in \mathcal{S}_c$ we have*

$$\operatorname{Var}(x) = |x(0)| + |x(1)| + 2 \sum_{\tau \in \operatorname{supp}(x)\setminus\{0,1\}} |x(\tau)|. \tag{1.1.12}$$

In particular, $x \in BV$ if and only if the series in (1.1.12) converges.

Proof. Consider an arbitrary partition $0 = t_0 < \ldots < t_n = 1$ of $[0,1]$. Then

$$\begin{aligned}\sum_{j=1}^{n} |x(t_{j-1}) - x(t_j)| &\leq \sum_{j=1}^{n} \big(|x(t_{j-1})| + |x(t_j)|\big) = |x(0)| + |x(1)| + 2\sum_{j=1}^{n-1} |x(t_j)| \\ &\leq |x(0)| + |x(1)| + 2 \sum_{\tau \in \operatorname{supp}(x)\setminus\{0,1\}} |x(\tau)|.\end{aligned}$$

This shows the "$\leq$"-part in (1.1.12).
For the reverse inequality let $n \in \mathbb{N}$ with $n \leq \#(\operatorname{supp}(x)\setminus\{0,1\})$ be fixed, where $\#A$ denotes the number of elements in a set A, and pick numbers $t_2, t_4, t_6, \ldots, t_{2n} \in \operatorname{supp}(x)$ in strictly increasing order. Set $t_0 := 0$ and $t_{2n+2} := 1$, and pick for each $j \in \{0,\ldots,n\}$

numbers $t_{2j+1} \in (t_{2j}, t_{2j+2}) \setminus \operatorname{supp}(x)$ which is possible as $\operatorname{supp}(x)$ is countable. Then

$$\begin{aligned}\operatorname{Var}(x) &\geq \sum_{j=1}^{2n+2} |x(t_{j-1}) - x(t_j)| = \sum_{j=0}^{n} |x(t_{2j})| + \sum_{j=1}^{n+1} |x(t_{2j})| \\ &= |x(0)| + |x(1)| + 2\sum_{j=1}^{n} |x(t_{2j})|. \end{aligned} \tag{1.1.13}$$

If $\operatorname{supp}(x)$ is finite we are done by putting $n = \#(\operatorname{supp}(x)\setminus\{0,1\})$, because then the right hand side of (1.1.13) coincides with the right hand side of (1.1.12). If $\operatorname{supp}(x)$ is infinite, we may let $n \to \infty$, and then the right hand side of (1.1.13) converges to the right hand side of (1.1.12) provided that for each n one picks the t_j appropriately. In any case, the proof is complete. ■

Formula (1.1.12) allows us to give an example of a function x that is nowhere monotone but yet of bounded variation [6].

Example 1.1.9. Let (q_n) be an enumeration of all rational numbers in $(0, 1)$. Define the function $x : [0, 1] \to \mathbb{R}$ by

$$x(t) = \begin{cases} 2^{-n} & \text{for } t = q_n, \\ 0 & \text{otherwise.} \end{cases}$$

Then x has countable support and according to formula (1.1.12) its variation is given by

$$\operatorname{Var}(x) = 2\sum_{n=1}^{\infty} 2^{-n} = 2.$$

However, x is monotone on no interval $[a, b] \subseteq [0, 1]$ with $a < b$. Since $\mathbb{Q}$ is dense in $\mathbb{R}$ we find $m, n \in \mathbb{N}$ such that $a < q_m < q_n < b$, and since $\mathbb{R}\setminus\mathbb{Q}$ is also dense in $\mathbb{R}$ we find some $c \in [0, 1]$ with $q_m < c < q_n$. But then $x(q_m), x(q_n) > 0 = x(c)$ which implies that x is not monotone on $[q_m, q_n]$ and hence also not on $[a, b]$. ♢

Interestingly, although BV-functions do not necessarily have any monotonicity behavior at all, they are generated by monotone functions [74].

Theorem 1.1.10 (Jordan's Decomposition Theorem). *A function $x : [0, 1] \to \mathbb{R}$ is of bounded variation if and only if it can be written as a difference $x = y - z$ of two increasing functions $y, z : [0, 1] \to \mathbb{R}$.*

Obviously, the set M of monotone functions on $[0, 1]$ is not a linear space. For instance, the sum of the two monotone functions $x(t) = t^2$ and $y(t) = 1-t$ is given by $(x+y)(t) = t^2 - t + 1$ which is not monotone on $[0, 1]$. Jordan's Decomposition Theorem shows $\operatorname{Span}(M) = BV$, that is, the linear hull of M is BV. Therefore, Theorem 1.1.10 can be understood as the dawn of the BV era. Moreover, it also shows that "nice" properties of monotone functions carry over to BV-functions. In particular, monotone and hence BV-functions have only at most countably many discontinuities, and each discontinuity is of first kind (jumps) or removable [163]. This means that for $x \in BV$ all the left-

and right-sided limits

$$\lim_{t\to t_0-} x(t) \text{ for } t_0 \in (0,1], \qquad \lim_{t\to t_0+} x(t) \text{ for } t_0 \in [0,1)$$

exist and are finite. Even better, by Lebesgue's Theorem, every monotone and hence BV-function is differentiable almost everywhere [86].

In general, functions which have no discontinuity of second kind are automatically bounded on compact intervals and usually called *regular*. So denoting by R the space of all regular functions on $[0,1]$, we have the inclusions

$$BV \subseteq R \subseteq B \cap L_\infty, \tag{1.1.14}$$

and both inclusions are strict. Note that each BV-function is indeed bounded, since the relation $|x(t)| \leq |x(0)| + |x(0) - x(t)| \leq |x(0)| + \mathrm{Var}(x)$ for each $t \in [0,1]$ implies

$$\|x\|_\infty \leq |x(0)| + \mathrm{Var}(x). \tag{1.1.15}$$

The characteristic function $\chi_{\mathbb{Q}\cap[0,1]}$ is bounded and measurable but not regular and thus shows that the second inclusion in (1.1.14) is strict. Formula (1.1.12) applied to functions with nonempty but finite support shows that BV and hence R contain discontinuous functions. However, there is an interesting interconnection between R and D: Since a Darboux function can have only essential discontinuities and a regular function can have only jump discontinuities, every regular Darboux function must be continuous. Conversely, every continuous function is clearly regular. Consequently,

$$R \cap D = C. \tag{1.1.16}$$

In particular, $BV \cap D = BV \cap C$. However, not every continuous and hence regular function belongs to BV, and this is why also the first inclusion in (1.1.14) is strict.

Example 1.1.11. The function $x = \varphi_{1,0,1}$, given by

$$x(t) = \begin{cases} t\sin\dfrac{1}{t} & \text{for } 0 < t \leq 1, \\ 0 & \text{for } t = 0, \end{cases} \tag{1.1.17}$$

is clearly continuous and hence regular as $|x(t)| \leq t$ for all $t \in [0,1]$. But x is not of bounded variation. To see this consider the points $0 < t_n < t_{n-1} < \ldots < t_1 < t_0 = 2/\pi < 1$ of the extremal points of x located at $t_j = 1/(j\pi + \pi/2)$. Then

$$\mathrm{Var}(x) \geq \sum_{j=1}^n |x(t_{j-1}) - x(t_j)| = \sum_{j=1}^n \left(t_{j-1} + t_j\right) = \frac{8}{\pi}\sum_{j=1}^n \frac{j}{4j^2-1} \geq \frac{2}{\pi}\sum_{j=1}^n \frac{1}{j}$$

which gets infinitely large as $n \to \infty$. ◇

The function x_3 in Figure 1.1.1 is also continuous and of unbounded variation. The functions x_1 and x_4 also have unbounded variation; only x_2 is of bounded variation.

In general, the function $\varphi_{\alpha,\beta,n}$ belongs to BV if and only if $\alpha > 1$. The proof for this general fact is given in our next result that summarizes the conditions on α, β and n in order to make $\varphi_{\alpha,\beta,n}$ a member of the classes B, C, D and BV. The special case $\beta = 0$ and $n = 1$ has been discussed comprehensively in [4]; we give the full proof here, since the generalization to arbitrary β and n will be of importance for the sequel.

Proposition 1.1.12. *For the functions $\varphi_{\alpha,\beta,n}$ the following statements are true.*

(a) $\varphi_{\alpha,\beta,n} \in B$ if and only if $\alpha \geq 0$.

(b) $\varphi_{\alpha,\beta,n} \in BV$ if and only if $\alpha > 1$.

(c) $\varphi_{\alpha,\beta,n} \in C$ if and only if $\alpha > 0$ and $\beta = 0$.

(d) $\varphi_{\alpha,\beta,n} \in D$ if and only if one of the following five cases is true.

(i) $\alpha > 0,\ \beta = 0$ and n is arbitrary,

(ii) $\alpha = 0,\ 0 \leq \beta \leq 1$ and n is even,

(iii) $\alpha = 0,\ -1 \leq \beta \leq 1$ and n is odd,

(iv) $\alpha < 0,\ \beta \geq 0$ and n is even,

(v) $\alpha < 0,\ \beta$ is arbitrary and n is odd.

Proof. (a) Note that for $\alpha < 0$ we have for $t_j := 1/(2j\pi + \pi/2)$ with $j \in \mathbb{N}_0$ that $\varphi_{\alpha,\beta,n}(t_j) = 1/(2j\pi + \pi/2)^\alpha \to \infty$ as $j \to \infty$. For $\alpha \geq 0$ we have on the other hand $|\varphi_{\alpha,\beta,n}(t)| \leq \max\{|\beta|, 1\}$ for all $t \in [0,1]$. Thus, (a) is established.

To prove (b), note that for $\alpha < 0$ we have that $\varphi_{\alpha,\beta,n} \notin BV$ because of (a). Thus we can assume that $\alpha \geq 0$ and that $\varphi_{\alpha,\beta,n}$ is bounded. For $t_j := 1/(j\pi + \pi/2)$ with $j \in \mathbb{N}_0$ we have

$$\mathrm{Var}(\varphi_{\alpha,\beta,n}) = |\varphi_{\alpha,\beta,n}(2/\pi) - \varphi_{\alpha,\beta,n}(1)| + \limsup_{t\to 0+} |\varphi_{\alpha,\beta,n}(t) - \beta| + \sum_{j=0}^{\infty} |\varphi_{\alpha,\beta,n}(t_j)|.$$

In particular, since $\varphi_{\alpha,\beta,n}$ is bounded, it is of bounded variation if and only if the series converges. We have

$$\sum_{j=0}^{\infty} |\varphi_{\alpha,\beta,n}(t_j)| = \sum_{j=0}^{\infty} t_j^\alpha = \sum_{j=0}^{\infty} \frac{1}{(j\pi + \pi/2)^\alpha},$$

and this series converges if and only if $\alpha > 1$. In this case, and only then, the function $\varphi_{\alpha,\beta,n}$ is of bounded variation.

(c) The function $\varphi_{\alpha,\beta,n}$ is clearly continuous at every point $t \in (0,1]$, so we only need to consider $t = 0$. Again by (a) we can assume that $\alpha \geq 0$, because for $\alpha < 0$ the function $\varphi_{\alpha,\beta,n}$ is unbounded and hence discontinuous. For $\alpha = 0$ consider the points $t_j := 2/(j\pi)$ for $j \in \mathbb{N}$ which converge to 0 as $j \to \infty$. Then $|\varphi_{0,\beta,n}(t_j)| = [1-(-1)^j]/2$ for all $j \in \mathbb{N}$, and the function $\varphi_{0,\beta,n}$ cannot be continuous at $t = 0$, no matter what n and β are.

For $\alpha > 0$, however, we have $|\varphi_{\alpha,\beta,n}(t)| \leq t^\alpha$ for $0 < t \leq 1$ and hence $\varphi_{\alpha,\beta,n}(t) \to 0$ as $t \to 0+$. This shows that $\varphi_{\alpha,\beta,n}$ is continuous at $t = 0$ if and only if $\beta = 0$.

(d) Again, $\varphi_{\alpha,\beta,n} \in D[\varepsilon, 1]$ for all $\varepsilon \in (0,1)$ and we only need to check what happens around $t = 0$. For even $n \in \mathbb{N}$ we have

$$\liminf_{t\to 0+} \varphi_{\alpha,\beta,n}(t) = 0 \qquad \text{and} \qquad \limsup_{t\to 0+} \varphi_{\alpha,\beta,n}(t) = \begin{cases} 0 & \text{for } \alpha > 0, \\ 1 & \text{for } \alpha = 0, \\ \infty & \text{for } \alpha < 0, \end{cases}$$

and for odd $n \in \mathbb{N}$,

$$\liminf_{t\to 0+} \varphi_{\alpha,\beta,n}(t) = \begin{cases} 0 & \text{for } \alpha > 0, \\ -1 & \text{for } \alpha = 0, \\ -\infty & \text{for } \alpha < 0 \end{cases} \qquad \text{and} \qquad \limsup_{t\to 0+} \varphi_{\alpha,\beta,n}(t) = \begin{cases} 0 & \text{for } \alpha > 0, \\ 1 & \text{for } \alpha = 0, \\ \infty & \text{for } \alpha < 0. \end{cases}$$

From these estimates (d) is an immediate consequence. ■

Note that the conditions in (a) and (b) do not depend on β and n, while the one in (c) is independent of n.

The two examples x_2 and x_3 in Figure 1.1.1 show that there is no inclusion between the two spaces C and BV. Nevertheless, C and BV are connected in the sense that the set NBV of all normalized functions $x^\# - x(0)$, where the *right regularization* $x^\#$ of a function $x \in BV$ is defined by

$$x^\#(t) = \begin{cases} \lim_{s\to t+} x(s) & \text{for } 0 \leq t < 1, \\ x(1) & \text{for } t = 1, \end{cases} \tag{1.1.18}$$

can be considered as the dual of C [6, Theorem 4.31], where the duality is established by a Riemann-Stieltjes integral.

One can show that BV, equipped with the norm

$$\|x\| = |x(0)| + \operatorname{Var}(x) \tag{1.1.19}$$

is a Banach space which is continuously embedded into B, by (1.1.15). One can even show that BV as well as C and B are closed under multiplication, that is, the spaces BV, C and B are algebras. In addition, C is also closed under forming reciprocals of functions x that have no zeros. Indeed, if $x \in C$ has no zeros, then it is bounded away from zero by the Permanence Principle. But then $1/x$ is also continuous. An analogue of this fact is not true in B and BV.

Example 1.1.13. The function

$$x(t) = \begin{cases} t & \text{for } 0 < t \leq 1, \\ 1 & \text{for } t = 0 \end{cases}$$

belongs to BV and hence also to B, but its reciprocal

$$\frac{1}{x}(t) = \begin{cases} 1/t & \text{for } 0 < t \leq 1, \\ 1 & \text{for } t = 0 \end{cases}$$

does not, since it is unbounded near $t = 0$. ◇

In contrast to C to ensure that $1/x \in BV$ or $1/x \in B$ we have to assume explicitly that $x \in BV$ is bounded away from zero, that is, $\text{supp}_\delta(x) = [0, 1]$ for some $\delta > 0$. In this case the estimate

$$\text{Var}(1/x) \leq \delta^{-2} \, \text{Var}(x)$$

is true.

However, the spaces BV, C and B are not only algebras, their norms are in a sense submultiplicative. In general, if a normed algebra $(X, \|\cdot\|)$ satisfies an estimate of the form

$$\|xy\| \leq c \, \|x\| \, \|y\| \quad \text{for } x, y \in X$$

with some constant $c > 0$ independent of x and y, then we call X a *Banach algebra* if $(X, \|\cdot\|)$ is a Banach space; in addition, we say that X is *normalized* if $c = 1$. For instance, the spaces C and B are both normalized Banach algebras with respect to the supremum norm $\|\cdot\|_\infty$ given in (1.1.3), whereas $(BV, \|\cdot\|_\infty)$ also equipped with the supremum norm is a normalized algebra, but not a Banach algebra. That BV when equipped with the supremum norm is not complete may be seen by the following example.

Example 1.1.14. For each $n \in \mathbb{N}$ let $x_n : [0, 1] \to \mathbb{R}$ be defined by

$$x_n(t) = \begin{cases} 1/j & \text{for } t = 1/(2j), j \in \{1, \ldots, n\}, \\ 0 & \text{otherwise.} \end{cases}$$

Then each x_n belongs to BV according to Proposition 1.1.8 as it has finite support. Moreover, (x_n) converges uniformly to the function $x : [0, 1] \to \mathbb{R}$, defined by

$$x(t) = \begin{cases} 1/j & \text{for } t = 1/(2j), j \in \mathbb{N}, \\ 0 & \text{otherwise,} \end{cases}$$

because

$$\|x_n - x\|_\infty = \frac{1}{n+1} \quad \text{for all } n \in \mathbb{N}.$$

In particular, (x_n) is a Cauchy sequence with respect to the supremum norm. However, the limit x does not belong to BV. On the one hand, x has countable support $\text{supp}(x) = \{1/2, 1/4, 1/6, 1/8, \ldots\}$, but on the other hand, by Proposition 1.1.8 its variation is

$$\text{Var}(x) = \sum_{j=1}^{\infty} \frac{2}{j} = \infty.$$

Therefore, $x \notin BV$, and hence $(BV, \|\cdot\|_\infty)$ is not complete. 

The most function spaces X we investigate in this thesis are real Banach algebras when endowed with a norm of the form $\|x\| = |x(0)| + \Phi(x)$, where $\Phi : X \to [0,\infty)$ is a seminorm on X. The space BV with the norm (1.1.19) is an example of such a space. Most of the time one can replace $\|\cdot\|$ by an equivalent norm $\|\cdot\|_X$ which then makes X even a normalized Banach algebra. Here come the details.

Proposition 1.1.15. *Let $(X, \|\cdot\|)$ be a Banach algebra with $X \subseteq B$ and norm*

$$\|x\| = |x(0)| + \Phi(x),$$

where $\Phi : X \to [0,\infty)$ is a seminorm satisfying

$$\Phi(xy) \le \alpha\,\Phi(y)\,\|x\|_\infty + \alpha\,\Phi(x)\,\|y\|_\infty \quad \text{for } x, y \in X \tag{1.1.20}$$

for some constant $\alpha \ge 1$ independent of x and y. Then X, equipped with the norm

$$\|x\|_X := \alpha\,\|x\|_\infty + \Phi(x),$$

is a normalized Banach algebra, i.e. $\|xy\|_X \le \|x\|_X\,\|y\|_X$ for all $x, y \in X$. Moreover, in case $X \hookrightarrow B$ both norms are equivalent.

Proof. Since Φ is a seminorm and $\|\cdot\|_\infty$ is a norm, $\|\cdot\|_X$ is a norm on X for any α. Moreover,

$$\begin{aligned}
\|xy\|_X &= \alpha\,\|xy\|_\infty + \Phi(xy) \le \alpha\,\|x\|_\infty\,\|y\|_\infty + \alpha\Phi(y)\,\|x\|_\infty + \alpha\Phi(x)\,\|y\|_\infty \\
&\le \alpha^2\,\|x\|_\infty\,\|y\|_\infty + \alpha\Phi(y)\,\|x\|_\infty + \alpha\Phi(x)\,\|y\|_\infty + \Phi(x)\Phi(y) \\
&= \big(\alpha\,\|x\|_\infty + \Phi(x)\big)\big(\alpha\,\|y\|_\infty + \Phi(y)\big) \\
&= \|x\|_X\,\|y\|_X\,.
\end{aligned}$$

Finally, if $X \hookrightarrow B$, there is some $c > 0$ such that $\|x\|_\infty \le c\,\|x\|$ for all $x \in X$. Then

$$\begin{aligned}
\|x\| &= |x(0)| + \Phi(x) \le \|x\|_\infty + \Phi(x) \le \alpha\,\|x\|_\infty + \Phi(x) = \|x\|_X\,, \\
\|x\|_X &= \alpha\,\|x\|_\infty + \Phi(x) \le \alpha c\,\|x\| + \|x\| \le (1+\alpha c)\,\|x\|\,,
\end{aligned}$$

which completes the proof. ■

For our space BV, the function $\Phi(x) := \mathrm{Var}(x)$ is a seminorm on BV satisfying the hypotheses of Proposition 1.1.15 with $\alpha = 1$. Accordingly, the space BV is a normalized Banach algebra when equipped with the equivalent norm

$$\|x\|_{BV} := \|x\|_\infty + \mathrm{Var}(x). \tag{1.1.21}$$

In this thesis we will primarily consider this norm for BV.
Besides $BV \hookrightarrow B$ we also have the embedding $BV \hookrightarrow L_\infty$, since each function in BV is measurable as it is bounded and has only countably many discontinuities, and the norms satisfy $\|x\|_{L_\infty} \le \|x\|_\infty \le \|x\|_{BV}$.

As we have seen, the class C is not a subclass of BV. However, the class AC of *absolutely continuous* functions is.

Definition 1.1.16. A function $x : [0,1] \to \mathbb{R}$ is said to be absolutely continuous, if it has the following property: For each $\varepsilon > 0$ there exists $\delta > 0$ such that for each finite collection $([a_j, b_j])_{1\le j\le n}$ of nonoverlapping[3] intervals $[a_j, b_j] \subseteq [0,1]$ the implication

$$\sum_{j=1}^{n} |a_j - b_j| \le \delta \qquad \Longrightarrow \qquad \sum_{j=1}^{n} |x(a_j) - x(b_j)| \le \varepsilon \tag{1.1.22}$$

holds.

It is well known that *absolutely continuous* functions can also be characterized as follows.

Theorem 1.1.17. *A function $x : [0,1] \to \mathbb{R}$ belongs to AC if and only if x is almost everywhere differentiable and its derivative x' belongs to L_1 and satisfies the Fundamental Theorem of Calculus, namely the identity*

$$x(t) - x(s) = \int_s^t x'(\tau)\,\mathrm{d}\tau \quad \textit{for } s,t \in [0,1]. \tag{1.1.23}$$

As said, it is easy to show that

$$AC \subseteq BV \tag{1.1.24}$$

and this inclusion is also strict, because every absolutely continuous function is continuous, but there are discontinuous functions of bounded variation.

Example 1.1.18. The characteristic function $x = \chi_{\{0\}}$ has bounded variation and is discontinuous at $t = 0$. Therefore, it cannot be absolutely continuous. Indeed, it does not satisfy (1.1.22): For $\varepsilon = 1/2$ and any $\delta > 0$ we have for $[a_1, b_1] = [0, \delta]$ that $|a_1 - b_1| = \delta$, but $|x(a_1) - x(b_1)| = 1 > \varepsilon$. ◊

Since every AC-function is continuous, one might ask if the inclusion (1.1.24) can be replaced by $AC = C \cap BV$. This is not true, and here is a prominent counterexample.

Example 1.1.19. Write every $t \in [0,1]$ in a ternary representation

$$t = \sum_{j=1}^{\infty} \frac{a_j(t)}{3^j} \quad \text{with } a_j(t) \in \{0,1,2\}, \tag{1.1.25}$$

and define the *Cantor set* $\mathcal{C} := \{t \in [0,1] \mid a_j(t) \in \{0,2\} \text{ for all } j \in \mathbb{N}\}$. It can be shown that $\mathcal{C}$ is independent of the ternary representation of t which may be not unique in general, and that $\mathcal{C}$ is a compact uncountable null set.
Moreover, denote by $N(t)$ the smallest j such that $a_j(t) = 1$ if it exists. If not, put $N(t) = \infty$. The *Cantor function* $c : [0,1] \to \mathbb{R}$ is then defined by

$$c(t) = \frac{1}{2^{N(t)}} + \frac{1}{2} \sum_{j=1}^{N(t)-1} \frac{a_j(t)}{2^j}.$$

[3]Nonoverlapping means here that the intersection of two intervals contains at most one point.

It can be shown that the definition of c is independent of the ternary representation of t, that is, c is well-defined, and that c is the unique continuous extension of the function $\tilde{c} : \mathcal{C} \to \mathbb{R}$ which satisfies

$$\tilde{c}(t) = \frac{1}{2}\sum_{j=1}^{\infty} \frac{a_j(t)}{2^j} \quad \text{for } t \in \mathcal{C}.$$

Moreover, c is increasing and continuous and hence belongs to $C \cap BV$. However, c is also differentiable almost everywhere with $c' = 0$, because c is constant outside of $\mathcal{C}$; in particular, $c(\mathcal{C}) = [0,1]$. By Theorem 1.1.17 it cannot be absolutely continuous, since it cannot fulfill the relation (1.1.23). ♢

Thus, functions $x \in AC$ apart from being continuous and of bounded variation must have a third property. Indeed, this property is called the *Luzin property* and states that x must map null sets into null sets. The *Vitali Banach Zaretskij Theorem* then yields that a function $x \in C \cap BV$ is absolutely continuous if and only if it satisfies the Luzin property [76]. The Cantor function from Example 1.1.19 does not satisfy this property, since $\mathcal{C}$ is a null set, but $c(\mathcal{C}) = [0,1]$ is not.

In contrast to BV-functions it is easier for AC-functions to calculate the variation explicitly. This is because of the following result.

Theorem 1.1.20. *For a function $x \in AC$ we have*

$$\operatorname{Var}(x) = \int_0^1 |x'(t)| \,\mathrm{d}t = \|x'\|_{L^1}. \tag{1.1.26}$$

Formula (1.1.26) geometrically means that x considered as a curve or path in $\mathbb{R}$ has length $\operatorname{Var}(x)$. Thus, functions of bounded variation are sometimes called *rectifiable.* Of course, from a geometric point of view this makes more sense in higher dimensions. The conditions (1.1.11) and (1.1.22) also make sense for functions attaining values in a normed space $(X, \|\cdot\|_X)$. One then just has to replace the absolute values by the norm $\|\cdot\|_X$. In this case Theorem 1.1.20 still holds true. In this thesis, however, we will only consider real-valued functions.

Before we pass to classes of functions with more regularity let us mention the following curiosity which shows that modifying the partitions or interval collections in the Definitions 1.1.7 and 1.1.16 leads to unexpected phenomena. Recall that BV is defined by the condition (1.1.11) where the supremum is taken over all *ordered* partitions of $[0,1]$. Instead, let us define the space SBV to consist of all functions $x : [0,1] \to \mathbb{R}$ satisfying (1.1.11) where the supremum is now taken over all *not necessarily ordered* collections of points $t_0, t_1, \ldots, t_{n-1}, t_n \in [0,1]$ which satisfy $\sum_{j=1}^{n} |t_{j-1} - t_j| \leq 1$ instead. We call the elements of SBV *functions of super bounded variation.* Clearly, $SBV \subseteq BV$. The following example shows that the inclusion is strict.

Example 1.1.21. Consider the function $x : [0,1] \to \mathbb{R}$, $t \mapsto \sqrt{t}$, and the collection t_0, t_1, ..., $t_{2k-1} \in [0,1]$ for $k \in \mathbb{N}$, defined by

$$t_0 = 0, \quad t_1 = \frac{3}{\pi^2}, \quad t_2 = 0, \quad t_3 = \frac{3}{\pi^2}\frac{1}{4}, \quad \ldots, \quad t_{2k-2} = 0, \quad t_{2k-1} = \frac{3}{\pi^2 k^2}.$$

Then

$$\sum_{j=1}^{2k-1} |t_{j-1} - t_j| \le 2\sum_{j=1}^{k} t_{2j-1} = \frac{6}{\pi^2}\sum_{j=1}^{k}\frac{1}{j^2} \le \frac{6}{\pi^2}\sum_{j=1}^{\infty}\frac{1}{j^2} = 1,$$

but

$$\sum_{j=1}^{2k-1} |x(t_{j-1}) - x(t_j)| \ge \sum_{j=1}^{k}\left|\sqrt{t_{2j-1}}\right| = \frac{\sqrt{3}}{\pi}\sum_{j=1}^{k}\frac{1}{j}$$

gets infinitely large as $k \to \infty$. Thus, $x \notin SBV$. However, as an increasing function, x belongs to BV. ◊

The next result characterizes SBV and states that SBV coincides with the space *Lip* of *Lipschitz continuous* functions, that is, functions $x : [0,1] \to \mathbb{R}$ for which there exists some $L > 0$ such that

$$|x(s) - x(t)| \le L|s-t| \quad \text{for } s,t \in [0,1]. \tag{1.1.27}$$

In this sense the following theorem provides another way of how to define *Lip*.

Theorem 1.1.22. *The equality $SBV = Lip$ is true.*

Proof. The inclusion $Lip \subseteq SBV$ is trivial. So assume that $x \in SBV$ and fix numbers $a, b \in [0,1]$ with $a \ne b$. Choose $n \in \mathbb{N}$ so that

$$n \le \frac{1}{|a-b|} \le 2n \tag{1.1.28}$$

and define the numbers $t_0 := a, t_1 := b, t_3 := a, t_4 := b, \ldots$, that is, $t_n := a$ if n is even and $t_n := b$ if n is odd. Then, by the first inequality in (1.1.28),

$$\sum_{j=1}^{n} |t_{j-1} - t_j| = n|a-b| \le 1.$$

Since $x \in SBV$ there is some $M > 0$ independent of a, b and n such that

$$\sum_{j=1}^{n} |x(t_{j-1}) - x(t_j)| \le M.$$

By the second inequality in (1.1.28) we get

$$|x(a) - x(b)| = \frac{1}{n}\sum_{j=1}^{n} |x(t_{j-1}) - x(t_j)| \le \frac{M}{n} \le 2M|a-b|,$$

and since a, b are arbitrary and M does not depend on a or b, $x \in Lip$. ■

We remark that a similar phenomenon occurs in Definition 1.1.16. If we there drop the assumption that the intervals are mutually nonoverlapping, then we also end up with the space *Lip* [6, Exercise 3.8].

Even if BV-functions can be quite chaotic and may have many discontinuities, the following result is very remarkable. It says that all discontinuities of a BV-function can be "smoothed out" by a suitable change of variables.

Theorem 1.1.23. *A function $x : [0,1] \to \mathbb{R}$ belongs to BV if and only if it may be represented as a composition $x = y \circ z$, where $y \in Lip$ with $\mathrm{lip}(y) \leq 1$ and $z : [0,1] \to [0,1]$ is increasing.*

Theorem 1.1.23 provides another way of how to define *BV*-functions. A proof can be found in [58].

It is well known that the space *Lip* is a linear space. For a function $x \in Lip$ we define its *Lipschitz constant* by

$$\mathrm{lip}(x) = \sup_{\substack{0 \leq s,t \leq 1 \\ s \neq t}} \frac{|x(s) - x(t)|}{|s - t|}. \tag{1.1.29}$$

For functions $x : [a,b] \to \mathbb{R}$ we write $\mathrm{lip}(x,[a,b])$ instead of $\mathrm{lip}(x)$. Similar notions are defined for $Lip(\mathbb{R})$.

One can easily show that lip is a seminorm on *Lip* and that the space *Lip* when endowed with the norm $\|x\| = |x(0)| + \mathrm{lip}(x)$ is a Banach algebra. Moreover, lip satisfies (1.1.20) with $\alpha = 1$, and so Proposition 1.1.15 ensures that the norm

$$\|x\|_{Lip} := \|x\|_\infty + \mathrm{lip}(x) \tag{1.1.30}$$

makes *Lip* a normalized Banach algebra.

Sometimes, functions $x : \mathbb{R} \to \mathbb{R}$ belong to a given function space X only when restricted to an interval. In this case, we write

$$X_{loc}(\mathbb{R}) := \Big\{x : \mathbb{R} \to \mathbb{R} \mid x \in X[a,b] \text{ for any interval } [a,b] \subseteq \mathbb{R}\Big\}.$$

One prominent example of such a space is the family

$$Lip_{loc}(\mathbb{R}) = \{x : \mathbb{R} \to \mathbb{R} \mid x \in Lip[a,b] \text{ for any interval } [a,b] \subseteq \mathbb{R}\}$$

of *locally Lipschitz continuous* functions that play an important role in the theory of *BV*-functions. For instance, the composition $g \circ x$ of a function $g : \mathbb{R} \to \mathbb{R}$ with any function $x \in BV$ is again a *BV*-function if and only if $g \in Lip_{loc}(\mathbb{R})$; we will investigate this and related phenomena in the Chapters 2 and 4 in much more detail.

Observe that the spaces $Lip_{loc}(\mathbb{R})$ and $Lip(\mathbb{R})$ have to be strictly distinguished. For instance, the function $t \to t^2$ for $t \in \mathbb{R}$ belongs to $Lip_{loc}(\mathbb{R})$ but not to $Lip(\mathbb{R})$. Similarly, convergence of a function sequence (x_n) in $Lip_{loc}(\mathbb{R})$ now cannot be considered with respect to the norm (1.1.30) anymore. Instead, we say that (x_n) converges in $Lip_{loc}(\mathbb{R})$ if and only if (x_n) converges in $Lip[a,b]$ for any two real numbers $a < b$.

Example 1.1.24. The functions $x_n : \mathbb{R} \to \mathbb{R}$, defined by $x_n(t) = \max\{|t| - n, 0\}$ are Lipschitz continuous throughout $\mathbb{R}$ with $\mathrm{lip}(x_n, \mathbb{R}) = 1$ for all $n \in \mathbb{N}$. However, $x_n(t) = 0$ for all $t \in [-n,n]$. This implies that $x_n \in Lip_{loc}(\mathbb{R})$ with $\mathrm{lip}(x_n,[a,b]) = 0$ for all $[a,b] \subseteq [-n,n]$ and $n \in \mathbb{N}$. We conclude that (x_n) converges in $Lip_{loc}(\mathbb{R})$ to $\mathbb{0}$, but diverges in $Lip(\mathbb{R})$. ♢

It is clear that *Lip* is a subspace of *AC*. One could ask if the characterization given in Theorem 1.1.17 still holds if one replaces $x' \in L_1$ with the requirement that x' is (bounded and) Riemann integrable. Such functions are due to the formula (1.1.23) also Lipschitz continuous with $\mathrm{lip}(x) = \|x'\|_{L_\infty}$. However, not every Lipschitz continuous function has a Riemann integrable derivative, even if the function is differentiable everywhere.

Example 1.1.25. In 1881 Vito Volterra constructed in [149] an everywhere differentiable function $v : [0,1] \to \mathbb{R}$ with $v(0) = v(1) = 0$ and a bounded (and hence Lebesgue integrable) but not Riemann integrable derivative. But this function v is then clearly Lipschitz continuous. ♢

Moreover, there are functions that are absolutely continuous but not Lipschitz continuous.

Example 1.1.26. The function $x : [0,1] \to \mathbb{R}$, defined by $x(t) = \sqrt{t}$, is absolutely continuous which can be shown easily with the help of Theorem 1.1.20. But it is not Lipschitz continuous, since otherwise we had

$$|x(0) - x(t)| = \sqrt{t} \leq Lt$$

and hence $1/L \leq \sqrt{t}$ for some $L > 0$ and all $t \in (0,1]$ which is not possible. ♢

Thus, Lipschitz continuous functions are precisely absolutely continuous functions with essentially bounded derivative.

In fact, the square root function from Example 1.1.26 is only Hölder continuous with exponent 1/2; this means that it satisfies an estimate of the form

$$|x(s) - x(t)| \leq L|s-t|^\alpha \quad \text{for } s,t \in [0,1] \tag{1.1.31}$$

with $\alpha = 1/2$. Functions x satisfying (1.1.31) for arbitrary $0 < \alpha \leq 1$ are called *Hölder continuous with exponent* α, and we write Lip_α for the space of such functions. Observe that this definition makes sense for $0 < \alpha \leq 1$ only, because for $\alpha \leq 0$ the condition (1.1.31) would not necessarily imply that x is continuous, and for $\alpha > 1$ any function satisfying (1.1.31) must be constant.

Even if *AC*-functions are differentiable only almost everywhere, formula (1.1.23) allows us to recover some regularity properties from their derivatives. This we will be very useful in Section 5.1.

Lemma 1.1.27. *Let $x : [a,b] \to \mathbb{R}$ be absolutely continuous, and let $D \subseteq [a,b]$ be the points of differentiability of x. Then the following statements hold.*

(a) If $x'|_D$ is uniformly continuous, then $x \in C^1[a,b]$ and x' is uniformly continuous on $[a,b]$.

(b) If $x'|_D$ is Lipschitz continuous, then $x \in C^1[a,b]$ and x' is Lipschitz continuous on $[a,b]$.

(c) If $x'|_D$ is constant, then x is affine on $[a,b]$.

(d) If $x'|_D$ is zero, then x is constant on $[a,b]$.

Proof. (a) By [142, Section 13, Theorem D] the restriction $x'|_D : D \to \mathbb{R}$ has a (unique) uniformly continuous extension $y : [a,b] \to \mathbb{R}$. Using formula (1.1.23) we obtain

$$x(t) = x(a) + \int_a^t x'(\tau)\,\mathrm{d}\tau = x(a) + \int_a^t y(\tau)\,\mathrm{d}\tau \quad \text{for } t \in [a,b].$$

But the Fundamental Theorem of Calculus now says that x indeed belongs to $C^1[a,b]$ and $x' = y$ is uniformly continuous on $[a,b]$.
(b) As shown in (a), the function x belongs to $C^1[a,b]$ and $x' = y$ is uniformly continuous, where y is again the unique uniformly continuous extension of $x'|_D$. In order to show that y is even Lipschitz continuous, fix $s,t \in [a,b]$. Since D is dense in $[a,b]$ we find sequences (s_n) and (t_n) in D converging to s and t, respectively. Since $x'|_D$ is Lipschitz continuous on D, we find some $L > 0$ independent of s and t such that

$$|y(s_n) - y(t_n)| = |x'(s_n) - x'(t_n)| \leq L|s_n - t_n| \quad \text{for all } n \in \mathbb{N}.$$

Letting $n \to \infty$ yields due to the continuity of y that

$$|y(s) - y(t)| \leq L|s-t|.$$

But this shows nothing else than that $x' = y$ is indeed Lipschitz continuous on $[a,b]$.
(c) By (a) we conclude that x is continuously differentiable on $[a,b]$. Thus, if $x'(t) = c$ for all $t \in D$, then $x' \equiv c$ throughout $[a,b]$, because D is dense and x' is continuous. This implies that x is affine.
(d) follows immediately from (c), because in the proof of (c) we can now put $c = 0$. ■

Note that in Lemma 1.1.27 we cannot get a better result than (a), that is, we cannot replace "uniformly continuous" by "continuous" there. The first example illustrating this is a function that is even Lipschitz continuous but not everywhere differentiable.

Example 1.1.28. The function $x : [0,1] \to \mathbb{R}$, $t \mapsto \max\{2t, 1\}$, is absolutely (even Lipschitz) continuous on $[0,1]$ and differentiable at every point of the union $D = [0,1/2) \cup (1/2,1]$ with $x' = 2\chi_{(1/2,1]}$. In particular, x' is continuous at every point of D. However, x is not continuously differentiable. Note that x' is not uniformly continuous on D, since

$$\Big|x'(1/2-\delta) - x'(1/2+\delta)\Big| = 2$$

for all $\delta \in (0,1/2)$. ♢

The second example is a function that is everywhere differentiable but has a discontinuous yet Lebesgue integrable derivative.

Example 1.1.29. The function $x : [0,1] \to \mathbb{R}$, defined by

$$x(t) = \begin{cases} t^2 \cos\dfrac{1}{t} - 2\displaystyle\int_0^t s\cos\frac{1}{s}\,\mathrm{d}s & \text{for } 0 < t \leq 1, \\ 0 & \text{for } t = 0, \end{cases}$$

is differentiable with

$$x'(t) = \sin\frac{1}{t} = \varphi_{0,0,1}(t) \quad \text{for } 0 < t \leq 1,$$

where the notation is borrowed from (1.1.1). At $t = 0$ we obtain with the help of de L'Hospital's rule

$$x'(0) = \lim_{t\to 0+} \frac{x(t) - x(0)}{t} = \lim_{t\to 0+} \left(t\cos\frac{1}{t} - \frac{2}{t}\int_0^t s\cos\frac{1}{s}\,\mathrm{d}s\right) = \lim_{t\to 0+} t\cos\frac{1}{t} = 0,$$

and consequently $x' = \varphi_{0,0,1}$ on $[0,1]$; in particular, x' is measurable and $|x'| \leq 1$ on $[0,1]$ which shows $x' \in L_\infty$. However, by Proposition 1.1.12 (c), the function $x' = \varphi_{0,0,1}$ is not continuous (at $t = 0$). ♢

Finally, the last classes we discuss here in this section are the classes C^n of n-times continuously differentiable functions on $[0,1]$, where we set $C^0 = C$. As usual, limits at the boundary points of $[0,1]$ are considered to be one-sided. It is clear that

$$C^n \subseteq C^m \quad \text{for } m \leq n,$$

and the inclusions are strict for $m < n$ which is shown by the following example.

Example 1.1.30. For $n \in \mathbb{N}_0$ consider the function $x_n : [0,1] \to \mathbb{R}$, defined by

$$x_n(t) = \frac{1}{(n+1)!}\left(t - \frac{1}{2}\right)^n \left|t - \frac{1}{2}\right|,$$

where we agree on $0^0 = 1$. Then it is straightforward to show that x_n is n-times continuously differentiable with k-th derivative

$$x_n^{(k)} = x_{n-k} \quad \text{for } 0 \leq k \leq n.$$

In particular, the n-th derivative

$$x_n^{(n)}(t) = \left|t - \frac{1}{2}\right| \quad \text{for } 0 \leq t \leq 1$$

is continuous, but not differentiable. This shows $x_n \in C^n \backslash C^{n+k}$ for all $n, k \in \mathbb{N}$. ♢

With the help of the Mean Value Theorem it is easily shown that

$$C^1 \subseteq Lip.$$

However, this inclusion is also strict, as the function $x_0(t) = |t - 1/2|$ from Example 1.1.30 shows.

In total, we have the chain of strict inclusions

$$C^n \subsetneq C^1 \subsetneq Lip \subsetneq AC \subsetneq C \cap BV \subsetneq BV \subsetneq R \subsetneq B \cap L_\infty \subsetneq L_\infty \subsetneq L_p \qquad (1.1.32)$$

for $n \in \mathbb{N}$ with $n > 1$ and $p \in \mathbb{R}$ with $p > 1$.

1.2 Classes of Functions of Generalized Bounded Variation

In this section we introduce spaces of functions of generalized bounded variation that will be studied in this thesis. Since those variations are more complicated to define than the Jordan variation to construct examples and counterexamples we will consider not the "smooth" functions $\varphi_{\alpha,\beta,n}$ defined in (1.1.1) but the following "jump" functions instead. For a real sequence (α_j) we define the function $\mathfrak{J}_{(\alpha_j)} : [0,1] \to \mathbb{R}$ by

$$\mathfrak{J}_{(\alpha_j)}(t) = \begin{cases} \alpha_j & \text{if } t = \frac{1}{2j} \text{ for some } j \in \mathbb{N}, \\ 0 & \text{otherwise.} \end{cases} \tag{1.2.1}$$

It is clear that $\mathfrak{J}_{(\alpha_j)}$ belongs to L_∞ and hence to all Lebesgue spaces for any sequence (α_j) and to either of the spaces C and D only if $\alpha_j = 0$ for all $j \in \mathbb{N}$. So for these spaces the functions (1.2.1) are of no interest. However, $\mathfrak{J}_{(\alpha_j)} \in B$ if and only if (α_j) is bounded, and $\mathfrak{J}_{(\alpha_j)} \in BV$ if and only if $\sum_{j=1}^\infty \alpha_j$ converges absolutely. To be more precise, $\operatorname{supp}(\mathfrak{J}_{(\alpha_j)}) \subseteq \{\frac{1}{2}, \frac{1}{4}, \frac{1}{6}, \ldots\}$ and by Proposition 1.1.8,

$$\operatorname{Var}\left(\mathfrak{J}_{(\alpha_j)}\right) = 2\sum_{j=1}^\infty |\alpha_j|. \tag{1.2.2}$$

For further reference, we summarize this as

Corollary 1.2.1. *The function $\mathfrak{J}_{(\alpha_j)}$ belongs to BV if and only if*

$$\sum_{j=1}^\infty |\alpha_j| < \infty.$$

Its variation can be calculated explicitly by (1.2.2).

As a special case we write for $\alpha \in \mathbb{R}$

$$\mathfrak{J}_\alpha := \mathfrak{J}_{(a_j)} \quad \text{for } \alpha_j = \frac{1}{j^\alpha} \text{ and all } j \in \mathbb{N}. \tag{1.2.3}$$

Then Corollary 1.2.1 says that $\mathfrak{J}_\alpha \in BV$ if and only if $\alpha > 1$. In this case, (1.2.2) yields

$$\operatorname{Var}(\mathfrak{J}_\alpha) = 2\sum_{j=1}^\infty \frac{1}{j^\alpha} = 2\zeta(\alpha),$$

where ζ denotes the Riemann zeta function. In particular, $\operatorname{Var}(\mathfrak{J}_2) = 2\zeta(2) = \pi^2/3$.

Since functions with countable support will be of great importance later on, we put a particular emphasis in investigating such functions in this section.

The Wiener Variation

We start with a self-evident generalization of Definition 1.1.7.

Definition 1.2.2. For a function $x : [0,1] \to \mathbb{R}$ and a real number $p \geq 1$ we call the possibly infinite number

$$\operatorname{Var}_p(x) = \sup_P \sum_{j=1}^{n} |x(t_{j-1}) - x(t_j)|^p \tag{1.2.4}$$

the *Wiener variation of x on* $[0,1]$, where the supremum is taken over all finite partitions $P : 0 = t_0 < \ldots < t_n = 1$ of $[0,1]$. If $\operatorname{Var}_p(x) < \infty$, we say that x has bounded Wiener variation and write $x \in WBV_p$.

Note that for $p = 1$ the space WBV_p reduces to BV. Thus, unless otherwise stated we always tacitly assume $p > 1$.

We test Definition 1.2.2 on functions with countable support. Naively, our intuition might make us believe that there is a perfect analogue to formula (1.1.12), namely something like

$$\operatorname{Var}_p(x) = |x(0)|^p + |x(1)|^p + 2 \sum_{\tau \in \operatorname{supp}(x) \setminus \{0,1\}} |x(\tau)|^p \tag{1.2.5}$$

for functions $x \in \mathcal{S}_c$. However, this is not true, and one indication for this might be that according to Proposition 1.1.8 we had $\operatorname{Var}_p(x) = \operatorname{Var}\left(|x|^p\right)$ which looks weird. Let us make this a little more explicit.

Example 1.2.3. Let us consider the function $x := \mathfrak{J}_{(\alpha_j)}$ with $\alpha_1 = 1, \alpha_2 = -1$ and $\alpha_j = 0$ for all $j \geq 3$. Then the right hand side of (1.2.5) becomes 4. However, if we take the partition $t_0 := 0, t_1 := 1/4, t_2 := 1/2, t_3 := 1$, then

$$\operatorname{Var}_p(x) \geq |x(0) - x(\tfrac{1}{4})|^p + |x(\tfrac{1}{4}) - x(\tfrac{1}{2})|^p + |x(\tfrac{1}{2}) - x(1)|^p = 2 + 2^p$$

which is larger than 4 for $p > 1$. Thus, formula (1.2.5) cannot be true in general.

A little more scrutiny shows $\operatorname{Var}_p(x) = 2 + 2^p$. Indeed, for an arbitrary partition $0 = t_0 < \ldots < t_n$ there are two possibilities: Either there exists some $k \in \{1, \ldots, n-1\}$ with $\frac{1}{4} = t_{k-1} < t_k = \frac{1}{2}$, or there is no such k. If such a k exists, then

$$\begin{aligned}\sum_{j=1}^{n} |x(t_{j-1}) - x(t_j)|^p &= \sum_{j=1}^{k-1} |x(t_{j-1}) - x(t_j)|^p + |x(t_{k-1}) - x(t_k)|^p + |x(t_k) - x(t_{k+1})|^p \\ &\quad + \sum_{j=k+2}^{n} |x(t_{j-1}) - x(t_j)|^p \\ &= 2 + 2^p.\end{aligned}$$

If not, then for each $k \in \{1, \ldots, n\}$ we have $t_{k-1} \neq \frac{1}{4}$ or $t_k \neq \frac{1}{2}$. But in this case,

$$\sum_{j=1}^{n} |x(t_{j-1}) - x(t_j)|^p \leq 4 \leq 2 + 2^p.$$

Consequently, since $\operatorname{Var}_p(x)$ is the supremum of all such sums, $\operatorname{Var}_p(x) = 2 + 2^p$. $\diamondsuit$

In contrast to Proposition 1.1.8 we can now not calculate the Wiener variation of a function with countable support explicitly but get at least some estimates from above and below.

Proposition 1.2.4. *For $x \in \mathcal{S}_c$ we have*

$$\operatorname{Var}\left(|x|^p\right) = \operatorname{Var}_p\left(|x|\right) \leq \operatorname{Var}_p(x) \leq 2^{p-1} \operatorname{Var}\left(|x|^p\right), \tag{1.2.6}$$

where

$$\operatorname{Var}\left(|x|^p\right) = |x(0)|^p + |x(1)|^p + 2 \sum_{\tau \in \operatorname{supp}(x) \setminus \{0,1\}} |x(\tau)|^p$$

according to (1.1.12). In particular, $x \in WBV_p$ if and only if $|x|^p \in BV$.

Proof. To see this first note that $t \mapsto |t|^p$ is convex and thus superadditive; in particular, $|a-b|^p \leq |a^p - b^p|$ for all $a, b \geq 0$. From this it follows that $\operatorname{Var}_p(|x|) \leq \operatorname{Var}(|x|^p)$. By considering the same special partition as in the proof of Proposition 1.1.8 one can show that also $\operatorname{Var}_p(|x|) \geq \operatorname{Var}(|x|^p)$. This shows the equality in (1.2.6).
Since $\big||a| - |b|\big| \leq |a-b|$ for all $a, b \in \mathbb{R}$, also the first inequality in (1.2.6) is proven.
For the right inequality consider an arbitrary partition $0 = t_0 < \ldots < t_n = 1$ of $[0,1]$. Again, since $t \mapsto |t|^p$ is convex, we have $|a-b|^p \leq 2^{p-1}\big(|a|^p + |b|^p\big)$ for all $a, b \in \mathbb{R}$. This implies

$$\begin{aligned}
\sum_{j=1}^{n} \left|x(t_{j-1}) - x(t_j)\right|^p &\leq 2^{p-1} \sum_{j=1}^{n} \left(|x(t_{j-1})|^p + |x(t_j)|^p\right) \\
&\leq 2^{p-1} \left(|x(0)|^p + |x(1)|^p + 2 \sum_{\tau \in \operatorname{supp}(x) \setminus \{0,1\}} |x(\tau)|^p\right) = 2^{p-1} \operatorname{Var}\left(|x|^p\right),
\end{aligned}$$

again by Proposition 1.1.8. This yields the claim. ∎

For our functions $\mathfrak{J}_{(\alpha_j)}$ this means the following.

Corollary 1.2.5. *The function $\mathfrak{J}_{(\alpha_j)}$ satisfies*

$$2 \sum_{j=1}^{\infty} |\alpha_j|^p \leq \operatorname{Var}_p\left(\mathfrak{J}_{(\alpha_j)}\right) \leq 2^p \sum_{j=1}^{\infty} |\alpha_j|^p; \tag{1.2.7}$$

in particular, it belongs to WBV_p if and only if the series in (1.2.7) converges.

Example 1.2.3 has shown that we cannot expect equality in either of these two inequalities in (1.2.7). However, observe that (1.2.7) contains (1.2.2) for $p = 1$.

For functions with countable support we deduce from (1.2.6) for $p = 1$ that $\operatorname{Var}(|x|) = \operatorname{Var}(x)$. However, even for $p = 1$ the right inequality in formula (1.2.6) is false for functions x having uncountable support.

Example 1.2.6. The function $x := 2\chi_{\mathbb{Q} \cap [0,1]} - \mathbb{1}$ is clearly of unbounded variation on $[0,1]$, but $|x|$ is equal to $\mathbb{1}$ and thus has variation 0. Note that $\operatorname{supp}(x) = [0,1]$. ◇

One can show that the equivalence $x \in BV \Leftrightarrow |x| \in BV$ is true for *continuous* functions [6].

In can be shown that WBV_p when endowed with the norm

$$\|x\| = |x(0)| + \mathrm{Var}_p(x)^{1/p} \tag{1.2.8}$$

is a Banach algebra, and that this norm satisfies the estimate (1.1.20) with $\Phi(x) = \mathrm{Var}_p(x)^{1/p}$ and $\alpha = 1$. Therefore, by Proposition 1.1.15, WBV_p together with the norm

$$\|x\|_{WBV_p} := \|x\|_\infty + \mathrm{Var}_p(x)^{1/p} \tag{1.2.9}$$

is a normalized Banach algebra.

It is clear that $BV \subseteq WBV_p$ for any $p \geq 1$. More generally, we have

$$BV \subseteq WBV_p \subseteq WBV_q \subseteq R \qquad \text{for } 1 \leq p \leq q. \tag{1.2.10}$$

All those inclusions are in fact continuous embeddings. For instance, for the last inclusion we have

$$|x(t)| \leq |x(0)| + \left(|x(t) - x(0)|^p\right)^{1/p} \leq |x(0)| + \mathrm{Var}_p(x)^{1/p}$$

and hence

$$\|x\|_\infty \leq |x(0)| + \mathrm{Var}_p(x)^{1/p} \leq \|x\|_{WBV_p}; \tag{1.2.11}$$

in particular, the two norms in (1.2.8) and (1.2.9) are equivalent by Proposition 1.1.15. That for $1 < p < q$ the inclusions in (1.2.10) are strict can now also be seen with the help of the functions $\mathfrak{J}_\alpha$ introduced in (1.2.3). Corollary 1.2.5 says that $\mathfrak{J}_\alpha$ belongs to WBV_p if and only if $\alpha p > 1$. In particular, for $1 < p < q$ we have $\mathfrak{J}_{1/p} \in (R \cap WBV_q) \backslash WBV_p$. In addition to the inclusions in (1.2.10) we also have

$$Lip_{1/p} \subseteq WBV_p \quad \text{for } p \geq 1,$$

which is a generalization of the inclusion $Lip \subseteq BV$.

In this context the following analogue to Theorem 1.1.23 is noteworthy.

Theorem 1.2.7. *A function $x : [0,1] \to \mathbb{R}$ belongs to WBV_p if and only if it may be represented as a composition $x = y \circ z$, where $y \in Lip_{1/p}$ and $z : [0,1] \to [0,1]$ is increasing.*

The Young Variation

Even more general than the Wiener variation is the Young variation which has been introduced 1937 by Laurence Chisholm Young in [160, 161]. The weighting function $t \mapsto |t|^p$ that has been used to weight the absolute values $|x(t_{j-1}) - x(t_j)|$ in Definition 1.2.2 is now replaced by a general convex function.

Definition 1.2.8. A function $\varphi : [0, \infty) \to [0, \infty)$ is said to be a Young function (or φ-function) if it is convex and such that $\varphi(t) = 0$ if and only if $t = 0$.

Note that according to this definition every Young function is continuous, strictly increasing and so that $\varphi(t) \to \infty$ as $t \to \infty$. Moreover, due to convexity, $\varphi(st) \leq s\varphi(t)$ for all $s \in [0, 1], t \in [0, \infty)$, as well as $\varphi(st) \geq s\varphi(t)$ for all $s \in [1, \infty), t \in [0, \infty)$.

Using the notion of a Young function the Young variation is now defined as follows.

Definition 1.2.9. For a function $x : [0, 1] \to \mathbb{R}$ and a Young function $\varphi : [0, \infty) \to [0, \infty)$ we call the possibly infinite number

$$\mathrm{Var}_{\varphi}(x) = \sup_{P} \sum_{j=1}^{n} \varphi\Big(|x(t_{j-1}) - x(t_j)|\Big)$$

the *Young variation (or φ-variation) of x on* $[0, 1]$, where the supremum is taken over all finite partitions $P : 0 = t_0 < \ldots < t_n = 1$ of $[0, 1]$.
Finally, we denote by

$$YBV_{\varphi} := \{x : [0, 1] \to \mathbb{R} \mid \mathrm{Var}_{\varphi}(\lambda x) < \infty \text{ for some } \lambda > 0\}$$

the space of functions of bounded Young variation.
Observe that $\varphi(t) = t^p$ for $p \geq 1$ is a Young function, and in this case, $YBV_{\varphi} = WBV_p$.

As in the Propositions 1.1.8 and 1.2.4 we will also for the Young variation consider functions with countable support.

Proposition 1.2.10. *Let φ be an arbitrary Young function. For $x \in \mathcal{S}_c$ we have*

$$\mathrm{Var}\left(\varphi(|x|)\right) = \mathrm{Var}_{\varphi}\left(|x|\right) \leq \mathrm{Var}_{\varphi}(x) \leq 2^{-1}\,\mathrm{Var}\left(\varphi(2|x|)\right), \tag{1.2.12}$$

where

$$\mathrm{Var}\left(\varphi(\lambda|x|)\right) = \varphi(\lambda|x(0)|) + \varphi(\lambda|x(1)|) + 2 \sum_{\tau \in \mathrm{supp}(x) \setminus \{0,1\}} \varphi\Big(\lambda|x(\tau)|\Big) \quad \text{for } \lambda > 0$$

according to (1.1.12). In particular, $x \in YBV_{\varphi}$ if and only if $\varphi\Big(\lambda|x|\Big) \in BV$ for some $\lambda > 0$.

Proof. We proceed as in the proof of Proposition 1.2.4. First note that φ is convex and hence superadditive; in particular, $\varphi(|a - b|) \leq |\varphi(a) - \varphi(b)|$ for all $a, b \geq 0$. From this it follows that $\mathrm{Var}_{\varphi}\left(|x|\right) \leq \mathrm{Var}\left(\varphi(|x|)\right)$. By considering the same special partition

as in the proof of Proposition 1.1.8 one can show that also $\mathrm{Var}_\varphi\left(|x|\right) \geq \mathrm{Var}\left(\varphi(|x|)\right)$. This shows the equality in (1.2.12).
Since φ is increasing and $\Big||a|-|b|\Big| \leq |a-b|$ for all $a, b \in \mathbb{R}$, also the inequality in the middle of (1.2.12) is proven.
For the remaining inequality consider an arbitrary partition $0 = t_0 < \ldots < t_n = 1$ of $[0,1]$. Again, since φ is increasing and convex we have $\varphi(|a-b|) \leq 2^{-1}\big[\varphi(2|a|)+\varphi(2|b|)\big]$ for all $a, b \in \mathbb{R}$. This implies

$$\begin{aligned}\sum_{j=1}^{n}\varphi\Big(|x(t_{j-1})-x(t_j)|\Big) &\leq 2^{-1}\sum_{j=1}^{n}\Big[\varphi\Big(2|x(t_{j-1})|\Big)+\varphi\Big(2|x(t_j)|\Big)\Big]\\ &= 2^{-1}\left(\varphi\Big(2|x(0)|\Big)+\varphi\Big(2|x(1)|\Big)+2\sum_{j=1}^{n-1}\varphi\Big(2|x(t_j)|\Big)\right)\\ &\leq 2^{-1}\left(\varphi\Big(2|x(0)|\Big)+\varphi\Big(2|x(1)|\Big)+2\sum_{\tau\in\mathrm{supp}(x)\setminus\{0,1\}}\varphi\Big(2|x(\tau)|\Big)\right)\\ &= 2^{-1}\,\mathrm{Var}\left(\varphi(2|x|)\right),\end{aligned}$$

where the last equality comes from (1.1.12). Thus, also the right inequality in (1.2.12) is true, and this completes the proof. ■

For our functions $\mathfrak{J}_{(\alpha_j)}$ this means the following.

Corollary 1.2.11. *The function $\mathfrak{J}_{(\alpha_j)}$ satisfies*

$$2\sum_{j=1}^{\infty}\varphi\Big(\lambda|\alpha_j|\Big) \leq \mathrm{Var}_\varphi\left(\lambda\mathfrak{J}_{(\alpha_j)}\right) \leq \sum_{j=1}^{\infty}\varphi\Big(2\lambda|\alpha_j|\Big) \quad \textit{for } \lambda > 0. \tag{1.2.13}$$

Moreover, it belongs to YBV_φ if and only if at least one of the two series in (1.2.13) converges for some $\lambda > 0$.

Proposition 1.2.10 explains why we have not defined the space YBV_φ by the set

$$YBV_\varphi^* := \{x : [0,1] \to \mathbb{R} \mid \mathrm{Var}_\varphi(x) < \infty\}$$

as we have done so for the other variations. The reason is that YBV_φ^* is not a linear space, since it is not closed under multiplication with scalars.

Example 1.2.12. The function $\varphi : [0,\infty) \to [0,\infty)$, defined by

$$\varphi(t) = \begin{cases} t\exp(-1/t) & \text{for } t > 0,\\ 0 & \text{for } t = 0,\end{cases}$$

is a Young function. Putting $\alpha_1 := 0$ and $\alpha_j := \frac{1}{4\log(j)}$ for $j \in \mathbb{N}, j \geq 2$, the function $x := \mathfrak{J}_{(\alpha_j)}$, given by (1.2.1), is nonnegative and has countable support. Moreover, by (1.1.12) we have with $\tau_j = 1/(2j)$,

$$\begin{aligned}\mathrm{Var}\left(\varphi(\lambda x)\right) &= 2\sum_{j=2}^{\infty}\varphi\Big(\lambda x(\tau_j)\Big) = 2\lambda\sum_{j=2}^{\infty}x(\tau_j)\exp\left(\frac{-1}{\lambda x(\tau_j)}\right)\\ &= \frac{\lambda}{2}\sum_{j=2}^{\infty}\frac{1}{j^{4/\lambda}\log(j)}.\end{aligned} \tag{1.2.14}$$

For $\lambda = 2$ the series in (1.2.14) converges which means that $\mathrm{Var}\left(\varphi(2x)\right) < \infty$. Thus, by (1.2.12), also $\mathrm{Var}_\varphi(x) < \infty$, that is, $x \in YBV_\varphi^*$. However, for $\lambda = 4$ the series in (1.2.14) diverges which means that $\mathrm{Var}\left(\varphi(4x)\right) = \infty$. Again by (1.2.12) we get that $\mathrm{Var}_\varphi(4x) = \infty$ which implies $4x \notin YBV_\varphi^*$. ◊

In general, one only has $YBV_\varphi = \mathrm{Span}(YBV_\varphi^*)$, that is, YBV_φ is the linear hull of YBV_φ^*. However, if φ satisfies a so-called *δ_2-condition*, that is,

$$\limsup_{t\to 0+} \frac{\varphi(2t)}{\varphi(t)} < \infty, \tag{1.2.15}$$

then YBV_φ^* is a linear space. In fact, the δ_2-condition is not only sufficient but also necessary for this [121]. This is exactly analogous to the properties of Orlicz spaces. Observe that the function φ in Example 1.2.12 does not satisfy a δ_2-condition, since

$$\limsup_{t\to 0+} \frac{\varphi(2t)}{\varphi(t)} = 2 \limsup_{t\to 0+} \exp\left(\frac{1}{2t}\right) = \infty.$$

The δ_2-condition, however, has also some other benefits, because it makes explicit calculations easier. For φ satisfying a δ_2-condition, the function $M : [0,\infty] \to [0,\infty]$, set by $M(0) := 0$, $M(\infty) := \infty$ and

$$M(T) := \sup_{0<t\le T} \frac{\varphi(2t)}{\varphi(t)} \quad \text{for } T > 0 \tag{1.2.16}$$

is well-defined and increasing. We then have

$$\mathrm{Var}_\varphi(2x) \le M\left(2\left\|x\right\|_\infty\right) \mathrm{Var}_\varphi(x) \quad \text{for } x \in YBV_\varphi, \tag{1.2.17}$$

and Proposition 1.2.10 reduces to the following familiar form.

Corollary 1.2.13. *Let φ be a Young function satisfying a δ_2-condition. For $x \in \mathcal{S}_c$ we have*

$$\mathrm{Var}\left(\varphi(|x|)\right) = \mathrm{Var}_\varphi\left(|x|\right) \le \mathrm{Var}_\varphi(x) \le 2^{-1} M\left(2\left\|x\right\|_\infty\right) \mathrm{Var}\left(\varphi(|x|)\right), \tag{1.2.18}$$

where M is as in (1.2.16) and

$$\mathrm{Var}\left(\varphi(|x|)\right) = \varphi(|x(0)|) + \varphi(|x(1)|) + 2 \sum_{\tau\in\mathrm{supp}(x)\setminus\{0,1\}} \varphi\left(|x(\tau)|\right)$$

according to (1.1.12). In particular, $x \in YBV_\varphi$ if and only if $\varphi \circ |x| \in BV$.

Note that for the special Young function $\varphi(t) = t^p$ we have $M(T) = 2^p$ for all $T > 0$ and hence Corollary 1.2.13 reduces to Proposition 1.2.4.

In order to equip YBV_φ with a norm, we consider the set $\{x \in YBV_\varphi \mid \mathrm{Var}_\varphi(x) \le 1\}$ which is absorbing, balanced and convex. Thus, the Minkowski functional

$$\mathfrak{M} : YBV_\varphi \to [0,\infty), \; x \mapsto \inf\{\lambda > 0 \mid \mathrm{Var}_\varphi(x/\lambda) \le 1\} \tag{1.2.19}$$

is a seminorm on YBV_φ [116] which exhibits the following properties.

Proposition 1.2.14. *If three functions $x, y, z : [0,1] \to \mathbb{R}$ with $x, y \in YBV_\varphi$ for some constants $\alpha, \beta \geq 0$ and each $s, t \in [0,1]$ satisfy the estimate*

$$|z(s) - z(t)| \leq \alpha|x(s) - x(t)| + \beta|y(s) - y(t)|,$$

then $z \in YBV_\varphi$ with $\mathfrak{M}(z) \leq \alpha\,\mathfrak{M}(x) + \beta\,\mathfrak{M}(y)$.

We do not give the proof here because we will prove a much more general result in Lemma 1.2.26 below. Nevertheless, from Proposition 1.2.14 follows that the Minkowski functional satisfies

$$\mathfrak{M}(xy) \leq \|x\|_\infty \mathfrak{M}(y) + \|y\|_\infty \mathfrak{M}(x)$$

for all $x, y \in YBV_\varphi$, because

$$\Big|x(s)y(s) - x(t)y(t)\Big| \leq \|x\|_\infty |y(s) - y(t)| + \|y\|_\infty |x(s) - x(t)|$$

holds for any $s, t \in [0,1]$. In particular, this shows that the space YBV_φ is closed under multiplication. Finally, equipped with the norm

$$\|x\| = |x(0)| + \mathfrak{M}(x), \tag{1.2.20}$$

the space YBV_φ is a Banach algebra which becomes normalized by Proposition 1.1.15 when (1.2.20) is replaced by

$$\|x\|_{YBV_\varphi} := \|x\|_\infty + \mathfrak{M}(x). \tag{1.2.21}$$

Similarly to BV the reciprocal of a function $x \in YBV_\varphi$ also belongs to YBV_φ if x is bounded away from zero. More precisely, if $\operatorname{supp}_\delta(x) = [0,1]$ for some $\delta > 0$, then

$$\operatorname{Var}_\varphi(1/x) \leq \operatorname{Var}_\varphi(x/\delta^2);$$

in particular,

$$\operatorname{Var}_p(1/x) \leq \delta^{-2p} \operatorname{Var}_p(x).$$

For the special case $\varphi(t) = t^p$ it is easily verified that

$$\|x\|_{YBV_\varphi} = \|x\|_\infty + \operatorname{Var}_p(x)^{1/p} = \|x\|_{WBV_p}.$$

Since the norm $\|\cdot\|_{YBV_\varphi}$ is defined via the unhandy Minkowski functional, it is difficult to estimate norms of YBV_φ-functions. If a δ_2-condition is assumed, we can consider the variations themselves instead.

Proposition 1.2.15. *Let φ be a Young function and let (x_n) be a sequence in YBV_φ. The following statements hold.*

(a)

$$\left(\forall \lambda > 0 : \lim_{n\to\infty} \operatorname{Var}_\varphi(\lambda x_n) = 0\right) \iff \lim_{n\to\infty} \mathfrak{M}(x_n) = 0.$$

(b) If φ satisfies a δ_2-condition and (x_n) is a bounded sequence in B, then

$$\lim_{n\to\infty} \mathrm{Var}_\varphi(x_n) = 0 \quad \Longleftrightarrow \quad \lim_{n\to\infty} \mathfrak{M}(x_n) = 0.$$

Proof. Part (a) has been proven in [120]. For (b) note that we only have to prove "$\Rightarrow$", because the converse follows from (a). To this end, fix $\varepsilon > 0$ and pick $m \in \mathbb{N}$ so large that $2^{-m} \le \varepsilon$. Let (x_n) be a sequence in YBV_φ which is bounded in B with $\mathrm{Var}_\varphi(x_n) \to 0$ as $n \to \infty$. Then there are $N \in \mathbb{N}$ and $C > 0$ such that $\|x_n\|_\infty \le C/2$ and $\mathrm{Var}_\varphi(x_n) \le M(2^m C)^{-m}$ for all $n \ge N$, where M denotes the function in (1.2.16). For a partition $0 = t_0 < \ldots < t_k = 1$ of $[0, 1]$ we obtain $|x_n(t_{j-1}) - x_n(t_j)| \le 2\,\|x_n\|_\infty \le C$, and from that we get

$$\begin{aligned}\sum_{j=1}^k \varphi\Big(2^m|x_n(t_{j-1}) - x_n(t_j)|\Big) &\le M(2^m C)^m \sum_{j=1}^k \varphi\Big(|x_n(t_{j-1}) - x_n(t_j)|\Big)\\ &\le M(2^m C)^m \,\mathrm{Var}_\varphi(x_n).\end{aligned}$$

Consequently, $\mathrm{Var}_\varphi(2^m x_n) \le M(2^m C)^m \,\mathrm{Var}_\varphi(x_n) \le 1$ and hence $\mathfrak{M}(x_n) \le 2^{-m} \le \varepsilon$ for all $n \ge N$. This shows $\mathfrak{M}(x_n) \to 0$ as $n \to \infty$. ■

Finally, in order to compare two spaces YBV_φ and YBV_ψ for two Young functions φ and ψ, we write $\psi \preceq \varphi$ if

$$\limsup_{t\to 0+} \frac{\psi(\lambda t)}{\varphi(t)} < \infty \qquad \text{for some } \lambda > 0, \tag{1.2.22}$$

and we also write $\psi \prec \varphi$ if $\psi \preceq \varphi$ and $\varphi \not\preceq \psi$. With this notation at hand it was shown in [42] that $YBV_\varphi \subseteq YBV_\psi$ if and only if $\psi \preceq \varphi$. In particular, two Young spaces YBV_φ and YBV_ψ coincide if and only if $\psi \preceq \varphi$ and $\varphi \preceq \psi$, and so $\psi \prec \varphi$ is equivalent to the strict inclusion $YBV_\varphi \subsetneq YBV_\psi$. For instance, the functions $\varphi(t) = t$ and $\psi(t) = t^2$ are both Young functions with $\psi \prec \varphi$. This means that $YBV_\varphi = BV \subsetneq YBV_\psi = WBV_2$, in accordance with (1.2.10).
In general, the chain of strict inclusions

$$BV \subsetneq YBV_\varphi \subsetneq YBV_\psi \subsetneq R \tag{1.2.23}$$

is true for $\psi \prec \varphi \prec \iota$, where $\iota(t) := t$ and $YBV_\iota = BV$, and it can be shown that all inclusions are continuous embeddings. For instance, for $x \in YBV_\psi$ and $\lambda > 0$ with $\mathrm{Var}_\psi(x/\lambda) \le 1$ we have

$$\begin{aligned}|x(t)| &\le |x(0)| + \lambda\psi^{-1}\Big(\psi(|x(t) - x(0)|/\lambda)\Big) \le |x(0)| + \lambda\psi^{-1}\Big(\mathrm{Var}_\psi(x/\lambda)\Big)\\ &\le |x(0)| + \lambda\psi^{-1}(1).\end{aligned}$$

Taking the infimum over all such λ gives $|x(t)| \le |x(0)| + \psi^{-1}(1)\mathfrak{M}(x)$ and hence

$$\|x\|_\infty \le |x(0)| + \psi^{-1}(1)\mathfrak{M}(x) \le \max\{1, \psi^{-1}(1)\}\,\|x\|_{YBV_\psi}\,; \tag{1.2.24}$$

in particular, the two norms (1.2.20) and (1.2.21) are equivalent by Proposition 1.1.15. Note that if $\psi(t) = t^p$ we have $\psi^{-1}(1) = 1$ and hence (1.2.24) reduces to (1.2.11).

Due to the immense generality of Young functions it is much more difficult to prove that all inclusions in (1.2.23) are strict if $\psi \prec \varphi \prec \iota$. But since we need a similar argument later in Section 3.2 we give here a proof. First note that it is easy to find a decreasing sequence (β_j) of positive numbers β_j tending to 0 such that $\sum_{j=1}^{\infty} \psi(\beta_j) = \infty$. Set $\alpha_j := \sqrt{\beta_j}$ for all $j \in \mathbb{N}$ and fix $\lambda > 0$. Since $\alpha_j \to 0$ as $j \to \infty$ there is some $n \in \mathbb{N}$ such that $\lambda \geq \alpha_j$ for all $j \geq n$. This implies

$$\sum_{j=1}^{\infty} \psi(\lambda \alpha_j) \geq \sum_{j=n}^{\infty} \psi(\alpha_j^2) = \sum_{j=n}^{\infty} \psi(\beta_j) = \infty.$$

According to Corollary 1.2.11 the function $\mathfrak{J}_{(\alpha_j)}$ does not belong to YBV_ψ yet it belongs to R as (α_j) converges to 0. This shows that the last inclusion in (1.2.24) is strict. For the first and middle inclusion we would like to find a real sequence (α_j) in such a way that $\mathfrak{J}_{(\alpha_j)} \in YBV_\psi \backslash YBV_\varphi$. According to Corollary 1.2.11 this would be the case if we can arrange (α_j) so that

$$\sum_{j=1}^{\infty} \psi(\alpha_j) < \infty \qquad \text{and} \qquad \sum_{j=1}^{\infty} \varphi(\lambda \alpha_j) = \infty \quad \text{for all } \lambda > 0. \tag{1.2.25}$$

The following quite technical Lemma provides a solution and we show how to apply it after its proof.

Lemma 1.2.16. *Let $\Phi : (0, \infty) \times [0, \infty) \to [0, \infty)$ satisfy the following conditions.*

(i) For each $t \geq 0$ the function $\Phi(\cdot, t)$ is increasing.

(ii) For each $\lambda > 0$ we have $\limsup_{t \to 0+} \dfrac{\Phi(\lambda, t)}{t} = \infty$.

Then for each $\alpha > 0$ there exists a sequence (t_j) of positive real numbers such that for all fixed $\lambda > 0$

$$\sum_{j=1}^{\infty} \Phi(\lambda, t_j) = \infty \qquad \textit{and} \qquad \sum_{j=1}^{\infty} t_j \leq \alpha.$$

Proof. First assume that for each $\lambda > 0$

$$m(\lambda) := \limsup_{t \to 0+} \Phi(\lambda, t) > 0. \tag{1.2.26}$$

Then pick a monotonically decreasing sequence (λ_j) converging to zero such that

$$\sum_{j=1}^{\infty} m(\lambda_j) = \infty.$$

For each $j \in \mathbb{N}$ we find due to (1.2.26) some $0 < t_j \leq \alpha/2^j$ so that $\Phi(\lambda_j, t_j) \geq m(\lambda_j)/2$. Then

$$\sum_{j=1}^{\infty} t_j \leq \alpha \sum_{j=1}^{\infty} \frac{1}{2^j} = \alpha.$$

For fixed $\lambda > 0$ there is some $k \in \mathbb{N}$ such that $\lambda \geq \lambda_j$ for each $j \geq k$. Then with the help of (i) we obtain

$$\sum_{j=1}^{\infty} \Phi(\lambda, t_j) \geq \sum_{j=k}^{\infty} \Phi(\lambda_j, t_j) \geq \frac{1}{2} \sum_{j=k}^{\infty} m(\lambda_j) = \infty,$$

as desired.

Now assume that (1.2.26) does not hold. Then there exists some $\lambda_0 > 0$ such that

$$\lim_{t \to 0+} \Phi(\lambda_0, t) = 0.$$

By replacing $\Phi(s,t)$ by $\Phi(s/\lambda_0, t)$ we can assume that $\lambda_0 = 1$. Moreover, we get from assumption (i)

$$\lim_{t \to 0+} \Phi(\lambda, t) = 0 \qquad \text{for all } \lambda \in (0,1]. \tag{1.2.27}$$

Define

$$\beta := \frac{2}{\alpha} + 2.$$

We are going to define two sequences (n_k) in $\mathbb{N}$ and (τ_k) in $(0, \infty)$ recursively as follows: We start with $n_1 := 1$ and choose $\tau_1 > 0$ so small that

$$1 \geq \Phi(1, \tau_1) > \beta \tau_1.$$

This is possible because of (1.2.27) and (ii). Once n_k and τ_k have been constructed, we first choose $n_{k+1} \in \mathbb{N}$ so that

$$\frac{2^k}{\Phi(1/n_k, \tau_k)} + n_k - 1 < n_{k+1} \leq \frac{2^k}{\Phi(1/n_k, \tau_k)} + n_k.$$

Afterwards we choose τ_{k+1} so small that

$$1 \geq \Phi(1/n_{k+1}, \tau_{k+1}) > \beta^{k+1} \tau_{k+1}$$

holds. Again, this is possible because of the conditions (1.2.27) and (ii).

In total, this construction ensures, that (n_k) and (τ_k) satisfy the relations

$$1 \geq \Phi(1/n_k, \tau_k) > \beta^k \tau_k \qquad \text{and} \tag{1.2.28}$$

$$\frac{2^k}{\Phi(1/n_k, \tau_k)} + n_k - 1 < n_{k+1} \leq \frac{2^k}{\Phi(1/n_k, \tau_k)} + n_k \qquad \text{for all } k \in \mathbb{N}; \tag{1.2.29}$$

in particular,

$$n_{k+1} - n_k \geq 2^k - 1 \geq 1 \tag{1.2.30}$$

and $n_k \to \infty$ as $k \to \infty$.

Rearranging (1.2.29) gives

$$\Phi(1/n_k, \tau_k) > \frac{2^k}{n_{k+1} - n_k + 1} \qquad \text{and} \qquad \frac{2^k}{n_{k+1} - n_k} \geq \Phi(1/n_k, \tau_k). \tag{1.2.31}$$

Using the first inequality of (1.2.31) leads to

$$\frac{1}{n_{k+1}-n_k}=\frac{1}{2^k}\cdot\frac{2^k}{n_{k+1}-n_k}<\frac{1}{2^k}\cdot\frac{n_{k+1}-n_k+1}{n_{k+1}-n_k}\cdot\Phi(1/n_k,\tau_k), \tag{1.2.32}$$

and using (1.2.30) yields

$$\frac{1}{2^k}\cdot\frac{n_{k+1}-n_k+1}{n_{k+1}-n_k}=\frac{1}{2^k}\left(1+\frac{1}{n_{k+1}-n_k}\right)\leq\frac{1}{2^k}\left(1+\frac{1}{2^k-1}\right)=\frac{1}{2^k-1}\leq 1.$$

This and (1.2.32) gives

$$\frac{1}{n_{k+1}-n_k}<\Phi(1/n_k,\tau_k). \tag{1.2.33}$$

Finally, if we compare the second inequality of (1.2.31) with (1.2.33), we find - taking (1.2.28) into account - by the Intermediate Value Theorem some $s_k\in[1,2]$ such that

$$\Phi(1/n_k,\tau_k)=\frac{s_k^k}{n_{k+1}-n_k}\geq\beta^k\tau_k. \tag{1.2.34}$$

We now define the sequence (t_j) by

$$t_j=\tau_k \qquad \text{for } n_k\leq j\leq n_{k+1}-1 \text{ and } j,k\in\mathbb{N}.$$

Fix $\lambda>0$. Then there is some $\ell\in\mathbb{N}$ such that $\lambda\geq 1/n_k$ for all $k\geq\ell$. By using the equality in (1.2.34) and (i) we obtain

$$\begin{aligned}\sum_{j=1}^{\infty}\Phi(\lambda,t_j)&=\sum_{k=1}^{\infty}\sum_{j=n_k}^{n_{k+1}-1}\Phi(\lambda,\tau_k)\geq\sum_{k=\ell}^{\infty}\sum_{j=n_k}^{n_{k+1}-1}\Phi(1/n_k,\tau_k)\\&=\sum_{k=\ell}^{\infty}(n_{k+1}-n_k)\Phi(1/n_k,\tau_k)=\sum_{k=\ell}^{\infty}s_k^k\geq\sum_{k=\ell}^{\infty}1=\infty\end{aligned}$$

which shows that $\sum_{j=1}^{\infty}\Phi(\lambda,t_j)=\infty$. On the other hand, by using the inequality in (1.2.34) we get similarly

$$\sum_{j=1}^{\infty}t_j=\sum_{k=1}^{\infty}(n_{k+1}-n_k)\tau_k\leq\sum_{k=1}^{\infty}\frac{s_k^k}{\beta^k}\leq\sum_{k=1}^{\infty}\left(\frac{2}{\beta}\right)^k=\alpha$$

which shows that $\sum_{j=1}^{\infty}t_j\leq\alpha$. This completes the proof. ■

We are going to show in the following how to use Lemma 1.2.16 to achieve (1.2.25). Let φ and ψ be two Young functions satisfying $\varphi\not\preceq\psi$ which means

$$\limsup_{t\to 0+}\frac{\varphi(\lambda t)}{\psi(t)}=\infty \quad \text{for all } \lambda>0. \tag{1.2.35}$$

Since ψ is a homeomorphism of $[0,\infty)$ with $\psi(0)=0$ we can substitute with $s=\psi(t)$ and obtain

$$\limsup_{s\to 0+}\frac{\varphi\big(\lambda\psi^{-1}(s)\big)}{s}=\infty \quad \text{for all } \lambda>0. \tag{1.2.36}$$

We now define $\Phi : (0,\infty) \times [0,\infty) \to [0,\infty)$ by

$$\Phi(\lambda, t) := \varphi\big(\lambda\psi^{-1}(t)\big).$$

Since ψ^{-1} and φ are increasing, condition (i) of Lemma 1.2.16 is met, and (1.2.36) takes care of condition (ii). Consequently, Lemma 1.2.16 delivers us a sequence (t_j) of positive real numbers such that

$$\sum_{j=1}^{\infty} t_j \leq 1 \qquad \text{and} \qquad \sum_{j=1}^{\infty} \Phi(\lambda, t_j) = \infty \quad \text{for all } \lambda > 0.$$

Substituting again $\alpha_j := \psi^{-1}(t_j)$, we obtain a sequence (α_j) of positive real numbers such that

$$\sum_{j=1}^{\infty} \psi(\alpha_j) \leq 1 \qquad \text{and} \qquad \sum_{j=1}^{\infty} \varphi(\lambda\alpha_j) = \infty \quad \text{for all } \lambda > 0. \tag{1.2.37}$$

But Corollary 1.2.11 now says that the function $\mathfrak{J}_{(\alpha_j)}$ belongs to $YBV_\psi \backslash YBV_\varphi$, as desired.

Later in Section 3.2 we will use the same idea to investigate multiplier sets in spaces of functions of bounded Young variation.

The Waterman Variation

Another generalized type of variation is that of Waterman which has been introduced in 1972 by Daniel Waterman [151].

Definition 1.2.17. A sequence $\Lambda = (\lambda_j)$ of positive real numbers λ_j is called a Waterman sequence if it is decreasing, converging to 0 as $j \to \infty$ and so that $\sum_{j=1}^{\infty} \lambda_j = +\infty$.

A particularly simple Waterman sequence is given by $\lambda_j = 1/j^q$ for $0 < q \leq 1$, and the special case $\lambda_j = 1/j$ was originally considered by Waterman.

Definition 1.2.18. The possibly infinite number

$$\mathrm{Var}_\Lambda(x) = \sup \sum_{j=1}^{\infty} \lambda_j |f(a_j) - f(b_j)|,$$

where the supremum is taken over all countably infinite collections $([a_j, b_j])$ of mutually nonoverlapping closed subintervals of $[0,1]$, is called the Waterman variation (or Λ-variation) of the function x over the interval $[0,1]$.

Finally, we denote by $\Lambda BV := \{x : [0,1] \to \mathbb{R} \mid \mathrm{Var}_\Lambda(x) < \infty\}$ the space of functions of bounded Waterman variation.

Observe that this definition remains the same if only finite interval collections are considered.

For the special sequence $\lambda_j = 1/j^q$ we write $\Lambda_q BV$ instead of ΛBV. Functions in $\Lambda_1 BV$ are of particular interest and called functions of bounded *harmonic variation.*

The first big difference in Definition 1.2.18 compared to the Definitions 1.1.7, 1.2.2 and 1.2.9 is that we here use (finite or infinite) collections of intervals. However, there is another possibility to calculate the Waterman variation using partitions.

Proposition 1.2.19. *Let $\Lambda = (\lambda_j)$ be a Waterman sequence and $x : [0,1] \to \mathbb{R}$ be any function. Then*

$$\mathrm{Var}_\Lambda(x) = \sup_{\sigma,P} \sum_{j=1}^{n} \lambda_{\sigma(j)} |x(t_{j-1}) - x(t_j)|, \tag{1.2.38}$$

where the supremum is taken over all partitions $P : 0 = t_0 < \ldots < t_n = 1$ of $[0,1]$ and all permutations σ of $\mathbb{N}$.

Proof. Let us denote the quantity on the right hand side of (1.2.38) by v. We need to show that $\mathrm{Var}_\Lambda(x) = v$.

Let σ be a permutation of $\mathbb{N}$ and let $0 = t_0 < \ldots < t_n = 1$ be a partition of $[0,1]$. Define $[a_{\sigma(j)}, b_{\sigma(j)}] := [t_{j-1}, t_j]$ for $j = 1, \ldots, n$ and $[a_{\sigma(j)}, b_{\sigma(j)}] := [0,0] = \{0\}$ for $j > n$. The collection $([a_j, b_j])$ then consists of mutually disjoint subintervals of $[0,1]$, and we obtain

$$\begin{aligned}\sum_{j=1}^{n} \lambda_{\sigma(j)} |x(t_{j-1}) - x(t_j)| &= \sum_{j=1}^{n} \lambda_{\sigma(j)} |x(a_{\sigma(j)}) - x(b_{\sigma(j)})| \le \sum_{j=1}^{\infty} \lambda_j |x(a_j) - x(b_j)| \\ &\le \mathrm{Var}_\Lambda(x).\end{aligned}$$

Taking the supremum with respect to the partitions and permutations on the left hand side yields $v \le \mathrm{Var}_\Lambda(x)$.

For the reverse inequality take a countably infinite collection $([a_j, b_j])$ of mutually disjoint intervals and fix $n \in \mathbb{N}$. Then the first n intervals can be ordered in the sense that there is some permutation τ of $\{1, \ldots, n\}$ such that

$$b_{\tau(j)} \le a_{\tau(j+1)} \quad \text{for } j \in \{1, \ldots, n-1\}.$$

We now define the permutation σ of $\mathbb{N}$ by $\sigma(2j) := \tau(j)$ and $\sigma(2j-1) := n + j$ for $j \in \{1, \ldots, n\}$ and $\sigma(j) := j$ for $j > 2n$. We also define a partition $0 =: t_0 \le \ldots \le t_{2n+1} := 1$ by $t_{2j} := b_{\tau(j)}$ and $t_{2j-1} := a_{\tau(j)}$ for $j \in \{1, \ldots, n\}$. Then,

$$\begin{aligned}\sum_{j=1}^{n} \lambda_j |x(a_j) - x(b_j)| &= \sum_{j=1}^{n} \lambda_{\tau(j)} |x(a_{\tau(j)}) - x(b_{\tau(j)})| = \sum_{j=1}^{n} \lambda_{\sigma(2j)} |x(t_{2j-1}) - x(t_{2j})| \\ &\le \sum_{j=1}^{2n} \lambda_{\sigma(j)} |x(t_{j-1}) - x(t_j)| \le v.\end{aligned}$$

Letting $n \to \infty$ yields

$$\sum_{j=1}^{\infty} \lambda_j |x(a_j) - x(b_j)| \le v,$$

and taking the supremum with respect to the collections of intervals gives $\mathrm{Var}_\Lambda(x) \le v$. This completes the proof. ∎

We now test Definition 1.2.18 on functions with countable support. More generally speaking, for arbitrary functions with countable support we have an analogue to the Propositions 1.2.4 and 1.2.10.

Proposition 1.2.20. *Let $\Lambda = (\lambda_j)$ be a Waterman sequence. For $x \in \mathcal{S}_c$ we have*

$$\sup_{\sigma,n} \sum_{\substack{\text{pairwise distinct} \\ \tau_1,\dots,\tau_n \in \text{supp}(x)}} \lambda_{\sigma(j)}|x(\tau_j)| \leq \text{Var}_\Lambda(x) \leq 2 \sup_{\sigma,n} \sum_{\substack{\text{pairwise distinct} \\ \tau_1,\dots,\tau_n \in \text{supp}(x)}} \lambda_{\sigma(j)}|x(\tau_j)|, \qquad (1.2.39)$$

where the supremum is taken over all natural numbers $n \leq \#\,\text{supp}(x)$ and all permutations σ of $\mathbb{N}$.
In particular, $x \in \Lambda BV$ if and only if the supremum in (1.2.39) is finite.

Proof. For the left inequality in (1.2.39) we assume for simplicity that $\text{supp}(x) \subseteq (0,1)$; the other case is similar. Fix $n \leq \#\,\text{supp}(x)$, pick numbers $\tau_1, \dots, \tau_{n-1} \in \text{supp}(x)$, bring them in increasing order and relabel them by $t_2, t_4, t_6, \dots, t_{2n-2}$. Set $t_0 := 0$ and $t_{2n} := 1$, and pick for each $j \in \{1, \dots, n\}$ numbers $t_{2j-1} \in (t_{2j-2}, t_{2j}) \setminus \text{supp}(x)$. Fix a permutation σ of $\mathbb{N}$ and define another permutation ϱ of $\mathbb{N}$ by $\varrho(2j) = \sigma(j)$ and $\varrho(2j-1) = \sigma(n+j)$ for $j \in \{1, \dots, n\}$ and by $\varrho(j) = \sigma(j)$ for $j > 2n$. Then, by Proposition 1.2.19,

$$\begin{aligned}
\text{Var}_\Lambda(x) &\geq \sum_{j=1}^{2n} \lambda_{\varrho(j)}|x(t_{j-1}) - x(t_j)| \\
&= \sum_{j=0}^{n-1} \lambda_{\varrho(2j+1)}|x(t_{2j}) - x(t_{2j+1})| + \sum_{j=1}^{n} \lambda_{\varrho(2j)}|x(t_{2j-1}) - x(t_{2j})| \\
&= \sum_{j=1}^{n-1} \left(\lambda_{\varrho(2j+1)} + \lambda_{\varrho(2j)}\right)|x(\tau_j)| \geq \sum_{j=1}^{n-1} \lambda_{\sigma(j)}|x(\tau_j)|.
\end{aligned}$$

Taking the supremum with respect to σ and the partitions we obtain, again with Proposition 1.2.19, the left inequality in (1.2.39).
We now show the right inequality. To do this let $n \leq \#\,\text{supp}(x)$, and consider an arbitrary finite collection $([a_j, b_j])_{1 \leq j \leq n}$ of nonoverlapping subintervals of $[0,1]$ and any permutation σ of $\mathbb{N}$. We have

$$\begin{aligned}
\sum_{j=1}^{n} \lambda_j|x(a_j) - x(b_j)| &\leq \sum_{j=1}^{n} \lambda_j|x(a_j)| + \sum_{j=1}^{n} \lambda_j|x(b_j)| \\
&\leq \sup_\sigma \sum_{j=1}^{n} \lambda_{\sigma(j)}|x(a_j)| + \sup_\sigma \sum_{j=1}^{n} \lambda_{\sigma(j)}|x(b_j)| \\
&\leq 2 \sup_{\sigma,n} \sum_{\substack{\text{pairwise distinct} \\ \tau_1,\dots,\tau_n \in \text{supp}(x)}} \lambda_{\sigma(j)}|x(\tau_j)|.
\end{aligned}$$

By taking the supremum on the left hand side with respect to interval collections we obtain the remaining inequality. ■

For our functions $\mathfrak{J}_{(\alpha_j)}$ defined by (1.2.1) this reads as follows.

Corollary 1.2.21. *Let $\Lambda = (\lambda_j)$ be a Waterman sequence. The function $\mathfrak{J}_{(\alpha_j)}$ satisfies*

$$\sup_\sigma \sum_{j=1}^{\infty} \lambda_{\sigma(j)}|\alpha_j| \leq \text{Var}_\Lambda\left(\mathfrak{J}_{(\alpha_j)}\right) \leq 2 \sup_\sigma \sum_{j=1}^{\infty} \lambda_{\sigma(j)}|\alpha_j|; \qquad (1.2.40)$$

in particular, it belongs to ΛBV if and only if the supremum in (1.2.40) is finite.

In particular, the function $\mathfrak{J}_\alpha = \mathfrak{J}_{(1/j^\alpha)}$ introduced in (1.2.3) belongs to ΛBV if and only if $\sum_{j=1}^{\infty} \lambda_j / j^\alpha$ converges which is possible only for $\alpha > 0$. The supremum over permutations σ in Corollary 1.2.21 can then be dropped, because the two sequences (λ_j) and $(1/j^\alpha)$ are both decreasing for $\alpha \geq 0$ and hence the supremum is attained if $\sigma(j) = j$ for all $j \in \mathbb{N}$.

However, in general it is not sufficient to drop the permutation σ in (1.2.39) or in (1.2.40) and simply consider

$$\mathrm{Var}_\Lambda^*(x) := \sum_{j=1}^{\infty} \lambda_j |x(\tau_j)| \tag{1.2.41}$$

for functions having countable support, because (1.2.41) depends on how $\mathrm{supp}(x)$ is actually enumerated. In particular, $\mathrm{Var}_\Lambda^*(x)$ is simply not well-defined as an expression that is supposed to only depend on x. The following example illustrates this.

Example 1.2.22. Let $\Lambda = (\lambda_j) = (\alpha_j)$ be defined by $\lambda_j := \alpha_j := 1/\sqrt{j}$ for $j \in \mathbb{N}$, and let $x := \mathfrak{J}_{(\alpha_j)}$. If $\mathrm{supp}(x) = \{\frac{1}{2}, \frac{1}{4}, \dots\} = \{\tau_1, \tau_2, \dots\}$ is given in decreasing order, then

$$\mathrm{Var}_\Lambda^*(x) = \sum_{j=1}^{\infty} \lambda_j \alpha_j = \sum_{j=1}^{\infty} \frac{1}{j} = \infty.$$

However, we now show that $\mathrm{Var}_\Lambda^*(x)$ may be finite if the τ_j are ordered "properly". We first define $P := \{2^k \mid k \in \mathbb{N}_0\}$ to be the set of all powers of 2, and write the set $\mathbb{N} \backslash P = \{k_1, k_2, k_3, \dots\} = \{3, 5, 6, 7, 9, \dots\}$ in increasing order. We now define an injective function $\sigma_1 : \mathbb{N} \backslash P \to P$ as follows. First, pick $\sigma_1(k_1) = \sigma_1(3) \in P$ so large that $\lambda_3 \lambda_{\sigma_1(3)} \leq 2^{-3}$, and then pick the numbers $\sigma_1(k_j) \in P$ successively so large that they exceed $\sigma_1(k_{j-1})$ and satisfy $\lambda_{k_j} \lambda_{\sigma_1(k_j)} \leq 2^{-k_j}$. This is possible since $\lambda_k \to 0$ as $k \to \infty$. We then have

$$\lambda_j \lambda_{\sigma_1(j)} \leq 2^{-j} \quad \text{for all } j \in \mathbb{N} \backslash P.$$

Since P and $\mathbb{N} \backslash \sigma_1(\mathbb{N} \backslash P)$ have the same cardinality, we may find a bijective function $\sigma_2 : P \to \mathbb{N} \backslash \sigma_1(\mathbb{N} \backslash P)$. The function $\sigma : \mathbb{N} \to \mathbb{N}$, defined by

$$\sigma(j) = \begin{cases} \sigma_1(j) & \text{for } j \in \mathbb{N} \backslash P, \\ \sigma_2(j) & \text{for } j \in P, \end{cases}$$

is a permutation of $\mathbb{N}$. If we now define $\tilde{\tau}_j := \tau_{\sigma(j)}$, we obtain

$$\begin{aligned} \mathrm{Var}_\Lambda^*(x) &= \sum_{j=1}^{\infty} \lambda_j x(\tilde{\tau}_j) = \sum_{j \in \mathbb{N} \backslash P} \lambda_j \lambda_{\sigma_1(j)} + \sum_{j \in P} \lambda_j \lambda_{\sigma_2(j)} \leq \sum_{j \in \mathbb{N} \backslash P} 2^{-j} + \sum_{k \in \mathbb{N}_0} \lambda_{2^k} \\ &\leq \sum_{j \in \mathbb{N}} 2^{-j} + \sum_{k \in \mathbb{N}_0} \sqrt{2}^{-k} = 3 + \sqrt{2} < \infty. \end{aligned}$$

Thus, if $\mathrm{supp}(x)$ is enumerated in a clumsy way, Var_Λ^* cannot decide whether x has bounded Λ-variation. ◇

In total, one could understand Var_Λ^* to be a "variation" of $x \in \mathcal{S}_c$ with respect to one particular enumeration of $\mathrm{supp}(x)$, whereas $\mathrm{Var}_\Lambda(x)$ denotes the supremum of all such variations.

One can show that $\mathrm{Var}_\Lambda(\cdot)$ is a seminorm which makes ΛBV a Banach space when equipped with the norm

$$\|x\| = |x(0)| + \mathrm{Var}_\Lambda(x). \tag{1.2.42}$$

For two functions $x, y \in \Lambda BV$ and a collection of nonoverlapping intervals $([a_j, b_j])_{j\in\mathbb{N}}$ one obtains the estimates

$$\begin{aligned}\sum_{j=1}^{\infty}\lambda_j|x(a_j)y(a_j) - x(b_j)y(b_j)| &\le \sum_{j=1}^{\infty}\lambda_j\Big(\|x\|_\infty|y(a_j) - y(b_j)| + \|y\|_\infty|x(a_j) - x(b_j)|\Big)\\ &= \|x\|_\infty\sum_{j=1}^{\infty}\lambda_j|y(a_j) - y(b_j)| + \|y\|_\infty\sum_{j=1}^{\infty}\lambda_j|x(a_j) - x(b_j)|\\ &\le \|x\|_\infty\mathrm{Var}_\Lambda(y) + \|y\|_\infty\mathrm{Var}_\Lambda(x)\end{aligned}$$

from which

$$\mathrm{Var}_\Lambda(xy) \le \|x\|_\infty\mathrm{Var}_\Lambda(y) + \|y\|_\infty\mathrm{Var}_\Lambda(x)$$

follows. In particular, ΛBV is even an algebra and, when equipped with the norm

$$\|x\|_{\Lambda BV} := \|x\|_\infty + \mathrm{Var}_\Lambda(x), \tag{1.2.43}$$

a normalized Banach algebra by Proposition 1.1.15. Again, as we have seen for the Jordan, Wiener and Young variation, if $x \in \Lambda BV$ is bounded away from zero, that is, $\mathrm{supp}_\delta(x) = [0,1]$ for some $\delta > 0$, then $1/x \in \Lambda BV$ with

$$\mathrm{Var}_\Lambda(1/x) \le \delta^{-2}\,\mathrm{Var}_\Lambda(x).$$

We are now going to compare two Waterman spaces ΓBV and ΛBV for two arbitrary Waterman sequences $\Gamma = (\gamma_j)$ and $\Lambda = (\lambda_j)$. It was shown in the paper [129] that $\Lambda BV \subseteq \Gamma BV$ if and only if

$$\limsup_{n\to\infty}\frac{\sum_{j=1}^{n}\gamma_j}{\sum_{j=1}^{n}\lambda_j} < \infty; \tag{1.2.44}$$

we will write $\Gamma \preceq \Lambda$ in this case. Similarly as for Young functions we write $\Gamma \prec \Lambda$ if $\Gamma \preceq \Lambda$ and $\Lambda \not\preceq \Gamma$. In general, one has the chain of inclusions

$$BV \subseteq \Lambda BV \subseteq \Gamma BV \subseteq R. \tag{1.2.45}$$

for $\Gamma \preceq \Lambda$, where all inclusions are continuous embeddings. For instance, we have

$$|x(t)| \le |x(0)| + |x(0) - x(t)| \le |x(0)| + \gamma_1^{-1}\,\mathrm{Var}_\Gamma(x)$$

and hence

$$\|x\|_\infty \le |x(0)| + \gamma_1^{-1}\operatorname{Var}_\Gamma(x) \le \max\{1, 1/\gamma_1\}\,\|x\|_{\Gamma BV} \tag{1.2.46}$$

which shows on the one hand the right embedding in (1.2.45) and on the other hand with the help of Proposition 1.1.15 that the two norms (1.2.42) and (1.2.43) are equivalent.

The left and right inclusion in (1.2.45) are strict, and for $\Gamma \prec \Lambda$ also the middle is. To show this, the following auxiliary result is helpful and will also be needed later in Section 3.2.

Lemma 1.2.23. *Let (a_j) and (b_j) be sequences of positive real numbers such that*

$$\sum_{j=1}^\infty a_j = \infty = \sum_{j=1}^\infty b_j \quad \text{and} \quad \limsup_{n\to\infty} \frac{\sum_{j=1}^n a_j}{\sum_{j=1}^n b_j} = \infty.$$

Then for each $\alpha > 0$ there exists a monotonically decreasing sequence (u_j) of positive real numbers tending to zero such that $\sum_{j=1}^\infty a_j u_j$ remains divergent, whereas $\sum_{j=1}^\infty b_j u_j$ becomes convergent with limit α.

Before we turn to the proof let us quickly mention why this result might be surprising. If we define $A_n := \sum_{j=1}^n a_j$, $B_n := \sum_{j=1}^n b_j$ and $\gamma_n := A_n/B_n$, then $A_n = \gamma_n B_n$ and $\limsup_{n\to\infty} \gamma_n = \infty$. If (a_j) is bounded, then A_n grows at most linearly. Therefore,

$$\lim_{n\to\infty} \frac{\gamma_n}{n} = \lim_{n\to\infty} \frac{\frac{1}{n}A_n}{B_n} = 0,$$

which means that γ_n grows slower than linearly. Therefore, A_n grows only a little faster than B_n; nevertheless, there is always a suitable sequence (u_j) such that $\sum_{j=1}^\infty a_j u_j$ diverges while $\sum_{j=1}^\infty b_j u_j$ converges to any positive number we choose!

Proof of Lemma 1.2.23. Let A_n and B_n be defined as before, that is,

$$A_n := \sum_{j=1}^n a_j \quad \text{and} \quad B_n := \sum_{j=1}^n b_j \qquad \text{for } n \in \mathbb{N},$$

and set $A_0 := B_0 := 0$. Then the assertion states that $\limsup_{n\to\infty} \frac{A_n}{B_n} = \infty$, and from this we obtain

$$\limsup_{n\to\infty} \frac{A_n - A_m}{B_n - B_m} = \infty$$

for each fixed $m \in \mathbb{N}_0$. This is why we can find a strictly increasing sequence (n_k) of nonnegative integers with $n_1 := 0$ such that

$$B_{n_{k+2}} - B_{n_{k+1}} \ge \frac{1}{2}\Big(B_{n_{k+1}} - B_{n_k}\Big) \qquad \text{and}$$

$$\frac{A_{n_{k+1}} - A_{n_k}}{B_{n_{k+1}} - B_{n_k}} \ge 2^k \qquad \text{for all } k \in \mathbb{N}.$$

Writing $\mathcal{A}_k := A_{n_{k+1}} - A_{n_k}$ and $\mathcal{B}_k := B_{n_{k+1}} - B_{n_k}$, this reads

$$\mathcal{B}_{k+1} \geq \frac{1}{2}\mathcal{B}_k, \tag{1.2.47}$$

$$\frac{\mathcal{A}_k}{\mathcal{B}_k} \geq 2^k \tag{1.2.48}$$

for all $k \in \mathbb{N}$. We now define

$$u_j := \frac{\alpha}{2^k \mathcal{B}_k} \qquad \text{for } n_k + 1 \leq j \leq n_{k+1} \text{ and } j, k \in \mathbb{N}.$$

Then we obtain from (1.2.47) that (u_j) is monotonically decreasing. Moreover, with the help of (1.2.48),

$$\sum_{j=1}^{\infty} a_j u_j = \sum_{k=1}^{\infty} \sum_{j=n_k+1}^{n_{k+1}} a_j u_j = \sum_{k=1}^{\infty} \frac{\alpha}{2^k \mathcal{B}_k} \sum_{j=n_k+1}^{n_{k+1}} a_j = \sum_{k=1}^{\infty} \frac{\alpha}{2^k} \frac{\mathcal{A}_k}{\mathcal{B}_k} \geq \alpha \sum_{k=1}^{\infty} 1 = \infty,$$

and this proves that $\sum_{j=1}^{\infty} a_j u_j$ diverges. On the other hand we get similarly

$$\sum_{j=1}^{\infty} b_j u_j = \sum_{k=1}^{\infty} \sum_{j=n_k+1}^{n_{k+1}} b_j u_j = \sum_{k=1}^{\infty} \frac{\alpha}{2^k \mathcal{B}_k} \sum_{j=n_k+1}^{n_{k+1}} b_j = \sum_{k=1}^{\infty} \frac{\alpha}{2^k} \frac{\mathcal{B}_k}{\mathcal{B}_k} = \alpha \sum_{k=1}^{\infty} \frac{1}{2^k} = \alpha,$$

which shows that $\sum_{j=1}^{\infty} b_j u_j$ converges to α.
Finally, since

$$\alpha \geq \sum_{j=1}^{n} b_j u_j \geq B_n u_n$$

and $B_n \to \infty$, we must have $\lim_{j\to\infty} u_j = 0$, as desired. ■

With the help of Lemma 1.2.23 we can now show that all inclusions in (1.2.45) are strict for $\Gamma \prec \Lambda$. For the first inclusion put $a_j := 1$ and $b_j := \lambda_j$ for $j \in \mathbb{N}$ in Lemma 1.2.23. Then

$$\limsup_{n\to\infty} \frac{\sum_{j=1}^{n} a_j}{\sum_{j=1}^{n} b_j} = \limsup_{n\to\infty} \frac{n}{\sum_{j=1}^{n} \lambda_j} = \infty.$$

Consequently, Lemma 1.2.23 gives us a decreasing sequence (u_j) of positive real numbers such that

$$\sum_{j=1}^{n} u_j = \infty \qquad \text{and} \qquad \sum_{j=1}^{n} \lambda_j u_j < \infty.$$

But this means that the function $\mathfrak{J}_{(u_j)}$ does not belong to BV by Proposition 1.1.8. But since the u_j are in decreasing order, we have for any permutation σ of $\mathbb{N}$ that

$$\sum_{j=1}^{\infty} \lambda_{\sigma(j)} u_j \leq \sum_{j=1}^{\infty} \lambda_j u_j < \infty;$$

in particular, Proposition 1.2.20 ensures $\mathfrak{J}_{(u_j)} \in \Lambda BV$, and so the first inclusion in (1.2.45) is indeed strict. To show that also the second inclusion is strict is similar; we will skip this because we use a very similar argument in more detail in Section

3.2. For the last inclusion simply consider $\mathfrak{J}_{(1/\Gamma_j)}$ for a Waterman sequence (γ_j), where $\Gamma_j := \gamma_1 + \ldots + \gamma_j$. Then $\mathfrak{J}_{(1/\Gamma_j)}$ belongs to R, since $\Gamma_j \to \infty$ as $j \to \infty$, but it does not belong to ΓBV by the Abel-Dini-Theorem [50, 77].

One might wonder why we did not give an instance of a function that belongs to R but to none of the spaces ΓBV for any Waterman sequence Γ. The reason is the following remarkable relation which shows how "close" the Waterman spaces are to the space BV of functions of bounded Jordan variation and to the space R of regular functions. It says that

$$BV = \bigcap_{\Lambda} \Lambda BV \qquad \text{and} \qquad R = \bigcup_{\Lambda} \Lambda BV, \tag{1.2.49}$$

where the intersection and union are taken over all Waterman sequences Λ. In particular, every continuous function belongs to a suitable Waterman space!
Let us remark that there is a link between the Wiener space WBV_p, the Young space YBV_φ and the special Waterman space $\Lambda_q BV$. One may show that for $p > 1$ and $1 - 1/p < q \leq 1$ the inclusion $WBV_p \subseteq \Lambda_q BV$ holds, whereas the reverse inclusion $\Lambda_q BV \subseteq WBV_p$ holds for $p > 1$ and $0 < q \leq 1 - 1/p$. Similarly, it can be proved that $YBV_\varphi \subseteq \Lambda_q BV$ holds for an arbitrary Young function φ, if $\varphi^{1-q} \in L_1$.

The Riesz Variation

The last space we are dealing with is the space of functions of bounded variation in the sense of Riesz which has been introduced in 1910 by Frigyes Riesz in [135, 136].

Definition 1.2.24. Let $p \in (1, \infty)$ and let x be a real-valued function defined on $[0, 1]$. The possibly infinite number

$$\mathrm{RVar}_p(x) = \sup_P \sum_{j=1}^{n} \left| \frac{x(t_{j-1}) - x(t_j)}{t_{j-1} - t_j} \right|^p (t_j - t_{j-1}),$$

where the supremum is taken over all finite partitions $P : 0 = t_0 < \ldots < t_n = 1$ of $[0, 1]$, is called the Riesz variation of the function x over $[0, 1]$.
Moreover, we denote by $RBV_p := \{x : [0, 1] \to \mathbb{R} \mid \mathrm{RVar}_p(x) < \infty\}$ the space of functions of bounded Riesz variation.

Recall that in [135] Riesz proved the following characterization of RBV_p.

Theorem 1.2.25 (Riesz). *The class RBV_p for $1 < p < \infty$ coincides with the class AC of absolutely continuous functions with derivatives in the Lebesgue space L_p. Moreover, in this case the Riesz variation can be calculated explicitly by the formula*

$$\mathrm{RVar}_p(x) = \|x'\|_{L_p}^p = \int_0^1 |x'(t)|^p \, dt. \tag{1.2.50}$$

In particular, functions with countable support which play an important role in the spaces WBV_p, YBV_φ and ΛBV behave calmly in RBV_p for $p > 1$. Indeed, such a function belongs to RBV_p for $p > 1$ if and only if it is zero everywhere. This implies that also $\mathfrak{J}_{(\alpha_j)} \in RBV_p$ for $p > 1$ if and only if $\alpha_j = 0$ for all $j \in \mathbb{N}$.

For $p = 1$, the Riesz variation formally reduces to the Jordan variation, but the formula (1.2.50) is then no longer true, as we have seen in (1.1.24) and the example thereafter. However, it still holds for the class of all absolutely continuous functions which is a subclass of BV, and then we again end up with Theorem 1.1.20. We agree that whenever we use the symbol RBV_p we tacitly assume $p > 1$, unless stated otherwise.

Formula (1.2.50) also implies that $\mathrm{RVar}_p(x)^{1/p}$ is a seminorm satisfying

$$\mathrm{RVar}_p(xy)^{1/p} \leq \|x\|_\infty \mathrm{RVar}_p(y)^{1/p} + \|y\|_\infty \mathrm{RVar}_p(x)^{1/p},$$

and hence that RBV_p is an algebra which is complete when equipped with the norm

$$\|x\| = |x(0)| + \mathrm{RVar}_p(x)^{1/p}. \tag{1.2.51}$$

By Proposition 1.1.15, RBV_p becomes a normalized Banach algebra with respect to the norm

$$\|x\|_{RBV_p} := \|x\|_\infty + \mathrm{RVar}_p(x)^{1/p} \tag{1.2.52}$$

which is equivalent to (1.2.51). This is because

$$|x(t)| \leq |x(0)| + |x(t) - x(0)| \leq |x(0)| + \left(\frac{|x(t) - x(0)|^p}{t^{p-1}} \right)^{1/p} \leq |x(0)| + \mathrm{RVar}_p(x)^{1/p}.$$

Thus,

$$\|x\|_\infty \leq |x(0)| + \mathrm{RVar}_p(x)^{1/p} \leq \|x\|_{RBV_p}. \tag{1.2.53}$$

Similar to the space C of continuous functions the reciprocal $1/x$ of a function $x \in RBV_p$ belongs again to RBV_p if x has no zeros, because then it must be bounded away from zero automatically. In this case, $\mathrm{supp}_\delta(x) = [0,1]$ for some $\delta > 0$, and we obtain

$$\mathrm{RVar}_p(1/x) \leq \delta^{-2p} \mathrm{RVar}_p(x).$$

There is a remarkable interconnection between the Riesz spaces RBV_p and the Sobolev space $W^{1,p} = W^{1,p}[0,1]$: It can be shown that a function $x : [0,1] \to \mathbb{R}$ belongs to $W^{1,p}$ for $1 < p < \infty$ if and only if there is some $y \in AC$ with $y' \in L_p$ such that $x = y$ almost everywhere [56]. Consequently, $x \in W^{1,p}$ if and only if x agrees almost everywhere with a function in RBV_p. This means that RBV_p consists precisely of the continuous representatives of $W^{1,p}$. In this sense Riesz introduced Sobolev spaces, at least in the scalar case, around 25 years prior to Sobolev.

Since the Lebesgue spaces are decreasing with respect to p, that is, $L_p \subsetneq L_q$ for $1 < q < p < \infty$ (and even for $1 = q < p \leq \infty$) we immediately obtain the first three strict inclusions in

$$Lip \subsetneq RBV_p \subsetneq RBV_q \subsetneq AC \subsetneq BV. \tag{1.2.54}$$

In this sense, the formal extension $RBV_\infty := Lip$ would be reasonable. The last inclusion has been investigated in (1.1.24) and the example thereafter.

A handy summary

As we have seen, all our BV-norms have the form $\|x\|_X^* = |x(0)| + \Phi_X(x)$ respectively $\|x\|_X = \|x\|_\infty + \Phi_X(x)$, where Φ_X is the corresponding seminorm. Since we do not want to overburden the reader with cumbersome case distinctions we sometimes use a general approach working for all BV-spaces. Table 1.2.1 summarizes which seminorm belongs to which space.

Table 1.2.1: BV-spaces and their seminorms.

$X =$	BV	WBV_p	YBV_φ	ΛBV	RBV_p
$\Phi_X =$	Var	$\mathrm{Var}_p^{1/p}$	$\mathfrak{M}$	Var_Λ	$\mathrm{RVar}_p^{1/p}$

Using these notations the following auxiliary result is a generalization of Proposition 1.2.14.

Lemma 1.2.26. *Let X be one of the spaces BV, WBV_p, YBV_φ, ΛBV or RBV_p. Let $n \in \mathbb{N}$, $x_1, \ldots, x_n \in X$ and $\alpha_1, \ldots, \alpha_n \geq 0$, and let $x : [0,1] \to \mathbb{R}$ be a function satisfying*

$$|x(s) - x(t)| \leq \sum_{j=1}^{n} \alpha_j |x_j(s) - x_j(t)| \quad \text{for all } s, t \in [0,1]. \tag{1.2.55}$$

Then $x \in X$ and

$$\Phi_X(x) \leq \sum_{j=1}^{n} \alpha_j \Phi_X(x_j).$$

Proof. We first take care of the case $X = YBV_\varphi$ which includes the cases $X = BV$ and $X = WBV_p$. The statement is clearly true if $\alpha_1 = \ldots = \alpha_n = 0$, so we can assume that $\alpha_l > 0$ for at least one $l \in \{1, \ldots, n\}$. For each $k \in \{1, \ldots, n\}$ let $\lambda_k > 0$ be so that $\mathrm{Var}_\varphi(x_k/\lambda_k) \leq 1$. For $\nu := \alpha_1\lambda_1 + \ldots + \alpha_n\lambda_n > 0$ and $\gamma_k := \alpha_k\lambda_k/\nu \in [0,1]$ we have $\alpha_k/\nu = \gamma_k/\lambda_k$ and $\gamma_1 + \ldots + \gamma_n = 1$, and thus we obtain from the convexity of φ that for any partition $0 = t_0 < \ldots < t_m = 1$ of $[0,1]$,

$$\begin{aligned}
\sum_{j=1}^{m} \varphi\left(\frac{|x(t_{j-1}) - x(t_j)|}{\nu}\right) &\leq \sum_{j=1}^{m} \varphi\left(\sum_{k=1}^{n} \frac{\alpha_k |x_k(t_{j-1}) - x_k(t_j)|}{\nu}\right) \\
&= \sum_{j=1}^{m} \varphi\left(\sum_{k=1}^{n} \frac{\gamma_k |x_k(t_{j-1}) - x_k(t_j)|}{\lambda_k}\right) \leq \sum_{k=1}^{n} \gamma_k \left[\sum_{j=1}^{m} \varphi\left(\frac{|x_k(t_{j-1}) - x_k(t_j)|}{\lambda_k}\right)\right] \\
&\leq \sum_{k=1}^{n} \gamma_k \,\mathrm{Var}_\varphi(x_k/\lambda_k) \leq 1.
\end{aligned}$$

Consequently, $\mathfrak{M}(x) \leq \nu = \alpha_1\lambda_1 + \ldots \alpha_n\lambda_n$, and taking the infimum over all such λ_k yields the claim.

For $X = \Lambda BV$ fix a permutation σ of $\mathbb{N}$ and a partition $0 = t_0 < \ldots < t_m = 1$ of $[0,1]$. Then from (1.2.55) we get

$$\begin{aligned}\sum_{j=1}^{m} \lambda_{\sigma(j)}|x(t_{j-1}) - x(t_j)| &\leq \sum_{j=1}^{m} \lambda_{\sigma(j)} \left(\sum_{k=1}^{n} \alpha_k |x_k(t_{j-1}) - x_k(t_j)|\right) \\ &= \sum_{k=1}^{n} \alpha_k \sum_{j=1}^{m} \lambda_{\sigma(j)}|x_k(t_{j-1}) - x_k(t_j)| \\ &\leq \sum_{k=1}^{n} \alpha_k \operatorname{Var}_\Lambda(x_k) = \sum_{k=1}^{n} \alpha_k \Phi_{\Lambda BV}(x_k),\end{aligned}$$

where the last inequality comes from Proposition 1.2.19. Taking the supremum with respect to σ and partitions we obtain by another application of Proposition 1.2.19 that

$$\Phi_{\Lambda BV}(x) = \operatorname{Var}_\Lambda(x) \leq \alpha_1 \Phi_{\Lambda BV}(x_1) + \ldots + \alpha_n \Phi_{\Lambda BV}(x_n),$$

as claimed.

Finally, for $X = RBV_p$, we get from (1.2.55) that

$$\begin{aligned}\left(\sum_{j=1}^{m} \frac{|x(t_{j-1}) - x(t_j)|^p}{|t_{j-1} - t_j|^{p-1}}\right)^{1/p} &\leq \left(\sum_{k=1}^{n} \frac{1}{|t_{j-1} - t_j|^{p-1}} \left(\sum_{j=1}^{m} \alpha_k |x_k(t_{j-1}) - x_k(t_j)|\right)^p\right)^{1/p} \\ &= \left(\sum_{j=1}^{m} \left(\sum_{k=1}^{n} \frac{\alpha_k |x_k(t_{j-1}) - x_k(t_j)|}{|t_{j-1} - t_j|^{(p-1)/p}}\right)^p\right)^{1/p}.\end{aligned}$$

With the help of Minkowski's inequality this can be further estimated by

$$\begin{aligned}&\sum_{k=1}^{n} \left(\sum_{j=1}^{m} \left(\frac{\alpha_k |x_k(t_{j-1}) - x_k(t_j)|}{|t_{j-1} - t_j|^{(p-1)/p}}\right)^p\right)^{1/p} \leq \sum_{k=1}^{n} \alpha_k \left(\sum_{j=1}^{m} \frac{|x_k(t_{j-1}) - x_k(t_j)|^p}{|t_{j-1} - t_j|^{p-1}}\right)^{1/p} \\ &\leq \sum_{k=1}^{n} \alpha_k \operatorname{RVar}_p(x_k)^{1/p}.\end{aligned}$$

Taking the supremum with respect to partitions, the desired estimate follows and completes the proof. ■

We remark that it is not sufficient to replace (1.2.55) by the simpler estimate

$$|x(t)| \leq \alpha_1 |x_1(t)| + \ldots + \alpha_n |x_n(t)| \quad \text{for } t \in [0,1],$$

because then x may have unbounded variation. Here is an example.

Example 1.2.27. The characteristic function $x = \chi_{\mathbb{Q}\cap[0,1]}$ does not belong to BV yet satisfies $|x(t)| \leq 1 = |\mathbb{1}(t)|$ for all $t \in [0,1]$ and $\mathbb{1} \in BV$. ◇

Since $RBV_p \subseteq AC$ for $p > 1$, see (1.2.54), one cannot expect that the space RBV_p contains some Hölder space Lip_γ for a suitable $\gamma < 1$. In fact, one may show that there are functions which belong to each Hölder space Lip_γ for $\gamma < 1$, but not to BV. Consequently, such functions cannot belong to the smaller space RBV_p for any $p > 1$. A comparison of our results on the Wiener space WBV_p and the Riesz space RBV_p shows that these spaces have quite different properties.

- The space RBV_p is decreasing in p, while the space WBV_p is increasing in p.
- The space RBV_p is contained in C for any $p > 1$, while the space WBV_p contains BV and hence also discontinuous functions.
- The space RBV_p is contained in BV, while the space WBV_p contains functions of unbounded Jordan variation.
- The space WBV_p contains all Hölder continuous functions for $\gamma \leq 1/p$, while the space RBV_p contains functions which are not Hölder continuous for any γ.

Table 1.2.2 gives an overview about when our functions $\mathfrak{J}_{(\alpha_j)}$ and their special cases $\mathfrak{J}_\alpha = \mathfrak{J}_{(1/j^\alpha)}$ for $\alpha \in \mathbb{R}$ introduced in (1.2.1) and (1.2.3) belong to certain BV-spaces.

Table 1.2.2: The functions $\mathfrak{J}_{(\alpha_j)}$ and $\mathfrak{J}_\alpha$ and when they belong to our BV-spaces.

X	$\mathfrak{J}_{(\alpha_j)} \in X$ if and only if	$\mathfrak{J}_\alpha \in X$ if and only if
BV	$\sum_{j=1}^\infty \lvert\alpha_j\rvert < \infty$	$\alpha > 1$
WBV_p	$\sum_{j=1}^\infty \lvert\alpha_j\rvert^p < \infty$	$\alpha p > 1$
YBV_φ	$\exists \lambda > 0 : \sum_{j=1}^\infty \varphi\big(\lambda\lvert\alpha_j\rvert\big) < \infty$	$\exists \lambda > 0 : \sum_{j=1}^\infty \varphi\left(\frac{\lambda}{j^\alpha}\right) < \infty$
ΛBV	$\sup_\sigma \sum_{j=1}^\infty \lambda_{\sigma(j)}\lvert\alpha_j\rvert < \infty$	$\sum_{j=1}^\infty \frac{\lambda_j}{j^\alpha} < \infty$
RBV_p	$\forall j \in \mathbb{N} : \alpha_j = 0$	never

Helly's Selection Principle

We end this introductory chapter with an important result that all the BV spaces discussed so far have in common.

Theorem 1.2.28 (Helly's Selection Principle). *Let X be one of the spaces BV, WBV_p, YBV_φ, ΛBV or RBV_p, and let (x_n) be a bounded sequence in X. Then (x_n) has a subsequence that converges pointwise to some function $x \in X$.*

Proof. The proof for BV was given in [71], for YBV_φ and therefore also for WBV_p in [121], and for ΛBV in [152]. The proof for RBV_p follows from that for BV. Indeed, if (x_n) is a bounded sequence in RBV_p for $p > 1$, then it is also a bounded sequence in BV by Jensen's inequality and hence possesses a subsequence that converges pointwise to $x \in BV$. We call this subsequence again (x_n) and show that x in fact belongs to RBV_p. To see this fix a partition $0 = t_0 < \ldots < t_m = 1$ of $[0, 1]$. Since (x_n) is bounded in RBV_p, there is some M such that

$$\sum_{j=1}^m \left|\frac{x_n(t_{j-1}) - x_n(t_j)}{t_{j-1} - t_j}\right|^p (t_j - t_{j-1}) \leq \mathrm{RVar}_p(x_n) \leq M \quad \text{for all } n \in \mathbb{N}.$$

Letting $n \to \infty$ yields

$$\sum_{j=1}^{m} \left| \frac{x(t_{j-1}) - x(t_j)}{t_{j-1} - t_j} \right|^p (t_j - t_{j-1}) \leq M$$

and hence $\mathrm{RVar}_p(x) \leq M$. ■

Helly's Selection Principle may be viewed as a certain counterpart of the Arzelà-Ascoli compactness criterion for BV-functions. Note that in general we cannot expect to find a subsequence that converges in the BV-norm, because BV is infinite dimensional and hence not compact, even for quite simple sequences.

Example 1.2.29. The functions $x_n := \chi_{\{1/(2n)\}}$ form a bounded sequence in BV and converge pointwise to $\mathbb{0}$. But because of $\|x_n\|_{BV} = 3$ for all $n \in \mathbb{N}$ there is no subsequence converging in the BV-norm to $\mathbb{0}$. ◇

One might think that since regular functions have only countably many discontinuities a variant of Helly's Selection Principle might be true also in the space R. Unfortunately, it is not.

Example 1.2.30. Let (q_n) be an enumeration of all rational numbers in $\mathbb{Q} \cap (0,1)$. The functions $x_n := \chi_{\{q_1,\ldots,q_n\}}$ with $n \in \mathbb{N}$ form a bounded sequence in R that cannot have a pointwise convergent subsequence with limit in R. Indeed, let $(x_{n_k})_k$ be any subsequence of (x_n) that converges pointwise to some function $x \in B$. Then $x = \chi_{\{q_{n_1},q_{n_2},q_{n_3},\ldots\}}$, but x cannot belong to R. To see this note that the sequence $(q_{n_k})_k$ is bounded and hence has a subsequence converging to some $r \in [0,1]$; we name this subsequence $(q_{n_k})_k$ again and assume without loss of generality that $r \in [0,1)$. Then $x(q_{n_k}) = 1$ for all $k \in \mathbb{N}$ and hence $\limsup_{t\to r+} x(t) = 1$. But since $\mathbb{Q}$ is countable, $\liminf_{t\to r+} x(t) = 0$. Consequently, $x \notin R$. ◇

But even for simple sequences, a convergence of a subsequence stronger than pointwise can in general not be achieved.

The following result is intermediate between Theorem 1.2.28 and Example 1.2.30 and thus acts "between" BV and R. It illustrates once more the difference between these spaces.

Theorem 1.2.31. *Let (x_n) be a bounded sequence in R. Assume that (x_n) has uniformly bounded ε-variation, that is, for each $\varepsilon > 0$,*

$$\sup_{n\in\mathbb{N}} \Big(\inf \big\{ \mathrm{Var}(y) \mid y \in BV, \|x_n - y\|_\infty \leq \varepsilon \big\} \Big) < \infty. \tag{1.2.56}$$

Then (x_n) has a subsequence that converges pointwise to some function $x \in R$.

The proof and many more results in this direction about regular functions can be found in [63]. We remark that the sequence (x_n) of Example 1.2.30 is bounded yet not of uniformly bounded ε-variation. Indeed, for $\varepsilon = 1/4$, any $y \in BV$ and fixed $n \in \mathbb{N}$ with $\|x_n - y\|_\infty \leq 1/4$ we have $|x_n(q_j) - y(q_j)| = |1 - y(q_j)| \leq 1/4$ for $j \in \{1,\ldots,n\}$ and

$|x_n(r) - y(r)| = |y(r)| \leq 1/4$ for any $r \in [0,1]\setminus\mathbb{Q}$. This means that $|y(q_j) - y(r)| \geq 1/2$ for those j and r. But since the q_j are countable, $\operatorname{Var}(y) \geq n$ which shows

$$\inf \Big\{ \operatorname{Var}(y) \mid y \in BV, \|x_n - y\|_\infty \leq \varepsilon \Big\} \geq n.$$

Consequently, the supremum in (1.2.56) is infinite.

Chapter 2

Functions with Primitive

In this second chapter we will analyze the class of all functions which have a primitive with respect to its size and its relations to other function classes. Apart from recalling known and discussing new results we put a particular emphasis on examples and counterexamples.

Interested and mindful students may ask the following question during a first semester course:

Can one characterize those real functions which have a primitive?

In other words: Can we tell by looking at a function $x : [0,1] \to \mathbb{R}$ whether there is a differentiable function $f : [0,1] \to \mathbb{R}$ with $f'(t) = x(t)$ for each $t \in [0,1]$? Here, the limits for $f'(0)$ and $f'(1)$ are considered to be one-sided. In what follows we will denote the class of all such functions x by the symbol Δ.

An elementary answer to this problem is, at least to the best of our knowledge, not known. Of course, there are easy conditions which are either only necessary or only sufficient and which also arise naturally in first semester courses. For instance, every continuous function belongs to Δ, and every function from Δ is a Darboux function by the Theorem of Darboux [72, 78]. With the notions introduced in Section 1.2 apart from the inclusion (1.1.4) we have

$$C \subseteq \Delta \subseteq D. \tag{2.0.1}$$

Thus, Δ is situated between continuous and Darboux functions.

The goal of this chapter is an analysis of the class Δ with respect to its size, its algebraic properties and its relation to other function classes. For most examples the functions $\varphi_{\alpha,\beta,n}$ introduced in (1.1.1) will serve as a key ingredient.
For instance, the function x given in Example 1.1.29 is differentiable with $x' = \varphi_{0,0,1}$. In other words, $\varphi_{0,0,1} \in \Delta$. However, by Proposition 1.1.12 (c), the function $\varphi_{0,0,1}$ is discontinuous. This shows that the first inclusion in (2.0.1) is strict. We now show by means of another example that also the second inclusion in (2.0.1) is strict.

Example 2.0.1. By Proposition 1.1.12 (d), the function $x := \varphi_{0,0,2}$ belongs to D which is intuitively clear. However, x has no primitive on the entire interval $[0,1]$. In fact, x has a primitive on $(0,1]$, to wit the function

$$f(t) = \frac{t}{2} + \frac{t^2}{4}\sin\frac{2}{t} - \frac{1}{2}\int_0^t s\sin\frac{2}{s}\,\mathrm{d}s,$$

which is continuously extendable by $f(0) = 0$. Moreover, this extension is even differentiable at $t = 0$ with

$$f'(0) = \lim_{t\to 0+}\frac{f(t)-f(0)}{t} = \frac{1}{2} - \lim_{t\to 0+}\frac{1}{2t}\int_0^t s\sin\frac{2}{s}\,\mathrm{d}s = \frac{1}{2} - \frac{1}{2}\lim_{t\to 0+} t\sin\frac{2}{t} = \frac{1}{2}$$

which, however, does not coincide with $\varphi_{0,0,2}(0)$. Although the function f is differentiable on $[0,1]$, and $f' = x$ on $(0,1]$, it is still no primitive for x on $[0,1]$, because it has the "wrong value" at $t = 0$.

If g was a primitive of $x = \varphi_{0,0,2}$ on $[0,1]$ with $g(1) = f(1)$, then g would coincide with f on $(0,1]$ by the Fundamental Theorem of Calculus. By continuity, $f = g$ on $[0,1]$ and hence $1/2 = f'(0) = g'(0) = \varphi_{0,0,2}(0) = 0$ which is not possible. Consequently, $\varphi_{0,0,2} \notin \Delta$. ◊

In Proposition 2.1.5 below, we will add another node to Proposition 1.1.12, namely a characterization of those parameters $(\alpha, \beta, n) \in \mathbb{R}^2 \times \mathbb{N}$ for which $\varphi_{\alpha,\beta,n}$ belongs to Δ. We then have in detail

> $\varphi_{\alpha,\beta,n} \in \Delta$ *if and only if one of the following three cases is true.*
>
> *(i)* $\alpha > 0$, $\beta = 0$ *and* n *is even,*
>
> *(ii)* $\alpha = 0$, $\beta = \frac{1}{\pi}\int_0^\pi \sin^n t\,\mathrm{d}t$ *and* n *is even,*
>
> *(iii)* $\alpha > -1$, $\beta = 0$ *and* n *is odd.*

This shows again that the f in Example 2.0.1 could not be a primitive of $x = \varphi_{0,0,2}$, because the correct value at $t = 0$ should have been

$$\frac{1}{\pi}\int_0^\pi \sin^2 t\,\mathrm{d}t = \frac{1}{2}$$

which was also calculated explicitly.

As we have seen in Section 1.2 the function classes C, B and BV are linear spaces, but D is not, as was shown in Example 1.1.2. However, since differentiation is a linear operator, the class Δ is also a linear space. To be more precise, if $x, y \in \Delta$ with primitives f and g, respectively, then $x + \lambda y \in \Delta$ with primitive $f + \lambda g$ for any $\lambda \in \mathbb{R}$.

By that version of the Fundamental Theorem of Calculus which is taught in every first semester calculus course a primitive can be reconstructed from its derivative by integration. But this additionally needs the Riemann or Lebesgue integrability of the derivative which in general is not given automatically. Here is an example of a function that is differentiable everywhere, but its derivative is integrable neither in the sense of Lebesgue nor in the sense of Riemann.

Example 2.0.2. The function $x : [0,1] \to \mathbb{R}$, defined by

$$x(t) = \begin{cases} t^2 \cos \dfrac{1}{t^2} & \text{for } 0 < t \le 1, \\ 0 & \text{for } t = 0, \end{cases}$$

is differentiable on $[0,1]$, and its derivative is given by

$$x'(t) = \begin{cases} 2t \cos \dfrac{1}{t^2} + \dfrac{2}{t} \sin \dfrac{1}{t^2} & \text{for } 0 < t \le 1, \\ 0 & \text{for } t = 0. \end{cases}$$

However, x' is not Lebesgue integrable: The term $2t \cos t^{-2}$ is continuous and hence Lebesgue integrable, but the second term $\frac{2}{t} \sin \frac{1}{t^2}$ is not. Indeed, the substitution $t = \sqrt{s}$ would lead to the integral

$$\int_0^1 \frac{2}{t} \sin \frac{1}{t^2} \, \mathrm{d}t = \int_0^1 \frac{1}{s} \sin \frac{1}{s} \, \mathrm{d}s = \int_0^1 \varphi_{-1,0,1}(s) \, \mathrm{d}s.$$

But it is easy to show that $\varphi_{\alpha,\beta,n}$ belongs to L_1 if and only if $\alpha > -1$. However, one can also show (and we will do so in Proposition 2.1.5 below) that x' is improperly Riemann integrable.
In addition to the function x we consider again Volterra's function $v : [0,1] \to \mathbb{R}$ introduced in Example 1.1.25. Remember that v is differentiable with $v(1) = 0$ such that v' is bounded (and hence Lebesgue integrable) but not Riemann integrable on $[0,1]$. The function $x + v$ now is an example of a differentiable function the derivative of which is neither (improperly) Riemann nor Lebesgue integrable. ◊

As the last example has shown in order to reconstruct a function from its derivative neither the Riemann nor the Lebesgue integral is the right choice. The next and first section of this chapter is therefore dedicated to another type of integration which is a little more powerful.

2.1 The Kurzweil-Henstock Integral

It is tempting to recover a differentiable function by integrating its derivative. However, Example 2.0.2 has shown that neither the Riemann nor the Lebesgue integral can do that in a proper way. But the following notion of integration can.

Definition 2.1.1. A finite collection of pairs $(\tau_j, [t_{j-1}, t_j])_{1 \le j \le n}$ consisting of real points $\tau_j \in [t_{j-1}, t_j]$ and closed intervals $[t_{j-1}, t_j]$ with $0 = t_0 < \ldots < t_n = 1$ is called a *tagged partition of* $[0,1]$. For a function $\gamma : [0,1] \to (0, \infty)$ such a tagged partition is called *γ-fine* if

$$0 < |t_{j-1} - t_j| \le 2\gamma(\tau_j)$$

holds for all $j \in \{1, \ldots, n\}$.
Finally, a function $x : [0,1] \to \mathbb{R}$ is *Kurzweil-Henstock integrable (KH-integrable)* on $[0,1]$ if there is some number $A \in \mathbb{R}$ with the following property: For each $\varepsilon > 0$

there exists some function $\gamma : [0,1] \to (0,\infty)$ such that every γ-fine tagged partition $(\tau_j, [t_{j-1}, t_j])_{1\le j\le n}$ satisfies

$$\left|\sum_{j=1}^{n} x(\tau_j)(t_j - t_{j-1}) - A\right| \le \varepsilon.$$

In this case, we write

$$\int_0^1 x(t)\,\mathrm{d}t := A$$

and name this number the *Kurzweil-Henstock integral* (KH-integral) of x over $[0,1]$. The family of all functions which are KH-integrable over $[0,1]$ will be denoted by KH.

The KH-integral has all common properties such as monotonicity and linearity with respect to the integrand as well as additivity with respect to the domain of integration [68]. A nice and accessible introduction into integration theory using only the KH-integral may be found in the recent book [61].

The KH-integral is a true generalization of the Lebesgue integral in the following sense: Every improperly Riemann integrable function and every Lebesgue integrable function is also KH-integrable. The converse, however, is not true which is again shown by the functions defined in (1.1.1):

Example 2.1.2. The function $\varphi_{-1,0,1}$ is improperly Riemann integrable on $[0,1]$ and so also KH-integrable, but not Lebesgue integrable as we have already seen in Example 2.0.2. Hence, the function $x := \varphi_{-1,0,1} + \chi_{\mathbb{Q}\cap[0,1]}$ is KH-integrable yet neither improperly Riemann nor Lebesgue integrable. ◇

One can show that a function x is Lebesgue integrable if and only if both x and $|x|$ are KH-integrable; the function $|\varphi_{-1,0,1}|$ may serve as an example of a (nonnegative) function that is not KH-integrable.

In contrast to the Riemann integral there is no sense in considering improper KH-integrals. Indeed, if $x : [0,1] \to \mathbb{R}$ is KH-integrable on every interval $[c,1]$ for $c \in (0,1)$, and if the limit

$$\lim_{c\to 0+} \int_c^1 x(t)\,\mathrm{d}t$$

exists and is finite, then x is KH-integrable on all of $[0,1]$, and its KH-integral is given by the formula

$$\int_0^1 x(t)\,\mathrm{d}t = \lim_{c\to 0+} \int_c^1 x(t)\,\mathrm{d}t.$$

A similar reasoning works for the left sided limit. A proof can be found in [68].

With the help of the KH-integral, one has an improved version of the classical Fundamental Theorem of Calculus. In fact, the following pointwise version is true.

Theorem 2.1.3. *For $x : [0,1] \to \mathbb{R}$ the following statements hold.*

(a) If $x \in KH$, then the function $f : [0,1] \to \mathbb{R}$, given by

$$f(t) := \int_0^t x(s)\,\mathrm{d}s, \tag{2.1.1}$$

is differentiable at $t \in [0,1]$ if and only if the limit

$$\lim_{\delta\to 0} \frac{1}{\delta}\int_t^{t+\delta} x(s)\,\mathrm{d}s$$

exists and is finite. In this case, its value coincides with $f'(t)$.

(b) If $x \in KH$ and if $x(t) = x(t+)$ for some $t \in [0,1)$, then

$$x(t) = \lim_{\delta\to 0+} \frac{1}{\delta}\int_t^{t+\delta} x(s)\,\mathrm{d}s. \tag{2.1.2}$$

A similar result is true in the case $x(t) = x(t-)$ for $t \in (0,1]$.

(c) If $f : [0,1] \to \mathbb{R}$ is a differentiable function with $f'(t) = x(t)$ for all $t \in [0,1]$, then $x \in KH$ with

$$f(t) - f(s) = \int_s^t x(\tau)\,\mathrm{d}\tau \quad \textit{for all } s,t \in [0,1].$$

(d) A function $x : [0,1] \to \mathbb{R}$ possesses a primitive if and only if $x \in KH$ and

$$x(t) = \lim_{\delta\to 0} \frac{1}{\delta}\int_t^{t+\delta} x(s)\,\mathrm{d}s \tag{2.1.3}$$

holds for any $t \in [0,1]$. The unique *primitive f of x satisfying $f(0) = 0$ is then given by (2.1.1).*

Proof. Part (a) follows immediately from the equality

$$\frac{f(t+\delta) - f(t)}{\delta} = \frac{1}{\delta}\int_t^{t+\delta} x(s)\,\mathrm{d}s,$$

which holds for all $t \in [0,1]$ and $\delta \in \mathbb{R}\backslash\{0\}$ with $0 \le t + \delta \le 1$.

For (b) fix $\varepsilon > 0$ and $t \in [0,1)$ so that $x(t) = x(t+)$. Then there is some $\eta > 0$ with $x(t) - \varepsilon \le x(s) \le x(t) + \varepsilon$ for $t \le s \le t + \eta$. For $0 < \delta < \eta$ we deduce that

$$\delta\big(x(t) - \varepsilon\big) = \int_t^{t+\delta} \big(x(t) - \varepsilon\big)\,\mathrm{d}s \le \int_t^{t+\delta} x(s)\,\mathrm{d}s \le \int_t^{t+\delta} \big(x(t) + \varepsilon\big)\,\mathrm{d}s = \delta\big(x(t) + \varepsilon\big),$$

and so

$$\limsup_{\delta\to 0+} \left| \frac{1}{\delta}\int_t^{t+\delta} x(s)\,\mathrm{d}s - x(t) \right| \le \varepsilon.$$

Since $\varepsilon > 0$ was chosen arbitrarily, (2.1.2) has been established.

Finally, a proof for (c) may be found in [68].

Part (d) is only a reformulation of the other statements. ■

Observe that a combination of (b) and (d) yields the well-known fact that every continuous function has a primitive.

Theorem 2.1.3 answers completely the aforementioned problem of characterizing functions with primitives. In particular, part (d) shows that in order to prove $f \in \Delta$ one has to check two things: First, one needs to prove that the function is KH-integrable, and second, one has to calculate the limit in (2.1.3) at each point and show that it actually coincides with the corresponding value of the function. We will do exactly this a couple times later on.

The problem of deciding whether a function has a primitive without using any too technical tools is an old one. We give here two further characterizations of the functions in Δ.

The first criterion uses a refinement of the notion of KH-integrability and works without the limit (2.1.3). Let $t_0, \ldots, t_n, \tau_1, \ldots \tau_n$ points in $[0,1]$ with $t_0 = 0$, $t_n = 1$ and $\tau_j \in [t_{j-1}, t_j]$ for $1 \leq j \leq n$. Moreover, let $\gamma : [0,1] \to (0, \infty)$ be an arbitrary function. We call the partition induced by $t_0, \ldots, t_n$ *γ-super fine* if $0 < |t_j - t_{j-1}| \leq 2\gamma(\tau_j)$ for all $j = 1, \ldots, n$.

We call a function $x : [0,1] \to \mathbb{R}$ *SKH-integrable* on $[0,1]$ if there is some number $A \in \mathbb{R}$ with the following property: For all $\varepsilon > 0$ and $c > 0$ there is a function $\gamma : [0,1] \to (0, \infty)$ such that for all γ-super fine partitions $(\tau_j, [t_{j-1}, t_j])$ the implication

$$\sum_{j=1}^{n} |t_j - t_{j-1}| \leq c \quad \Longrightarrow \quad \left| \sum_{j=1}^{n} x(\tau_j)(t_j - t_{j-1}) - A \right| \leq \varepsilon$$

holds. In this case, we call the number A the *super KH-integral* of x over $[0,1]$.

Let us make some comments to this definition. The subtle difference to Definition 2.1.1 is on the one hand that here the points t_j do not have to be ordered as $t_0 < t_1 < \ldots < t_{n-1} < t_n$, but that on the other hand this must be compensated by restricting the total length of the intervals $[t_{j-1}, t_j]$ by c. In other words, the variation of the (finite) sequence (t_j) must not exceed c. This condition makes no sense for $c < 1$ because then the points t_j cannot satisfy $t_0 = 0$ and $t_n = 1$. For $c \geq 1$ it means that, roughly speaking, the points t_j may not jump back and forth too often. We now have

Theorem 2.1.4. *A function $x : [0,1] \to \mathbb{R}$ belongs to Δ if and only if x is SKH-integrable.*

The first one who proved the "if"-part of this theorem was Robbins in 1943 [137]; his proof is short and elementary yet covers only continuous functions. Only recently in 2012, Thomson generalized the arguments and gave a proof for the version stated here that is also true for not necessarily continuous functions [146].

Since the tagged partitions $(\tau_j, [t_{j-1}, t_j])_{1 \leq j \leq n}$ of $[0,1]$ have to satisfy stronger requirements than those in Definition 2.1.1, it is clear that an SKH-integrable function is "better" than just an KH-integrable function. Indeed, Theorem 2.1.4 shows that SKH encapsulates both the KH-integrability and the limit condition (2.1.3) which is needed for the existence of a primitive according to Theorem 2.1.3 (d).

Moreover, the difference between the requirements of Definition 2.1.1 and Theorem 2.1.4 reminds of the definition of absolutely continuous functions, see Definition 1.1.16: There, one uses finitely many nonoverlapping subintervals of $[0,1]$. If "nonoverlapping" is dropped, then one ends up at the smaller class of Lipschitz continuous functions.

The second criterion we give is one attempt to characterize the functions in Δ without an integral. For instance, Freiling posed the following result [64]: A function ψ which maps any interval $I \subseteq [0,1]$ to a real number $\psi(I)$ is called *additive* if for any two nonoverlapping closed intervals $I, J \subseteq [0,1]$ we have

$$\psi(I \cup J)|I \cup J| = \psi(I)|I| + \psi(J)|J|,$$

where $|I|$ denotes the length of I. In [64] the author shows that a function $x : [0,1] \to \mathbb{R}$ belongs to Δ if and only if there is an additive interval function ψ such that

$$\lim_{I \to t} \psi(I) = x(t) \qquad \text{for } 0 \leq t \leq 1.$$

Here, the notion $I \to t$ means that for any sequence (I_n) of intervals $I_n \subseteq [0,1]$ with $|I_n| \to 0$ and $t \in I_n$, the quantities $\psi(I_n)$ converge to $x(t)$ as $n \to \infty$. Our Theorem 2.1.3 delivers such an interval function in virtue of

$$\psi([a,b]) = \frac{1}{b-a} \int_a^b x(s)\,\mathrm{d}s$$

for $0 \leq a < b \leq 1$ and $\psi(I) = x(t)$ for $I = [t,t]$. Freiling himself admits in [64] that any such interval function hides some kind of integral and therefore does not really represent a new tool to characterize Δ. We will mention other attempts below.

Let us come back to Theorem 2.1.3. The integral in (2.1.3) can be seen as an average value of x along the interval $[t, t+\delta]$. Accordingly, part (d) of Theorem 2.1.3 says that x has a primitive if and only if x is in the mean around t equal to $x(t)$. In particular, x can have only essential discontinuities, as, for instance, $x = \varphi_{\alpha,\beta,n}$ for $\alpha \leq 0$ at $t = 0$ shows. Moreover, if x oscillates around $x(t)$, then this must happen in a more or less symmetric manner. Part (b) of the following proposition illustrates that behavior for our oscillatory functions (1.1.1). Here, the integral

$$\sigma_n := \int_0^\pi \sin^n t\,\mathrm{d}t = \begin{cases} 2\dfrac{(n-1)!!}{n!!} & \text{for odd } n, \\[2ex] \pi\dfrac{(n-1)!!}{n!!} & \text{for even } n, \end{cases} \tag{2.1.4}$$

will be of particular importance, where $n!!$ denotes the double factorial which is defined by

$$n!! = \prod_{k=0}^{\lceil \frac{n}{2} \rceil - 1} (n-2k) \quad \text{for } n \in \mathbb{N}$$

and gives the product of all integers from 1 up to n that have the same parity. We then have

Proposition 2.1.5. *For the functions $\varphi_{\alpha,\beta,n}$ from (1.1.1) the following relations hold.*

(a) $\varphi_{\alpha,\beta,n} \in KH$ if and only if $\alpha > -1$ and n is even, or if $\alpha > -2$ and n is odd.

(b) $\varphi_{\alpha,\beta,n} \in \Delta$ if and only if one of the following three cases is true.

(i) $\alpha > 0$, $\beta = 0$ and n is even,

(ii) $\alpha = 0$, $\beta = \sigma_n/\pi$, and n is even,

(iii) $\alpha > -1$, $\beta = 0$ and n is odd.

Proof. (a) First, let n be even. Due to $\varphi_{\alpha,\beta,n} \geq 0$ the KH-integrability is equivalent to the Lebesgue integrability. For $\alpha > -1$ the function $\varphi_{\alpha,\beta,n}$ is dominated by the Lebesgue integrable function $t \mapsto t^\alpha$ and therefore is Lebesgue integrable itself. For $\alpha \leq -1$ the substitution $t = 1/s$ leads to

$$\begin{aligned}\int_0^1 t^\alpha \sin^n \frac{1}{t}\,\mathrm{d}t &\geq \int_0^1 \frac{1}{t}\sin^n \frac{1}{t}\,\mathrm{d}t = \int_1^\infty \frac{1}{s}\sin^n s\,\mathrm{d}s \geq \sum_{j=1}^{\infty}\int_{j\pi}^{(j+1)\pi}\frac{1}{s}\sin^n s\,\mathrm{d}s\\ &\geq \sum_{j=1}^{\infty}\frac{1}{(j+1)\pi}\int_{j\pi}^{(j+1)\pi}\sin^n s\,\mathrm{d}s = \frac{\sigma_n}{\pi}\sum_{j=1}^{\infty}\frac{1}{j+1} = \infty,\end{aligned}$$

where we have used the shortcut (2.1.4). In this case, $\varphi_{\alpha,\beta,n}$ cannot be KH-integrable.

Now, let n be odd and $\alpha > -2$. With the same substitution as above we come to

$$\int_0^1 t^\alpha \sin^n \frac{1}{t}\,\mathrm{d}t = \int_1^\infty \frac{1}{s^{\alpha+2}}\sin^n s\,\mathrm{d}s. \tag{2.1.5}$$

Let $u > 2\pi$ and $k \in \mathbb{N}$ be so that $2k\pi \leq u < 2(k+1)\pi$. Then

$$\int_1^u \frac{1}{s^{\alpha+2}}\sin^n s\,\mathrm{d}s = \left\{\int_1^{2\pi} + \sum_{j=1}^{k-1}\int_{2j\pi}^{2(j+1)\pi} + \int_{2k\pi}^{u}\right\}\frac{1}{s^{\alpha+2}}\sin^n s\,\mathrm{d}s. \tag{2.1.6}$$

Due to

$$\begin{aligned}\int_{2j\pi}^{(2j+1)\pi}\frac{1}{s^{\alpha+2}}\sin^n s\,\mathrm{d}s &\leq \frac{1}{(2j\pi)^{\alpha+2}}\int_{2j\pi}^{(2j+1)\pi}\sin^n s\,\mathrm{d}s = \frac{\sigma_n}{(2j\pi)^{\alpha+2}},\\ \int_{2j\pi}^{(2j+1)\pi}\frac{1}{s^{\alpha+2}}\sin^n s\,\mathrm{d}s &\geq \frac{1}{\big((2j+1)\pi\big)^{\alpha+2}}\int_{2j\pi}^{(2j+1)\pi}\sin^n s\,\mathrm{d}s = \frac{\sigma_n}{\big((2j+1)\pi\big)^{\alpha+2}}\end{aligned}$$

and

$$\begin{aligned}\int_{(2j+1)\pi}^{2(j+1)\pi}\frac{1}{s^{\alpha+2}}\sin^n s\,\mathrm{d}s &\leq \frac{1}{(2(j+1)\pi)^{\alpha+2}}\int_{(2j+1)\pi}^{2(j+1)\pi}\sin^n s\,\mathrm{d}s = -\frac{\sigma_n}{\big(2(j+1)\pi\big)^{\alpha+2}},\\ \int_{(2j+1)\pi}^{2(j+1)\pi}\frac{1}{s^{\alpha+2}}\sin^n s\,\mathrm{d}s &\geq \frac{1}{\big((2j+1)\pi\big)^{\alpha+2}}\int_{(2j+1)\pi}^{2(j+1)\pi}\sin^n s\,\mathrm{d}s = -\frac{\sigma_n}{\big((2j+1)\pi\big)^{\alpha+2}}\end{aligned}$$

we obtain

$$0 \le \int_{2j\pi}^{2(j+1)\pi} \frac{1}{s^{\alpha+2}} \sin^n s \,\mathrm{d}s \le \frac{\sigma_n}{(2j\pi)^{\alpha+2}} - \frac{\sigma_n}{\big(2(j+1)\pi\big)^{\alpha+2}} \tag{2.1.7}$$

and hence

$$\begin{aligned}\sum_{j=1}^{k-1} \int_{2j\pi}^{2(j+1)\pi} \frac{1}{s^{\alpha+2}} \sin^n s \,\mathrm{d}s &\le \sum_{j=1}^{k-1} \left(\frac{\sigma_n}{(2j\pi)^{\alpha+2}} - \frac{\sigma_n}{\big(2(j+1)\pi\big)^{\alpha+2}} \right) \\ &= \frac{\sigma_n}{(2\pi)^{\alpha+2}} - \frac{\sigma_n}{\big(2k\pi\big)^{\alpha+2}}.\end{aligned}$$

Since $\alpha > -2$, this implies that the series in (2.1.6) converges. Moreover, the last integral in (2.1.6) tends to 0 as $u \to \infty$, because then also $k \to \infty$. Consequently, $\varphi_{\alpha,\beta,n}$ is KH-integrable (even improperly Riemann integrable) in this case.
For $\alpha \le -2$ and $m \in \mathbb{N}$ we have

$$\left| \int_{m\pi}^{(m+1)\pi} \frac{1}{s^{\alpha+2}} \sin^n s \,\mathrm{d}s \right| = \int_{m\pi}^{(m+1)\pi} \frac{1}{s^{\alpha+2}} \big|\sin^n s\big| \,\mathrm{d}s \ge \int_{m\pi}^{(m+1)\pi} \big|\sin^n s\big| \,\mathrm{d}s = \sigma_n > 0.$$

Consequently, the limit

$$\lim_{u\to\infty} \int_1^u \frac{1}{s^{\alpha+2}} \sin^n s \,\mathrm{d}s$$

does not exist in $\mathbb{R}$ and so $\varphi_{\alpha,\beta,n}$ cannot be KH-integrable.

(b) We will combine part (a) with Theorem 2.1.3 (d). To this end, let n be even. By part (a), the function $\varphi_{\alpha,\beta,n}$ is only KH-integrable for $\alpha > -1$ and hence cannot have a primitive for $\alpha \le -1$ by Theorem 2.1.3 (d). Since $\varphi_{\alpha,\beta,n}$ is continuous on every interval $[\varepsilon, 1]$ it suffices according to Theorem 2.1.3 that the limit condition

$$\lim_{\delta\to 0+} \frac{1}{\delta} \int_0^\delta \varphi_{\alpha,\beta,n}(t) \,\mathrm{d}t = \varphi_{\alpha,\beta,n}(0) = \beta$$

holds precisely for the claimed values of α and β. For $\alpha \ge 0$ we obtain similarly as in part (a) with the substitution $t = 1/s$ for $\delta > 0$ the estimates

$$0 \le \int_0^\delta t^\alpha \sin^n \frac{1}{t} \,\mathrm{d}t = \int_{1/\delta}^\infty \frac{1}{s^{\alpha+2}} \sin^n s \,\mathrm{d}s \le \int_{1/\delta}^\infty \frac{1}{s^{\alpha+2}} \,\mathrm{d}s = \frac{\delta^{\alpha+1}}{\alpha+1}.$$

For $\alpha > 0$ it follows that

$$0 \le \frac{1}{\delta} \int_0^\delta t^\alpha \sin^n \frac{1}{t} \,\mathrm{d}t \le \frac{\delta^\alpha}{\alpha+1} \longrightarrow 0 \qquad \text{as } \delta \to 0+,$$

and this is why in this case $\varphi_{\alpha,\beta,n}$ has a primitive if and only if $\beta = 0$.
On the other hand, for $-1 < \alpha \le 0$ and $m \in \mathbb{N}$ we get

$$\begin{aligned}\int_0^{1/(m\pi)} t^\alpha \sin^n \frac{1}{t} \,\mathrm{d}t &= \int_{m\pi}^\infty \frac{1}{t^{\alpha+2}} \sin^n t \,\mathrm{d}t = \sum_{j=m}^\infty \int_{j\pi}^{(j+1)\pi} \frac{1}{t^{\alpha+2}} \sin^n t \,\mathrm{d}t \\ &\ge \frac{1}{\pi^{\alpha+2}} \sum_{j=m}^\infty \frac{1}{(j+1)^{\alpha+2}} \int_{j\pi}^{(j+1)\pi} \sin^n t \,\mathrm{d}t = \frac{\sigma_n}{\pi^{\alpha+2}} \sum_{j=m+1}^\infty \frac{1}{j^{\alpha+2}} \\ &\ge \frac{\sigma_n}{\pi^{\alpha+2}} \int_{m+1}^\infty \frac{1}{t^{\alpha+2}} \,\mathrm{d}t = \frac{\sigma_n}{\pi^{\alpha+2}(\alpha+1)(m+1)^{\alpha+1}},\end{aligned} \tag{2.1.8}$$

where we have used (2.1.4) again. Analogously, one can show for $m \geq 2$ that

$$\int_0^{1/(m\pi)} t^\alpha \sin^n \frac{1}{t}\,\mathrm{d}t = \int_{m\pi}^\infty \frac{1}{t^{\alpha+2}} \sin^n t\,\mathrm{d}t \leq \frac{\sigma_n}{\pi^{\alpha+2}(\alpha+1)(m-1)^{\alpha+1}}. \tag{2.1.9}$$

From (2.1.8) we deduce that for $-1 < \alpha < 0$,

$$\limsup_{\delta\to 0+} \frac{1}{\delta} \int_0^\delta t^\alpha \sin^n \frac{1}{t}\,\mathrm{d}t = \infty,$$

which means that $\varphi_{\alpha,\beta,n}$ cannot have a primitive in this case.

For $\alpha = 0$, $m \in \mathbb{N}$, $0 < \delta < 1$ and $\delta(m-1)\pi \leq 1 \leq \delta m\pi$ we have again with the substitution $t = 1/s$,

$$\int_0^\delta \sin^n \frac{1}{t}\,\mathrm{d}t = \int_{1/\delta}^\infty \frac{1}{s^2} \sin^n s\,\mathrm{d}s = \int_{1/\delta}^{m\pi} \frac{1}{s^2} \sin^n s\,\mathrm{d}s + \int_{m\pi}^\infty \frac{1}{s^2} \sin^n s\,\mathrm{d}s.$$

From (2.1.8) and (2.1.9) we obtain for $m \geq 2$,

$$\frac{\sigma_n}{\pi}\frac{m-1}{m+1} \leq \frac{1}{\delta}\int_{m\pi}^\infty \frac{1}{s^2} \sin^n s\,\mathrm{d}s \leq \frac{\sigma_n}{\pi}\frac{m}{m-1}.$$

Since $m \to \infty$ for $\delta \to 0+$, it follows that

$$\lim_{\delta\to 0+} \frac{1}{\delta}\int_{m\pi}^\infty \frac{1}{s^2} \sin^n s\,\mathrm{d}s = \frac{\sigma_n}{\pi}.$$

Moreover, since

$$0 \leq \frac{1}{\delta}\int_{1/\delta}^{m\pi} \frac{1}{s^2} \sin^n s\,\mathrm{d}s \leq \delta\sigma_n \longrightarrow 0 \qquad \text{as } \delta \to 0+,$$

we end in total at

$$\lim_{\delta\to 0+} \frac{1}{\delta}\int_0^\delta \sin^n \frac{1}{t}\,\mathrm{d}t = \frac{\sigma_n}{\pi}.$$

In this case, $\varphi_{\alpha,n,\beta}$ has a primitive if and only if $\beta = \sigma_n/\pi$, as claimed.

We now handle the case when n is odd. By part (a), the function $\varphi_{\alpha,\beta,n}$ is KH-integrable only for $\alpha > -2$ and hence cannot have a primitive for $\alpha \leq -2$ by Theorem 2.1.3 (d). In addition, $\varphi_{\alpha,\beta,n}$ is continuous on every interval $[\varepsilon, 1]$, by Theorem 2.1.3 (d) it again suffices to show that the limit condition

$$\lim_{\delta\to 0+} \frac{1}{\delta}\int_0^\delta \varphi_{\alpha,\beta,n}(t)\,\mathrm{d}t = \varphi_{\alpha,\beta,n}(0) = \beta$$

is satisfied precisely for $\alpha > -1$ and $\beta = 0$. For $\alpha > -2$ we get with the substitution $t = 1/s$ for $0 < \delta < 1$,

$$\int_0^\delta t^\alpha \sin^n \frac{1}{t}\,\mathrm{d}t = \int_{1/\delta}^\infty \frac{1}{s^{\alpha+2}} \sin^n s\,\mathrm{d}s.$$

For $\alpha > -1$ we deduce for $2(m-1)\pi \le 1/\delta \le 2m\pi$ and $m \in \mathbb{N}$,

$$\frac{1}{\delta}\int_0^\delta t^\alpha \sin^n \frac{1}{t}\,\mathrm{d}t = \frac{1}{\delta}\left\{\int_{1/\delta}^{2m\pi} + \sum_{j=m}^{\infty}\int_{2j\pi}^{2(j+1)\pi}\right\}\frac{1}{s^{\alpha+2}}\sin^n s\,\mathrm{d}s. \tag{2.1.10}$$

Since

$$\frac{1}{\delta}\left|\int_{1/\delta}^{2m\pi}\frac{1}{s^{\alpha+2}}\sin^n s\,\mathrm{d}s\right| \le 2\delta^{\alpha+1}\sigma_n,$$

the first term in (2.1.10) goes to 0 as $\delta \to 0+$. For the sum in (2.1.10) we get from (2.1.7),

$$\begin{aligned}0 &\le \frac{1}{\delta}\sum_{j=m}^{\infty}\int_{2j\pi}^{2(j+1)\pi}\frac{1}{s^{\alpha+2}}\sin^n s\,\mathrm{d}s \le \frac{1}{\delta}\sum_{j=m}^{\infty}\left(\frac{\sigma_n}{(2j\pi)^{\alpha+2}} - \frac{\sigma_n}{\big(2(j+1)\pi\big)^{\alpha+2}}\right)\\ &= \frac{\sigma_n}{\delta(2m\pi)^{\alpha+2}} \le \delta^{\alpha+1}\sigma_n,\end{aligned}$$

and this implies that also the entire sum in (2.1.10) tends to 0 as $\delta \to 0+$. Consequently, in this case, $\varphi_{\alpha,\beta,n}$ has a primitive if and only if $\beta = 0$.

Let $-2 < \alpha \le -1$. Introducing the shortcut

$$a_j := \int_{j\pi}^{(j+1)\pi}\frac{1}{s^{\alpha+2}}|\sin^n s|\,\mathrm{d}s \quad \text{for } j \in \mathbb{N}_0,$$

the sequence (a_j) decreases to 0 as $j \to \infty$. Moreover, due to the convexity of the function $t \mapsto 1/t^{\alpha+2}$ we get the estimate

$$\begin{aligned}a_j + a_{j+2} &= \int_{j\pi}^{(j+1)\pi}\frac{1}{s^{\alpha+2}}|\sin^n s|\,\mathrm{d}s + \int_{(j+2)\pi}^{(j+3)\pi}\frac{1}{s^{\alpha+2}}|\sin^n s|\,\mathrm{d}s\\ &= \int_{(j+1)\pi}^{(j+2)\pi}\frac{1}{(t-\pi)^{\alpha+2}}|\sin^n(t-\pi)|\,\mathrm{d}t + \int_{(j+1)\pi}^{(j+2)\pi}\frac{1}{(t+\pi)^{\alpha+2}}|\sin^n(t+\pi)|\,\mathrm{d}t\\ &= \int_{(j+1)\pi}^{(j+2)\pi}\left(\frac{1}{(t-\pi)^{\alpha+2}} + \frac{1}{(t+\pi)^{\alpha+2}}\right)|\sin^n t|\,\mathrm{d}t\\ &\ge 2\int_{(j+1)\pi}^{(j+2)\pi}\frac{1}{t^{\alpha+2}}|\sin^n t|\,\mathrm{d}t = 2a_{j+1}\end{aligned}$$

which implies

$$a_j - a_{j+1} \ge a_{j+1} - a_{j+2} \quad \text{for all } j \in \mathbb{N}_0.$$

By a generalization of the error estimates for alternating series proved in [36] we have

$$\frac{a_m}{2} \le \left|\sum_{j=m}^{\infty}(-1)^j a_j\right| \le \frac{a_{m-1}}{2} \quad \text{for all } m \in \mathbb{N}.$$

Since we have for $m \in \mathbb{N}$,

$$\int_0^{1/(m\pi)} t^\alpha \sin^n\frac{1}{t}\,\mathrm{d}t = \sum_{j=m}^{\infty}\int_{j\pi}^{(j+1)\pi}\frac{1}{s^{\alpha+2}}\sin^n s\,\mathrm{d}s = \sum_{j=m}^{\infty}(-1)^j a_j$$

we obtain for even $m \in \mathbb{N}$,

$$\int_0^{1/(m\pi)} t^\alpha \sin^n \frac{1}{t}\,\mathrm{d}t = \sum_{j=m}^{\infty}(-1)^j a_j \geq \frac{a_m}{2} \geq \frac{\sigma_n}{2\big[(m+1)\pi\big]^{\alpha+2}}$$

which in turn implies

$$\limsup_{\delta\to 0+} \frac{1}{\delta}\int_0^\delta t^\alpha \sin^n \frac{1}{t}\,\mathrm{d}t \geq \frac{\sigma_n}{2}.$$

Similarly, one can show that

$$\liminf_{\delta\to 0+} \frac{1}{\delta}\int_0^\delta t^\alpha \sin^n \frac{1}{t}\,\mathrm{d}t \leq -\frac{\sigma_n}{2}.$$

Thus, by Theorem 2.1.3 (d) the function $\varphi_{\alpha,\beta,n}$ cannot have a primitive in this case. ∎

Note that the case $\alpha = 0$ for even n in Proposition 2.1.5 (b) is special. Here we can see that a function x has a primitive only if x oscillates at t around $x(t)$ so that big amplitudes compensate each other.

Since every continuous function has a primitive, Theorem 2.1.3 ensures the inclusions

$$C \subseteq \Delta \subseteq KH, \tag{2.1.11}$$

which supplement the inclusions given in (2.0.1). The function $\varphi_{\alpha,0,1}$ for $-2 < \alpha \leq -1$ shows, that the second inclusion is strict and that the inclusion $D \subseteq KH$ does not hold, while any function with finite support shows that the inclusion $KH \subseteq D$ is also false.

In addition, Proposition 2.1.5 (b) shows that for fixed α and n there is at most one possibility for β to guarantee that $\varphi_{\alpha,\beta,n}$ has a primitive. Behind this observation hides a more general result which can be stated as follows.

Theorem 2.1.6. *If the two functions $x, y \in \Delta$ agree almost everywhere, then they agree in fact everywhere.*

Proof. According to Theorem 2.1.3 we have that $x, y \in KH$, and by Theorem 9.9 of [68] it follows that

$$\int_I x(s)\,\mathrm{d}s = \int_I y(s)\,\mathrm{d}s$$

holds for any interval $I \subseteq [0,1]$. From Theorem 2.1.3 (d) it then follows that

$$x(t) = \lim_{\delta\to 0+}\frac{1}{\delta}\int_t^{t+\delta} x(s)\,\mathrm{d}s = \lim_{\delta\to 0+}\frac{1}{\delta}\int_t^{t+\delta} y(s)\,\mathrm{d}s = y(t)$$

for every $t \in [0,1]$. ∎

Theorem 2.1.6 shows that we cannot change the values of a function in Δ at any point we please, because then the function will no longer belong to Δ; we have seen this already in Example 2.0.1. However, it is not surprising that the condition (a) in Proposition 2.1.5 does not depend on β, because it is an integrability condition.

Our considerations suggest that this phenomenon is also true for Darboux functions, because if we change a function only at a single point it looses its intermediate value property in general. The following example illustrates that this suspicion is not true.

Example 2.1.7. For arbitrary $\beta \in \mathbb{R}$ let us denote by x_β the function

$$x_\beta(t) := \varphi_{-1/2,\beta,1}(t) = \begin{cases} \dfrac{1}{\sqrt{t}} \sin \dfrac{1}{t} & \text{for } 0 < t \leq 1, \\ \beta & \text{for } t = 0. \end{cases}$$

By Proposition 1.1.12 (d) (v) this function x_β is a Darboux function for every $\beta \in \mathbb{R}$. Moreover, x_β and x_γ agree on $(0,1]$, but in case $\beta \neq \gamma$ certainly not on $[0,1]$. In particular, this shows that the assumption $x, y \in \Delta$ in Theorem 2.1.6 cannot be dropped, because by Proposition 2.1.5 (b) the function x_β belongs to Δ for $\beta = 0$, but not for any $\beta \neq 0$. ◊

An important problem in the theory of real functions is to determine whether a given class of functions is closed under uniform convergence. The probably simplest examples are the classes B and C, because every first semester student learns that the uniform limit of a sequence of bounded respectively continuous functions on a compact interval is again bounded respectively continuous. We will see in Section 6.1 that besides uniform convergence there are further weaker types of convergence with similar continuity preserving properties.

A little more surprising might be that the same is true for the class Δ. A quite elementary proof for this fact which uses the Mean Value Theorem for differentiable functions can be found in [147]. We give here an alternative proof as a direct application of our main Theorem 2.1.3 (d).

Theorem 2.1.8. *Let (x_n) be a sequence of real-valued functions that converges on $[0,1]$ uniformly to some function x. If each x_n belongs to Δ, then so does x.*

Proof. By Theorem 2.1.3 (d), every x_n belongs to KH and satisfies condition (2.1.3). Due to the uniform convergence, we find for fixed $\varepsilon > 0$ some $N \in \mathbb{N}$ such that for all $t \in [0,1]$ and $n \geq N$ we have

$$x_n(t) - \varepsilon \leq x(t) \leq x_n(t) + \varepsilon.$$

In particular, x is measurable as a limit of measurable functions and bounded from above and below by the KH-integrable functions x_n and so also KH-integrable by Theorem 5.11 of [87]. For $n \geq N$ and $t, t+\delta \in [0,1]$ with $\delta \neq 0$,

$$\frac{1}{\delta}\int_t^{t+\delta} x_n(s)\,\mathrm{d}s - \varepsilon \leq \frac{1}{\delta}\int_t^{t+\delta} x(s)\,\mathrm{d}s \leq \frac{1}{\delta}\int_t^{t+\delta} x_n(s)\,\mathrm{d}s + \varepsilon.$$

Letting $\delta \to 0$ yields together with Theorem 2.1.3 (d) that

$$x_n(t) - \varepsilon \leq \liminf_{\delta\to 0} \frac{1}{\delta}\int_t^{t+\delta} x(s)\,\mathrm{d}s \leq \limsup_{\delta\to 0} \frac{1}{\delta}\int_t^{t+\delta} x(s)\,\mathrm{d}s \leq x_n(t) + \varepsilon,$$

and letting $n \to \infty$ afterwards gives

$$x(t) - \varepsilon \leq \liminf_{\delta \to 0} \frac{1}{\delta} \int_t^{t+\delta} x(s)\,\mathrm{d}s \leq \limsup_{\delta \to 0} \frac{1}{\delta} \int_t^{t+\delta} x(s)\,\mathrm{d}s \leq x(t) + \varepsilon.$$

Since $\varepsilon > 0$ was chosen arbitrarily, x satisfies indeed (2.1.3) and thus belongs to the class Δ again by Theorem 2.1.3 (d). ■

One could ask whether Theorem 2.1.8 remains true if the uniform convergence is replaced by one of the other aforementioned weaker forms of convergence. We will consider such types in more detail and answer this question in Section 6.1.

As mentioned above, there were further attempts in the literature to characterize the functions in Δ without using integrals; see [43, 53, 115], for instance. It is reasonable to try to find a characterization by looking at sub- and superlevel sets of the form

$$S_\alpha(x) := \{t \in [0,1] \mid x(t) < \alpha\} \quad \text{and} \quad P_\alpha(x) := \{t \in [0,1] \mid x(t) > \alpha\}. \tag{2.1.12}$$

Characterizing a class $\mathcal{A}$ of function on $[0,1]$ then means that there is a system $\mathcal{S}$ of subsets of $\mathbb{R}$ such that

$$x \in \mathcal{A} \iff \forall \alpha \in \mathbb{R} : \; S_\alpha(x), P_\alpha(x) \in \mathcal{S}. \tag{2.1.13}$$

For some classes $\mathcal{A}$ this is easy to do. For instance, a function $x : [0,1] \to \mathbb{R}$ is continuous if and only if the sets in (2.1.12) are all open (in this case, $\mathcal{S}$ consists of all open sets in $\mathbb{R}$), monotone if and only if the sets in (2.1.12) are all intervals (in this case, $\mathcal{S}$ consists of all intervals in $\mathbb{R}$) and measurable if and only if the sets in (2.1.12) are all measurable (in this case, $\mathcal{S}$ consists of all Lebesgue measurable subsets of $\mathbb{R}$). However, it is not possible to find a set system $\mathcal{S}$ for the class $\mathcal{A} = \Delta$ such that (2.1.13) is fulfilled.

Example 2.1.9. The function $x := \varphi_{0,0,1} + \mathbb{1}$ is nonnegative and belongs to Δ according to Proposition 2.1.5 (b). By Example 2.0.1 we get that $x^2(t) = \varphi_{0,0,2}(t) + 2\varphi_{0,0,1}(t) + 1$ does not belong to Δ.

Assume now that there is a set system $\mathcal{S}$ for $\mathcal{A} = \Delta$ such that (2.1.13) is fulfilled. Then $S_\alpha(x), P_\alpha(x) \in \mathcal{S}$ for any $\alpha \in \mathbb{R}$; in particular, $\emptyset = S_{-1}(x) \in \mathcal{S}$ and $[0,1] = P_{-1}(x) \in \mathcal{S}$. We now fix $\beta \in \mathbb{R}$. If $\beta \geq 0$, then

$$S_\beta(x^2) = S_{\sqrt{\beta}^2}(x^2) = S_{\sqrt{\beta}}(x) \in \mathcal{S} \qquad \text{and} \qquad P_\beta(x^2) = P_{\sqrt{\beta}^2}(x^2) = P_{\sqrt{\beta}}(x) \in \mathcal{S}.$$

If $\beta < 0$ we have

$$S_\beta(x^2) = \emptyset \in \mathcal{S} \qquad \text{and} \qquad P_\beta(x^2) = [0,1] \in \mathcal{S}.$$

Thus, we have shown that $S_\beta(x^2), P_\beta(x^2) \in \mathcal{S}$ for any β. According to (2.1.13) this would imply that $x^2 \in \Delta$, a contradiction. ◇

Generally speaking, if $h : \mathbb{R} \to \mathbb{R}$ is a strictly increasing homeomorphism of $\mathbb{R}$ and $S_\alpha(x)$ belongs to some class $\mathcal{S}$ for any $\alpha \in \mathbb{R}$, then $S_\alpha(h \circ x) = S_{h^{-1}(\alpha)}(x)$ also has to belong to $\mathcal{S}$ for any $\alpha \in \mathbb{R}$, and a similar reasoning is true for the sets P_α. Thus, any function class $\mathcal{A}$ satisfying (2.1.13) for some set system $\mathcal{S}$ must be closed under outer compositions with strictly increasing homeomorphisms. As Example 2.1.9 has shown, the class Δ is not closed under the strictly increasing homeomorphism $h(u) = u$ for $u < 0$ and $h(u) = u^2$ for $u \geq 0$; we will revisit this and related problems later in Section 2.3.

In view of this last example, the following result is particularly surprising. Its proof can be found in [132].

Theorem 2.1.10. *A function $x : \mathbb{R} \to \mathbb{R}$ belongs to $\Delta(\mathbb{R})$ if and only if for each set $E \subseteq \mathbb{R}$ there is a function $y \in \Delta(\mathbb{R})$ with $x^{-1}(E) = y^{-1}(E)$.*

Thus, a function x has a primitive on $\mathbb{R}$ if its preimages locally look like those of another function with primitive. This last result is somewhat circular, because in order to establish the existence of a primitive, we use the existence of a primitive of another function. In other words: A function which does not have a primitive produces the wrong kind of preimages; this observation was refined in [43].

We have seen that the Theorems 2.1.3 and 2.1.4 fully characterize the functions in the class Δ. Moreover, Δ exhibits some pleasant and unpleasant structural properties. For instance, Δ is closed under summation and uniform convergence, but it is not closed under multiplication and composition, as Example 2.1.9 has shown. Therefore, the following problems naturally arise.

Problem 2.1.11. *Find a necessary and sufficient condition on $g : [0, 1] \to \mathbb{R}$ such that $xg \in \Delta$ for all $x \in \Delta$.*

Problem 2.1.12. *Find a necessary and sufficient condition on $g : \mathbb{R} \to \mathbb{R}$ such that $g \circ x \in \Delta$ for all $x \in \Delta$.*

Problem 2.1.13. *Find a necessary and sufficient condition on $g : [0, 1] \to [0, 1]$ such that $x \circ g \in \Delta$ for all $x \in \Delta$.*

The function g in any of these problems is in a sense universal: It has to withstand the multiplication respectively outer composition respectively inner composition by any function $x \in \Delta$. We will discuss these and related problems in the next sections of this chapter, and start with Problem 2.1.11.

2.2 Products of Derivatives

While we have given a full characterization of the class Δ of functions on $[0,1]$ having a primitive, we will now discuss the structural properties of Δ in more detail.
Since differentiation is a linear procedure, the sum of two functions of Δ again belongs to Δ. However, Example 2.1.9 has shown that the product of two functions in Δ may not belong to Δ anymore. We give here another example which might be more interesting, since one factor will be continuous.

Example 2.2.1. The functions

$$x(t) = \varphi_{-1/2,0,1}(t) = \begin{cases} \dfrac{1}{\sqrt{t}} \sin \dfrac{1}{t} & \text{for } 0 < t \leq 1, \\ 0 & \text{for } t = 0 \end{cases} \tag{2.2.1}$$

and

$$y(t) = \varphi_{1/2,0,1}(t) = \begin{cases} \sqrt{t} \sin \dfrac{1}{t} & \text{for } 0 < t \leq 1, \\ 0 & \text{for } t = 0 \end{cases} \tag{2.2.2}$$

both have a primitive f respectively g by Proposition 2.1.5 (b), namely

$$f(t) = t\sqrt{t} \cos \frac{1}{t} - \frac{3}{2} \int_0^t \sqrt{s} \cos \frac{1}{s} \, ds \qquad \text{and} \qquad g(t) = \int_0^t \sqrt{s} \sin \frac{1}{s} \, ds.$$

The product xy, however, is the function $\varphi_{0,0,2}$ which does not have a primitive, as we have seen in Example 2.0.1 and in Proposition 2.1.5 (b). ◇

Recall that Problem 2.1.11 asks for a universal *multiplier* for Δ, that is, for a function $g : [0,1] \to \mathbb{R}$ such that $xg \in \Delta$ for all $x \in \Delta$. Example 2.2.1 has shown that $g \in C$ is too weak for being a multiplier for Δ. Of course, as usual we can give stronger conditions on at least one of the factors $x, y \in \Delta$ to ensure $xy \in \Delta$. Such conditions will be given in the sequel[1]. Our first pair of conditions also sets the first bridge for the interconnection between Δ and BV.

Theorem 2.2.2. *For $x, y : [0,1] \to \mathbb{R}$ the following statements hold.*

(a) If $x \in \Delta$ and $y \in BV \cap C$, then $xy \in \Delta$.

(b) If $x \in \Delta \cap B$ and $y \in C$, then $xy \in \Delta$.

A proof of Theorem 2.2.2 using the approximation theorem of Weierstrass can be found in [11]. We will give another proof later for the more general result in Theorem 2.2.12, but use only the KH-integral.

Part (a) of Theorem 2.2.2 says that any $y \in BV \cap C$ is a multiplier for Δ. Moreover, Example 2.2.1 shows that we cannot drop the assumption $y \in BV$ in (a) and also not $x \in B$ in (b). On the other hand, Theorem 2.2.2 allows us to find other multipliers

[1]Compare to [6], Exercises 1.38, 1.41, 1.44, 1.45.

g for Δ. For instance, it suffices that g is continuous and injective, because then it is strictly monotone and thus belongs to BV. Even simpler, the condition $g \in C^1$ is sufficient, because $C^1 \subseteq BV \cap C$. If, in addition to $g \in C^1$, one requires x to have a bounded primitive, then $xg \in \Delta$ can be deduced from Theorem 2.2.2 (b) instead of (a). Indeed, if f is a bounded primitive of x, then from the product rule we obtain $xg = f'g = (fg)' - fg' \in \Delta$, since g' is continuous.

In order to tackle Problem 2.1.11 in a more systematic way, it is reasonable to formally introduce the class of multipliers in a more general setting. For a function space X of real-valued functions on $[0,1]$ we denote by

$$X/X := \{g : [0,1] \to \mathbb{R} \mid xg \in X \text{ for all } x \in X\} \tag{2.2.3}$$

the *multiplier set* of X. So our main interest is how the set Δ/Δ looks like. A description of X/X for other spaces is also interesting. The next chapter is dedicated to this problem in much more generality. In particular, we will investigate X/X and related more general multiplier sets if the underlying function spaces are one of the BV-type spaces introduced in Chapter 1. But for now, let us stay with X/X for some classes X that have been considered so far. If X contains the constant function $\mathbb{1}$, then clearly

$$X/X \subseteq X, \tag{2.2.4}$$

and in case that X is not closed under multiplication the inclusion in (2.2.4) is strict. Note that X/X is always closed under multiplication: Indeed, if $g, h \in X/X$ are given and $x \in X$ is fixed, then $xg \in X$ and hence $xgh \in X$ which shows that then $gh \in X/X$.

For the classical spaces $X \in \{C, B, BV\}$ we clearly have equality in (2.2.4), and Theorem 2.2.2 (a) shows that $BV \cap C \subseteq \Delta/\Delta$. The sets Δ/Δ, KH/KH and D/D are far more difficult to describe; we will discuss them in the sequel.

We start with KH/KH. From the Hölder Inequality we get that the multiplier set for L_1 is L_∞. However, L_∞ is not the multiplier set for KH.

Example 2.2.3. The function

$$g(t) := \operatorname{sign} \varphi_{0,0,1}(t) = \begin{cases} -1 & \text{if } \sin(1/t) < 0, \\ 0 & \text{if } \sin(1/t) = 0, \\ 1 & \text{if } \sin(1/t) > 0 \end{cases}$$

belongs to L_∞. But for $x = \varphi_{-1,0,1} \in KH$ we have $xg = |x| \notin KH$, as we have seen after Example 2.1.2. Thus, $g \notin KH/KH$. ◇

The class KH/KH was fully described in [87]. The author proved the equality

$$KH/KH = BV^*, \tag{2.2.5}$$

where BV^* denotes the space of all functions $x : [0,1] \to \mathbb{R}$ that are almost everywhere equal to some function in BV. This explains that g of Example 2.2.3 could not be a multiplier for KH, since it can be transformed into a BV-function only if one changes its values on a set of positive measure. From (2.2.5) and $BV \subseteq BV^*$ it also follows that for $x \in KH$ and $y \in BV$ we have $xy \in KH$. This has been proven for the first time by Chelidze and Dzhvarsheishvili [39]; a proof can also be found in [87] and [88]. Moreover, on every subinterval $[a,b]$ of $[0,1]$ with $0 \le a \le b \le 1$ a Hölder-type inequality holds, namely

$$\left| \int_a^b x(t)y(t)\,\mathrm{d}t \right| \le \|x\|_{KH[a,b]} \, \|y\|_{BV[a,b]} \, , \tag{2.2.6}$$

where

$$\|x\|_{KH[a,b]} := \sup_{a \le c \le d \le b} \left| \int_c^d x(t)\,\mathrm{d}t \right|$$

is a natural norm on $KH[a,b]$, and $\|\cdot\|_{BV[a,b]}$ is the usual norm for $BV[a,b]$, introduced in Chapter 1. With these tools we are now able to give the promised proof of Theorem 2.2.2.

Proof of Theorem 2.2.2. (a) Let $x \in \Delta$ and $y \in BV \cap C$. By Theorem 2.1.3 (c) we have $x \in KH$, and from (2.2.5) follows $xy \in KH$. In order to prove $xy \in \Delta$ we have to check the limit condition (2.1.3) (with x replaced by xy). To this end, fix $t \in [0,1)$; the case $t = 1$ is similar and will be omitted. Since $x(s)y(s) = [x(s) - x(t)]y(s) + x(t)y(s)$ for $s \in [0,1]$ we get from $y \in \Delta$ and (2.1.3) that

$$\begin{aligned} \lim_{\delta \to 0} \frac{1}{\delta} \int_t^{t+\delta} x(s)y(s)\,\mathrm{d}s &= \lim_{\delta \to 0} \frac{1}{\delta} \int_t^{t+\delta} [x(s) - x(t)]y(s)\,\mathrm{d}s + x(t) \lim_{\delta \to 0} \frac{1}{\delta} \int_t^{t+\delta} y(s)\,\mathrm{d}s \\ &= \lim_{\delta \to 0} \frac{1}{\delta} \int_t^{t+\delta} [x(s) - x(t)]y(s)\,\mathrm{d}s + x(t)y(t). \end{aligned}$$

Thus, we can assume that $x(t) = 0$. By (2.2.6) we obtain

$$\left| \int_t^{t+\delta} x(s)y(s)\,\mathrm{d}s \right| \le \sigma(\delta) \, \|y\|_{BV} \, ,$$

where we set

$$\sigma(\delta) := \sup_{[u,v] \subseteq [t,t+\delta]} \left| \int_u^v x(s)\,\mathrm{d}s \right| .$$

By Theorem 2.1.3 (a) and (d) the function $f : [t,1] \to \mathbb{R}$, defined by

$$f(u) = \int_t^u x(s)\,\mathrm{d}s,$$

is differentiable at $u = t$ with $f'(t) = x(t) = 0$. This implies that we find for each $\varepsilon > 0$ some $\eta > 0$ so that $|f(u)| \le \varepsilon |u - t|$ for $|u - t| \le \eta$. For $|\delta| \le \eta$ we get

$$\sigma(\delta) = \sup_{[u,v] \subseteq [t,t+\delta]} |f(u) - f(v)| \le \varepsilon \sup_{[u,v] \subseteq [t,t+\delta]} \big(|v - t| + |u - t| \big) \le 2\varepsilon\delta.$$

We deduce

$$\lim_{\delta \to 0} \frac{1}{\delta} \int_t^{t+\delta} x(s)y(s)\,\mathrm{d}s = 0,$$

which is exactly (2.1.3) for xy and proves (a).

For (b) assume that $x \in \Delta \cap B$ and $y \in C$. We approximate y uniformly by piecewise linear and continuous functions. For instance, define y_n on $[0,1]$ piecewise linear and continuous by $y_n(k2^{-n}) = y(k2^{-n})$ for $k \in \{0, \ldots, 2^n\}$ and $n \in \mathbb{N}$. Since y is continuous, the functions y_n converge indeed uniformly on $[0,1]$ to y, and since they are all piecewise linear, all of them belong to $BV \cap C$. By part (a), $xy_n \in \Delta$ for each $n \in \mathbb{N}$. Since x is bounded, the products xy_n converge also uniformly to xy, and finally Theorem 2.1.8 ensures $xy \in \Delta$. ■

Note that we have not used the continuity of y in our proof of part (a), but merely $y \in \Delta$. However, this is not a real generalization: Any BV-function possesses only jump discontinuities, but the Darboux property, which every function in Δ has, excludes them. Thus, $BV \cap \Delta = BV \cap C$. One could also argue that the proof of part (a) of Theorem 2.2.2 remains valid if one replaces y almost everywhere with another function $y \in BV \cap \Delta$. In other words, can we replace the assumption $y \in BV \cap C$ by the apparently weaker assumption $y \in BV^* \cap \Delta$? The answer is given by

Theorem 2.2.4. *The identity*

$$BV^* \cap \Delta = BV \cap C \tag{2.2.7}$$

holds.

Proof. Because of $BV \subseteq BV^*$ and $C \subseteq \Delta$ we only need to show the inclusion $BV^* \cap \Delta \subseteq BV \cap C$. To this end, let $x \in BV^*$ be a function with primitive. By definition of BV^* there is a function $h \in BV$ which agrees almost everywhere with x on $[0,1]$. The right regularization

$$h^{\#}(t) = \begin{cases} \lim\limits_{s \to t+} h(s) & \text{for } t \in [0,1), \\ x(1) & \text{for } t = 1 \end{cases}$$

has bounded variation, is right-continuous and almost everywhere equal to h, because it differs from h in the (at most countably many) points of discontinuity of h. In particular, $h^{\#}$ agrees almost everywhere with x and satisfies $h^{\#}(1) = x(1)$. As a BV-function, $h^{\#}$ is KH-integrable, and with the help of Theorem 2.1.3 (b) and (d) as well as [68, Theorem 9.9] it follows for $t \in [0,1)$ that

$$h^{\#}(t) = \lim_{\delta \to 0+} \frac{1}{\delta} \int_t^{t+\delta} h^{\#}(s)\,\mathrm{d}s = \lim_{\delta \to 0+} \frac{1}{\delta} \int_t^{t+\delta} x(s)\,\mathrm{d}s = x(t).$$

Since $h^{\#}(1) = x(1)$, the two functions $h^{\#}$ and x are in fact equal everywhere on $[0,1]$. Thus, x is a Darboux function of bounded variation and hence continuous. ■

In Theorem 2.2.4 the set $BV^* \cap \Delta$ cannot be replaced by the larger set $BV^* \cap D$. In [45], the author constructed a bounded upper semicontinuous Darboux function $g : [0,1] \to \mathbb{R}$ which is 0 almost everywhere but not everywhere. This function then lies within $BV^* \cap D$, but neither in BV nor in C. Moreover, this g can also not belong to Δ by Theorem 2.2.4. The same is shown by Theorem 2.1.6: If g had a primitive, then it would be identically zero.

The multipliers in KH/KH are completely described by (2.2.5). In the literature there can also be found some asymmetric conditions for being in KH/KH. For instance, the authors of [67] show using the Hölder space Lip_α the following: *If a primitive of $x \in \Delta$ belongs to Lip_α, and if $y \in Lip_\beta$, then $xy \in KH$ if $\alpha + \beta > 1$.* This description of KH/KH is independent of our characterization using BV and BV^*. In fact, $g \in Lip_\beta$ is not even necessary for g being in KH/KH, not even for $g \in \Delta/\Delta$, as the following example illustrates.

Example 2.2.5. The function $g : [0,1] \to \mathbb{R}$, defined by

$$g(t) = \begin{cases} \dfrac{1}{\log(2/t)} & \text{for } 0 < t \leq 1, \\ 0 & \text{for } t = 0, \end{cases}$$

is increasing and hence of bounded variation. By (2.2.5), g belongs to KH/KH. Moreover, g is also continuous, and so by Theorem 2.2.2 (a) it even belongs to Δ/Δ. However, a simple calculation shows that g does not belong to Lip_β for any $\beta > 0$. ◇

We now turn to the problem how the multipliers in Δ/Δ look like. First, the two classes Δ and Δ/Δ are different in structure: On the one hand and as we have seen above, Δ/Δ is an algebra, whereas Δ is not closed under multiplication by Example 2.1.9 or 2.2.1. On the other hand, Δ is closed under uniform convergence by Theorem 2.1.8. However, Δ/Δ is not. To see this, first note that since Δ/Δ is closed under multiplication, the assumption $g \in \Delta/\Delta$ implies $g \in \Delta$ and hence $g^2, g^3, \ldots, g^n \in \Delta/\Delta$ for all $n \in \mathbb{N}$. Therefore, as $g(t) = t$ and any constant function belong to Δ/Δ by Theorem 2.2.2 (a), also any polynomial is a member of Δ/Δ. By the Approximation Theorem of Weierstrass any continuous function can be uniformly approximated on $[0,1]$ by polynomials. Now, if Δ/Δ was closed under uniform convergence, then any continuous function would also belong to Δ/Δ. However, in Example 2.2.1 we have seen that the continuous function given in (2.2.2) serves as a counterexample.

From what we have seen so far, the class Δ/Δ must have the form

$$\Delta/\Delta = X \cap \Delta, \tag{2.2.8}$$

where X is a (for now unknown) class of functions which contains BV as a subclass. Conversely, in the next theorem we give a superclass of X.

Theorem 2.2.6. *The inclusion*

$$\Delta/\Delta \subseteq B \tag{2.2.9}$$

holds.

Proof. Let $g \in \Delta/\Delta$ be given; in particular, $g \in \Delta$ and hence $g \in KH$ by Theorem 2.1.3 (c), and thus g is Lebesgue measurable [68, Theorem 9.12 (c)]. We first show that

g is *essentially* bounded in the sense of the Lebesgue measure. Since the statements $g \in L_\infty$ and $g^2 \in L_\infty$ are equivalent and since Δ/Δ is closed under multiplication, we can assume that g is nonnegative throughout $[0,1]$. To invoke a contradiction, we assume that $g \notin L_\infty$ and construct some function $x \in \Delta$ with $xg \notin \Delta$; this proves the theorem. Because $g \notin L_\infty$, which means $\operatorname{ess\,sup} g([0,1]) = \infty$, we find a strictly monotone sequence (t_n) in $[0,1]$ which converges to some $c \in [0,1]$ and satisfies

$$\operatorname{ess\,sup} g([t_n, t_{n+1}]) \geq 5n \quad \text{for } n \in \mathbb{N}.$$

Without loss of generality we can assume that $c \in [0,1)$ and that (t_n) is strictly decreasing; the other cases are similar. Thus, we find $s_n \in [t_n, t_{n-1}]$ and measurable sets $A_n \subseteq [s_n, s_n + \mu_n]$ of measure $|A_n| = \mu_n/2 > 0$, so that

$$t_n < s_n - \mu_n < s_n < s_n + \mu_n < s_n + 2\mu_n < t_{n-1}$$

and $g(t) \geq 4n$ for $t \in A_n$. We now define $\alpha_n > 0$ by

$$\alpha_n := \frac{t_n - t_{n+1}}{2n\mu_n} \quad \text{for } n \in \mathbb{N},$$

as well as the function $x : [0,1] \to \mathbb{R}$ by

$$x(t) = \begin{cases} 0 & \text{for } t \in \{0, c, 1\}, \\ 0 & \text{for } t \in \{t_n, s_n - \mu_n, s_n + 2\mu_n\}, n \in \mathbb{N}, \\ \alpha_n & \text{for } s_n \leq t \leq s_n + \mu_n, n \in \mathbb{N}, \\ \text{linear} & \text{otherwise.} \end{cases}$$

The following picture shows x on some interval $[t_n, t_{n-1}]$.

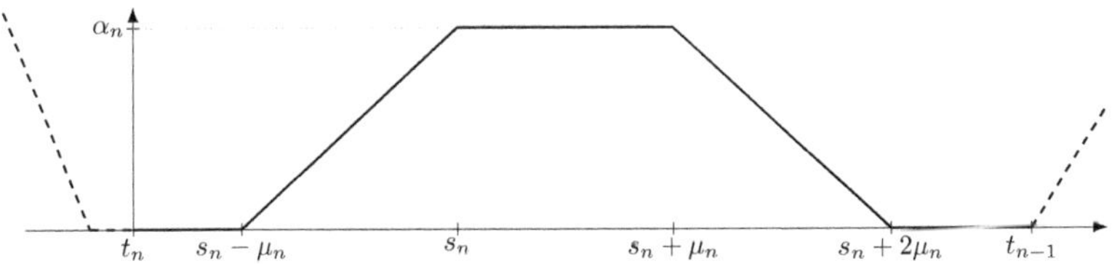

Figure 2.2.1: The function x on a subinterval.

The function x is KH-integrable on $[0,c]$ and on every interval $[d,1]$ for $d \in (c,1)$, because x is piecewise linear on $[0,1]\backslash\{c\}$. For such a d and $n \in \mathbb{N}$ with $t_n < d \leq t_{n-1}$ we have

$$\int_d^1 x(t)\,\mathrm{d}t \leq \int_{t_n}^1 x(t)\,\mathrm{d}t = \sum_{j=1}^n \int_{s_j-\mu_j}^{s_j+2\mu_j} x(t)\,\mathrm{d}t = 2\sum_{j=1}^n \mu_j\alpha_j = \sum_{j=1}^n \frac{t_j - t_{j+1}}{j} \leq t_1 - t_{n+1}.$$

Since the right hand side remains bounded for $d \to c+$ and since x is nowhere negative, we deduce that x is even Lebesgue integrable on $[c,1]$. Moreover,

$$\begin{aligned} \int_c^d x(t)\,\mathrm{d}t &\leq \sum_{j=n}^\infty \int_{t_j}^{t_{j-1}} x(t)\,\mathrm{d}t = \sum_{j=n}^\infty \int_{s_j-\mu_j}^{s_j+2\mu_j} x(t)\,\mathrm{d}t = 2\sum_{j=n}^\infty \mu_j\alpha_j \\ &= \sum_{j=n}^\infty \frac{t_j - t_{j+1}}{j} \leq \frac{t_n - c}{n}, \end{aligned}$$

and from that follows

$$\frac{1}{d-c}\int_c^d x(t)\,\mathrm{d}t \le \frac{t_n-c}{(d-c)n} \le \frac{1}{n}.$$

Since $d \to c+$ implies $n \to \infty$, we obtain

$$\lim_{d\to c+} \frac{1}{d-c}\int_c^d x(t)\,\mathrm{d}t = 0 = x(c),$$

and as $x(t) = 0$ for $0 \le t \le c$, Theorem 2.1.3 (d) shows indeed $x \in \Delta$.

It remains to show that $xg \notin \Delta$, and for that we again use Theorem 2.1.3 (d). If $xg \notin KH$, we are done by Theorem 2.1.3 (c). Therefore we assume $xg \in KH$ and have to show that the limit condition (2.1.3) is violated. By construction of x we have

$$\begin{aligned}\int_c^{t_n} x(t)g(t)\,\mathrm{d}t &= \sum_{j=n}^{\infty}\int_{t_{j+1}}^{t_j} x(t)g(t)\,\mathrm{d}t \ge \sum_{j=n}^{\infty}\int_{A_j} x(t)g(t)\,\mathrm{d}t \ge 2\sum_{j=n}^{\infty} j\mu_j\alpha_j \\ &= \sum_{j=n}^{\infty}(t_j - t_{j+1}) = t_n - c.\end{aligned}$$

However, from this estimates it follows that

$$\limsup_{\delta\to 0+} \frac{1}{\delta}\int_c^{c+\delta} x(t)g(t)\,\mathrm{d}t \ge 1 > 0 = x(0)g(0),$$

which means that (2.1.3) is indeed violated. We deduce $xg \notin \Delta$, contradicting the fact that $g \in \Delta/\Delta$.

Finally, we now show that g is really bounded. Due to $g \in \Delta/\Delta$ we have $g \in \Delta$, and since $g \ge 0$ and $g \in L_\infty$ we get $g(t) \le \|g\|_{L_\infty}$ for almost all $t \in [0,1]$. From Theorem 2.1.3 (d) in combination with [68, Theorem 9.9] we obtain

$$g(t) = \lim_{\delta\to 0} \frac{1}{\delta}\int_t^{t+\delta} g(t)\,\mathrm{d}t \le \lim_{\delta\to 0} \frac{1}{\delta}\int_t^{t+\delta} \|g\|_{L_\infty}\,\mathrm{d}t = \|g\|_{L_\infty}$$

for each $t \in [0,1]$ and consequently $g \in B$, as claimed. ■

In the proof of the last theorem we have deduced $g \in B$ from $g \in \Delta \cap L_\infty$. This deduction does not work anymore if we replace $g \in \Delta \cap L_\infty$ by $g \in D \cap L_\infty$. We show that there is a Darboux function which is zero almost everywhere (and in particular essentially bounded with L_∞-norm 0), but attains every real number in every proper real interval (and so is dramatically unbounded). This example is similar to one found in the introduction of the book [147], in which the authors construct a Darboux function that maps any subinterval $[a,b]$ onto the entire interval $[0,1]$ but has no primitive on subintervals. Our example is an unbounded version of this idea.[2]

[2]Compare to Example 3.1.9.

Example 2.2.7. Let $I_n = [a_n, b_n] \subseteq [0,1]$ with $a_n, b_n \in \mathbb{Q}$ and $a_n < b_n$ for $n \in \mathbb{N}$ be an enumeration of all proper subintervals of $[0,1]$ with rational end points. Inductively, one can extract pairwise disjoint Cantor sets $C_n \subseteq I_n$ such that each has Lebesgue measure zero. Since each such set C_n has the cardinality of $\mathbb{R}$, there are bijective functions $g_n : C_n \to \mathbb{R}$ for each $n \in \mathbb{N}$. We put

$$g(t) := \begin{cases} g_n(t) & \text{for } t \in C_n, n \in \mathbb{N}, \\ 0 & \text{otherwise.} \end{cases}$$

Now, if an interval $[a,b] \subseteq [0,1]$ with $a < b$ is given, then since $\mathbb{Q}$ is dense in $\mathbb{R}$ we find some $m \in \mathbb{N}$ satisfying $I_m \subseteq [a,b]$. From $g(I_m) \supseteq g_m(C_m) = \mathbb{R}$ we deduce $g([a,b]) = \mathbb{R}$; in particular, g is a Darboux function that attains every real number in any proper real interval. Because all the sets C_n have measure zero, their union has measure zero, too, and g is indeed zero almost everywhere. By Theorem 2.1.6, g cannot have a primitive on some subinterval of $[0,1]$. ◇

We are now going to answer the question on how the functions in Δ/Δ look like. In order to do so, let us consider the following class of functions.

$$\overline{BV} := \left\{ g : [0,1] \to \mathbb{R} \mid \forall t \in [0,1] : \limsup_{\delta \to 0} \operatorname{Var}(g, [t+\delta, t+2\delta]) < \infty \right\}. \tag{2.2.10}$$

This class is situated between BV and L_1, that is, the inclusions

$$BV \subseteq \overline{BV} \subseteq L_1 \tag{2.2.11}$$

hold, where the first is obvious, and the second is shown in our next

Proposition 2.2.8. *Every function in $\overline{BV}$ is Lebesgue integrable.*

Proof. First note that any $x \in \overline{BV}$ is (Lebesgue) measurable since it can have only countably many points of discontinuity. In order to show that x is Lebesgue integrable it suffices to show that x is locally Lebesgue integrable on $[0,1]$, that is, $x \in L_1([t_0 - \delta, t_0 + \delta] \cap [0,1])$ for all $t_0 \in [0,1]$ and suitable $\delta > 0$ depending on t_0. We only show this for $t_0 = 0$, the other cases are similar.
Since $x \in \overline{BV}$ there are $M > 0$ and $N \in \mathbb{N}$ such that

$$\operatorname{Var}\left(x, I_n\right) \le M \quad \text{for } n \ge N,$$

where $I_n := [2^{-n-1}, 2^{-n}]$. For any $t \in I_n$ we have $|x(t)| \le |x(2^{-n})| + \operatorname{Var}(x, I_n) \le |x(2^{-n})| + M$; in particular, $|x(2^{-n-1})| \le M + |x(2^{-n})|$ for all $n \ge N$. This implies $|x(2^{-n})| \le (n-N)M + m$ for all $n \ge N$, where $m := |x(2^{-N})|$, and thus

$$|x(t)| \le (n - N + 1)M + m \le nM + m \quad \text{for } t \in I_n, n \ge N.$$

We obtain

$$\begin{aligned} \int_0^{2^{-N}} |x(t)| \, dt &= \sum_{n=N}^{\infty} \int_{2^{-n-1}}^{2^{-n}} |x(t)| \, dt \le \sum_{n=N}^{\infty} \left(2^{-n} - 2^{-n-1}\right)\left(nM + m\right) \\ &= 2^{-N}(m + M + MN) \end{aligned}$$

which shows that x is indeed Lebesgue integrable on $[0, 2^{-N}]$. ■

The characteristic function $\chi_{\mathbb{Q}\cap[0,1]}$ which belongs to $L_1\backslash\overline{BV}$ shows that the second inclusion in (2.2.11) is strict. Our next example proves that also the first inclusion in (2.2.11) is strict. Moreover, it also illustrates that in contrast to BV the class $\overline{BV}$ is not contained in B.

Example 2.2.9. The function

$$x(t) = \begin{cases} \log(t) & \text{for } 0 < t \leq 1, \\ 0 & \text{for } t = 0 \end{cases}$$

is unbounded and so certainly not in BV. Moreover, for $t > 0$ and $t+\delta, t+2\delta \in (0,1)$ we have

$$\begin{aligned} \operatorname{Var}(x, [t+\delta, t+2\delta]) &= |x(t+\delta) - x(t+2\delta)| = |\log(t+\delta) - \log(t+2\delta)| \\ &= \left|\log\frac{t+\delta}{t+2\delta}\right| \longrightarrow 0 \quad \text{as } \delta \to 0, \end{aligned}$$

and for $t = 0$ and $0 < \delta < 1/2$ we get

$$\operatorname{Var}(x, [\delta, 2\delta]) = |x(\delta) - x(2\delta)| = \left|\log\frac{1}{2}\right| = \log(2).$$

Consequently, $x \in \overline{BV}$. ◇

One could argue that any function $x \in \overline{BV} \cap D$ must be bounded, since such a function can have only finitely many discontinuities. Indeed, for any $t \in [0,1]$ the function x must be of bounded variation on each interval $[t+\delta, t+2\delta]$ for sufficiently small $\delta > 0$ and hence continuous there because of the Darboux property. This means that for each $t \in [0,1]$ the function x must be continuous on $[t-\delta, t+\delta]\backslash\{t\}$ for small $\delta > 0$, and thus can have only finitely many discontinuities as $[0,1]$ is compact. However, the following example shows that even if $x \in \overline{BV} \cap D$ is discontinuous at only one point, then it may still be unbounded.

Example 2.2.10. Define the sequence of natural numbers $(k_n)_{n\in\mathbb{N}_0}$ by

$$k_0, k_1, k_2, k_3, \ldots = 0, 1, 0, 1, 2, 1, 0, 1, 2, 3, 2, 1, 0, 1, 2, 3, 4, 3, 2, 1, 0, \ldots,$$

which, for instance, can be constructed explicitly as follows. For $n \in \mathbb{N}_0$ we set

$$i_n := \left\lfloor \frac{\sqrt{1+4n}-1}{2} \right\rfloor, \qquad a_n := i_n(i_n+1) \qquad \text{and} \qquad b_n := (i_n+1)(i_n+2).$$

Then

$$k_n = \frac{b_n - a_n}{2} - \left|\frac{a_n + b_n}{2} - n\right|.$$

We now define the function $x : [0,1] \to \mathbb{R}$ by $x(0) = 0$, and piecewise linear and continuous on $(0,1]$ by $x(2^{-n}) = k_n$ for $n \in \mathbb{N}_0$. Here is a picture of x on $[2^{-12}, 1]$, where the t-axis has partially been scaled in a logarithmic manner to make the line segments more visible.

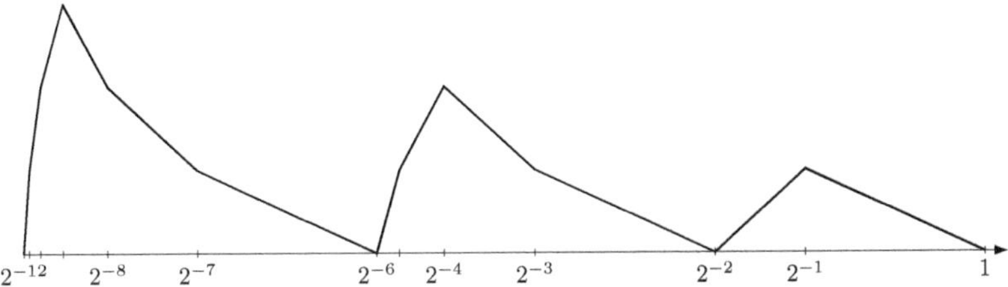

Figure 2.2.2: The function x on a partially logarithmic scale.

Then $x \in BV[a,1]$ for all $a \in (0,1)$ and $\mathrm{Var}(x,[2^{-(n+1)},2^{-n}]) = |k_{n+1} - k_n| = 1$ for all $n \in \mathbb{N}_0$, and this implies $x \in \overline{BV}$. To see this, note that for $\delta \in (0,1)$ we find some $n \in \mathbb{N}$ such that $2^{-n} \leq \delta \leq 2^{-n+1}$. Therefore, $2^{-n+1} \leq 2\delta \leq 2^{-n+2}$; we obtain

$$\mathrm{Var}(x,[\delta,2\delta]) \leq \mathrm{Var}(x,[2^{-n},2^{-n+2}]) = 2$$

and thus indeed $x \in \overline{BV}$.
Moreover, since x is continuous and nonnegative on $(0,1]$, and since $x([0,2^{-n}]) = [0,\infty)$, we conclude $x \in D$, but $x \notin B$. ♢

As we have seen, the inclusion $\overline{BV} \cap D \subseteq B$ is false. But it turns true if we replace D by the space Δ.

Proposition 2.2.11. *The inclusion*

$$\overline{BV} \cap \Delta \subseteq B$$

holds.

Proof. Fix $x \in \overline{BV} \cap \Delta$, and assume that x is unbounded. Then there is a sequence (t_n) in $[0,1]$ such that $|x(t_n)| \to \infty$ as $n \to \infty$. Since $[0,1]$ is compact, we can assume by passing to a suitable subsequence that (t_n) converges monotonically to some t_0. We suppose that (t_n) decreases to $t_0 \in [0,1)$ and that $x(t_n) \to \infty$; the other cases are similar. For each $n \in \mathbb{N}$ there is some $k \in \mathbb{N}$ such that $t_n \in J_k := [t_0 + 2^{-k}, t_0 + 2^{-k+1}]$. For any $s \in J_k$ we clearly have

$$x(t_n) \leq x(s) + \mathrm{Var}(x, J_k). \tag{2.2.12}$$

Since $x \in \Delta$, by Theorem 2.1.3 (d) x is also KH-integrable and satisfies the limit condition (2.1.3). Consequently, integrating (2.2.12) on both sides with respect to s gives

$$x(t_n) \leq \frac{1}{2^{-k}} \int_{t_0+2^{-k}}^{t_0+2^{-k+1}} x(s)\,\mathrm{d}s + \mathrm{Var}(x, J_k). \tag{2.2.13}$$

Since x belongs to $\overline{BV}$, there is some $M > 0$ such that

$$\limsup_{k\to\infty} \operatorname{Var}(x, J_k) = M < \infty, \tag{2.2.14}$$

and since x satisfies (2.1.3) at t_0 we get

$$\lim_{k\to\infty} \frac{1}{2^{-k}} \int_{t_0+2^{-k}}^{t_0+2^{-k+1}} x(s)\,\mathrm{d}s = x(t_0). \tag{2.2.15}$$

Letting $n \to \infty$ implies $k \to \infty$, and we obtain from (2.2.13), (2.2.14) and (2.2.15) that

$$\begin{aligned}\infty = \lim_{n\to\infty} x(t_n) &\le \limsup_{k\to\infty} \left(\frac{1}{2^{-k}} \int_{t_0+2^{-k}}^{t_0+2^{-k+1}} x(s)\,\mathrm{d}s + \operatorname{Var}(x, J_k) \right) \\ &= \lim_{k\to\infty} \frac{1}{2^{-k}} \int_{t_0+2^{-k}}^{t_0+2^{-k+1}} x(s)\,\mathrm{d}s + \limsup_{k\to\infty} \operatorname{Var}(x, J_k) = x(t_0) + M < \infty.\end{aligned}$$

But this is impossible, and the claim is proven. ∎

Proposition 2.2.11 implies that the function x constructed in Example 2.2.10 while having the Darboux property cannot possess a primitive. Moreover, the compactness of the domain of definition is important: The function $x = \log$ on $(0, 1]$ with $x(0) = 0$ from Example 2.2.9 belongs to $C^1(0, 1]$ and hence also to $D(0, 1]$, $\Delta(0, 1]$ and $\overline{BV}(0, 1]$, but not to $B(0, 1]$ and $BV(0, 1]$.

The specialty of the class $\overline{BV}$ is now that it is exactly that class X of functions we were searching for in (2.2.8). More precisely, we have the following result.

Theorem 2.2.12. *The identity*

$$\Delta/\Delta = \overline{BV} \cap \Delta$$

holds.

Fleissner was the first, who characterized Δ/Δ completely in the sense of Theorem 2.2.12 using an improper Stieltjes integral and a similar type of bounded variation which he called *distant bounded variation* [59]. Later, Mařík reformulated Fleissner's results and used only functions of bounded variation instead of Stieltjes integrals [111]. He identified the missing class X in (2.2.8) to be the class of those functions $g : [0, 1] \to \mathbb{R}$ which satisfy for all $t \in [0, 1]$,

$$\limsup_{n\to\infty} \left[\operatorname{Var}\left(g, \left[t + \frac{1}{2^n}, t + \frac{2}{2^n}\right]\right) + \operatorname{Var}\left(g, \left[t - \frac{2}{2^n}, t - \frac{1}{2^n}\right]\right) \right] < \infty. \tag{2.2.16}$$

Mařík together with Bruckner and Weil then used the condition

$$\limsup_{n\to\infty} \left[\operatorname{Var}\left(g, \left[t + \frac{1}{n}, t + \frac{2}{n}\right]\right) + \operatorname{Var}\left(g, \left[t - \frac{2}{n}, t - \frac{1}{n}\right]\right) \right] < \infty \tag{2.2.17}$$

instead of (2.2.16) [24]. However, it is easily seen that the classes characterized by (2.2.16) and (2.2.17) in fact coincide with the class $\overline{BV}$. This was also recognized by

Mařík who gave an alternative proof for Theorem 2.2.12 and his original reformulation of Fleissner's result in the paper [113].

We summarize some of the ideas of Mařík's first proof [111] to give a proof for the inclusion "$\supseteq$" in Theorem 2.2.12. This will give another impression of how useful Theorem 2.1.3 is for those kind of arguments. The other inclusion is more difficult and technical to prove; we refer the reader to the literature mentioned above.

Proof of "$\supseteq$" of Theorem 2.2.12. Let $g \in \overline{BV} \cap \Delta$ and $x \in \Delta$; we have to show that $xg \in \Delta$. To this end, fix $t \in [0, 1)$; as in our proof of Theorem 2.2.2 (a) we can assume that $x(t) = 0$. For a given $\varepsilon > 0$ we find some $\eta > 0$ such that the function

$$f(u) := \int_t^u x(s)\,\mathrm{d}s$$

satisfies the estimate $|f(u)| \leq \varepsilon|u-t|$ for $|t-u| \leq \eta$. Writing $I_j := [t+2^{-(j+1)}, t+2^{-j}]$, we find due to $g \in \overline{BV} \cap \Delta$ and Proposition 2.2.11 some $M > 0$ and $N \in \mathbb{N}$ with

$$g(s) + \mathrm{Var}(g, I_j) \leq M \quad \text{for } j \geq N, s \in I_j.$$

From (2.2.6) follows $xg \in KH(I_j)$ for all $j \geq N$, as well as

$$\left|\int_{I_j} x(s)g(s)\,\mathrm{d}s\right| \leq M s_j, \tag{2.2.18}$$

where we put similarly to (2.2.6)

$$s_j := \sup_{[u,v]\subseteq I_j} \left|\int_u^v x(s)\,\mathrm{d}s\right|.$$

From the estimate $|f(u)| \leq \varepsilon|u-t|$ we obtain

$$s_j = \sup_{[u,v]\subseteq I_j} |f(u) - f(v)| \leq \frac{\varepsilon}{2^{j-1}}.$$

For $2^{-(n+1)} \leq \delta < 2^{-n} < \varepsilon$ we have using (2.2.18)

$$\left|\int_t^{t+\delta} x(s)g(s)\,\mathrm{d}s\right| \leq \sum_{j=n}^{\infty} \left|\int_{I_j} x(s)g(s)\,\mathrm{d}s\right| \leq M\sum_{j=n}^{\infty} s_j \leq M\varepsilon\sum_{j=n}^{\infty}\frac{1}{2^{j-1}} = \frac{4M\varepsilon}{2^n} \leq 8M\delta\varepsilon;$$

in particular, $xg \in KH[t, t+\delta]$. We deduce

$$\limsup_{\delta\to 0+} \frac{1}{\delta}\left|\int_t^{t+\delta} x(s)g(s)\,\mathrm{d}s\right| \leq 8M\varepsilon,$$

and since $\varepsilon > 0$ had been chosen arbitrarily, we obtain

$$\lim_{\delta\to 0+} \frac{1}{\delta}\int_t^{t+\delta} x(s)g(s)\,\mathrm{d}s = 0 = x(t)g(t).$$

Similarly, one can show that $xg \in KH[t-\delta, t]$ for $t \in (0,1]$ and sufficiently small $\delta > 0$, as well as

$$\lim_{\delta \to 0-} \frac{1}{\delta} \int_t^{t+\delta} x(s)g(s)\,\mathrm{d}s = 0 = x(t)g(t).$$

In total, we have shown that xg is KH-integrable in a neighborhood of any point $t \in [0,1]$ and that it satisfies (2.1.3) there. Since $[0,1]$ is compact, $xg \in KH$, and an application of Theorem 2.1.3 (d) yields indeed $xg \in \Delta$. ■

Fleissner's and Mařík's results encapsulated in Theorem 2.2.12 show, as mentioned above, that the missing class in (2.2.8) is $X = \overline{BV}$. We check this result immediately in case of our test functions $\varphi_{\alpha,\beta,n}$ defined in (1.1.1).

Proposition 2.2.13. *For the functions (1.1.1) the following relations hold.*

(a) $\varphi_{\alpha,\beta,n} \in \overline{BV}$ *if and only if* $\alpha \geq 1$.

(b) $\varphi_{\alpha,\beta,n} \in \Delta/\Delta$ *if and only if* $\alpha \geq 1$ *and* $\beta = 0$.

Proof. To prove (a) first note that $\varphi_{\alpha,\beta,n} \in BV$ if and only if $\alpha > 1$ by Proposition 1.1.12 (b). Therefore, we focus on the case when $\alpha \leq 1$. Due to $\varphi_{\alpha,\beta,n} \in BV[\varepsilon,1] \subseteq \overline{BV}[\varepsilon,1]$ for every $\varepsilon \in (0,1)$, we only need to investigate those triples $(\alpha,\beta,n) \in (-\infty,1] \times \mathbb{R} \times \mathbb{N}$ for which the condition

$$\limsup_{\delta \to 0+} \operatorname{Var}(\varphi_{\alpha,\beta,n},[\delta,2\delta]) < \infty$$

is fulfilled. To this end, fix $\delta \in (0,1/2)$. For $t_m := 1/(m\pi + \frac{\pi}{2})$ with $m \in \mathbb{N}$ and $t_m \leq 2\delta \leq t_{m-1}$ we have $t_{2m+1} \leq \delta \leq t_{2m-2}$. As the t_m run through the local maxima and minima of $\varphi_{\alpha,\beta,n}$ we get for $\alpha = 1$

$$\begin{aligned}
\operatorname{Var}\left(\varphi_{1,\beta,n},[\delta,2\delta]\right) &\leq \operatorname{Var}\left(\varphi_{1,\beta,n},[t_{2m+1},t_{m-1}]\right) = \sum_{j=m-1}^{2m+1} |\varphi_{1,\beta,n}(t_j)| = \sum_{j=m-1}^{2m+1} t_j \\
&\leq t_{m-1} + \int_{m-1}^{2m+1} \frac{1}{\pi t + \frac{\pi}{2}}\,\mathrm{d}t \\
&= \frac{1}{m\pi - \frac{\pi}{2}} + \frac{1}{\pi}\log\left(\frac{4m+3}{2m-1}\right) \longrightarrow \frac{\log 2}{\pi} \qquad \text{as } m \to \infty.
\end{aligned}$$

Because $\delta \to 0+$ implies $m \to \infty$, we deduce that

$$\limsup_{\delta \to 0+} \operatorname{Var}\left(\varphi_{1,\beta,n},[\delta,2\delta]\right) \leq \frac{\log 2}{\pi} < \infty,$$

and hence $\varphi_{1,\beta,n} \in \overline{BV}$.

For $0 \leq \alpha < 1$ one shows similarly

$$\begin{aligned}
\operatorname{Var}\left(\varphi_{\alpha,\beta,n}[\delta,2\delta]\right) &\geq \operatorname{Var}\left(\varphi_{\alpha,\beta,n},[t_{2m-2},t_m]\right) = \sum_{j=m}^{2m-2} |\varphi_{\alpha,\beta,n}(t_j)| = \sum_{j=m}^{2m-2} t_j^\alpha \\
&\geq \int_m^{2m-1} \frac{1}{(\pi t + \frac{\pi}{2})^\alpha}\,\mathrm{d}t \geq \frac{1}{(2\pi)^\alpha}\int_m^{2m-1} \frac{1}{t^\alpha}\,\mathrm{d}t \\
&= \frac{(2m-1)^{1-\alpha} - m^{1-\alpha}}{(2\pi)^\alpha(1-\alpha)} \longrightarrow \infty \qquad \text{as } m \to \infty.
\end{aligned}$$

Since again $\delta \to 0+$ implies $m \to \infty$, we obtain this time

$$\limsup_{\delta \to 0+} \operatorname{Var}\left(\varphi_{\alpha,\beta,n}, [\delta, 2\delta]\right) = \infty$$

and hence $\varphi_{\alpha,\beta,n} \notin \overline{BV}$.

Finally, we have for $\alpha < 0$,

$$\begin{aligned}\operatorname{Var}(\varphi_{\alpha,\beta,n}, [\delta, 2\delta]) \geq \operatorname{Var}(\varphi_{\alpha,\beta,n}, [t_{2m-2}, t_m]) &= \sum_{j=m}^{2m-2} |\varphi_{\alpha,\beta,n}(t_j)| = \sum_{j=m}^{2m-2} t_j^{\alpha} \\ &= \sum_{j=m}^{2m-2} (j\pi + \pi/2)^{|\alpha|} \geq (m\pi + \pi/2)^{|\alpha|} \geq m^{|\alpha|},\end{aligned}$$

and so again $\varphi_{\alpha,\beta,n} \notin \overline{BV}$. Consequently, part (a) is proven.

A proof of (b) is now immediate. Just combine part (a), Theorem 2.2.12 and Proposition 2.1.5 (b). ■

Proposition 2.2.13 allows us to provide another example showing that the first inclusion in (2.2.11) is strict. Indeed, combining Proposition 1.1.12 (b) with Proposition 2.2.13 (a) shows that $\varphi_{1,\beta,n} \in \overline{BV} \setminus BV$ and, in addition, $\varphi_{1,0,n} \in (\Delta/\Delta)\backslash BV$. So there are functions of unbounded variation that still belong to Δ/Δ. At first glance it might be surprising that Δ/Δ contains also discontinuous functions of unbounded variation; we construct such a function in the following

Example 2.2.14. For $n \in \mathbb{N}$ put $\mu_n := 3^{1-n}/4$ and $t_n := 2^{-n}$, and define $g : [0,1] \to \mathbb{R}$ by

$$g(t) = \begin{cases} 0 & \text{for } t \in \{0, 1\}, \\ 0 & \text{for } t \in \{t_n, t_n + 2\mu_n\}, n \in \mathbb{N}, \\ 1 & \text{for } t = t_n + \mu_n, n \in \mathbb{N}, \\ \text{linear} & \text{otherwise,} \end{cases}$$

see the following graphic.

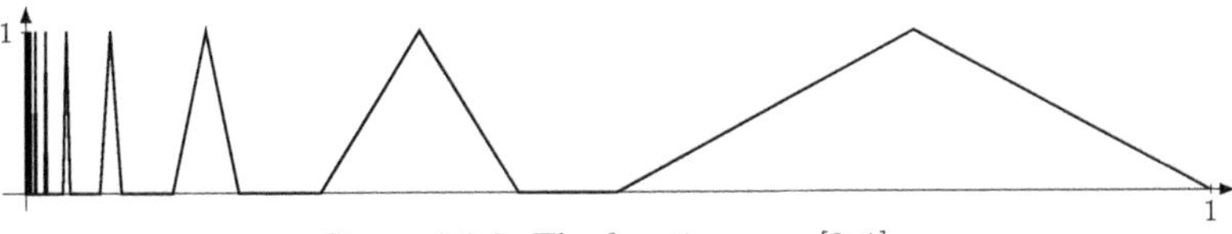

Figure 2.2.3: The function g on $[0, 1]$.

Then g is continuous and piecewise linear on $(0, 1]$ and attains values only in $[0, 1]$. However, g is discontinuous at $t = 0$, and of unbounded variation near 0, as

$$\operatorname{Var}(g, [t_{n+k}, t_n]) = 2k \quad \text{for } k \in \mathbb{N}.$$

As a measurable, bounded and nonnegative function, g is KH-integrable on $[0,1]$ and satisfies for $t_{n+1} \le \delta \le t_n$ the estimates

$$\frac{1}{\delta}\int_0^\delta g(t)\,\mathrm{d}t \le \frac{1}{t_{n+1}}\int_0^{t_n} g(t)\,\mathrm{d}t = 2^{n+1}\frac{3}{4}\sum_{j=n+1}^{\infty}\frac{1}{3^j} = \frac{3}{4}\left(\frac{2}{3}\right)^n \longrightarrow 0 \qquad \text{as } n \to \infty.$$

Since $\delta \to 0+$ implies $n \to \infty$, we obtain

$$\lim_{\delta\to 0+}\frac{1}{\delta}\int_0^\delta g(t)\,\mathrm{d}t = 0 = g(0),$$

and so g has a primitive by Theorem 2.1.3 (d).

For $t_{n+1} \le \delta \le t_n$ we have $t_n \le 2\delta \le t_{n-1}$, and this yields

$$\mathrm{Var}(g,[\delta,2\delta]) \le \mathrm{Var}(g,[t_{n+1},t_{n-1}]) = 4;$$

in particular,

$$\limsup_{\delta\to 0+}\mathrm{Var}(g,[\delta,2\delta]) \le 4 < \infty,$$

which means nothing but $g \in \overline{BV}$. Theorem 2.2.12 now says $g \in \Delta/\Delta$. ◊

The zigzag function in Example 2.2.14 visualizes the subtle difference between the classes Δ and Δ/Δ. In order to make a function g with peaks belong to Δ, its peaks have to get slim sufficiently quickly, and in order to ensure $g \in \Delta/\Delta$, the positions of those peaks have to go to zero sufficiently fast. Roughly speaking, Δ is responsible for the width of the peaks, and Δ/Δ for their placement.

We thus have characterized the multiplier sets X/X for $X \in \{B, BV, C, KH, \Delta\}$. As mentioned at the beginning of this chapter, the class D/D is also hard to handle. We will investigate this class in much more detail in the next chapter, but we still give some structural properties here and put a particular emphasis on its relation to Δ.

We first show by means of an example that D is also not closed under multiplication.

Example 2.2.15. In Example 1.1.2 we constructed two functions $x, y \in D$, the sum of which does not lie within D. From that we easily obtain a corresponding example for the product, namely the two functions $\tilde{x} = \exp\circ x$ and $\tilde{y} = \exp\circ y$. They are Darboux functions, but their product $\tilde{x}\tilde{y} = \exp\circ(x+y) = \exp\circ\chi_{\{0\}}$ is not. ◊

Even more interesting is the following example [24] which shows that the product of two functions of Δ does not necessarily belong to D anymore.

Example 2.2.16. Let $x, y : [0,1] \to \mathbb{R}$ be defined by

$$x(t) = \pi\max\left\{\varphi_{0,1/\pi,1}(t), 0\right\} \qquad \text{and} \qquad y(t) = \pi\min\left\{\varphi_{0,-1/\pi,1}(t), 0\right\}.$$

Then both x and y are Lebesgue integrable and continuous on $(0,1]$. Similarly as in the proof of Proposition 2.1.5 one can show that both x and y have primitives. On the other hand, $xy = -\chi_{\{0\}}$ is not even a Darboux function. ◊

By replacing y by $-y$ in Example 2.2.15 we see that the fraction x/y of two functions $x, y \in D$ (with $y(t) \neq 0$ for $t \in [0,1]$) also does not necessarily lie in D. A similar modification for an analogue of Example 2.2.16, however, is not possible. Hruška has shown in [11], that the quotient x/y of two functions $x, y \in \Delta$ (again with $y(t) \neq 0$ for $t \in [0,1]$) surprisingly always belongs to D!

Because of our discussion of (2.2.4) and Example 2.2.15 we conclude that the strict inclusion

$$D/D \subsetneq D \tag{2.2.19}$$

must be true, and so the class D/D is certainly smaller than D. How small D/D really is will be shown in Theorem 3.1.8 in the next chapter, according to which D/D contains only constant functions!

In Figure 2.2.4 below we give a summary of identities and inclusions of the most important function spaces that have been considered in this chapter. Here, $A \longrightarrow B$ means $A \subsetneq B$.

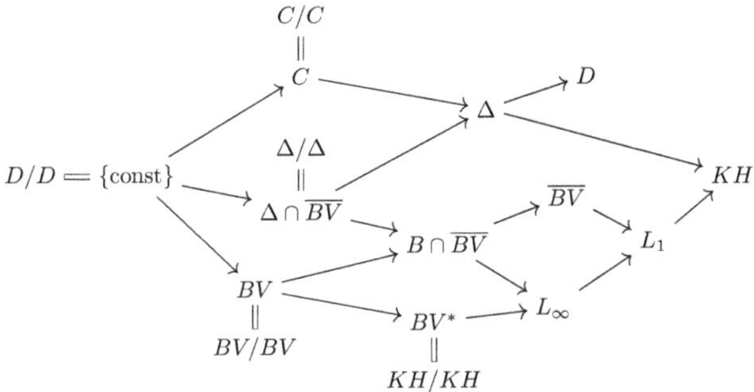

Figure 2.2.4: Inclusions between classical sets.

The following example shows that none of the somewhat exotic sets $\overline{BV}$ and BV^* which are both strict supersets of BV includes the other. The same is true if one replaces $\overline{BV}$ by the smaller class $B \cap \overline{BV}$.

Example 2.2.17. By Proposition 1.1.12 (a) and (b) and Proposition 2.2.13 (a) the functions $\varphi_{1,\beta,n}$ belong to $B \cap \overline{BV}$ for arbitrary $\beta \in \mathbb{R}$ and $n \in \mathbb{N}$, but not to BV and hence also not to BV^*. However, the function

$$x(t) := \begin{cases} n & \text{for } t = 1/n, n \in \mathbb{N}, \\ 0 & \text{otherwise} \end{cases}$$

belongs to BV^*, but not to $\overline{BV}$ let alone to B, as $x(1/n) = n \to \infty$ and

$$\operatorname{Var}\left(x, \left[\frac{1}{2n}, \frac{1}{n}\right]\right) \geq \left|x\left(\frac{1}{2n}\right) - x\left(\frac{1}{n}\right)\right| = n$$

show. ◇

In Table 2.2.1 below we gather all conditions on $(\alpha, \beta, n) \in \mathbb{R}^2 \times \mathbb{N}$ for which the functions $\varphi_{\alpha,\beta,n}$ from (1.1.1) belong to a certain class of functions.

Table 2.2.1: Conditions under which $\varphi_{\alpha,\beta,n}$ belongs to certain function classes.

$\varphi_{\alpha,\beta,n} \in$	if α	and β	and n
B	≥ 0	arbitrary	arbitrary
C	> 0	$= 0$	arbitrary
D	> 0	$= 0$	arbitrary
	$= 0$	$\in [0, 1]$	even
	$= 0$	$\in [-1, 1]$	odd
	< 0	≥ 0	even
	< 0	arbitrary	odd
BV	> 1	arbitrary	arbitrary
BV^*	> 1	arbitrary	arbitrary
$\overline{BV}$	≥ 1	arbitrary	arbitrary
Δ	> 0	$= 0$	even
	$= 0$	$= \frac{1}{\pi} \int_0^\pi \sin^n t\, dt$	even
	> -1	$= 0$	odd
Δ/Δ	≥ 1	$= 0$	arbitrary
L_1	> -1	arbitrary	arbitrary
KH	> -1	arbitrary	even
	> -2	arbitrary	odd
KH/KH	> 1	arbitrary	arbitrary

We close this section with a question that is important for applications: *How does the algebra $Alg(\Delta)$ generated by Δ look like?* Of course, this question makes sense only because the product of two functions with primitive does not necessarily have a primitive.

It is easy to see that any function $x \in \Delta$ belongs to the class $\mathcal{B}_1$ of *Baire-1 functions* which are those functions that can be written as a pointwise limit of continuous functions. Indeed, if f is a primitive of $x \in \Delta$, then

$$x(t) = f'(t) = \lim_{n\to\infty} n\big[f(t + 1/n) - f(t)\big] \quad \text{for } 0 \leq t \leq 1,$$

where f has been continuously extended to $[0, 2]$ by $f(t) = f(1) + x(1)(t - 1)$ for $t \in (1, 2]$.

Thus, $\Delta \subseteq \mathcal{B}_1$, and since $\mathcal{B}_1$ is an algebra, we even have $Alg(\Delta) \subseteq \mathcal{B}_1$. A characterization of those Baire-1 functions which can be expressed as a product of two (or, more general, finitely many) functions from Δ is to the best of our knowledge not known; we only know that not every Baire-1 function can be written that way. Here is an example of such a function, that even belongs to $\mathcal{B}_1 \cap D$.

Example 2.2.18. By the Propositions 1.1.12 and 2.1.5 it is immediate that the function $p : [0,1] \to \mathbb{R}$, defined by

$$p(t) := 2 + \varphi_{0,-1,1}(t) = \begin{cases} 2 + \sin \dfrac{1}{t} & \text{for } 0 < t \leq 1, \\ 1 & \text{for } t = 0, \end{cases}$$

is positive and belongs to $\mathcal{B}_1 \cap D$. However, by Proposition 2.1.5 (b) it has no primitive. We now show that p cannot be written as a product of two functions from Δ.
Assume the opposite, that is, $p = xy$ for some $x, y \in \Delta$. Then both x and y are Darboux functions, and since p is positive everywhere on $[0,1]$, both x and y are either positive or negative throughout $[0,1]$, since a sign change of x or y implies that p has a zero. Moreover, since $1 \leq p \leq 3$ on $[0,1]$, we get from the Cauchy Schwarz Inequality for $0 < \delta \leq 1$,

$$\left(\frac{1}{\delta}\int_0^\delta p(t)\,\mathrm{d}t\right)^2 \leq \frac{3}{\delta^2}\left(\int_0^\delta \sqrt{x(t)y(t)}\,\mathrm{d}t\right)^2 \leq 3\left(\frac{1}{\delta}\int_0^\delta x(t)\,\mathrm{d}t\right)\left(\frac{1}{\delta}\int_0^\delta y(t)\,\mathrm{d}t\right).$$

Letting $\delta \to 0+$ yields together with Theorem 2.1.3 (d) and Proposition 2.1.5 (b),

$$4 \leq 3x(0)y(0) = 3p(0) = 3,$$

a contradiction. 

In view of Example 2.2.18 the following result of Preiss proved in [131] is remarkable.

Theorem 2.2.19. *The identity*

$$\mathcal{B}_1 = \{xy + z \mid x, y, z \in \Delta\} \tag{2.2.20}$$

is true, that is, every Baire-1 function p can be written as a sum $p = xy + z$ of a product of two derivatives and another derivative. If p is bounded, then the functions x, y, z can be chosen to be bounded, as well. In particular, $Alg(\Delta) = \mathcal{B}_1$.

Theorem 2.2.19 is indeed astonishing: Although there are many Baire-1 functions p which cannot be expressed as a product of two functions form Δ it suffices to add a suitable function from Δ to change that. We illustrate this with the following

Example 2.2.20. Let $p : [0,1] \to \mathbb{R}$ be defined as in Example 2.2.18; we already know that p cannot be written as a product of two derivatives. On the other hand, we have seen in Example 2.2.16 that $-\chi_{\{0\}} = xy$ with suitable $x, y \in \Delta$. Finally the function

$$z(t) := 2 + \varphi_{0,0,1}(t) = \begin{cases} 2 + \sin \dfrac{1}{t} & \text{for } 0 < t \leq 1, \\ 2 & \text{for } t = 0 \end{cases}$$

has a primitive, because in contrast to p it has the correct value at $t = 0$. In total,

$$p = -\chi_{\{0\}} + z = xy + z,$$

in accordance with Theorem 2.2.19. ◇

The determination of $Alg(X)$ for a given function class X reminds of the well-known problem from Linear Algebra to determine the linear hull $\mathrm{Span}(X)$ of a given set of vectors X which itself is not a vector space. For instance, if M is the set of monotone functions on $[0,1]$, then $\mathrm{Span}(M) = BV$, and so this problem can be seen as the dawn of the BV era. For $X \in \{B, BV, C, \Delta, KH\}$ this problem is not of major interest, as $\mathrm{Span}(X) = X$ in these cases. However, in view of Example 1.1.2, the question on how $\mathrm{Span}(D)$ looks like, is interesting indeed. The answer is very surprising and given in the following theorem. Since with Example 2.2.7 we now have all tools at hand we give here a short proof based on an idea presented in the book [22].

Theorem 2.2.21. *The linear hull* $\mathrm{Span}(D)$ *comprises all functions, since every function* $x : [0,1] \to \mathbb{R}$ *can be written as the sum* $x = y + z$ *of two Darboux functions* $y, z \in D$.

Proof. Let $x : [0,1] \to \mathbb{R}$ be an arbitrary function, and let g be the function constructed in Example 2.2.7. We put

$$h(t) := \begin{cases} \log|g(t)| & \text{if } g(t) \neq 0, \\ 0 & \text{if } g(t) = 0, \end{cases}$$

as well as

$$y(t) := \begin{cases} h(t) & \text{if } g(t) \geq 0, \\ x(t) - h(t) & \text{if } g(t) < 0 \end{cases} \qquad \text{and} \qquad z(t) := \begin{cases} x(t) - h(t) & \text{if } g(t) \geq 0, \\ h(t) & \text{if } g(t) < 0. \end{cases}$$

The two functions y and z then attain every real number in every proper interval. To see this, fix $[a,b] \subseteq [0,1]$ with $a < b$ and $\xi \in \mathbb{R}$. Since g attains every real number in $[a,b]$ there is some $t \in [a,b]$ such that $g(t) = e^{\xi} > 0$. Then $h(t) = \xi$ and hence $y(t) = h(t) = \xi$. Similarly, one shows that z also attains ξ by replacing e^{ξ} by $-e^{\xi}$. Thus, the functions y and z are Darboux functions. The identity $x = y + z$ is clear. ■

It seems that the first one who mentioned the result of Theorem 2.2.21 was Lindenbaum [90]. Other proofs of Theorem 2.2.21 were also given by Sierpiński [141] and Fast [57].

2.3 Compositions of Derivatives

In this section we discuss the composition $x \circ y$ of two functions x and y and are particularly interested in the case when x and y come from Δ, provided that x is defined on the range of y. To be more specific, recall that Problem 2.1.12 asks for a universal *perturbation* for Δ, that is, for a function $g : \mathbb{R} \to \mathbb{R}$ such that $g \circ x \in \Delta$ for all $x \in \Delta$. Similar to the multiplier problem investigated in Section 2.2 we approach this problem more rigorously and introduce

$$\Pi(X) := \{g : \mathbb{R} \to \mathbb{R} \mid g \circ x \in X \text{ for all } x \in X\}, \tag{2.3.1}$$

where X is a set of real-valued functions defined on $[0,1]$. In other words, the class $\Pi(X)$ is the largest class of possible outer perturbations that do not leave X. Of course, we are most interested in finding $\Pi(\Delta)$.
Analogously, Problem 2.1.13 asks for a universal *substitution* for Δ, that is, for a function $g : [0,1] \to [0,1]$ such that $x \circ g \in \Delta$ for all $x \in \Delta$. We similarly introduce the set

$$\Sigma(X) := \{g : [0,1] \to [0,1] \mid x \circ g \in X \text{ for all } x \in X\}, \tag{2.3.2}$$

where X is as before. In other words, the class $\Sigma(X)$ is the largest class of possible inner substitutions that do not leave X. Again, the class $\Sigma(\Delta)$ is of particular interest for us.
In general, if X contains all[3] affine functions $x(t) = at + b$ for arbitrary $a, b \in \mathbb{R}$, then

$$\Pi(X) \subseteq X_{loc}(\mathbb{R}), \tag{2.3.3}$$

and if the identity function $x(t) = t$ is contained in X, then

$$\Sigma(X) \subseteq X. \tag{2.3.4}$$

For instance, it is obvious that the classes B, C and D are "closed under composition" in the following sense:

$$\begin{aligned}
\Pi(B) &= B_{loc}(\mathbb{R}), & \Sigma(B) &= \{g : [0,1] \to [0,1]\}, \\
\Pi(C) &= C(\mathbb{R}), & \Sigma(C) &= \{g : [0,1] \to [0,1] \mid g \in C\}, \\
\Pi(D) &= D_{loc}(\mathbb{R}), & \Sigma(D) &= \{g : [0,1] \to [0,1] \mid g \in D\},
\end{aligned}$$

or less mathematical and more as a mnemonic:

$$\Pi(B) = \Sigma(B) = B, \qquad \Pi(C) = \Sigma(C) = C, \qquad \Pi(D) = \Sigma(D) = D.$$

The classes BV, Δ and KH, however, are not closed under outer composition; we summarize three counterexamples in the following

Example 2.3.1. (a) Let $g : \mathbb{R} \to \mathbb{R}$ and $x : [0,1] \to \mathbb{R}$ be defined by $g(u) = \sqrt{|u|}$ and $x(t) = \varphi_{2,0,4}(t)$. Then we have $g \in BV_{loc}(\mathbb{R})$ and $x \in BV$ by Proposition 1.1.12 (b), but by the same Proposition $g \circ x = \varphi_{1,0,2} \notin BV$.

(b) Let $g : \mathbb{R} \to \mathbb{R}$ and $x : [0,1] \to \mathbb{R}$ be defined by $g(u) = u^2$ and $x(t) = \varphi_{0,0,1}(t)$. Then we have $g \in \Delta_{loc}(\mathbb{R})$ and $x \in \Delta$ by Example 1.1.29, but by Example 2.0.1 on the other hand $g \circ x = \varphi_{0,0,2} \notin \Delta$.

(c) Let $g : \mathbb{R} \to \mathbb{R}$ and $x : [0,1] \to \mathbb{R}$ be defined by $g(u) = u^2$ and $x(t) = \varphi_{-1/2,0,1}(t)$. Then we have $g \in KH_{loc}(\mathbb{R})$ and $x \in KH$ by Proposition 2.1.5 (a), but by the same Proposition $g \circ x = \varphi_{-1,0,2} \notin KH$. ◇

[3]In fact, it suffices to guarantee that affine functions of the form $x(t) = 2at - a$ for infinitely many arbitrarily large $a > 0$ are contained in X.

Sometimes, describing $\Pi(X)$ is not so obvious. For example, Josephy showed in 1987 in [75] that the perturbation set $\Pi(BV)$ precisely contains locally Lipschitz continuous functions, that is

$$\Pi(BV) = Lip_{loc}(\mathbb{R}).$$

The function $g(u) = u^2$, for instance, belongs to $\Pi(BV)$, but the function $g(u) = \sqrt{|u|}$ does not. This explains Example 2.3.1 (a).

We will now talk about the set $\Pi(\Delta)$ which is most important for us in this section. Recall that we have seen that Δ is not an algebra, because it is not closed under multiplication. The parts (b) and (c) in Example 2.3.1 illustrated that even in case $g \in C^\infty(\mathbb{R})$ from $x \in \Delta$ (respectively $x \in KH$) it does not necessarily follow that $g \circ x \in \Delta$ (respectively $g \circ x \in KH$); we therefore have $C^\infty \not\subseteq \Pi(\Delta)$ and also $C^\infty \not\subseteq \Pi(KH)$. While we do not know how exactly $\Pi(KH)$ looks like, the class $\Pi(\Delta)$ is known and will be given in Theorem 2.3.3 below. But first a remark is in order. As Example 2.3.1 (b) shows, a continuous perturbation g may transform a function x with primitive into a function $g \circ x$ without a primitive. If we replace the condition $x \in \Delta$ with the condition $x \in \Delta/\Delta$ (which is stronger, because the inclusion $\Delta/\Delta \subsetneq \Delta$ is strict), we get the following asymmetric result.

Theorem 2.3.2. *From $g \in C(\mathbb{R})$ and $x \in \Delta/\Delta$ it follows that $g \circ x \in \Delta$.*

Proof. By Theorem 2.2.6 the function x is bounded, that is, $|x(t)| \leq M$ for some $M > 0$ and all $t \in [0, 1]$. By the Approximation Theorem of Weierstrass we find a sequence (p_n) of polynomials which converges on $[-M, M]$ uniformly to g. Then the sequence $(p_n \circ x)$ converges uniformly on $[-M, M]$ to $g \circ x$. As we have seen after Example 2.2.5, each composition $p_n \circ x$ belongs to Δ/Δ, and since Δ/Δ is a subset of Δ, also to Δ itself. But then Theorem 2.1.8 ensures $g \circ x \in \Delta$, as claimed. ∎

The quite surprising solution for the perturbation problem for $X = \Delta$ is contained in the following theorem the proof of which can be found in [13].

Theorem 2.3.3. *A function $g : \mathbb{R} \to \mathbb{R}$ belongs to $\Pi(\Delta)$ if and only if g is affine, that is, g has the form $g(u) = au + b$ for some fixed numbers $a, b \in \mathbb{R}$.*

Theorem 2.3.3 explains why the class Δ remains not invariant even under the extremely smooth nonaffine function $g(u) = u^2$ of Example 2.3.1 (b). We can formulate this result more drastically as follows: *If g is not affine, then we always find some function x with primitive so that $g \circ x$ has no primitive.* Moreover, from this it also follows that we cannot decide whether a function x belongs to Δ by just looking at its level sets (2.1.12). Indeed, if we compose $x : [0, 1] \to \mathbb{R}$ with the nonaffine smooth function $g(u) = u^3$, then the sets (2.1.12) for x and $g \circ x$ are the same, while x may have a primitive, whereas $g \circ x$ may not.

We now turn to a discussion of Problem 2.1.13 and try to find (2.3.2) for several classes X, especially for $X = \Delta$. Within the upcoming Theorems 2.3.4 and 2.3.5 we give some sufficient conditions for this case. In the first theorem we denote by BV^1 the set of all differentiable functions $x : [0,1] \to \mathbb{R}$ with $x' \in BV$. Note that every function from $x \in BV^1$ automatically belongs to C^1, since by a theorem of Darboux, x' is a Darboux function, and hence can have only essential discontinuities which, however, cannot occur, as $x' \in BV$ can have only jump discontinuities. Consequently, we have the surprising inclusion $BV^1 \subseteq C^1$, whereas the analogous inclusion $BV \subseteq C$ is of course not true.

Theorem 2.3.4. *Let $g : [0,1] \to [0,1]$ be in BV^1 with $g'(t) \neq 0$ for all $t \in [0,1]$. Then $x \circ g \in \Delta$ for all $x \in \Delta$.*

Proof. Let $x \in \Delta$, and let f be a primitive of x. If we define

$$\tilde{x} := (f \circ g)' = (x \circ g)g' \qquad \text{and} \qquad \tilde{y} := \frac{1}{g'}, \tag{2.3.5}$$

then we have $\tilde{x} \in \Delta$ (according to our construction), and $\tilde{y} \in C \cap BV$, because g' is bounded away from zero. Due to

$$x \circ g = f' \circ g = \frac{(f \circ g)'}{g'} = \tilde{x}\tilde{y}$$

we can apply Theorem 2.2.2 (a) to $\tilde{x}$ and $\tilde{y}$ and obtain $x \circ g \in \Delta$, as claimed. ■

Theorem 2.3.4 is a simple consequence of Theorem 2.2.2 (a). If we use part (b) of Theorem 2.2.2 instead, we get new information yet can only take bounded functions into account.

Theorem 2.3.5. *Let $g : [0,1] \to [0,1]$ be in C^1 with $g'(t) \neq 0$ for all $t \in [0,1]$. Then $x \circ g \in \Delta$ for all $x \in \Delta \cap B$.*

Proof. Fix $x \in \Delta \cap B$, let f be a primitive of x and define $\tilde{x}$ and $\tilde{y}$ as in (2.3.5). This time, $\tilde{x} \in \Delta$ is bounded, since x is bounded, and g' as a continuous function is also bounded away from zero which ensures that $\tilde{y}$ is also continuous. Therefore, the same calculation as in Theorem 2.3.4 shows that from Theorem 2.2.2 (b), applied to $\tilde{x}$ and $\tilde{y}$, the claim follows. ■

The Theorems 2.3.4 and 2.3.5 raise the question whether the condition $g \in C^1$ with $g' \neq 0$ is sufficient, that is, whether one can replace $g \in BV^1$ by $g \in C^1$ in Theorem 2.3.4 or $x \in \Delta \cap B$ by $x \in \Delta$ in Theorem 2.3.5. In fact, a much more general result is true. In [112] and also in [132] the authors gave a detailed yet very technical characterization of $\Sigma(\Delta)$.

Theorem 2.3.6. *Let g be an increasing homeomorphism of $[0,1]$, and let $\gamma : [0,1] \to \mathbb{R}$ be a function satisfying*

$$\underline{D}g^{-1}(t) \leq \gamma(t) \leq \overline{D}g^{-1}(t) \quad \textit{for } 0 \leq t \leq 1, \tag{2.3.6}$$

where $\underline{D}g$ and $\overline{D}g$ denote the respective lower and upper Dini derivatives of g. Then $g \in \Sigma(\Delta)$ if and only if

$$\limsup_{\delta \to 0} \frac{1}{g^{-1}(t+\delta) - g^{-1}(t)} \int_t^{t+\delta} \operatorname{Var}\big(\gamma, [t,s]\big)\, \mathrm{d}s < \infty$$

is true for all $t \in [0,1]$.

In the light of Theorem 2.3.6 the hypothesis $g \in BV^1$ in Theorem 2.3.4 which might look artificial at first glance makes now sense. Indeed, together with the assumptions made in Theorem 2.3.4 the function g is even a diffeomorphism, and we can but $\gamma(t) = (g^{-1})'(t)$ in (2.3.6) which then satisfies the integral criterion of Theorem 2.3.6.

By the way, the substitution set $\Sigma(BV)$ is also known and was identified in [75], as well. We call a function $g : [0,1] \to [0,1]$ *pseudo-monotone* if there is some natural number $N \in \mathbb{N}$ such that the preimage $g^{-1}[a,b]$ of any interval $[a,b] \subseteq [0,1]$ can be written as a union of at most N intervals. Of course, every monotone function is pseudo-monotone, and it can be shown that any pseudo-monotone function is of bounded variation [75]. The converse, however, is not true, and here is an Example illustrating this.

Example 2.3.7. By Proposition 1.1.12 (b), the function $g := \varphi_{2,0,2}$ is of bounded variation, but because of the identity

$$g^{-1}(\{0\}) = \left\{0, \frac{1}{\pi}, \frac{1}{2\pi}, \frac{1}{3\pi}, \ldots\right\}$$

it is not pseudo-monotone. ◇

In [75] the author proves that $\Sigma(BV)$ precisely contains all pseudo-monotone functions. The function g of Example 2.3.7 therefore belongs to $BV \backslash \Sigma(BV)$. We will come back to these functions in Section 4.2.

The following Table 2.3.1 gives an overview about the sets X/X, $\Pi(X)$ and $\Sigma(X)$ (as far as we know) for the most important classes X.

Table 2.3.1: X/X, $\Pi(X)$ and $\Sigma(X)$ for important classes X.

X	C	B	BV	Δ	D	KH
X/X	C	B	BV	$\Delta \cap \overline{BV}$	constant	BV^*
$\Pi(X)$	C	B	locally Lipschitz	affine	D	???
$\Sigma(X)$	C	B	pseudo-monotone	Theorem 2.3.6	D	???

In the context of the perturbation and substitution of functions several problems concerning the decomposition of given functions arise. Recall that from the theory of BV-functions the following results are known.

(a) A function $x : [0,1] \to \mathbb{R}$ belongs to BV if and only if x can be written as a composition $x = g \circ y$, where $y : [0,1] \to [0,1]$ is increasing and $g : [0,1] \to \mathbb{R}$ is Lipschitz continuous with $\text{lip}(g) \leq 1$.

(b) For every function $x \in BV \cap C$ there is a function $g : [0,1] \to [0,1]$ such that $x \circ g$ is differentiable at every point of $[0,1]$.

The first result shows that a given BV-function can be decomposed into two much better BV-functions, whereas the second result means that a given continuous BV-function can be improved by a suitable substitution. Analogously, for the classes Δ and Δ/Δ questions for similar decompositions may be asked.

Problem 2.3.8. *Can one find classes $X \subseteq \Delta$, such that for every function $x \in X$ there is a substitution $g : [0,1] \to [0,1]$ so that $x \circ g$ belongs to Δ or Δ/Δ? For instance, is this possible for $X = D$?*

We remark that Problem 2.3.8 has been answered partially by Maximoff in 1947 for $X = D \cap \mathcal{B}_1$ [114]. He showed that every Baire-1 function $x : [0,1] \to \mathbb{R}$ with the Darboux property can be transformed by an increasing homeomorphism $g : [0,1] \to [0,1]$ so that $x \circ g$ has a primitive.
An analogue of Maximoff's result cannot be true for functions which belong to D but not to $\mathcal{B}_1$. For instance, let x be the Darboux function from Example 2.2.7 that attains all real values on any proper real interval. Then x is nowhere continuous and therefore no $\mathcal{B}_1$-function [128]. Now, if $g : [0,1] \to [0,1]$ is any increasing homeomorphism, then g is strictly increasing and by the Intermediate Value Theorem maps proper intervals onto proper intervals. Consequently, the function $x \circ g$ also attains all real values on every proper interval and hence cannot be continuous anywhere. Thus, again, $x \circ g$ cannot be a Baire-1 function and, in particular, cannot have a primitive.

Chapter 3

Multiplier Spaces

In Example 2.1.9 we have seen that the class Δ of all real-valued functions having a primitive on $[0,1]$ is not closed under multiplication, and it therefore seemed natural to search for functions g with the property that, for each x which is a derivative, the product xg is a derivative, as well. According to (2.2.3) we denoted the set of those functions by Δ/Δ and gave a characterization in Theorem 2.2.12. We will now continue the discussion of X/X in both a more general and a more rigorous way.

Given two function classes X and Y of real-valued functions defined on $[0,1]$, can we identify all functions g such that the product xg belongs to Y whenever x belongs to X? Such a function g is said to be a *multiplier of the set Y over the set X*.

Definition 3.0.1. Let X and Y be two sets of functions $[0,1] \to \mathbb{R}$. We call the set

$$Y/X := \{g : [0,1] \to \mathbb{R} \mid xg \in Y \text{ for all } x \in X\} \tag{3.0.1}$$

the *multiplier set of Y over X*.

The following properties of Y/X are immediate consequences of the definition.

Proposition 3.0.2. *Let X, X_1, X_2, Y, Y_1, Y_2 be sets of real-valued functions on $[0,1]$. The following statements are true.*

(a) If $Y_1 \subseteq Y_2$, then $Y_1/X \subseteq Y_2/X$.

(b) If $X_1 \subseteq X_2$, then $Y/X_2 \subseteq Y/X_1$.

(c) If $\mathbb{1} \in X$, then $Y/X \subseteq Y$.

(d) If $X \subseteq Y$ and Y is closed under multiplication, then $Y \subseteq Y/X$.

(e) If $X \subseteq Y$ and X is closed under multiplication, then $X \subseteq Y/X$.

(f) If $\mathbb{1} \in X \cap Y$ and Y is closed under multiplication, then $Y/X = Y$ if and only if $X \subseteq Y$.

Although Proposition 3.0.2 follows immediately from the definition of Y/X, let us quickly justify some of the results. For instance, if $Y_1 \subseteq Y_2$ and $g \in Y_1/X$, then

$xg \in Y_1$ for all $x \in X$. But since $Y_1 \subseteq Y_2$ we also have that $xg \in Y_2$ for all $x \in X$, and this shows $g \in Y_2/X$ and hence (a). Part (b) is proven similarly, and (c) is clear. For (d) fix $g \in Y$ and $x \in X$. Since $X \subseteq Y$ we also have $x \in Y$, and since Y is closed under multiplication, $xg \in Y$. This shows $g \in Y/X$. Similarly, one can prove (e). For (f) note that if $X \subseteq Y$, then $Y/X = Y$ follows from (c) and (d). If, however, $X \nsubseteq Y$, then there exists some function x belonging to X but not to Y. Since $\mathbb{1}$ belongs to Y, but $x\mathbb{1} = x$ does not, we have $\mathbb{1} \notin Y/X$ and hence $Y \nsubseteq Y/X$. This shows (f).

As we have already seen in Section 2.2, the explicit calculation of Y/X for given sets X and Y, in some cases is quite easy, in other cases - even if $X = Y$ - surprisingly difficult, and sometimes even leads to some kind of degeneracy if $X \nsubseteq Y$. For example, the Hölder inequality for Lebesgue spaces implies

$$L_q/L_p = \begin{cases} L_{pq/(p-q)} & \text{for } 1 \le q \le p < \infty, \\ \{\mathbb{0}\ a.e.\} & \text{for } 1 \le p < q < \infty; \end{cases}$$

in particular, $L_p/L_p = L_\infty$. For the sake of completeness we will give a short proof of this well-known relation in Theorem 3.2.11.

More generally, the analogous problem has also been solved for Orlicz spaces (see e.g. [3, 12, 127], and references therein). The first results have been obtained in the 1960ies and state for example that, if L_Φ, L_Ψ denote two Orlicz spaces defined by Young functions Φ and Ψ, respectively, then

$$L_\Psi/L_\Phi = L_\infty \qquad \text{if } 0 < \limsup_{u\to\infty} \frac{\Psi(\lambda u)}{\Phi(u)} < \infty \quad \text{for all } \lambda > 0;$$

in particular, $L_\Phi/L_\Phi = L_\infty$. For further references regarding multipliers in Lebesgue-, Orlicz and other more abstract spaces we refer the reader to the papers of Lech Maligranda and his coauthors [79, 80, 81, 97, 98, 99, 100].

In this chapter we will discuss the classes Y/X when X and Y are classical spaces like B, C, D, Δ and BV as well as other BV-type spaces like WBV_p, YBV_φ, ΛBV and RBV_p.
For our classical classes B and C apart from the obvious identities $B/B = B$ and $C/C = C$ it is straightforward to show that

$$B/C = B \quad \text{and} \quad C/B = \{\mathbb{0}\};$$

we will prove this at the beginning of the next section.
For the class Δ of functions having a primitive that we have excessively studied in Chapter 2 it is much harder to determine multiplier space Y/X if X or Y is equal to Δ. As we have seen in Theorem 2.2.12 even the class Δ/Δ is not so easy to describe. We will discuss such classes in more detail in the next Section 3.1 and compare them to BV. Section 3.2 is dedicated to functions of generalized bounded variation. To be precise we will compare the classical spaces to each of the spaces YBV_φ, ΛBV and RBV_p. Moreover, we will characterize $\Gamma BV/\Lambda BV$ for two arbitrary Waterman

sequences Γ and Λ, YBV_ψ/YBV_φ for two arbitrary Young functions ψ and φ, and RBV_q/RBV_p for two arbitrary exponents $1 \leq p, q < \infty$. For the definitions of these spaces we refer the reader to Section 1.2.

However, there are still open problems even for classical spaces; in particular, the class D which is closely related to Δ and has been investigated in Chapter 2 causes many difficulties. Right after (2.2.19) we promised to prove that D/D contains only constant functions; this will be done in Theorem 3.1.8 below. Moreover, a complete characterization of the multiplier sets D/C and D/Δ - to our knowledge - is not known; we discuss them in more detail. Since we can present only partial results regarding these classes, we obtain as a consequence only partial results for D/RBV_p and similar classes in Section 3.2.

3.1 Multipliers in Classical Spaces

Recall from (2.0.1) that the (strict) inclusions $C \subsetneq \Delta \subsetneq D$ hold which is sometimes helpful in calculating multiplier sets involving one of these classes.

Our main interest of this section is to characterize Y/X for $Y, X \in \{B, C, D, \Delta\}$. Note that since B and C are algebras with $\mathbb{1}$ and $C \subseteq B$, by Proposition 3.0.2 (f) we immediately get $C/C = C$, $B/B = B$ and $B/C = B$.
Our first result shows that if X compared to Y is too large, then Y/X only contains the zero function. This is true for $Y \in \{C, D, \Delta\}$ and certain spaces X:

Proposition 3.1.1. *Let X be a class of real-valued functions on $[0,1]$ that contains all characteristic functions of singletons.*[1] *Then for $g : [0,1] \to \mathbb{R}$ the following statements are equivalent.*

(a) $g = \mathbb{0}$. *(b)* $g \in C/X$. *(c)* $g \in \Delta/X$. *(d)* $g \in D/X$.

Proof. Obviously, (a) implies (b). Since $C \subseteq \Delta \subseteq D$ we obtain from Proposition 3.0.2 (a) the inclusions $C/X \subseteq \Delta/X \subseteq D/X$, which show the implications "(b)⇒(c)" and "(c)⇒(d)". For "(d)⇒(a)" fix $g \in D/X$ and $t \in [0,1]$. Then $x := \chi_{\{t\}} \in X$ and hence $xg = \chi_{\{t\}} g(t) \in D$ which is possible only if $g(t) = 0$. Since t was arbitrarily chosen, $g = \mathbb{0}$. ∎

For instance, if we choose $X = B$ in Proposition 3.1.1, then we obtain $C/B = \Delta/B = D/B = \{\mathbb{0}\}$. However, the example $C/C = C$ shows that we cannot drop the assumption that X contains characteristic functions of singletons.

Our next result characterizes B/D and B/Δ. Recall that the symbols $\mathcal{S}_f$ and $\mathcal{S}_c$ denote the classes of real-valued functions on $[0,1]$ with finite and countable support, respectively.

[1] Such spaces will become very important in Section 4.1 for the investigation of the linear multiplication operator, see Definition 4.1.2 and below.

Theorem 3.1.2. *For $g : [0,1] \to \mathbb{R}$ the following statements are equivalent.*

(a) $g \in \mathcal{S}_f$. *(b)* $g \in B/D$. *(c)* $g \in B/\Delta$.

Proof. If $\operatorname{supp}(g)$ is finite, then $\operatorname{supp}(xg)$ is also finite for any function $x : [0,1] \to \mathbb{R}$; in particular, $xg \in B$ for all $x : [0,1] \to \mathbb{R}$, and this shows "(a)⇒(b)". Since $\Delta \subseteq D$, it follows from Proposition 3.0.2 (b) that $B/D \subseteq B/\Delta$ which shows "(b)⇒(c)".
For the remaining part "(c)⇒(a)" assume that $\operatorname{supp}(g)$ is infinite. Then there is a sequence (t_n) in $\operatorname{supp}(g)$ of distinct numbers converging to some $t \in [0,1]$. Without loss of generality we may assume that the sequence (t_n) is strictly decreasing, $0 < t_n < 1$ for all $n \in \mathbb{N}$ and $t = 0$, because the argument for the general case is basically the same. Choose $\varepsilon_n > 0$ so that

$$\varepsilon_n \le \frac{g(t_n)}{n}\left(t_{n+1}^2 - t_{n+2}^2\right), \tag{3.1.1}$$

$$t_{n+1} + \varepsilon_{n+1} < t_n - \varepsilon_n \qquad \text{for all } n \in \mathbb{N}. \tag{3.1.2}$$

Let us define the function $x : [0,1] \to \mathbb{R}$ as follows: Let $x(0) = x(1) = 0$,

$$x(t_n) = n/g(t_n),$$
$$x(t_n - \varepsilon_n) = x(t_n + \varepsilon_n) = 0 \quad \text{for } n \in \mathbb{N},$$

and let x be piecewise linear and continuous otherwise. Then x is well-defined by (3.1.2) and nonnegative and continuous on the interval $(0,1]$, and for $t_{n+1} < \delta \le t_n$ we have by (3.1.1),

$$0 \le \int_0^\delta x(t)\,\mathrm{d}t \le \sum_{j=n}^{\infty} \varepsilon_j x(t_j) = \sum_{j=n}^{\infty} j\varepsilon_j/g(t_j) \le \sum_{j=n+1}^{\infty} \left(t_j^2 - t_{j+1}^2\right) = t_{n+1}^2 \le \delta^2.$$

In particular, x is Lebesgue and hence KH-integrable on $[0,1]$ with

$$\lim_{\delta \to 0+} \frac{1}{\delta}\int_0^\delta x(t)\,\mathrm{d}t = 0 = x(0).$$

Since x is continuous on $(0,1]$, it follows that $x \in \Delta$ by Theorem 2.1.3 (b) and (d). But $x(t_n)g(t_n) = n$ for all $n \in \mathbb{N}$ showing $xg \notin B$ and hence $g \notin B/\Delta$. ■

As a consequence we get a result similar to Proposition 3.1.1: If Y compared to X is a huge class of bounded functions, then the multiplier spaces Y/D and Y/Δ are again very small.

Corollary 3.1.3. *Let $Y \subseteq B$ be a class of real-valued functions on $[0,1]$ that contains all functions with finite support. Then for $g : [0,1] \to \mathbb{R}$ the following statements are equivalent.*

(a) $g \in \mathcal{S}_f$. *(b)* $g \in Y/D$. *(c)* $g \in Y/\Delta$.

Proof. If supp(g) is finite, then supp(xg) is also finite; in particular, $xg \in \mathcal{S}_f \subseteq Y$ for any $x : [0,1] \to \mathbb{R}$, and this shows "(a)⇒(b)". Since $\Delta \subseteq D$, it follows from Proposition 3.0.2 (b) that $Y/D \subseteq Y/\Delta$ which shows "(b)⇒(c)". Since $Y \subseteq B$, Proposition 3.0.2 (a) yields $Y/\Delta \subseteq B/\Delta$. But Theorem 3.1.2 gives that B/Δ and therefore also Y/Δ contains only functions with finite support. Consequently, "(c)⇒(a)" is proven. ∎

For instance, if we take $Y = B$, then Corollary 3.1.3 is exactly Theorem 3.1.2. However, the relation $\Delta/\Delta = \overline{BV} \cap \Delta$ of Theorem 2.2.12 shows that we cannot drop the assumption that Y contains all functions with finite support.

One of the most difficult problems in the framework of classical function spaces seems to be a characterization of D/D which we will handle in the following.
If one replaces "multiplication" by "summation", the following result due to Radaković is known and was proven in [133].

Theorem 3.1.4. *Let $g : [0,1] \to \mathbb{R}$ be such that $x + g \in D$ for all $x \in D$. Then g is constant.*

Several authors claimed that Radaković proved in [133] the same statement for products. Others stated that the assertion for products can easily be deduced from Theorem 3.1.4 by taking logarithms [20, 23, 66, 123], but none of them proves this. We will show in the sequel that the product version actually can be deduced from Theorem 3.1.4, but its proof is, at least in our opinion, not as trivial as it might appear at first glance. We start by treating a special case first, namely we assume that a multiplier $g \in D/D$ has no zeros in $[0,1]$ at all. Then, due to the Darboux property, g must be either everywhere positive or everywhere negative on $[0,1]$. Taking logarithms as suggested then indeed yields that g must degenerate to a constant. In fact, the following slightly more general result is true.

Lemma 3.1.5. *Let $0 \leq a < b \leq 1$ and $g \in D/D$ be so that $g(t) \neq 0$ for all $t \in (a,b)$. Then g is constant on $[a,b]$.*

Proof. Fix $g \in D/D$. Then $g \in D$ by Proposition 3.0.2 (c), and due to the Darboux property we have that g is either strictly positive or strictly negative on (a,b). We assume the first, the latter case is similar.
Pick $\delta > 0$ so small that $\delta < (b-a)/2$. Then $[a+\delta, b-\delta] \subseteq (a,b)$, and we can define $h : [0,1] \to [a+\delta, b-\delta]$ by $h(t) = (b-a-2\delta)t + a + \delta$. Then h is linear and maps $[0,1]$ bijectively onto $[a+\delta, b-\delta]$ with linear inverse. Moreover, the function $G := g \circ h$ satisfies $G(t) > 0$ for all $t \in [0,1]$ and still belongs to D/D. To see this, fix $F \in D$ and define $x : [0,1] \to \mathbb{R}$ by

$$x(t) = \begin{cases} F(0) & \text{for } t \in [0, a+\delta), \\ F(h^{-1}(t)) & \text{for } t \in [a+\delta, b-\delta], \\ F(1) & \text{for } t \in (b-\delta, 1]. \end{cases}$$

Then $x \in D$, $F = x \circ h$, and $FG = (xg) \circ h$. Since $g \in D/D$, we have $xg \in D$ and hence $FG \in D$ as claimed.

For any $x \in D$ we have $\exp \circ x \in D$ and now obtain $x+\log G = \log(e^x G) \in D$. Theorem 3.1.4 yields that the Darboux function $\log G$ and hence G is constant. But this means that g is constant on $[a+\delta, b-\delta]$. Since δ can be chosen arbitrarily small, g is constant on (a,b), and due to the Darboux property it is constant even on $[a,b]$. ■

In particular, if $g \in D/D$ has no zeros in $[0,1]$ at all, Lemma 3.1.5 shows, that it follows indeed from Theorem 3.1.4 that g is constant. But if g has zeros, things are not at all so obvious.
However, the general product case follows immediately from Theorem 2 of [123]. We give an alternative proof and show how the product result can indeed be deduced from Theorem 3.1.4, for which we need some technicalities that will be given in the following.

Definition 3.1.6. The symbol $\mathfrak{c} := \operatorname{card}(\mathbb{R})$ denotes the cardinality of the set $\mathbb{R}$ of real numbers. Let $\mathcal{I}$ be any system of subintervals of $[0,1]$. We call a set $E \subseteq [0,1]$ $\mathfrak{c}$-*dense with respect to* $\mathcal{I}$ if $\operatorname{card}(E \cap I) = \mathfrak{c}$ for all $I \in \mathcal{I}$.

For instance, $[0,1]$ itself is $\mathfrak{c}$-dense with respect to any system of nondegenerate intervals, whereas $\mathbb{Q} \cap [0,1]$ is never. A similar version of the following auxiliary result has been proven in the paper [124].

Lemma 3.1.7. *Let $E \subseteq [0,1]$ and let $\mathcal{I} = (I_n)_{n\in N}$, $N \subseteq \mathbb{N}$, be the countable system of all closed and nondegenerate subintervals of $[0,1]$ with rational end points. If E is $\mathfrak{c}$-dense with respect to $\mathcal{I}$, then there are pairwise disjoint sets $(A_n)_{n\in N}$ such that $A_n \subseteq E \cap I_n$ and $\operatorname{card}(A_n) = \mathfrak{c}$ for all $n \in N$.*

Proof. Let $\mathcal{J}$ be the system of all open subintervals of $[0,1]$ which intersect E. We first prove that any set $F \subseteq [0,1]$ is $\mathfrak{c}$-dense with respect to $\mathcal{I}$ if and only if F is $\mathfrak{c}$-dense with respect to the system $\mathcal{J}$.
To see this assume that F is $\mathfrak{c}$-dense with respect to $\mathcal{I}$ and fix $J \in \mathcal{J}$. Then there is some $t \in J \cap E$. Since J is open, we find some $I \in \mathcal{I}$ such that $t \in I \subseteq J$. Since F is $\mathfrak{c}$-dense with respect to $\mathcal{I}$, it follows that $\mathfrak{c} = \operatorname{card}(F \cap I) \leq \operatorname{card}(F \cap J)$ and therefore $\operatorname{card}(F \cap J) = \mathfrak{c}$. This shows that F is also $\mathfrak{c}$-dense with respect to $\mathcal{J}$.
For the opposite direction suppose that F is $\mathfrak{c}$-dense with respect to $\mathcal{J}$ and fix $I \in \mathcal{I}$. Since E is $\mathfrak{c}$-dense with respect to $\mathcal{I}$, there is some $t \in E \cap I^\circ$. But then $I^\circ \in \mathcal{J}$, and since F is $\mathfrak{c}$-dense with respect to $\mathcal{J}$ we obtain $\operatorname{card}(F \cap I) \geq \operatorname{card}(F \cap I^\circ) = \mathfrak{c}$ which implies $\operatorname{card}(F \cap I) = \mathfrak{c}$. Consequently, F is also $\mathfrak{c}$-dense with respect to $\mathcal{I}$.
In particular, E is $\mathfrak{c}$-dense with respect to $\mathcal{J}$. Thus, by [124], there are pairwise disjoint sets $B_1, B_2, B_3, \ldots \subseteq E$, each of which being $\mathfrak{c}$-dense with respect to $\mathcal{J}$ (and therefore also with respect to $\mathcal{I}$), such that

$$E = \bigcup_{n\in\mathbb{N}} B_n.$$

Define $A_n := B_n \cap I_n \subseteq E \cap I_n$ for $n \in N$. Then $A_n \subseteq B_n$ for all $n \in N$; in particular, the A_n are pairwise disjoint. Moreover, since each B_n is $\mathfrak{c}$-dense with respect to $\mathcal{I}$, $\operatorname{card}(A_n) = \operatorname{card}(B_n \cap I_n) = \mathfrak{c}$, as desired. ■

We are now ready to prove

Theorem 3.1.8. *A function $g : [0,1] \to \mathbb{R}$ belongs to D/D, if and only if g is constant.*

Proof. Obviously, any constant function belongs to D/D, so it remains to prove the converse. To do this fix $g \in D/D$. First, let us note that then $g \in D$ by Proposition 3.0.2 (c). Since D/D is closed under multiplication we have $g^2 \in D/D$. Moreover, because of $g \in D$, showing that g is constant is equivalent to showing that g^2 is constant. Therefore, we can assume that $g \geq 0$.
Let

$$Z := g^{-1}(\{0\}) = \{t \in [0,1] \mid g(t) = 0\}$$

be the set of zeros of g. If $Z = \emptyset$, that is, $g(t) > 0$ for all $t \in [0,1]$, then it follows immediately from Lemma 3.1.5 that g is constant.
Assume now that $Z \neq \emptyset$. We would like to show that $Z = [0,1]$ which means that $g = \mathbb{0}$ on $[0,1]$. The proof for this will be divided into two parts. In the first part we show that Z is a closed subset of $[0,1]$, and in the second part we show that Z is a dense subset of $[0,1]$.
To prove the first part assume that Z is not closed. Then there exists some $c \in [0,1]$ with $g(c) > 0$ and a sequence (t_n) of elements of Z converging to c. Let us consider the set

$$E := \big\{t \in [0,1] \mid g(t) < g(c)/2\big\}.$$

We claim that for each interval $[a,b] \subseteq [0,1]$ either $E \cap [a,b] = \emptyset$ or $\operatorname{card}(E \cap [a,b]) = \mathfrak{c}$. To see this, let us fix $[a,b] \subseteq [0,1]$ and assume that $E \cap [a,b] \neq \emptyset$. Let $s \in [a,b] \cap E$. If $g(t) < g(c)/2$ for all $t \in [a,b]$, then $E \cap [a,b] = [a,b]$ and our claim is proved. If, however, $g(\tau) \geq g(c)/2$ for some $\tau \in [a,b]$, then, since the function g is a Darboux function, it attains all real numbers between $g(s) < g(c)/2$ and $g(\tau) \geq g(c)/2$ on $[a,b]$; in particular, $\operatorname{card}(E \cap [a,b]) = \mathfrak{c}$, which again proves our claim.
Now, let $(I_n)_{n\in\mathbb{N}}$ be the countable collection of all closed and nondegenerate subintervals of $[0,1]$ with rational end points. Define

$$N := \{n \in \mathbb{N} \mid E \cap I_n \neq \emptyset\}.$$

As shown above, $\operatorname{card}(E \cap I_n) = \mathfrak{c}$ for each $n \in N$, that is, E is $\mathfrak{c}$-dense with respect to $(I_n)_{n\in N}$. Therefore, by Lemma 3.1.7, we can choose pairwise disjoint sets $A_n \subseteq E \cap I_n$ and surjections $h_n : A_n \to (0,1]$ for all $n \in N$. Then the set $A := \bigcup_{n\in N} A_n$ clearly satisfies $A \subseteq E$.
Consider the function $h : [0,1] \to [0,1]$, defined by the formula

$$h(t) = \begin{cases} h_n(t) & \text{for } t \in A_n \text{ and } n \in N, \\ 0 & \text{for } t \in [0,1]\backslash A. \end{cases}$$

Since the sets A_n are pairwise disjoint, the function h is well-defined. We claim that $h \in D$. To see this, fix $[a,b] \subseteq [0,1]$. Assume that there exists some $m \in N$ such that $I_m \subseteq [a,b]$. Then

$$[0,1] \supseteq h([a,b]) \supseteq h(I_m) \supseteq h(A_m) = h_m(A_m) = (0,1],$$

and this shows that $h([a,b])$ is an interval.
Now assume that $I_k \not\subseteq [a,b]$ for all $k \in N$. Then $E \cap (a,b) = \emptyset$, since otherwise there was some $t \in E \cap (a,b)$, and therefore there was some $m \in \mathbb{N}$ such that $t \in I_m \subseteq (a,b)$. But then we would have $I_m \cap E \neq \emptyset$ which means that $m \in N$, contradicting the assumption that $I_k \not\subseteq [a,b]$ for all $k \in N$. Thus, we have $E \cap (a,b) = \emptyset$, that is $g(t) \geq g(c)/2$ for all $t \in (a,b)$. By Lemma 3.1.5 we infer that $g(t) = d$ for some $d \geq g(c)/2$ and all $t \in [a,b]$; in particular, $E \cap [a,b] = \emptyset$. Therefore, $h(t) = 0$ for all $t \in [a,b]$, because otherwise we would have $h(s) > 0$ for some $s \in [a,b] \cap A$, and since $A \subseteq E$ we had $s \in E \cap [a,b]$ which is not possible. This means $h([a,b]) = \{0\}$. Thus, we have shown that h maps closed intervals onto intervals which proves that indeed $h \in D$.
Since c is an accumulation point of Z we may assume (by passing to a subsequence, if necessary) that $t_n < c$ for all $n \in \mathbb{N}$ or $t_n > c$ for all $n \in \mathbb{N}$. We are going to investigate only the first case, because the second can be treated similarly. Let us define the function $x : [0,1] \to \mathbb{R}$ by the formula

$$x(t) = \begin{cases} h(t) & \text{for } 0 \leq t < c, \\ 1 & \text{for } c \leq t \leq 1. \end{cases}$$

We claim that $x \in D$. To see this, fix $[a,b] \subseteq [0,1]$ with $a < b$. If $b < c$, then $x([a,b]) = h([a,b])$ is an interval, because $h \in D$. If $c \leq a$, then obviously $x([a,b]) = \{1\}$. Now, let $a < c \leq b$. Since $t_n < c$ for all $n \in \mathbb{N}$ and $t_n \to c$, there exists some $m \in \mathbb{N}$ such that $a < t_m < c$ and $g(t_m) = 0$; in particular, $t_m \in E$ and there exists some $n \in N$ such that $t_m \in I_n \subseteq [a,c) \subseteq [a,b]$. But then

$$[0,1] \supseteq x([a,b]) \supseteq x([a,c]) \supseteq x(I_n) = h(I_n) \supseteq h(A_n) = h_n(A_n) = (0,1],$$

and again, $x([a,b])$ is an interval. Consequently, $x \in D$.
Now, we have

$$x(t)g(t) = \begin{cases} h(t)g(t) & \text{for } 0 \leq t < c, \\ g(t) & \text{for } c \leq t \leq 1. \end{cases}$$

Then $x(t)g(t) \leq g(c)/2$ for $t < c$ and $x(c)g(c) = g(c)$, showing that $xg \notin D$ and contradicting the fact that $g \in D/D$. Thus, Z is closed and the first part of the proof is completed.

We now pass to the second part of the proof in which we show that Z is dense. First note that Z is not empty by assumption. Assume that Z is not dense. Then there is an open nonempty interval $I \subseteq [0,1]$ such that $g(t) > 0$ for all $t \in I$. Fix $s \in I$ and define

$$a := s - \operatorname{dist}\Big((Z \cap [0,s]) \cup \{0\}, s\Big) \quad \text{and} \quad b := s + \operatorname{dist}\Big((Z \cap [s,1]) \cup \{1\}, s\Big).$$

Then $g(t) > 0$ for all $t \in (a,b)$ and hence, by Lemma 3.1.5, $g(t) = d$ for some $d > 0$ and all $t \in [a,b]$. However, since Z is closed and not empty, it follows that $a \in Z$ or

$b \in Z$, which implies $g(a) = 0$ or $g(b) = 0$. In any case we end up with $d = 0$, and this is our desired and final contradiction. ■

Note that Theorem 3.1.8 also shows that D is not closed under multiplication, since if it was, then Proposition 3.0.2 (f) implied $D/D = D$, a contradiction.

Lemma 3.1.7 can also be used to create well-known examples of functions which attain every real number on every subinterval of $[0, 1]$. We have given an example of such a function already in Example 2.2.7; we will present here another such function, but this time we prove its existence with the help of Lemma 3.1.7.

Example 3.1.9. Indeed, if we apply Lemma 3.1.7 on $E = [0, 1]$ we obtain pairwise disjoint sets $(A_n)_{n\in\mathbb{N}}$ such that $A_n \subseteq I_n$ and $\mathrm{card}(A_n) = \mathfrak{c}$ for all $n \in \mathbb{N}$, where (I_n) is an enumeration of all closed proper subintervals of $[0, 1]$ with rational end points. Due to $\mathrm{card}(A_n) = \mathfrak{c}$ there are surjections $\varphi_n : A_n \to \mathbb{R}$ for all $n \in \mathbb{N}$. Define $x : [0, 1] \to \mathbb{R}$ by

$$x(t) = \begin{cases} \varphi_n(t) & \text{for } t \in A_n \text{ and } n \in \mathbb{N}, \\ 0 & \text{otherwise.} \end{cases}$$

If $I \subseteq [0, 1]$ is any proper closed subinterval of $[0, 1]$, then there is some $n \in \mathbb{N}$ such that $A_n \subseteq I_n \subseteq I$. Since $x(A_n) = \varphi_n(A_n) = \mathbb{R}$ we obtain $x(I) = \mathbb{R}$, and so x has the desired property. In particular, $x \in D$, but x is continuous at no point of $[0, 1]$ and hence not even a Baire-1 function [128]. ◇

A consequence of the Theorems 3.1.2 and 3.1.8 is the following

Theorem 3.1.10. *Let $g : [0, 1] \to \mathbb{R}$. Then the following statements are equivalent.*

(a) $g = \mathbb{0}$.

(b) $g \in C/D$.

(c) $g \in C/\Delta$.

(d) $g \in \Delta/D$.

(e) $g \in C/B$.

(f) $g \in \Delta/B$.

(g) $g \in D/B$.

Proof. Obviously, $\mathbb{0}$ belongs to each of the given multiplier classes which shows that (a) implies all the other statements. Moreover, since $C \subseteq \Delta \subseteq D$ it follows from the parts (a) and (b) of Proposition 3.0.2 that $C/D \subseteq C/\Delta$ and $C/X \subseteq \Delta/X \subseteq D/X$ for $X \in \{D, \Delta, B\}$, which shows the implications "(b)⇒(c)", "(b)⇒(d)", "(e)⇒(f)" and "(f)⇒(g)".

For "(c)⇒(a)" fix $g \in C/\Delta$. Due to $C \subseteq B$ we get with the help of the parts (a) and (c) of Proposition 3.0.2 that $C/\Delta \subseteq B/\Delta$ and $C/\Delta \subseteq C$. Thus, $g \in C \cap B/\Delta$, and by Theorem 3.1.2, $\mathrm{supp}(g)$ is finite, which is possible only if $g = \mathbb{0}$.

For "(d)⇒(a)" fix $g \in \Delta/D$ and note that $\Delta/D \subseteq D/D$, due to $\Delta \subseteq D$ and Proposition 3.0.2 (a). Theorem 3.1.8 implies that g must be constant. The function $x := \varphi_{0,1,1}$, defined in (1.1.1), belongs to $D\setminus\Delta$ by the Propositions 1.1.12 and 2.1.5, and so does xg, unless $g = \mathbb{0}$.

The implication "(g)⇒(a)" is already covered by Proposition 3.1.1. ■

The Multiplier Spaces D/C and D/Δ

While we have given the characterization of Δ/Δ in Theorem 2.2.12 which was found by Fleissner and Mařík [59, 111], the class Δ/C was also identified by Mařík, who proved in [112] that g belongs to Δ/C if and only if

$$g \in \Delta \qquad \text{and} \qquad \limsup_{\tau \to t} \frac{1}{\tau - t} \int_t^\tau |g(s)|\, ds < \infty \quad \text{for } 0 \leq t \leq 1. \tag{3.1.3}$$

However, it is noteworthy that Theorem 2.2.2 together with Proposition 3.0.2 (c) and (e) tells us that

$$\Delta/(C \cap BV) = \Delta, \tag{3.1.4}$$

and Proposition 3.0.2 (b) says $\Delta/C \subseteq \Delta/(C \cap BV)$. Moreover, a comparison between (3.1.3) and (3.1.4) suggests that Δ/C is strictly smaller than $\Delta/(C \cap BV)$. Indeed, condition (3.1.3) forces $g \in \Delta/C$ to be Lebesgue integrable. Thus, any function $g \in \Delta$ that is not Lebesgue integrable belongs to $\Delta/(C \cap BV)$, but not to Δ/C; the derivative $g = x'$ of the function x of Example 2.0.2 is an instance of such a function.

The remaining two classes D/C and D/Δ are way more difficult to characterize, and their exact form is - at least to our knowledge - unknown. In the following we will discuss them in more detail and begin with D/C.
First note that $C \subseteq D/C \subseteq D$ by (c) and (e) of Proposition 3.0.2; we will show that both inclusions are strict in the sequel. In [122] and [144] it was shown that for each nonconstant continuous function x there is some function g_x with the following properties:

- g_x is a Darboux function,
- xg_x is not a Darboux function, and
- the set of points of discontinuity of g_x is meager.

In particular, such g_x cannot belong to D/C, and so, indeed, $D/C \subsetneq D$. Moreover, the function $\varphi_{0,\beta,1}$, defined in (1.1.1), belongs to $\Delta \backslash C$ for $\beta = 0$ by Proposition 1.1.12, and to $(D \cap \mathcal{B}_1) \backslash \Delta$ for $0 < |\beta| \leq 1$ by the Propositions 1.1.12 and 2.1.5, where $\mathcal{B}_1$ denotes the class of Baire-1 functions. This shows that we also have the strict inclusions $C \subsetneq \Delta \subsetneq D \cap \mathcal{B}_1$. It is therefore reasonable to ask whether the classes Δ or even $D \cap \mathcal{B}_1$ fit somewhere into the chain of inclusions $C \subseteq D/C \subseteq D$. In order to answer this question let us recall a characterization of the functions in $D \cap \mathcal{B}_1$ that has been given by Young in [162].

Theorem 3.1.11. *Let $g \in \mathcal{B}_1$. Then g belongs to $D \cap \mathcal{B}_1$ if and only if the following two requirements are satisfied.*

(i) For each $t \in (0, 1]$ there exists a sequence (s_n) in $[0, 1]$ converging to t such that $s_n < t$ for all $n \in \mathbb{N}$ and $\lim_{n\to\infty} g(s_n) = g(t)$.

(ii) For each $t \in [0, 1)$ there exists a sequence (t_n) in $[0, 1]$ converging to t such that $t < t_n$ for all $n \in \mathbb{N}$ and $\lim_{n\to\infty} g(t_n) = g(t)$.

With the help of Theorem 3.1.11 we now obtain

Proposition 3.1.12. *A function $g : [0,1] \to \mathbb{R}$ belongs to the class $(D \cap \mathcal{B}_1)/C$ if and only if $g \in D \cap \mathcal{B}_1$.*

Proof. By Proposition 3.0.2 (c), the inclusion $(D \cap \mathcal{B}_1)/C \subseteq D \cap \mathcal{B}_1$ holds, so it remains to show that the reverse implication also holds. To show this assume that $g \in D \cap \mathcal{B}_1$, and fix $x \in C$ and $t \in [0,1]$. Then $xg \in \mathcal{B}_1$. Without loss of generality let $t \in (0,1)$; the cases $t = 0$ and $t = 1$ are treated similarly. By Theorem 3.1.11, there exist sequences $(s_n), (t_n) \subseteq [0,1]$ both converging to t such that $s_n < t < t_n$ for all $n \in \mathbb{N}$ and

$$\lim_{n\to\infty} g(s_n) = g(t) = \lim_{n\to\infty} g(t_n).$$

Since $x \in C$,

$$\lim_{n\to\infty} x(s_n)g(s_n) = x(t)g(t) = \lim_{n\to\infty} x(t_n)g(t_n).$$

Again by Theorem 3.1.11 we have $xg \in D \cap \mathcal{B}_1$ showing that indeed $g \in (D \cap \mathcal{B}_1)/C$, as claimed. ∎

The last result simply says $(D \cap \mathcal{B}_1)/C = D \cap \mathcal{B}_1$; in particular $D \cap \mathcal{B}_1 \subseteq D/C$ by Proposition 3.0.2 (a), and for $g \in D \cap \mathcal{B}_1$ and $x \in C \subseteq \mathcal{B}_1$ we have that $xg \in \mathcal{B}_1$. This suggests that D/C is strictly larger than $D \cap \mathcal{B}_1$, which is indeed true and content of the following

Proposition 3.1.13. *There is a function $g \in D/C$ that does not belong to $\mathcal{B}_1$.*

Proof. In [57] it was proven that there exists a function $g_0 \in D$ such that for each $x \in C$ and each proper interval $I \subseteq [0,1]$ we have $(x + g_0)(I) = \mathbb{R}$. We now consider the function $g := e^{g_0}$ which also belongs to D, since exp is continuous.
Fix $x \in C$ and assume that $x(t) > 0$ for all $t \in I$ and some interval $I := [a,b] \subseteq [0,1]$ with $a < b$. Define $y : [0,1] \to \mathbb{R}$ by

$$y(t) := \begin{cases} x(a) & \text{for } t \in [0,a), \\ x(t) & \text{for } t \in I \\ x(b) & \text{for } t \in (b,1]. \end{cases}$$

Then $y \in C$ and $y(t) > 0$ for all $t \in [0,1]$. Moreover,

$$(xg)(I) = (yg)(I) = e^{\log(y)+g_0}(I) = (0,\infty).$$

If, however, $x(t) < 0$ for all $t \in I$, then a similar argument shows that

$$(xg)(I) = (-\infty, 0).$$

In particular, if we take $x = \mathbb{1}$, then g attains all positive real numbers in every interval. But then g cannot be an element of $\mathcal{B}_1$ as it is discontinuous everywhere [128].
Now, fix any compact nondegenerate subinterval I of $[0,1]$ and again some $x \in C$. Then $x(I) = [c,d]$ for some $c, d \in \mathbb{R}$ with $c \leq d$.

We now distinguish eight cases. If $0 < c = d$, then $(xg)(I) = cg(I) = c(0,\infty) = (0,\infty)$, if $c = d < 0$, then $(xg)(I) = c(0,\infty) = (-\infty,0)$, and for $c = 0 = d$ we have $(xg)(I) = \{0\}$. If $0 < c < d$, then $(xg)(I) = (0,\infty)$. If $c = 0 < d$, then I contains an interval J in which x attains only positive numbers. In this case we have $(xg)(I) = [0,\infty)$. A similar argument shows that if $c < 0 < d$, then $(xg)(I) = \mathbb{R}$, and if $c < 0 = d$, then $(xg)(I) = (-\infty,0]$. Finally, if $c < d < 0$, then we obtain $(xg)(I) = (-\infty,0)$. In either case, $(xg)(I)$ is an interval, and so $xg \in D$. But this means nothing else than $g \in D/C$, as desired. ■

Looking a little closer at the construction of g in the proof of Proposition 3.1.13 one might think that D/C consists of those functions which attain every (positive) real number in every interval. This is not true. In [133] was given an example of a function g attaining all real numbers in every nondegenerate subinterval of $[0,1]$ such that $x+g \notin D$ for some $x \in C$. Exponentiation gives a function $\tilde{g}$ attaining all positive real numbers in each nondegenerate subinterval of $[0,1]$ such that $x\tilde{g} \notin D$ for some $x \in C$ and hence $\tilde{g} \notin D/C$. In particular, this shows once again that D/C is a proper subset of D. Conclusively, we have

$$C \subsetneq \Delta \subsetneq D \cap \mathcal{B}_1 \subsetneq D/C \subsetneq D.$$

We now turn to D/Δ. Note that by (a) and (b) of Proposition 3.0.2 we have $\Delta/\Delta \subseteq D/\Delta \subseteq D/C$. In contrast to D/C the class $D \cap \mathcal{B}_1$ is not a subclass of D/Δ, and even worse neither is Δ. In Example 2.2.16 were given two functions in $\Delta \subseteq D \cap \mathcal{B}_1$ the product of which does not belong to D; in particular, $D/\Delta \subsetneq D$.

Again from (a) and (b) of Proposition 3.0.2 we get that $\mathcal{M} := (D \cap \mathcal{B}_1)/(D \cap \mathcal{B}_1)$ is a subclass of D/Δ. Here, the class $\mathcal{M}$ has been characterized more explicitly in [60]. Accordingly, $\mathcal{M}$ consists of all functions g in $D \cap \mathcal{B}_1$ such that if $s \in [0,1]$ is a point of right (left) discontinuity of g, then $g(s) = 0$, and there exists a sequence (t_n) in $[0,1]$ of zeros of g converging from the right (left) to s; in particular, any continuous function belongs to $\mathcal{M}$. Thus, both the class Δ/Δ and the class $\mathcal{M}$ are contained in D/Δ, although there is no inclusion between them.

Example 3.1.14. Recall that the functions $x = \varphi_{-1/2,0,1}$ and $y = \varphi_{1/2,0,1}$ which have been considered in Example 2.2.1 both belong to Δ, but their product $xy = \varphi_{0,0,2}$ does not, and so y is not an element of Δ/Δ which is also clear by Proposition 2.2.13. Moreover, y is continuous and hence lies in $\mathcal{M} \cap \Delta$.

On the other hand the function g from Example 2.2.14 is neither continuous nor of bounded variation but belongs to Δ/Δ with $g(0) = 0$. The function $\tilde{g} : [0,1] \to \mathbb{R}$, defined by $\tilde{g}(t) = g(t) + 1$, then still belongs to Δ/Δ but cannot belong to $\mathcal{M}$ since $\tilde{g}$ is discontinuous at $t = 0$ with $\tilde{g}(0) = 1 \neq 0$. ◊

In addition, this example also shows that both Δ/Δ and $\mathcal{M}$ are proper subsets of D/Δ, and that Δ is no subset of $\mathcal{M}$. Finally, $\mathcal{M}$ is also no subset of Δ, since the function $\varphi_{-1,0,1}$ belongs to $\mathcal{M}$, but not to Δ by Proposition 2.1.5.

The following Figure 3.1.1 summarizes inclusions of most function classes that have been considered so far. Here, $A \longrightarrow B$ means $A \subsetneq B$.

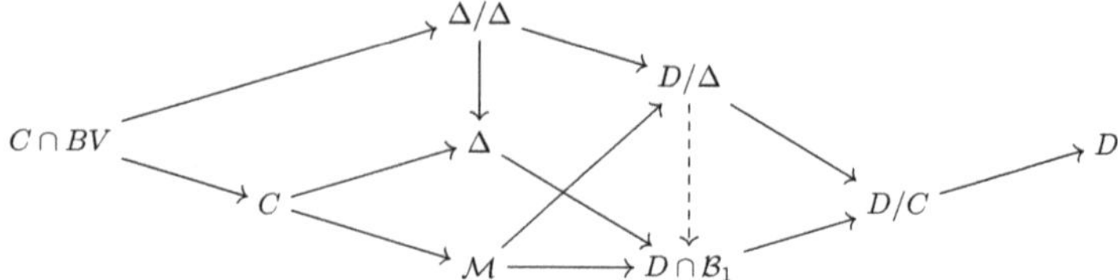

Figure 3.1.1: Inclusions between classical sets.

Other inclusions than those shown in this diagram do not hold. The dashed arrow indicates, however, that we were not able to decide wether the inclusion $D/\Delta \subseteq D \cap \mathcal{B}_1$ holds.

Multipliers in BV

In this last part of this section we are going to extend our considerations by comparing each of the classical classes C, Δ, D and B considered so far to the class BV of functions of bounded Jordan variation introduced in Definition 1.1.7.

First note that since both BV and B are algebras with $\mathbb{1}$ and linked by $BV \subseteq B$, we get by Proposition 3.0.2 (f) that $BV/BV = BV$, $B/B = B$ and $B/BV = B$. Our next result in this section characterizes BV/B and BV/C.

Theorem 3.1.15. *Let $g : [0,1] \to \mathbb{R}$. Then the following statements are equivalent.*

(a) $g \in BV \cap \mathcal{S}_c$. *(b) $g \in BV/B$.* *(c) $g \in BV/C$.*

Proof. "(a)⇒(b)": Assume first that $g \in BV$ and supp(g) is countable. Fix $x \in B$. Then there exists $M > 0$ such that $|x(t)| \leq M$ for all $t \in [0,1]$. It follows that $\operatorname{supp}(xg) \subseteq \operatorname{supp}(g)$, and by Proposition 1.1.8,

$$\operatorname{Var}(xg) \leq 2 \sum_{t \in \operatorname{supp}(g)} |x(t)g(t)| \leq 2M \sum_{t \in \operatorname{supp}(g)} |g(t)| \leq 2M \operatorname{Var}(g) < \infty,$$

showing that $xg \in BV$. Consequently, $g \in BV/B$.

"(b)⇒(c)": Since $C \subseteq B$ we get by Proposition 3.0.2 (b) that $BV/B \subseteq BV/C$.

"(c)⇒(a)": Assume $g \in BV/C$. Since $\mathbb{1} \in C$, Proposition 3.0.2 (c) immediately yields $g \in BV$. Now, suppose that supp(g) is uncountable. Then by Lemma 1.1.5 we find some $m > 0$ and a strictly monotone sequence (t_n) in $(0,1)$ converging to some $t \in [0,1]$ such that $|g(t_n)| \geq m$ for all $n \in \mathbb{N}$. Without loss of generality we may assume that the sequence (t_n) is strictly decreasing and $t = 0$. Let us pick $s_n \in (t_{n+1}, t_n)$ for $n \in \mathbb{N}$ and let us define $x : [0,1] \to \mathbb{R}$ piecewise linear and continuous on $[0,1]$ by $x(0) = x(1) = 0$,

$$x(s_n) = 0 \quad \text{and} \quad x(t_n) = 1/n \quad \text{for } n \in \mathbb{N}.$$

Then $x \in C$ and

$$\mathrm{Var}(xg) \geq \sum_{n=1}^{\infty} |x(t_n)g(t_n)| \geq m \sum_{n=1}^{\infty} \frac{1}{n} = \infty$$

which shows $xg \notin BV$ and contradicts $g \in BV/C$. ■

Finally, another application of Proposition 3.1.1 with $X = BV$ immediately gives $C/BV = \Delta/BV = D/BV = \{\mathbb{0}\}$. Moreover, Corollary 3.1.3, applied to $Y = BV$, yields $BV/D = BV/\Delta = \mathcal{S}_f$.

Table 3.1.1 below summarizes all multiplier classes considered so far. Note that we were not able to fully characterize the classes D/C and D/Δ.

Table 3.1.1: Multipliers in classical spaces.

$C/C = C$	$\Delta/C = (3.1.3)$	$D/C = ???$	$B/C = B$	$BV/C = BV \cap \mathcal{S}_c$
$C/\Delta = \{\mathbb{0}\}$	$\Delta/\Delta = \overline{BV} \cap \Delta$	$D/\Delta = ???$	$B/\Delta = \mathcal{S}_f$	$BV/\Delta = \mathcal{S}_f$
$C/D = \{\mathbb{0}\}$	$\Delta/D = \{\mathbb{0}\}$	$D/D = \mathbb{1}\mathbb{R}$	$B/D = \mathcal{S}_f$	$BV/D = \mathcal{S}_f$
$C/B = \{\mathbb{0}\}$	$\Delta/B = \{\mathbb{0}\}$	$D/B = \{\mathbb{0}\}$	$B/B = B$	$BV/B = BV \cap \mathcal{S}_c$
$C/BV = \{\mathbb{0}\}$	$\Delta/BV = \{\mathbb{0}\}$	$D/BV = \{\mathbb{0}\}$	$B/BV = B$	$BV/BV = BV$

It is now reasonable to ask whether the results showing in this table remain true if we replace BV by one of the other BV-spaces WBV_p, YBV_φ, ΛBV or RBV_p. Moreover, we are interested in investigating $\Gamma BV/\Lambda BV$ for two arbitrary Waterman sequences Γ and Λ, YBV_ψ/YBV_φ for two arbitrary Young functions ψ and φ and RBV_q/RBV_p for two arbitrary exponents $1 \leq p, q < \infty$. These and related problems will be discussed and answered in the next section.

3.2 Multipliers in Generalized BV-Spaces

In this section we are going to generalize the results of the previous section to spaces of functions of generalized bounded variation. Let us start with the spaces of functions of bounded Wiener and Young variation introduced in the Definitions 1.2.2 and 1.2.9, respectively. Multiplier sets for the Wiener spaces WBV_p have already been characterized in [40]. We will give here a more general argument for the Young spaces YBV_φ from which all results for Wiener spaces will follow.

Multipliers in WBV_p and YBV_φ

First note that since YBV_φ for an arbitrary Young function φ is an algebra with $\mathbb{1}$, we have $YBV_\varphi/YBV_\varphi = YBV_\varphi$ by Proposition 3.0.2 (f). Moreover, $YBV_\varphi \subseteq B$ implies $B/YBV_\varphi = B$, again by Proposition 3.0.2 (f).

Another application of Proposition 3.1.1 for $X = YBV_\varphi$ for some Young function φ immediately gives $C/YBV_\varphi = \Delta/YBV_\varphi = D/YBV_\varphi = \{\mathbb{0}\}$. However, if we apply Corollary 3.1.3 to $Y = YBV_\varphi$ instead, we obtain that the spaces YBV_φ/D and YBV_φ/Δ are the same and precisely consist of all functions having finite support, i.e. $YBV_\varphi/D = YBV_\varphi/\Delta = \mathcal{S}_f$.

In the sequel we are going to describe the multiplier spaces of two arbitrary spaces YBV_φ and YBV_ψ. Note that by Proposition 3.0.2 (f) we have $YBV_\psi/YBV_\varphi = YBV_\psi$ if and only if $YBV_\varphi \subseteq YBV_\psi$. Consequently, for the general case we need to know when one such space is contained in the other. But this has already been answered in (1.2.22) and below. Accordingly, if φ and ψ are two Young functions, then $YBV_\varphi \subseteq YBV_\psi$ if and only if $\psi \preceq \varphi$.

We will now turn to our main theorem concerning Young spaces which characterizes the multiplier spaces YBV_ψ/YBV_φ. For this we need the technical result from Lemma 1.2.16.

Theorem 3.2.1. *Let φ and ψ be two Young functions. Then the following statements hold.*

(a) If $\psi \preceq \varphi$, then $YBV_\psi/YBV_\varphi = YBV_\psi$.

(b) If $\psi \npreceq \varphi$, then $YBV_\psi/YBV_\varphi = YBV_\psi \cap \mathcal{S}_c$.

Proof. (a) By (1.2.22), the condition $\psi \preceq \varphi$ implies $YBV_\varphi \subseteq YBV_\psi$, and so the result follows immediately from Proposition 3.0.2 (f), as mentioned before.

(b) Assume first that $g \in YBV_\psi$ and $\operatorname{supp}(g) \subseteq \{t_j \mid j \in \mathbb{N}\}$ is countable, where we arbitrarily pick $t_j \in [0,1]\setminus\operatorname{supp}(g)$ for $j > \#\operatorname{supp}(g)$ if $\operatorname{supp}(g)$ is finite. Then $\operatorname{Var}_\psi(\lambda g) < \infty$ for some $\lambda > 0$. If $x \in YBV_\varphi$ is given, then x is bounded by some $M > 0$, say. Then $\operatorname{supp}(xg)$ is countable as well, and for $\mu := \lambda/(2M)$ we obtain by Proposition 1.2.10,

$$\operatorname{Var}_\psi(\mu x g) \leq \sum_{j=1}^{\infty} \psi(2\mu|(xg)(t_j)|) \leq \sum_{j=1}^{\infty} \psi(\lambda|g(t_j)|) \leq \operatorname{Var}_\psi(\lambda g) < \infty.$$

Thus, $xg \in YBV_\psi$ which shows $g \in YBV_\psi/YBV_\varphi$.

We now prove the converse, i.e. $YBV_\psi/YBV_\varphi \subseteq YBV_\psi \cap \mathcal{S}_c$. Due to $\psi \npreceq \varphi$ we have by (1.2.22) that

$$\limsup_{t\to 0+} \frac{\psi(\lambda t)}{\varphi(t)} = \infty \qquad \text{for all } \lambda > 0.$$

Since φ is an increasing homeomorphism of $[0,\infty)$, substituting $s = \varphi(t)$ and defining $\Phi(\lambda, s) := \psi\big(\lambda\varphi^{-1}(s)\big)$ leads to

$$\limsup_{s\to 0+} \frac{\Phi(\lambda, s)}{s} = \infty \qquad \text{for all } \lambda > 0.$$

Moreover, $\Phi(\cdot, s)$ is increasing for each fixed $s \geq 0$. Thus, we can apply Lemma 1.2.16 with $\alpha = 1$ and obtain a sequence (τ_j) in $(0, \infty)$ such that

$$\sum_{j=1}^{\infty} \Phi(\lambda, \tau_j) = \infty \quad \text{for all } \lambda > 0 \qquad \text{and} \qquad \sum_{j=1}^{\infty} \tau_j \leq 1.$$

The substitution $u_j := \varphi^{-1}(\tau_j)$ therefore yields a sequence (u_j) in $(0, \infty)$ which satisfies

$$\sum_{j=1}^{\infty} \psi(\lambda u_j) = \infty \quad \text{for all } \lambda > 0 \qquad \text{and} \qquad \sum_{j=1}^{\infty} \varphi(u_j) \leq 1.$$

Assume now that $g \in YBV_\psi/YBV_\varphi$, but $g \notin YBV_\psi \cap \mathcal{S}_c$. By Proposition 3.0.2 (c) we have $g \in YBV_\psi$, so $\operatorname{supp}(g)$ must be uncountable. By Lemma 1.1.5 we get some $m > 0$ and a sequence (t_j) in $(0, 1)$ of pairwise distinct numbers such that $|g(t_j)| \geq m$ for all $j \in \mathbb{N}$. Define the function $x : [0, 1] \to \mathbb{R}$ by $x(t_j) := u_j$ for all $j \in \mathbb{N}$ and $x(t) = 0$ otherwise. Then on the one hand, by Proposition 1.2.10,

$$\operatorname{Var}_\varphi(x/2) \leq \sum_{j=1}^{\infty} \varphi(|x(t_j)|) = \sum_{j=1}^{\infty} \varphi(u_j) \leq 1,$$

and so $x \in YBV_\varphi$. However, for each $\lambda > 0$ we get again from Proposition 1.2.10,

$$\operatorname{Var}_\psi(\lambda x g) \geq \sum_{j=1}^{\infty} \psi(\lambda |(xg)(t_j)|) \geq \sum_{j=1}^{\infty} \psi(\lambda m u_j) = \infty$$

which shows $xg \notin YBV_\psi$ and eventually $g \notin YBV_\psi/YBV_\varphi$, a contradiction. ∎

We are now going to compare YBV_ψ with the classical spaces B, D, Δ and C from the first section of this chapter. Here we have again an analogue of Theorem 3.1.15.

Theorem 3.2.2. *Let ψ be a Young function and let $g : [0, 1] \to \mathbb{R}$. Then the following statements are equivalent.*

(a) $g \in YBV_\psi \cap \mathcal{S}_c$. *(b) $g \in YBV_\psi/B$.* *(c) $g \in YBV_\psi/C$.*

Proof. For "(a)⇒(b)" assume that $g \in YBV_\psi$ and $\operatorname{supp}(g) \subseteq \{t_j \mid j \in \mathbb{N}\}$ is countable, where we again choose $t_j \in [0, 1] \setminus \operatorname{supp}(g)$ arbitrarily for $j > \#\operatorname{supp}(g)$ if $\operatorname{supp}(g)$ is finite; in particular, $\operatorname{Var}_\psi(\lambda g) < \infty$ for some $\lambda > 0$. Fix $x \in B$. Then x is bounded by some $M > 0$, say. Of course, we have $\operatorname{supp}(xg) \subseteq \operatorname{supp}(g)$. Therefore, for $\mu := \lambda/(2M)$ we obtain by Proposition 1.2.10,

$$\operatorname{Var}_\psi(\mu x g) \leq \sum_{j=1}^{\infty} \psi(2\mu |x(t_j) g(t_j)|) \leq \sum_{j=1}^{\infty} \psi(\lambda |g(t_j)|) \leq \operatorname{Var}_\psi(\lambda g) < \infty$$

which shows $xg \in YBV_\psi$ and hence $g \in YBV_\psi/B$.

Note that "(b)⇒(c)" immediately follows from Proposition 3.0.2 (b), since $C \subseteq B$.

We now prove the remaining part "(c)⇒(a)". To this end, assume that $g \in YBV_\psi$ has uncountable support. Then by Lemma 1.1.5 we find a strictly monotone sequence (t_j)

in $(0,1)$ converging to some $t \in [0,1]$ and a constant $m > 0$ such that $|g(t_j)| \geq m$ for all $j \in \mathbb{N}$, and we may assume that (t_j) is strictly decreasing and converges to $t = 0$. Pick $s_j \in (t_{j+1}, t_j)$ for all $j \in \mathbb{N}$. Note that since ψ is an increasing homeomorphism of $[0,\infty)$ with $\varphi(0) = 0$, the numbers

$$u_j := \sqrt{\psi^{-1}\left(\frac{1}{j}\right)} \quad \text{for } j \in \mathbb{N}$$

are well-defined and converge to 0 in such a way that

$$\sum_{j=1}^{\infty} \psi\left(u_j^2\right) = \sum_{j=1}^{\infty} \psi\left(\psi^{-1}(1/j)\right) = \sum_{j=1}^{\infty} \frac{1}{j} = \infty.$$

Since $u_j \to 0$ as $j \to \infty$, for each fixed $\lambda > 0$ there is some $N(\lambda) \in \mathbb{N}$ such that $\lambda \geq u_j$ for all $j \geq N(\lambda)$. Therefore, since ψ is increasing,

$$\sum_{j=1}^{\infty} \psi(\lambda u_j) \geq \sum_{j=N(\lambda)}^{\infty} \psi\left(u_j^2\right) = \infty.$$

Define $x : [0,1] \to \mathbb{R}$ piecewise linear and continuous by

$$x(0) := x(1) := 0, \quad x(t_j) := u_j \quad \text{and} \quad x(s_j) := 0 \qquad \text{for all } j \in \mathbb{N}.$$

Then $x \in C$ and for each $\lambda > 0$ we obtain

$$\mathrm{Var}_{\psi}(\lambda x g) \geq \sum_{j=1}^{\infty} \psi(\lambda |(xg)(t_j) - (xg)(s_j)|) = \sum_{j=1}^{\infty} \psi(\lambda |g(t_j) u_j|) \geq \sum_{j=1}^{\infty} \psi(\lambda m u_j) = \infty.$$

But this shows $xg \notin YBV_{\psi}$ and hence $g \notin YBV_{\psi}/C$. ■

Let us discuss some special cases of the Theorems 3.2.1 and 3.2.2. For $\varphi(t) = t^p$ and $\psi(t) = t^q$ for $1 \leq p, q < \infty$, the spaces YBV_{φ} and YBV_{ψ} precisely coincide with the spaces of functions of bounded Wiener variation, i.e. $YBV_{\varphi} = WBV_p$ and $YBV_{\psi} = WBV_q$. In this case, the condition $\psi \preceq \varphi$ is equivalent to $p \leq q$, and Theorem 3.2.1 reads as follows.

Corollary 3.2.3. *Let $1 \leq p, q < \infty$. Then the following statements hold.*

(a) If $p \leq q$, then $WBV_q/WBV_p = WBV_q$.

(b) If $p > q$, then $WBV_q/WBV_p = WBV_q \cap \mathcal{S}_c$.

In particular, for $p = 1 \leq q$ we obtain $WBV_q/BV = WBV_q$, and for $p > q = 1$ we get $BV/WBV_p = BV \cap \mathcal{S}_c$. More generally, if only one of the Young functions is replaced by $t \mapsto t^p$, then Theorem 3.2.1 yields the following four cases.

Corollary 3.2.4. *Let $1 \leq p, q < \infty$ and φ and ψ be Young functions. Then the following statements hold.*

(a) If $\limsup_{t\to 0+} \frac{\psi(t)}{t^p} < \infty$, *then* $YBV_\psi/WBV_p = YBV_\psi$.

(b) If $\limsup_{t\to 0+} \frac{\psi(t)}{t^p} = \infty$, *then* $YBV_\psi/WBV_p = YBV_\psi \cap \mathcal{S}_c$.

(c) If $\limsup_{t\to 0+} \frac{t^q}{\varphi(t)} < \infty$, *then* $WBV_q/YBV_\varphi = WBV_q$.

(d) If $\limsup_{t\to 0+} \frac{t^q}{\varphi(t)} = \infty$, *then* $WBV_q/YBV_\varphi = WBV_q \cap \mathcal{S}_c$.

Note that indeed Corollary 3.2.4 reduces to Corollary 3.2.3 for $\varphi(t) = t^p$ and $\psi(t) = t^q$. In general, if we put $p = 1$ in Corollary 3.2.4 (a) we obtain from the convexity of ψ and $\psi(0) = 0$ that

$$\limsup_{t\to 0+} \frac{\psi(t)}{t} \leq \limsup_{t\to 0+} \frac{t\psi(1)}{t} = \psi(1) < \infty$$

and hence $YBV_\psi/BV = YBV_\psi$. Similarly, if $\iota \npreceq \varphi$ for $\iota(t) = t$ being the identity function which is equivalent to

$$\limsup_{t\to 0+} \frac{t}{\varphi(t)} = \infty,$$

we get from (d) by putting $q = 1$ that $BV/YBV_\varphi = BV \cap \mathcal{S}_c$. Note that on the other hand the case $\iota \preceq \varphi$ leads to $YBV_\varphi = BV$ and hence to $BV/YBV_\varphi = BV$.

Moreover, the special case of Theorem 3.2.2 with $\psi(t) = t^q$ reads as follows.

Corollary 3.2.5. *Let $1 \leq q < \infty$ and $g : [0,1] \to \mathbb{R}$. Then the following statements are equivalent.*

(a) $g \in WBV_q \cap \mathcal{S}_c$. *(b)* $g \in WBV_q/B$. *(c)* $g \in WBV_q/C$.

Note that for $q = 1$ Corollary 3.2.5 reduces to Theorem 3.1.15.

Multipliers in ΛBV

We now continue our generalizations to the spaces of functions of bounded variation in the sense of Waterman introduced in Definition 1.2.18. We are particularly interested in characterizing $Y/\Lambda BV$ and $\Gamma BV/X$ for $X, Y \in \{B, C, D, \Delta, BV, \Lambda BV\}$, where Γ and Λ are arbitrary given Waterman sequences.

First notice that ΛBV is an algebra with $\mathbb{1}$, and so from Proposition 3.0.2 (f) we get $\Lambda BV/\Lambda BV = \Lambda BV$.

Recall that equation (1.2.49) says that not only $\Lambda BV \subseteq B$ holds for all Waterman sequences Λ, but also that each Waterman space comprises BV, and that each regular function belongs to at least one Waterman space. Because of that and Proposition 3.0.2 (f) we immediately get $\Gamma BV/BV = \Gamma BV$ and $B/\Lambda BV = B$.

Let us now consider an analogue to Theorem 3.1.15.

Theorem 3.2.6. *Let $g : [0,1] \to \mathbb{R}$ and Γ be a Waterman sequence. Then the following statements are equivalent.*

(a) $g \in \Gamma BV \cap \mathcal{S}_c$. *(b)* $g \in \Gamma BV/B$. *(c)* $g \in \Gamma BV/C$.

Proof. Throughout this proof let the Waterman sequence Γ be given by $\Gamma = (\gamma_j)$.
"(a)⇒(b)". Assume that $g \in \Gamma BV$ has countable support. Fix $x \in B$. Then there exists $M > 0$ such that $|x(t)| \leq M$ for any $t \in [0,1]$. Of course we have $\operatorname{supp}(xg) \subseteq \operatorname{supp}(g) \subseteq \{\tau_1, \tau_2, \ldots\}$ with $\tau_j \in \operatorname{supp}(g)$, where we again pick arbitrary $\tau_j \in [0,1] \setminus \operatorname{supp}(g)$ for $j > \#\operatorname{supp}(g)$ if $\operatorname{supp}(g)$ is finite. By Proposition 1.2.20 we have

$$\operatorname{Var}_\Gamma(xg) \leq 2 \sup_\sigma \sum_{j=1}^{\infty} \gamma_{\sigma(j)} |(xg)(\tau_j)| \leq 2M \sup_\sigma \sum_{j=1}^{\infty} \gamma_{\sigma(j)} |g(\tau_j)| \leq 2M \operatorname{Var}_\Gamma(g) < \infty,$$

where σ runs through all permutations of $\mathbb{N}$. Thus $xg \in \Gamma BV$ and hence $g \in \Gamma BV/B$.

"(b)⇒(c)" follows directly from Proposition 3.0.2 (b), since $C \subseteq B$.
"(c)⇒(a)". Assume now that $g \in \Gamma BV/C$. Certainly, $\mathbb{1} \in C$, and so $g \in \Gamma BV$ by Proposition 3.0.2 (c). Suppose now that $\operatorname{supp}(g)$ is uncountable. By Lemma 1.1.5 we obtain an $m > 0$ and some sequence (t_n) in $(0,1)$ of pairwise distinct terms such that $|g(t_n)| \geq m$ for all $n \in \mathbb{N}$. Without loss of generality we can assume that (t_n) is strictly decreasing and converging to 0. Pick $s_n \in (t_{n+1}, t_n)$ for $n \in \mathbb{N}$, and let $\Gamma_n := \gamma_1 + \ldots + \gamma_n$. Now, define $x \in C$ piecewise linear by $x(0) = x(1) = 0$, $x(s_n) = 0$ and $x(t_n) = 1/\Gamma_n$ for $n \in \mathbb{N}$. Let us define a finite collection of nonoverlapping subintervals of the interval $[0,1]$ by $[a_j, b_j] = [s_j, t_j]$ for $j \in \{1, \ldots, n\}$. Then

$$\operatorname{Var}_\Gamma(xg) \geq \sum_{j=1}^{n} \gamma_j |(xg)(a_j) - (xg)(b_j)| = \sum_{j=1}^{n} \gamma_j |(xg)(t_j)| \geq m \sum_{j=1}^{n} \frac{\gamma_j}{\Gamma_j}.$$

By a result of Abel and Dini [77],

$$\sum_{j=1}^{\infty} \frac{\gamma_j}{\Gamma_j} = \infty,$$

and hence $xg \notin \Gamma BV$, contradicting $g \in \Gamma BV/C$. ■

An application of Proposition 3.1.1 for $X = \Lambda BV$ for some Waterman sequence Λ immediately gives $C/\Lambda BV = \Delta/\Lambda BV = D/\Lambda BV = \{\mathbb{0}\}$. Moreover, if we apply Corollary 3.1.3 to $Y = \Lambda BV$ instead, we obtain $\Lambda BV/D = \Lambda BV/\Delta = \mathcal{S}_f$.

In the sequel we are going to describe the multiplier spaces of two arbitrary Waterman spaces ΓBV and ΛBV. Note that by Proposition 3.0.2 (f) we have $\Gamma BV/\Lambda BV = \Gamma BV$ if and only if $\Lambda BV \subseteq \Gamma BV$. Consequently, for the general case we need to know when one such space is contained in the other. But this has already been answered in the discussion around (1.2.44). Accordingly, if Γ and Λ are two Waterman sequences, then $\Lambda BV \subseteq \Gamma BV$ if and only if $\Gamma \preceq \Lambda$.
We are now in position to prove the following

Theorem 3.2.7. *Let Γ and Λ be two Waterman sequences. Then the following statements hold.*

(a) If $\Gamma \preceq \Lambda$, then $\Gamma BV/\Lambda BV = \Gamma BV$.

(b) If $\Gamma \npreceq \Lambda$, then $\Gamma BV/\Lambda BV = \Gamma BV \cap \mathcal{S}_c$.

Proof. Throughout this proof let $\Gamma = (\gamma_j)$ and $\Lambda = (\lambda_j)$.
Note that by (1.2.44) the assertion $\Gamma \preceq \Lambda$ is equivalent to $\Lambda BV \subseteq \Gamma BV$. Consequently, (a) follows immediately from Proposition 3.0.2 (f), as mentioned before.
For (b) assume first that $g \in \Gamma BV$ and $\operatorname{supp}(g) \subseteq \{t_1, t_2, \ldots\}$ is countable, where we pick arbitrary $t_j \in [0,1] \setminus \operatorname{supp}(g)$ for $j > \#\operatorname{supp}(g)$ if $\operatorname{supp}(g)$ is finite. If $x \in \Lambda BV$ is given, then x is bounded by some $M > 0$, say. Then $\operatorname{supp}(xg) \subseteq \operatorname{supp}(g)$, and we obtain by Proposition 1.2.20,

$$\operatorname{Var}_\Gamma(xg) \leq 2 \sup_\sigma \sum_{j=1}^{\infty} \gamma_{\sigma(j)} |x(t_j) g(t_j)| \leq 2M \sup_\sigma \sum_{j=1}^{\infty} \gamma_{\sigma(j)} |g(t_j)| \leq 2M \operatorname{Var}_\Gamma(g) < \infty,$$

and hence $xg \in \Gamma BV$ which shows that $g \in \Gamma BV/\Lambda BV$.
For the reverse implication assume that $\operatorname{supp}(g)$ is uncountable. By Lemma 1.1.5 there is a constant $m > 0$ and a sequence (t_j) in $(0,1)$ such that $|g(t_j)| \geq m$ for all $j \in \mathbb{N}$. Now apply Lemma 1.2.23 to $a_j = \gamma_j, b_j = \lambda_j$ and $\alpha = 1$ and obtain a monotonically decreasing sequence (u_j) in $(0,\infty)$ tending to zero such that

$$\sum_{j=1}^{\infty} \gamma_j u_j = \infty \qquad \text{and} \qquad \sum_{j=1}^{\infty} \lambda_j u_j = 1.$$

Define $x(t_j) := u_j$ and $x(t) = 0$ elsewhere. Then with the help of Proposition 1.2.20 we have

$$\operatorname{Var}_\Lambda(x) \leq 2 \sum_{j=1}^{\infty} \lambda_j u_j = 2,$$

that is, $x \in \Gamma BV$. Note that the sum on the right hand side of (1.2.39) is maximal for the given ordering of (λ_j) and (u_j) as both sequences are decreasing [6]. On the other hand, again by Proposition 1.2.20,

$$\operatorname{Var}_\Gamma(xg) \geq \sum_{j=1}^{\infty} \gamma_j |x(t_j) g(t_j)| \geq m \sum_{j=1}^{\infty} \gamma_j u_j = \infty$$

showing that $xg \notin \Gamma BV$ and hence $g \notin \Gamma BV/\Lambda BV$. That each function $g \in \Gamma BV/\Lambda BV$ belongs to ΓBV follows from Proposition 3.0.2 (c). This completes the proof. ∎

In the light of (1.2.49) it is reasonable to ask how the "limit" spaces $BV/\Lambda BV$, $\Gamma BV/BV$, $R/\Lambda BV$ and $\Gamma BV/R$ look like. With the help of Proposition 3.0.2 we obtain the following

Corollary 3.2.8. *Let Λ be a Waterman sequence. Then $BV/\Lambda BV = BV \cap \mathcal{S}_c$.*

Proof. Let $g \in BV$ have countable support. Then Theorem 3.1.15 yields $g \in BV/B$, and since $\Lambda BV \subseteq B$ we get by Proposition 3.0.2 (b) that $g \in BV/\Lambda BV$.

Suppose now that $g \in BV/\Lambda BV$ for $\Lambda = (\lambda_j)$; in particular, $g \in BV$ by Proposition 3.0.2 (c). The sequence $\Gamma = (\gamma_j)$, defined by $\gamma_j := \sqrt{\lambda_j}$, clearly is a Waterman sequence, and since $BV \subseteq \Gamma BV$ we obtain from Proposition 3.0.2 (a) that $g \in \Gamma BV/\Lambda BV$. However, since

$$\lim_{n\to\infty} \frac{\gamma_n}{\lambda_n} = \lim_{n\to\infty} \frac{\sqrt{\lambda_n}}{\lambda_n} = \lim_{n\to\infty} \frac{1}{\sqrt{\lambda_n}} = \infty,$$

we obtain with the help of the Stolz-Cesàro-Theorem [119] that $\Gamma \not\preceq \Lambda$. Finally, Theorem 3.2.7 yields that $\mathrm{supp}(g)$ is countable. ■

For the "flipped" space $\Gamma BV/BV$ note that we already know $\Gamma BV/BV = \Gamma BV$. Moreover, since $\Lambda BV \subseteq R$ we also obtain $R/\Lambda BV = R$ from Proposition 3.0.2 (f). Finally, since $C \subseteq R \subseteq B$, we have that $\Gamma BV/B \subseteq \Gamma BV/R \subseteq \Gamma BV/C$ by Proposition 3.0.2 (b), and Theorem 3.2.6 eventually shows that $\Gamma BV/R = \Gamma BV \cap \mathcal{S}_c$.

Let us now discuss some special cases of the Theorems 3.2.6 and 3.2.7. For $0 < p, q \leq 1$ we consider the two Waterman sequences $\Lambda_p = (1/n^p)$ and $\Lambda_q = (1/n^q)$ for which the the condition $\Lambda_q \preceq \Lambda_p$ is equivalent to $p \leq q$ (again by the Stolz-Cesàro-Theorem [119]). Consequently, Theorem 3.2.7 reads as follows.

Corollary 3.2.9. *Let $0 < p, q \leq 1$. Then the following statements hold.*

(a) If $p \leq q$, then $\Lambda_q BV/\Lambda_p BV = \Lambda_q BV$.

(b) If $p > q$, then $\Lambda_q BV/\Lambda_p BV = \Lambda_q BV \cap \mathcal{S}_c$.

Recall that the space $\Lambda_p BV$ is of particular interest for $p = 1$ and called the space of functions of bounded *harmonic variation*, abbreviated by the symbol $HBV = \Lambda_1 BV$. In particular, for $0 < p \leq 1 = q$ we obtain $HBV/\Lambda_p BV = HBV$, and for $p = 1 > q > 0$ we get $\Lambda_q BV/HBV = \Lambda_q BV \cap \mathcal{S}_c$, both as a consequence of Corollary 3.2.9.

Analogously, Theorem 3.2.6 then reads as follows.

Corollary 3.2.10. *Let $0 < q \leq 1$ and $g : [0,1] \to \mathbb{R}$. Then the following statements are equivalent.*

(a) $g \in \Lambda_q BV \cap \mathcal{S}_c$. *(b) $g \in \Lambda_q BV/B$.* *(c) $g \in \Lambda_q BV/C$.*

The spaces BV, WBV_p, YBV_φ and ΛBV have in common that they all contain functions of finite and countably infinite support. In the next and final subsection we consider multiplier spaces for the Riesz spaces RBV_p that do not have this property, as any function in RBV_p is automatically continuous for $p > 1$.

Multipliers in L_p and RBV_p

As a final generalization we now consider functions of bounded variation in the sense of Riesz as defined in Definition 1.2.24. Recall that for $1 < p < \infty$ the class RBV_p coincides with the class AC of absolutely continuous functions with derivatives in the space L_p. Moreover, in this case the Riesz variation can be calculated explicitly by the formula (1.2.50).
If we write

$$L'_p := \{x \in AC \mid x' \in L_p\} \quad \text{for } 1 \le p \le \infty,$$

then Riesz' result states that

$$L'_p = \begin{cases} AC & \text{for } p = 1, \\ RBV_p & \text{for } 1 < p < \infty, \\ Lip & \text{for } p = \infty, \end{cases}$$

whereas, by definition, $RBV_1 = BV \neq AC$.
Recall that RBV_p can be interpreted as the set of continuous representatives of the Sobolev space $W^{1,p}$; in particular, for each function $x \in W^{1,p}$ there is a function $\tilde{x} \in L'_p$ such that $x = \tilde{x}$ almost everywhere.
Thus, the Riesz spaces are closely related not only to Sobolev spaces but also to Lebesgue spaces. In order to find multipliers for Riesz spaces, we therefore need multipliers for Lebesgue spaces. As adumbrated at the beginning of this chapter, Hölder's inequality is of good use here. We will prove in a little more detail the following characterization of Lebesgue multipliers.

Theorem 3.2.11. *Let $1 \le p, q \le \infty$. Then*

$$L_q/L_p = \begin{cases} L_{\frac{pq}{p-q}} & \text{for } 1 \le q < p < \infty, \\ L_q & \text{for } 1 \le q < p = \infty, \\ L_\infty & \text{for } 1 \le p = q \le \infty, \\ \{\mathbb{0} \text{ a.e.}\} & \text{for } 1 \le p < q \le \infty. \end{cases} \tag{3.2.1}$$

Proof. We first prove for $1 \le p < \infty$ the identity

$$L_1/L_p = \begin{cases} L_{\frac{p}{p-1}} & \text{for } 1 < p < \infty, \\ L_\infty & \text{for } p = 1. \end{cases} \tag{3.2.2}$$

Note that for $1 < p < \infty$ the inclusion $L_{\frac{p}{p-1}} \subseteq L_1/L_p$ is a trivial consequence of Hölder's inequality. Indeed, if $g \in L_{\frac{p}{p-1}}$ and $x \in L_p$ are given, then Hölder's inequality guarantees immediately $xg \in L_1$ as $\frac{p}{p-1}$ is the Hölder conjugate to p. The same argument holds for $p = 1$ and shows $L_\infty \subseteq L_1/L_1$.
For the converse there is a little more to do. For fixed $g \in L_1/L_p$ with $1 \le p < \infty$ we also have $|g| \in L_1/L_p$. On L_p we define the linear functional $T : L_p \to \mathbb{R}$ by

$$Tx := \int_0^1 x(t)|g(t)| \,\mathrm{d}t$$

which is well-defined due to $|g| \in L_1/L_p$. We now show that T is bounded. If T is unbounded, there is a sequence (x_k) in L_p with $\|x_k\|_{L_p} = 1$ and $|Tx_k| \geq 3^k$ for all $k \in \mathbb{N}$. The function

$$x := \sum_{k=1}^{\infty} \frac{|x_k|}{2^k}$$

then belongs to L_p, because

$$\|x\|_{L_p} \leq \sum_{k=1}^{\infty} \frac{\|x_k\|_{L_p}}{2^k} = 1.$$

On the other hand, the function $x|g|$ does not belong to L_1, because by the Monotone Convergence Theorem,

$$\int_0^1 x(t)|g(t)|\,\mathrm{d}t = \sum_{k=1}^{\infty} \int_0^1 \frac{|x_k(t)|}{2^k}|g(t)|\,\mathrm{d}t \geq \sum_{k=1}^{\infty} \frac{|Tx_k|}{2^k} = \sum_{k=1}^{\infty} \frac{3^k}{2^k} = \infty.$$

But this contradicts $|g| \in L_1/L_p$. Consequently, T is a bounded linear functional on L_p and therefore has, by a well-known theorem of Riesz (see [153] for a proof), a representation of the form

$$Tx = \int_0^1 x(t)h(t)\,\mathrm{d}t$$

for some function $h \in L_q$, where q is the Hölder conjugate to p, that is, $q = \frac{p}{p-1}$ for $1 < p < \infty$ and $q = \infty$ for $p = 1$. Thus, $|g| = h$ almost everywhere which implies $g \in L_q$, as claimed. In total, formula (3.2.2) is established.

We are now going to deduce from (3.2.2) the remaining identities. Let $1 \leq q < p < \infty$ and set $r = \frac{p}{q}$ and $s = \frac{p}{p-q}$. Then $r, s > 1$ and $\frac{1}{r} + \frac{1}{s} = 1$. Fix $g \in L_{\frac{pq}{p-q}}$ and $x \in L_p$. Since $(|x|^q)^r = |x|^p \in L_1$ and $(|g|^q)^s = |g|^{\frac{pq}{p-q}} \in L_1$, we have $|x|^q \in L_r$ and $|g|^q \in L_s$, and from Hölder's inequality we obtain that $|xg|^q \in L_1$, hence $xg \in L_q$ and thus $g \in L_q/L_p$. The same argument works for $1 \leq p = q < \infty$. Finally, $L_\infty/L_\infty = L_\infty$ follows from Proposition 3.0.2 (f). This proves the first and third of the four identities in (3.2.1).

For the second note that for $1 \leq q < \infty$ the inclusion $L_q \subseteq L_q/L_\infty$ follows instantly from Hölder's inequality, and the inclusion $L_q/L_\infty \subseteq L_q$ is a consequence of Proposition 3.0.2 (c).

For the fourth and final identity in (3.2.1) fix $1 \leq p < q \leq \infty$ and $g \in L_q/L_p$ and assume that g is not zero almost everywhere. Then there is some measurable set $E \subseteq [0,1]$ of finite positive measure and some $m > 0$ such that $|g(t)| \geq m$ for all $t \in E$. For $x \in L_p$ we have $x \in L_p(E)$ and

$$\infty > \int_0^1 |x(t)g(t)|^q\,\mathrm{d}t \geq m^q \int_E |x(t)|^q\,\mathrm{d}t,$$

and consequently $x \in L_q(E)$. This shows $L_p(E) \subseteq L_q(E)$ and hence $p \geq q$, a contradiction. This proves $g = \mathbb{0}$ almost everywhere and completes the proof. ■

Before we proceed with a characterization of the multipliers RBV_q/RBV_p, let us remark another interpretation of Hölder's inequality. It states that if x and y belong to conjugate Lebesgue spaces, the product belongs to L_1. It is shown in [5] that this is precise in the following sense: If $1/p + 1/q = 1$ and $x \in L_p$, then

$$\|x\|_{L_p} = \sup\left\{\int_0^1 x(t)y(t)\,\mathrm{d}t \mid y \in L_q, \|y\|_{L_q} \le 1\right\}; \tag{3.2.3}$$

in particular, the norm of the linear functional

$$y \mapsto \int_0^1 x(t)y(t)\,\mathrm{d}t$$

is precisely the L_p-norm of x. In fact, one can show that the supremum in (3.2.3) is a maximum for $x \in L_p$ with $1 \le p < \infty$. The following example shows that for $p = \infty$ this is no longer true [5].

Example 3.2.12. The function $x : [0,1] \to \mathbb{R}$, $t \mapsto t$, belongs to L_∞ with $\|x\|_{L_\infty} = 1$. If any $y \in L_1$ with $0 < \|y\|_{L_1} \le 1$ is given, one can pick $\delta \in (0,1)$ so that

$$\int_0^{1-\delta} |y(t)|\,\mathrm{d}t \ge \frac{1}{2}\|y\|_{L_1},$$

as the Lebesgue integral is absolutely continuous with respect to the domain of integration. We now obtain

$$\begin{aligned}\left|\int_0^1 x(t)y(t)\,\mathrm{d}t\right| &\le \int_0^{1-\delta} t\,|y(t)|\,\mathrm{d}t + \int_{1-\delta}^1 t\,|y(t)|\,\mathrm{d}t \\ &\le (1-\delta)\int_0^{1-\delta} |y(t)|\,\mathrm{d}t + \int_{1-\delta}^1 |y(t)|\,\mathrm{d}t \\ &= \int_0^1 |y(t)|\,\mathrm{d}t - \delta\int_0^{1-\delta} |y(t)|\,\mathrm{d}t \le \left(1 - \frac{\delta}{2}\right)\|y\|_{L_1} < 1 = \|x\|_{L_\infty}.\end{aligned}$$

Consequently, the number $\|x\|_{L_\infty}$ and hence the supremum in (3.2.3) cannot be attained by values of the functional $\int_0^1 x(t)y(t)\,\mathrm{d}t$, as long as y satisfies $\|y\|_{L_1} \le 1$. ◇

We remark that there is another difference in this context, depending on the choice of p: In case $1 < p < \infty$, the supremum in (3.2.3) is attained by a unique function $y \in L_q$. However, for $p = 1$ or $p = \infty$, even if the supremum in (3.2.3) is attained by some function $y \in L_q$, this function must not be unique. This is illustrated by the next two examples [5].

Example 3.2.13. Consider the functions $x = y_1 = \chi_{[0,1/2]}$ and $y_2 = \mathbb{1}$. Then $x \in L_1$ and $y_1, y_2 \in L_\infty$ with $\|y_1\|_{L_\infty} = \|y_2\|_{L_\infty} = 1$. Moreover,

$$\|x\|_{L_1} = \frac{1}{2} = \int_0^1 x(t)y_1(t)\,\mathrm{d}t = \int_0^1 x(t)y_2(t)\,\mathrm{d}t,$$

but y_1 is not equal to y_2 almost everywhere. ◇

Example 3.2.14. Consider the functions $x = \mathbb{1}$, $y_1 = 2\chi_{[0,1/2]}$ and $y_2 = 2\chi_{[1/2,1]}$. Then $x \in L_\infty$ and $y_1, y_2 \in L_1$ with $\|y_1\|_{L_1} = \|y_2\|_{L_1} = 1$. Moreover,

$$\|x\|_{L_\infty} = 1 = \int_0^1 x(t)y_1(t)\,\mathrm{d}t = \int_0^1 x(t)y_2(t)\,\mathrm{d}t,$$

but y_1 is not equal to y_2 almost everywhere. ◇

We are now going to give a full characterization of the spaces RBV_q/RBV_p for arbitrary exponents $1 \leq p, q < \infty$ which is an immediate consequence of Theorem 3.2.11. Note that by Proposition 3.0.2 (f) we have $RBV_q/RBV_p = RBV_q$ if and only if $RBV_p \subseteq RBV_q$. Consequently, for the general case we need to know when one such space is contained in the other. Recall that similar to Lebesgue spaces we have $RBV_p \subseteq RBV_q$ if and only if $p \geq q$. In addition, note that $RBV_p \subseteq AC$ whenever $p > 1$. As a consequence, we obtain

Theorem 3.2.15. *Let $1 \leq p, q < \infty$. Then the following statements hold.*

(a) If $p \geq q$, then $RBV_q/RBV_p = RBV_q$.

(b) If $p < q$, then $RBV_q/RBV_p = \{\mathbb{0}\}$.

(c) If $q > 1$, then $RBV_q/AC = \{\mathbb{0}\}$.

Proof. (a) Since $p \geq q$, we have $RBV_p \subseteq RBV_q$, and (a) follows again immediately from Proposition 3.0.2 (f), as mentioned before.

For (b) with $p > 1$ fix $g \in RBV_q/RBV_p$ and $y \in L_p$. The function

$$x(t) = \int_0^t y(s)\,\mathrm{d}s$$

is then absolutely continuous with $x' = y \in L_p$ and hence belongs to RBV_p. From Proposition 3.0.2 (c) follows $g \in RBV_q$ and thus $g \in AC$ with $g' \in L_q$. Moreover, since g is a multiplier of RBV_q over RBV_p, we have $xg \in RBV_q$ and in particular $x'g + xg' \in L_q$. Because of $x \in AC \subseteq L_\infty$, the product xg' belongs to L_q and hence $x'g = yg$ must belong to L_q, as well. As this is true for all $y \in L_p$ we have shown that $g \in L_q/L_p$. By Theorem 3.2.11 it follows that g is zero almost everywhere, and since g is continuous, we conclude $g = \mathbb{0}$ everywhere, as desired.

For (c) repeat the argument used for (b) with $p = 1$ and RBV_p replaced by AC. Since $AC \subseteq BV$ we obtain the remaining part of (b) for $p = 1$ from (c) and Proposition 3.0.2 (b). This completes the proof. ∎

We remark that the parts (b) and (c) of Theorem 3.2.15 can also be proven without Theorem 3.2.11. Indeed, in order to prove (b) for $p > 1$ fix $g \in RBV_q/RBV_p$ for $p < q$ and assume that g is not identically zero. Because of $q > p \geq 1$ the function g is continuous, and there is some proper interval $[a, b] \subseteq (0, 1]$ and some constant $m > 0$ such that $|g(t)| \geq m$ for all $t \in [a, b]$. Now define

$$h(t) := \begin{cases} (t-a)^{-2/(p+q)} & \text{for } t \in (a, b], \\ 0 & \text{for } t \in [0, a] \cup (b, 1] \end{cases} \qquad \text{and} \qquad x(t) := \int_0^t h(s)\,\mathrm{d}s.$$

Then $h \in L_p \setminus L_q$, and therefore $x \in RBV_p \setminus RBV_q$. By definition of g, the product xg must belong to RBV_q. However, we obtain for $s \in (a, b]$,

$$\begin{aligned} \mathrm{RVar}_q(xg) &\geq (s-a)^{1-q}\,|(xg)(s)|^q \geq m^q (s-a)^{1-q}\,|x(s)|^q \\ &= \frac{m^q(p+q)}{p+q-2}(s-a)^{\frac{p-q}{p+q}}. \end{aligned} \tag{3.2.4}$$

Since $q > p$, the exponent $\frac{p-q}{p+q}$ in (3.2.4) is negative, and hence the right hand side in (3.2.4) goes to infinity as $s \to a+$. Consequently, $\mathrm{RVar}_q(xg) = \infty$ which contradicts $g \in RBV_q/RBV_p$. The same argument works with $p = 1$ to prove (c).

Let us add some comments to this result. First of all, although part (b) of Theorem 3.2.15 looks different than the corresponding results for the other BV-spaces (Theorem 3.2.1, Corollary 3.2.3 and Theorem 3.2.7), it still fits perfectly. Since in part (b) we assume $q > 1$, each function $g \in RBV_q/RBV_p$ must be continuous, and the only continuous function with countable support is the function $\mathbb{0}$.
Moreover, assertion (a) is of particular interest for $q = 1$, because then we obtain $BV/RBV_p = BV$ for all $p \in [1, \infty)$. In particular, for $p = 1$ we obtain again $BV/BV = BV$. On the other hand, from (b) for $p = 1$ we get that $RBV_q/BV = \{\mathbb{0}\}$ for all $q > 1$, and part (c) solves a conjecture that we made in [28].

We now compare RBV_q for fixed $q > 1$ to the classical spaces B, C, Δ and D, and also to Lip and AC; a first comparison of this kind was already given in Theorem 3.2.15 (c). First note that (1.2.54) implies

$$Lip \subsetneq RBV_q \subsetneq AC \subsetneq C \cap BV,$$

and we obtain the inclusions

$$Lip/C \subseteq RBV_q/C \subseteq AC/C \subseteq C \cap (BV/C)$$

and

$$Lip/BV \subseteq RBV_q/BV \subseteq AC/BV \subseteq C/BV$$

by Proposition 3.0.2 (a). But the only continuous function in BV/C is the zero function $\mathbb{0}$ by Theorem 3.1.15, and the only function in C/BV is also the zero function, as has been pointed out right after Theorem 3.1.15. Hence, $Lip/C = RBV_q/C = AC/C = \{\mathbb{0}\}$ and also $Lip/BV = RBV_q/BV = AC/BV = \{\mathbb{0}\}$, and since $C \subseteq B$ and $C \subseteq \Delta \subseteq D$ we obtain in total $Lip/X = RBV_q/X = AC/X = \{\mathbb{0}\}$ for $X \in \{B, BV, C, \Delta, D\}$ by Proposition 3.0.2 (b). Moreover, $RBV_q/Lip = RBV_q$ by Proposition 3.0.2 (f).

We are now considering Y/RBV_p for fixed $p > 1$ and $Y \in \{B, BV, C, \Delta, D, Lip, AC\}$. We already know that $BV/RBV_p = BV$. Also note that since $RBV_p \subseteq C \subseteq B$ we have $B/RBV_p = B$ and $C/RBV_p = C$ by Proposition 3.0.2 (f). Finally, since $\Delta/(C \cap BV) = \Delta$ by Theorem 2.2.2 and $RBV_p \subseteq C \cap BV$ we obtain $\Delta/RBV_p = \Delta$ by Proposition 3.0.2 (b) and (c). Similar relations hold for Y/Lip and Y/AC, namely $Y/Lip = Y/AC = Y$ for $Y \in \{C, \Delta, B, BV\}$. Moreover, since $RBV_p \subseteq AC$, Proposition 3.0.2 (f) yields $AC/RBV_p = AC$. Finally, $Lip/RBV_p \subseteq RBV_{2p}/RBV_p = \{\mathbb{0}\}$ by Proposition 3.0.2 (a) and Theorem 3.2.15 (b), and thus this gives us $Lip/RBV_p = \{\mathbb{0}\}$.
However, the class D/RBV_p and the analogous classes D/Lip and D/AC seem to be much more complex, and we do not know how they look like. In addition to the inclusions mentioned in Figure 3.1.1 at the end of Section 3.1, we only know that

$$D \cap \mathcal{B}_1 \subsetneq D/C \subseteq D/(C \cap BV) \subseteq D/AC \subseteq D/RBV_p \subseteq D/Lip \subseteq D,$$

and we believe that all of the given inclusions are strict.

The following tables summarize most of the multiplier classes under consideration. Table 3.2.1 compares the results from Section 3.1 to the results about Waterman spaces of Section 3.2.

Table 3.2.1: Multipliers in BV and ΛBV.

Jordan variation	Waterman variation
$C/BV = \{\mathbb{0}\}$	$C/\Lambda BV = \{\mathbb{0}\}$
$\Delta/BV = \{\mathbb{0}\}$	$\Delta/\Lambda BV = \{\mathbb{0}\}$
$D/BV = \{\mathbb{0}\}$	$D/\Lambda BV = \{\mathbb{0}\}$
$B/BV = B$	$B/\Lambda BV = B$
	$BV/\Lambda BV = BV \cap \mathcal{S}_c$
$BV/C = BV \cap \mathcal{S}_c$	$\Gamma BV/C = \Gamma BV \cap \mathcal{S}_c$
$BV/\Delta = \mathcal{S}_f$	$\Gamma BV/\Delta = \mathcal{S}_f$
$BV/D = \mathcal{S}_f$	$\Gamma BV/D = \mathcal{S}_f$
$BV/B = BV \cap \mathcal{S}_c$	$\Gamma BV/B = \Gamma BV \cap \mathcal{S}_c$
$BV/BV = BV$	$\Gamma BV/BV = \Gamma BV$
	$\Gamma BV/\Lambda BV = \Gamma BV$ for $\Gamma \preceq \Lambda$
	$\Gamma BV/\Lambda BV = \Gamma BV \cap \mathcal{S}_c$ for $\Gamma \not\preceq \Lambda$

Table 3.2.2 shows the new results from Section 3.2 about Wiener, Young and Riesz spaces.

Table 3.2.2: Multipliers in YBV_φ and RBV_p.

Young variation	Riesz variation ($1 < p, q < \infty$)
$C/YBV_\varphi = \{\mathbb{0}\}$	$C/RBV_p = C$
$\Delta/YBV_\varphi = \{\mathbb{0}\}$	$\Delta/RBV_p = \Delta$
$D/YBV_\varphi = \{\mathbb{0}\}$	$D/RBV_p = ???$
$B/YBV_\varphi = B$	$B/RBV_p = B$
$BV/YBV_\varphi = BV \cap \mathcal{S}_c$ for $\iota \not\preceq \varphi$	$BV/RBV_p = BV$
$YBV_\psi/C = YBV_\psi \cap \mathcal{S}_c$	$RBV_q/C = \{\mathbb{0}\}$
$YBV_\psi/\Delta = \mathcal{S}_f$	$RBV_q/\Delta = \{\mathbb{0}\}$
$YBV_\psi/D = \mathcal{S}_f$	$RBV_q/D = \{\mathbb{0}\}$
$YBV_\psi/B = YBV_\psi \cap \mathcal{S}_c$	$RBV_q/B = \{\mathbb{0}\}$
$YBV_\psi/BV = YBV_\psi$	$RBV_q/BV = \{\mathbb{0}\}$
$YBV_\psi/YBV_\varphi = YBV_\psi$ for $\psi \preceq \varphi$	$RBV_q/RBV_p = RBV_q$ for $q \leq p$
$YBV_\psi/YBV_\varphi = YBV_\psi \cap \mathcal{S}_c$ for $\psi \not\preceq \varphi$	$RBV_q/RBV_p = \{\mathbb{0}\}$ for $q > p$

Note that the results about Jordan, Waterman and Young variations are completely similar, whereas the results for the Riesz variation are slightly different. This is because functions in RBV_p for $p > 1$ are continuous. Moreover, similar to the class D/C we do not know how the classes D/RBV_p for $p > 1$ look like.

Finally, our last Table 3.2.3 below gives an overview about the other classes related to BV-type spaces.

Table 3.2.3: Multipliers in other classes.

Other classes ($1 < p, q < \infty$)		
$C/Lip = C$	$C/AC = C$	$RBV_q/Lip = RBV_q$
$\Delta/Lip = \Delta$	$\Delta/AC = \Delta$	$Lip/RBV_p = \{\mathbb{0}\}$
$D/Lip = ???$	$D/AC = ???$	$RBV_q/AC = \{\mathbb{0}\}$
$B/Lip = B$	$B/AC = B$	$AC/RBV_p = AC$
$BV/Lip = BV$	$BV/AC = BV$	$R/\Lambda BV = R$
$Lip/C = \{\mathbb{0}\}$	$AC/C = \{\mathbb{0}\}$	$\Lambda BV/R = \Lambda BV \cap \mathcal{S}_c$
$Lip/\Delta = \{\mathbb{0}\}$	$AC/\Delta = \{\mathbb{0}\}$	
$Lip/D = \{\mathbb{0}\}$	$AC/D = \{\mathbb{0}\}$	
$Lip/B = \{\mathbb{0}\}$	$AC/B = \{\mathbb{0}\}$	
$Lip/BV = \{\mathbb{0}\}$	$AC/BV = \{\mathbb{0}\}$	

Chapter 4

Linear Operators between BV-Spaces

The purpose of this chapter is to study several linear operators mainly in BV-spaces X and Y, where the symbols X and Y represent one of the spaces BV, WBV_p, YBV_φ, ΛBV or RBV_p introduced in Chapter 1. In detail we will consider

- the *multiplication operator* $M_g : X \to Y$, generated by a function $g : [0,1] \to \mathbb{R}$ and defined by

$$M_g x(t) = x(t)g(t), \tag{4.0.1}$$

- the *substitution operator* $S_g : X \to Y$, generated by a function $g : [0,1] \to [0,1]$ and defined by

$$S_g x(t) = x(g(t)), \tag{4.0.2}$$

- the *integral operator* $I_g : X \to Y$, generated by a function $g : [0,1] \times [0,1] \to \mathbb{R}$ and defined by

$$I_g x(t) = \int_0^1 g(t,s)x(s)\,\mathrm{d}s. \tag{4.0.3}$$

For all these operators we are particularly interested in analytic properties like acting conditions, continuity and compactness, but we will also investigate set-theoretic properties like injectivity, surjectivity and bijectivity.

4.1 Multiplication Operators

In this section we investigate the multiplication operator $M_g : X \to Y$, defined by

$$M_g x(t) := x(t)g(t) \quad \text{for } t \in [0,1],$$

where X and Y are linear spaces of real-valued functions on the interval $[0,1]$, and the generating function $g : [0,1] \to \mathbb{R}$ is given. In order to guarantee that M_g is well-defined, we have to make sure by imposing proper conditions on g that $M_g(X) \subseteq Y$.

That is, the product xg must belong to Y, whenever x belongs to X. Using the notation for multiplier sets

$$Y/X = \{g : [0,1] \to \mathbb{R} \mid xg \in Y \text{ for all } x \in X\}$$

introduced in (3.0.1) we have $M_g(X) \subseteq Y$ if and only if $g \in Y/X$.

As we have seen in the previous chapter, in some spaces, especially when $X = Y$, the classes Y/X are easy to find: For instance, we have seen in Section 3.1 that $B/B = B$ and $C/C = C$ which means for our multiplication operator nothing but $M_g(B) \subseteq B$ if and only if $g \in B$ and $M_g(C) \subseteq C$ if and only if $g \in C$.
Generally speaking, if $X = Y$ is closed under multiplication and contains $\mathbb{1}$ then $X/X = X$ by Proposition 3.0.2 (f) and therefore $M_g(X) \subseteq X$ if and only if $g \in X$. The spaces C and B clearly have both properties.
However, in other classes of functions which are not closed under addition or multiplication, a characterization of X/X or even Y/X can be much harder. As we have seen in Section 3.1 this is difficult especially if the class D of Darboux functions is involved. For instance, the class D/C satisfies the chain of inclusions

$$C \subsetneq \Delta \subsetneq D \cap \mathcal{B}_1 \subsetneq D/C \subsetneq D,$$

where $\mathcal{B}_1$ denotes the class of Baire-1 functions, but its exact characterization is - at least to our knowledge - unknown.
If X and Y are different spaces, the assumption $M_g(X) \subseteq Y$ can lead to a strong degeneracy of the generator g. Roughly speaking, this is true whenever X is "large" and Y is "small". For example, the classes C/B, C/D and D/B contain only $\mathbb{0}$ by Theorem 3.1.10. In terms of the multiplication operator this means that either of the inclusions $M_g(B) \subseteq C$, $M_g(D) \subseteq C$ and $M_g(B) \subseteq D$ is possible only for $g = \mathbb{0}$.

In this section we investigate M_g for the case when X and Y are some of the BV-type spaces BV, WBV_p, YBV_φ, ΛBV and RBV_p, introduced in Chapter 1. Since all five classes are closed under multiplication and contain the function $\mathbb{1}$, the multiplication operator from one such space into itself is always well-defined if and only if g itself is a member of that space. However, if X and Y are different spaces, the requirement of $M_g : X \to Y$ to be well-defined may again lead to some degeneracy. For instance, as we have seen in Corollary 3.2.3 and Theorem 3.2.15,

$$WBV_q/WBV_p = \begin{cases} WBV_q & \text{for } 1 \le p \le q, \\ WBV_q \cap \mathcal{S}_c & \text{for } 1 \le q < p, \end{cases} \tag{4.1.1}$$

$$RBV_q/RBV_p = \begin{cases} RBV_q & \text{for } 1 \le q \le p, \\ \{\mathbb{0}\} & \text{for } 1 \le p < q, \end{cases} \tag{4.1.2}$$

where $\mathcal{S}_c$ denotes the set of functions with countable support. According to the Theorems 3.2.1 and 3.2.7, similar relations hold in YBV_φ and ΛBV.

We are going to start our investigations by giving general criteria for injectivity, surjectivity and bijectivity for the multiplication operator $M_g : X \to Y$. Most of them can be expressed in terms of the support supp and supp_δ of the generating function g as defined in (1.1.8) and (1.1.9). Recall that if $\text{supp}(g)$ is uncountable, then $\text{supp}_\delta(g)$ is infinite for some $\delta > 0$. Conversely, if $\text{supp}_\delta(g)$ is countable for each $\delta > 0$, then $\text{supp}(g)$ is also countable as it is then a countable union of countable sets by (1.1.10).

Since M_g as a linear operator is injective if and only if its null space contains only the zero vector, we immediately obtain a criterion for injectivity.

Proposition 4.1.1. *For $g \in Y/X$, the operator $M_g : X \to Y$ is injective if and only if for each $x \in X \setminus \{\mathbb{0}\}$ there is some $t \in \text{supp}(g)$ such that $x(t) \neq 0$.*

This condition, however, is too broad and shows that an injectivity criterion in general does not only depend on g but also on X. In some cases, namely if the space X is sufficiently "large", the dependence on X is redundant. We give two such cases and introduce some terminology.

Definition 4.1.2. We say that a linear space X of real-valued functions on $[0, 1]$

- *separates points* if for each $t \in [0, 1]$ there is some $x \in X$ such that $x(t) \neq 0$,
- *strongly separates points* if X contains all characteristic functions of singletons,
- *uniformly separates points* if $X \subseteq C$ and if for each $t \in [0, 1]$ and each $\delta > 0$ there is some $x \in X$ such that $t \in \text{supp}(x) \subseteq [t - \delta, t + \delta]$.

Note that each space which separates points uniformly or strongly also separates points. Other relations, however, do not hold. For instance, the spaces C, B, BV, ΛBV, WBV_p, YBV_φ and RBV_p separate points. However, the spaces B, BV, WBV_p, YBV_φ and ΛBV separate points strongly, but not uniformly, whereas the spaces C and RBV_p separate points uniformly, but not strongly. Finally, the space of constant functions only separates points, but neither strongly nor uniformly.
With these definitions at hand, we obtain

Proposition 4.1.3. *For $g \in Y/X$ the following statements are true.*

(a) If $\text{supp}(g) = [0, 1]$, then the operator M_g is injective.

(b) If X strongly separates points, then the operator $M_g : X \to Y$ is injective if and only if $\text{supp}(g) = [0, 1]$.

(c) If X uniformly separates points, then the operator $M_g : X \to Y$ is injective if and only if $\overline{\text{supp}(g)} = [0, 1]$.

(d) If Y separates points, then the operator $M_g : X \to Y$ is surjective if and only if $\text{supp}(g) = [0, 1]$ and $1/g \in X/Y$.

(e) If Y separates points, then the operator $M_g : X \to Y$ is injective if it is surjective.

Proof. (a) Clearly, $xg = \mathbb{0}$ for $x \in X$ (if and) only if $x = \mathbb{0}$, and so the null space of M_g only contains $\mathbb{0}$ which implies the desired injectivity.

(b) Let M_g be injective and fix $t \in [0,1]$. Then $x := \chi_{\{t\}} \in X$, and since $x \neq \mathbb{0}$ and M_g is injective, $xg \neq \mathbb{0}$. But this is possible only if $g(t) \neq 0$. Since t was arbitrary, $g(t) \neq 0$ for all $t \in [0,1]$, i.e. $\operatorname{supp}(g) = [0,1]$.
Conversely, if $\operatorname{supp}(g) = [0,1]$, then M_g is injective by (a).

(c) Let M_g be injective and fix $t \in [0,1]$ such that $g(t) = 0$. For each $\delta > 0$ there is some $x \in X$ such that $t \in \operatorname{supp}(x) \subseteq [t-\delta, t+\delta]$. In particular, $x \neq \mathbb{0}$, and since M_g is injective, $xg \neq \mathbb{0}$. But since $\operatorname{supp}(xg) \subseteq \operatorname{supp}(x) \subseteq [t-\delta, t+\delta]$, there is some $s \in [t-\delta, t+\delta] \cap [0,1]$ such that $g(s) \neq 0$ and hence $s \in \operatorname{supp}(g)$. This shows that $\operatorname{supp}(g)$ is dense in $[0,1]$.
Conversely, if $\operatorname{supp}(g)$ is dense in $[0,1]$, then M_g is injective. To see this, fix $t \in [0,1]$ and $x \in X$ such that $xg = \mathbb{0}$. If $g(t) \neq 0$, then $x(t) = 0$. If $g(t) = 0$, then since $\operatorname{supp}(g)$ is dense in $[0,1]$, there is a sequence (t_n) in $[0,1]$ converging to t such that $g(t_n) \neq 0$ for each $n \in \mathbb{N}$. Then $x(t_n) = 0$ for each $n \in \mathbb{N}$ and due to continuity, $x(t) = 0$. Consequently, $x = \mathbb{0}$.

(d) Let M_g be surjective and fix $t \in [0,1]$. Since Y separates points there is some $y \in Y$ such that $y(t) \neq 0$. Since M_g is surjective we find some $x \in X$ such that $xg = y$; in particular, $g(t) \neq 0$, and consequently $g(t) \neq 0$ for all $t \in [0,1]$, as t was arbitrary. This shows $\operatorname{supp}(g) = [0,1]$. For any $y \in Y$ we again find $x \in X$ such that $xg = y$ and hence $y/g = x \in X$ which proves $1/g \in X/Y$.
For the converse assume that $\operatorname{supp}(g) = [0,1]$ and $1/g \in X/Y$. For $y \in Y$ the function $x := y/g$ belongs to X and satisfies $xg = y$, i.e. M_g is surjective.

(e) follows instantaneously from (a) and (d). ■

We remark that if $M_g : X \to X$ is surjective (or even bijective) and X does *not* separate points, then g may have zeros.

Example 4.1.4. Let $X = \{x \in BV \mid x(0) = 0\}$ and $g = \chi_{(0,1]}$. Then M_g maps X into itself and is bijective, because $M_g x = x$ for all $x \in X$. However, g obviously has a zero at $t = 0$. ♢

Note that this example is not contradictory to Proposition 4.1.3 (d), because X does not separate points. Even if the operator M_g is considered to be an operator from X into BV instead of X itself, then it does not contradict Proposition 4.1.3 (d), because in this case, M_g is not surjective anymore.

We now apply Proposition 4.1.3 to the spaces of our interest and obtain as a consequence the following result. Recall that according to our agreement after Definition 1.2.24 the symbol RBV_p implicitly means $p > 1$, unless otherwise stated. This implies that functions in RBV_p are automatically continuous, whereas functions in $RBV_1 = BV$ are usually not.

Corollary 4.1.5. *For $g \in Y/X$ the following statements are true.*

(a) If X is any of the spaces $B, BV, WBV_p, YBV_\varphi$ or ΛBV, then $M_g : X \to Y$ is injective if and only if $\operatorname{supp}(g) = [0,1]$.

(b) If X is any of the spaces C or RBV_p, then $M_g : X \to Y$ is injective if and only if $\overline{\operatorname{supp}(g)} = [0,1]$.

(c) If $Y \subseteq X \subseteq B$ and Y is any of the spaces $B, C, BV, WBV_p, YBV_\varphi, \Lambda BV$ or RBV_p, then $M_g : X \to Y$ is surjective if and only if $\inf_{t\in[0,1]} |g(t)| > 0$. *In this case, M_g is also injective.*

Proof. (a) follows from Proposition 4.1.3 (b), and (b) follows from part (c) of Proposition 4.1.3.

For (c) assume that M_g is surjective. Then from Proposition 4.1.3 (d) we obtain that $\operatorname{supp}(g) = [0,1]$ and $1/g \in X/Y$. Note that $\mathbb{1}$ belongs to all of the spaces B, BV, ΛBV, BV_φ, RBV_p and C, and we obtain $X/Y \subseteq X \subseteq B$ from Proposition 3.0.2 (c). Thus, $1/g$ is bounded, and this is possible only if $\inf_{t\in[0,1]} |g(t)| > 0$.
Conversely, assume that $\inf_{t\in[0,1]} |g(t)| > 0$. Then $\operatorname{supp}(g) = [0,1]$ and $1/g \in Y$. Since Y is closed under multiplication we have $Y \subseteq X/Y$ by Proposition 3.0.2 (e) and hence $1/g \in X/Y$. Again from Proposition 4.1.3 (d) we obtain that M_g is surjective, and that it is also injective follows from Proposition 4.1.3 (e). ■

In particular, for the special case $X = Y$, we have

Corollary 4.1.6. *The following statements are true.*

(a) If $X = Y$ is any of the spaces B, BV, WBV_p, YBV_φ or ΛBV, then $M_g : X \to X$ is injective if and only if $\operatorname{supp}(g) = [0,1]$.

(b) If $X = Y$ is any of the spaces C or RBV_p, then $M_g : X \to X$ is injective if and only if $\overline{\operatorname{supp}(g)} = [0,1]$.

(c) If $X = Y$ is any of the spaces B, C, BV, WBV_p, YBV_φ, ΛBV, or RBV_p, then $M_g : X \to X$ is surjective if and only if $\inf_{t\in[0,1]} |g(t)| > 0$.

(d) If $X = Y$ is any of the spaces B, C, BV, WBV_p, YBV_φ, ΛBV, or RBV_p, then $M_g : X \to X$ is bijective if and only if $\inf_{t\in[0,1]} |g(t)| > 0$. *In this case, $M_g^{-1} = M_{1/g}$.*

Note that for these spaces, a surjective multiplication operator $M_g : X \to X$ is always automatically injective and hence bijective. In this case, for the inverse operator the relation

$$M_g^{-1} = M_{1/g} \tag{4.1.3}$$

holds. However, this is no specialty of the particular space X, but of the operator itself. Indeed, if X is any space of real-valued functions on $[0,1]$ and $M_g : X \to X$ is

surjective, then it is always injective. To see this suppose by contradiction that there is some $x \in X$ with $M_g x = \mathbb{0}$, but $x \neq \mathbb{0}$. This means that there is some $t \in [0,1]$ such that $x(t) \neq 0$. Due to $xg = \mathbb{0}$ we get $g(t) = 0$. But since M_g is assumed to be surjective, we must find some $z \in X$ such that $M_g z = x$; in particular, $0 = z(t)g(t) = x(t) \neq 0$, a contradiction.
However, such an operator, even if it is bijective, does not have to satisfy the identity (4.1.3) anymore, because in the situation of Example 4.1.4 the function $1/g$ is not even defined.

We now turn to analytic properties of the multiplication operator $M_g : X \to X$. Here, we are particularly interested in continuity and compactness for X being one of the spaces BV, WBV_p, YBV_φ, ΛBV or RBV_p.
Recall that for a linear operator $L : X \to Y$ between two normed linear spaces $(X, \|\cdot\|_X)$ and $(Y, \|\cdot\|_Y)$ the operator norm is defined by

$$\|L\|_{X\to Y} = \sup_{x\neq \mathbb{0}} \frac{\|Lx\|_Y}{\|x\|_X} = \sup_{\|x\|_X=1} \|Lx\|_Y \,.$$

Then L is bounded if and only if $\|L\|_{X\to Y}$ is finite. Note that in these cases, as a linear operator, L is then even globally Lipschitz continuous.

We then have the following quite general result.

Proposition 4.1.7. *Let $(X, \|\cdot\|_X)$ and $(Y, \|\cdot\|_Y)$ be normed linear spaces of real-valued functions on $[0,1]$, where $X \hookrightarrow Y$ with embedding constant $c > 0$ and $\mathbb{1} \in X$ with $\|\mathbb{1}\|_X = 1$, and suppose that Y is a normalized algebra. Then, for $g \in Y/X$, the operator $M_g : X \to Y$ is bounded with*

$$\|g\|_Y \leq \|M_g\|_{X\to Y} \leq c\,\|g\|_Y \,.$$

Proof. Since $\mathbb{1} \in X$ we have $\|M_g\|_{X\to Y} \geq \|M_g \mathbb{1}\|_Y \, \|\mathbb{1}\|_X = \|g\|_Y$ which shows the first inequality.
Since $X \hookrightarrow Y$ with embedding constant $c > 0$, that is, $\|x\|_Y \leq c\,\|x\|_X$ for all $x \in X$, we have $\|xg\|_Y \leq \|x\|_Y \|g\|_Y \leq c\,\|x\|_X \|g\|_Y$, where we have used that Y is a normalized algebra in the first estimate. From this follows $\|M_g\|_{X\to Y} \leq c\,\|g\|_Y$ which shows the second inequality.
Finally, we have $g = g\mathbb{1} \in Y$, and this ensures that the upper and lower bounds for $\|M_g\|_{X\to Y}$ are finite. ∎

For the special case that $X = Y$ is one of our BV-spaces, we have $X/X = X$ and hence, since all those spaces contain the function $\mathbb{1}$ that has norm 1, we obtain the following

Corollary 4.1.8. *Let X be one of the spaces BV, WBV_p, YBV_φ, ΛBV or RBV_p. Then for $g \in X$ the operator $M_g : X \to X$ is well-defined and bounded with*

$$\|M_g\|_{X\to X} = \|g\|_X \,.$$

Proposition 4.1.7 allows us to compute the operator norm of M_g also for different combinations of BV-spaces. Let us give some sample results in this direction. For instance, we have for $1 \leq p \leq q < \infty$ that $WBV_p \hookrightarrow WBV_q$ with embedding constant 1, and thus Proposition 4.1.7 says

$$\|M_g\|_{WBV_p \to WBV_q} = \|g\|_{WBV_q} . \tag{4.1.4}$$

Similarly, we obtain from Proposition 4.1.7 for the Riesz spaces for $1 \leq q \leq p < \infty$,

$$\|M_g\|_{RBV_p \to RBV_q} = \|g\|_{RBV_q} . \tag{4.1.5}$$

However, the cases $p > q$ for the Wiener spaces and $q > p$ for the Riesz spaces are not covered by Proposition 4.1.7, even though the estimates (4.1.4) and (4.1.5) remain valid. Indeed, if $p > q$, then Corollary 3.2.3 yields that any $g \in WBV_q/WBV_p$ degenerates to a function in $WBV_q \cap \mathcal{S}_c$. Thus, for $x \in WBV_p$ we have $x \in B$ and

$$\|g\|_{WBV_q} \leq \|xg\|_{WBV_q} \leq \|x\|_\infty \|g\|_{WBV_q} \leq \|x\|_{WBV_p} \|g\|_{WBV_q}$$

and hence (4.1.4).
The case $q > p$ for the Riesz spaces is much more boring, because Theorem 3.2.15 then says that any $g \in RBV_q/RBV_p$ must be equal to $\mathbb{0}$, and then (4.1.5) is clearly true.

Other cases that are not covered by Proposition 4.1.7 either are sometimes also known. For instance, one can show with the help of Hölder's inequality that

$$\|M_g\|_{L_p \to L_q} = \|g\|_{L_{pq/(p-q)}} \qquad \text{for } p \geq q.$$

However, the operator M_g is not always bounded.

Example 4.1.9. Consider the space $C_0^1 := \{x \in C^1 \mid x(0) = 0\}$ equipped with the norm $\|\cdot\|_\infty$, and the function $g : [0,1] \to \mathbb{R}$, defined by

$$g(t) = \begin{cases} \dfrac{1}{t} & \text{for } 0 < t \leq 1, \\ 0 & \text{for } t = 0. \end{cases}$$

Then $g \in L_1/C_0^1$, because for $x \in C_0^1$ the function

$$x(t)g(t) = \begin{cases} \dfrac{x(t)}{t} & \text{for } 0 < t \leq 1, \\ 0 & \text{for } t = 0 \end{cases}$$

satisfies

$$\|M_g x\|_{L_1} = \int_0^1 |x(t)g(t)| \, \mathrm{d}t = \int_0^1 \frac{1}{t} \left| \int_0^t x'(s) \, \mathrm{d}s \right| \mathrm{d}t \leq \|x'\|_\infty .$$

Consequently, the operator $M_g : C_0^1 \to L_1$ is well-defined. However, M_g is not bounded. Indeed, the functions $x_n : [0,1] \to \mathbb{R}$, for $n \in \mathbb{N}$ defined by

$$x_n(t) = \begin{cases} 2nt - n^2t^2 & \text{for } 0 < t \leq 1/n, \\ 1 & \text{for } 1/n < t \leq 1, \end{cases}$$

belong to C_0^1 with $\|x_n\|_\infty = 1$ for all $n \in \mathbb{N}$, but the functions $M_g x_n$ satisfy

$$\|M_g x_n\|_{L_1} = \int_0^1 |M_g x_n(t)|\,\mathrm{d}t = \int_0^{1/n} \left(2n - n^2 t\right) \mathrm{d}t + \int_{1/n}^1 \frac{1}{t}\,\mathrm{d}t = \frac{3}{2} + \log(n)$$

and hence form an unbounded sequence in L_1. ♢

Note that the functions x_n constructed in Example 4.1.9 do not contradict Proposition 4.1.7, because $\mathbb{1} \notin C_0^1$, even though $C_0^1 \hookrightarrow L_1$ and $(C_0^1, \|\cdot\|_\infty)$ is a normalized algebra.

Let us now pass to compactness. A first criterion which yields compactness of a linear operator is that its range is finite dimensional. The following theorem states that the multiplication operator has finite dimensional range if and only if $\mathrm{supp}(g)$ is finite. This was first proven for $X = Y = BV$ and $X = Y = WBV_p$ in the recent papers [17] and [16], respectively, but it is also true in a more general setting. In fact, with the same idea we prove that the dimension of the range of M_g coincides with the number $\#\mathrm{supp}(g)$ of points at which g is not zero.

Theorem 4.1.10. *Let X be a linear space of real-valued functions on $[0,1]$ which separates points strongly, and let $g \in Y/X$. Then for $M_g : X \to Y$ we have* $\dim \mathrm{Im}(M_g) = \#\mathrm{supp}(g)$.

Proof. We first show the inequality $\dim \mathrm{Im}(M_g) \geq \#\mathrm{supp}(g)$ which is obviously true for $g = \mathbb{0}$. Thus we assume that $g \neq \mathbb{0}$ which implies $\#\mathrm{supp}(g) \geq 1$, and fix $n \in \mathbb{N}$ with $n \leq \#\mathrm{supp}(g)$. Then there are pairwise distinct numbers $t_1, \ldots, t_n \in \mathrm{supp}(g)$; in particular, $g(t_j) \neq 0$ for $1 \leq j \leq n$. The functions $y_j := \chi_{\{t_j\}} g$ belong to $\mathrm{Im}(M_g)$, since X contains all characteristic functions of singletons. For $j \in \{1, \ldots, n\}$ let $\lambda_j \in \mathbb{R}$ be so that

$$\sum_{j=1}^n \lambda_j y_j = \mathbb{0}.$$

By evaluating this equation at $t = t_k$ for each $k \in \{1, \ldots, n\}$, we get that

$$0 = \sum_{j=1}^n \lambda_j y_j(t_k) = \lambda_k \chi_{\{t_k\}} g(t_k) = \lambda_k g(t_k) \quad \text{for } 1 \leq k \leq n,$$

which implies $\lambda_k = 0$ for $1 \leq k \leq n$. Thus, $\{y_1, \ldots, y_n\}$ is a linearly independent subset of $\mathrm{Im}(M_g)$; in particular, $\dim \mathrm{Im}(M_g) \geq n$. Since this is true for each $n \leq \#\mathrm{supp}(g)$, we obtain $\dim \mathrm{Im}(M_g) \geq \#\mathrm{supp}(g)$.
In order to show the reverse inequality $\dim \mathrm{Im}(M_g) \leq \#\mathrm{supp}(g)$, we can assume that $\#\mathrm{supp}(g) < \infty$, because otherwise the inequality is clearly true. This time, let $n := \#\mathrm{supp}(g)$ and write $\mathrm{supp}(g) = \{t_1, \ldots, t_n\}$. Since X contains all characteristic functions of singletons, the functions $y_j := \chi_{\{t_j\}} g$ for $j \in \{1, \ldots, n\}$ form a subset of $\mathrm{Im}(M_g)$. Let $y \in \mathrm{Im}(M_g)$, that is, there is some $x \in X$ such that $y = xg$. Then

$$y = xy = \sum_{t \in [0,1]} x(t) g(t) \chi_{\{t\}} = \sum_{j=1}^n x(t_j) y_j$$

which shows that the linear hull of $\{y_1, \ldots, y_n\}$ contains y. But since y was chosen arbitrarily, it contains the entire range $\mathrm{Im}(M_g)$; in particular, $\dim \mathrm{Im}(M_g) \leq n = \#\mathrm{supp}(g)$, and this completes the proof. ■

Note that we cannot drop the word "strongly" in Theorem 4.1.10.

Example 4.1.11. Let $X = \mathbb{1}\mathbb{R}$ be the space of constant functions, let $Y = C$, and consider $M_g : X \to Y$, generated by $g(t) = t$. Then $\mathrm{supp}(g) = (0, 1]$ is even uncountable, but

$$\mathrm{Im}(M_g) = \Big\{y : [0, 1] \to \mathbb{R} \mid y(t) = at, a \in \mathbb{R}\Big\} = \mathrm{Span}\Big(\{g\}\Big)$$

is a one-dimensional subspace of C. ◇

We are now in position to prove the main result which provides a full characterization of those g that generate a compact multiplication operator $M_g : X \to X$. This result has been proven for the special case when $X = Y = WBV_p$ for $1 \leq p < \infty$ also in [16, 17].

Theorem 4.1.12. *Let X be one of the spaces BV, WBV_p, YBV_φ with $\varphi \in \delta_2$, or ΛBV, and let $g \in X$. Then $M_g : X \to X$ is compact if and only if* supp(g) *is countable.*

Proof. First note that for all the spaces under consideration we have $X/X = X$ and hence M_g is well-defined.
We first prove the theorem for $X = YBV_\varphi$ and some fixed Young function φ. This also implies the result for $X = BV$ and $X = WBV_p$ for $1 < p < \infty$. To this end, we begin by assuming that $\mathrm{supp}(g)$ is countable and show that $M_g : YBV_\varphi \to YBV_\varphi$ is compact. If $\mathrm{supp}(g)$ is finite, then M_g has finite dimensional range by Theorem 4.1.10 and hence is compact. If $\mathrm{supp}(g)$ is infinite, we can write $E := \{t_1, t_2, t_3, \ldots\} = \mathrm{supp}(g) \subseteq [0, 1]$. Setting $E_n := \{t_1, t_2, \ldots, t_n\}$, the functions $g_n := \chi_{E_n} g$ have finite support and thus belong to YBV_φ. By Theorem 4.1.10 the operators $M_{g_n} : YBV_\varphi \to YBV_\varphi$ have finite dimensional range and hence are compact. Moreover, since $g \in YBV_\varphi \subseteq B$ and $\varphi \in \delta_2$, the variation $\mathrm{Var}_\varphi(g)$ is finite. Finally, $g_n - g = -\chi_{E \setminus E_n} g$. We obtain with the help of Proposition 1.2.10,

$$\mathrm{Var}_\varphi\left(\frac{g_n - g}{2}\right) \leq \sum_{j=n+1}^{\infty} \varphi\Big(|g(t_j)|\Big). \tag{4.1.6}$$

Since, again by Proposition 1.2.10,

$$\sum_{j=n+1}^{\infty} \varphi\Big(|g(t_j)|\Big) \leq \sum_{j=1}^{\infty} \varphi\Big(|g(t_j)|\Big) \leq \mathrm{Var}_\varphi(g) < \infty,$$

the right hand side of (4.1.6) and therefore also $\mathrm{Var}_\varphi\Big((g_n - g)/2\Big)$ and hence $\mathrm{Var}_\varphi(g_n - g)$ (due to $\varphi \in \delta_2$) go to 0 as $n \to \infty$. By Proposition 1.2.15 (b), also $\mathfrak{M}(g_n - g) \to 0$ as $n \to \infty$. Additionally, by Proposition 1.2.10,

$$\varphi\Big(\|g_n - g\|_\infty\Big) = \varphi\left(\sup_{j>n} |g(t_j)|\right) = \sup_{j>n} \varphi\Big(|g(t_j)|\Big) \leq \sum_{j=n+1}^{\infty} \varphi\Big(|g(t_j)|\Big) \leq \mathrm{Var}_\varphi(g_n - g).$$

Consequently, $\varphi\big(\|g_n - g\|_\infty\big)$, hence $\|g_n - g\|_\infty$ and eventually $\|g_n - g\|_{YBV_\varphi}$ go to 0 as $n \to \infty$. Finally, by Corollary 4.1.8,

$$\|M_{g_n} - M_g\|_{YBV_\varphi} = \|g_n - g\|_{YBV_\varphi} \longrightarrow 0 \quad \text{as } n \to \infty,$$

and so M_g is compact, as well.

For the converse assume now that $\text{supp}(g)$ is uncountable which implies that $\text{supp}_\delta(g)$ is infinite for some $\delta > 0$. Then there is a sequence (t_n) in $\text{supp}_\delta(g)$ of pairwise disjoint points; in particular, $|g(t_n)| \geq \delta$ for all $n \in \mathbb{N}$. The functions $x_n := \chi_{\{t_n\}}$ form a bounded sequence in YBV_φ, but for $m, n \in \mathbb{N}$ with $m \neq n$ we have

$$\|M_g x_m - M_g x_n\|_{YBV_\varphi} \geq \|g \cdot (x_m - x_n)\|_\infty \geq \Big|g(t_n)\Big(x_m(t_n) - x_n(t_n)\Big)\Big| = |g(t_n)| \geq \delta$$

and hence $(M_g x_n)$ cannot have a convergent subsequence. Thus, M_g cannot be compact.

We now mimic this proof for $X = \Lambda BV$, where $\Lambda = (\lambda_j)$ is a fixed Waterman sequence; the ideas are the same. We first assume that $\text{supp}(g)$ is countable. If $\text{supp}(g)$ is finite, then M_g has finite dimensional range by Theorem 4.1.10 and hence is compact. If $\text{supp}(g)$ is infinite, we can write $E := \{t_1, t_2, t_3, \ldots\} = \text{supp}(g) \subseteq [0,1]$. Setting $E_n := \{t_1, t_2, \ldots, t_n\}$, the functions $g_n := \chi_{E_n} g$ have finite support and thus belong to ΛBV. By Theorem 4.1.10 the operators $M_{g_n} : \Lambda BV \to \Lambda BV$ have finite dimensional range and hence are compact. Moreover, $g_n - g = -\chi_{E \setminus E_n} g$, and thus by Proposition 1.2.20,

$$\text{Var}_\Lambda(g_n - g) \leq 2 \sup_\sigma \sum_{j=n+1}^{\infty} \lambda_{\sigma(j)} |g(t_j)|, \tag{4.1.7}$$

where the supremum is taken over all permutations σ of $\mathbb{N}$. Since

$$\sup_\sigma \sum_{j=n+1}^{\infty} \lambda_{\sigma(j)} |g(t_j)| \leq \sup_\sigma \sum_{j=1}^{\infty} \lambda_{\sigma(j)} |g(t_j)| \leq \text{Var}_\Lambda(g),$$

again by Proposition 1.2.20, the right hand side of (4.1.7) and therefore also $\text{Var}_\Lambda(g_n - g)$ goes to 0 as $n \to \infty$. Additionally,

$$\|g_n - g\|_\infty = \sup_{j>n} |g(t_j)| \leq \lambda_1^{-1} \sup_\sigma \sum_{j=n+1}^{\infty} \lambda_{\sigma(j)} |g(t_j)| \leq \lambda_1^{-1} \text{Var}_\varphi(g_n - g).$$

Consequently, $\|g_n - g\|_\infty$ and eventually $\|g_n - g\|_{\Lambda BV}$ go to 0 as $n \to \infty$. Finally, by Proposition 4.1.7,

$$\|M_{g_n} - M_g\|_{\Lambda BV} = \|g_n - g\|_{\Lambda BV} \longrightarrow 0 \quad \text{as } n \to \infty,$$

and so M_g is compact as well.

For the converse assume now that $\text{supp}(g)$ is uncountable which implies that $\text{supp}_\delta(g)$ is infinite for some $\delta > 0$. Then there is a sequence (t_n) in $\text{supp}_\delta(g)$ of pairwise disjoint points; in particular, $|g(t_n)| \geq \delta$ for all $n \in \mathbb{N}$. The functions $x_n := \chi_{\{t_n\}}$ form a bounded sequence in ΛBV, but for $m, n \in \mathbb{N}$ with $m \neq n$ we have

$$\|M_g x_m - M_g x_n\|_{\Lambda BV} \geq \|g \cdot (x_m - x_n)\|_\infty \geq \Big|g(t_n)\Big(x_m(t_n) - x_n(t_n)\Big)\Big| = |g(t_n)| \geq \delta > 0$$

and hence $(M_g x_n)$ cannot have a convergent subsequence. Thus, M_g cannot be compact. ■

For the Riesz spaces RBV_p we have a similar result. However, since each function in RBV_p is continuous, compactness of M_g leads to a stronger degeneracy.

Theorem 4.1.13. *For $g \in RBV_p$, the operator $M_g : RBV_p \to RBV_p$ is compact if and only if $g = \mathbb{0}$.*

Proof. If $g = \mathbb{0}$, the operator M_g is clearly compact. We now assume that g is not identically zero. Since g is continuous, there is some interval $[a, b] \subseteq [0, 1]$ with $a < b$ such that $g(t) \neq 0$ for all $t \in [a, b]$. The set

$$K := \Big\{x \in RBV_p \mid \forall t \in [0,1] \backslash [a,b] : x(t) = 0\Big\}$$

is a closed infinite dimensional subspace of RBV_p and hence complete. Obviously, $M_g(K) \subseteq K$, and we have in fact $M_g(K) = K$. To see this fix $y \in K$ and define $x : [0, 1] \to \mathbb{R}$ by

$$x(t) = \begin{cases} y(t)/g(t) & \text{for } t \in [a, b], \\ 0 & \text{for } t \in [0, 1] \backslash [a, b]. \end{cases}$$

Then clearly $M_g x = xg = y$. Since g is continuous on the compact set $[a, b]$, it is bounded away from zero which ensures $\mathrm{RVar}_p(x) < \infty$. Thus, the restriction $M_g|_K : K \to K$ is surjective and hence not compact, and so M_g is also not compact. ■

Note that Theorem 4.1.13 also provides an analogue for Theorem 4.1.10 in RBV_p: A multiplication operator $M_g : RBV_p \to RBV_p$ has finite dimensional range if and only if its generator degenerates to $g = \mathbb{0}$.

Recall that the essential norm of a bounded linear operator A between two Banach spaces X and Y is defined by

$$\|A\|_e := \inf\Big\{\, \|A - K\|_{X \to Y} \mid K : X \to Y \text{ linear and compact}\Big\} \tag{4.1.8}$$

and measures the distance from A to the closed subspace of compact linear operators. In particular, $\|A\|_e = 0$ if and only if A itself is compact. Our previous discussions suggest that in $X = Y = BV$ the equality

$$\|M_g\|_e = \inf\Big\{\delta > 0 \mid \mathrm{supp}_\delta(g) \text{ is finite}\Big\} \tag{4.1.9}$$

holds. We were not able to decide whether this conjecture is true. However, some partial results are possible. Recall that the *right regularization* $g^\#$ of a function $g \in BV$ is defined by

$$g^\#(t) = \begin{cases} \lim\limits_{s \to t+} g(s) & \text{for } t \in [0,1), \\ g(1) & \text{for } t = 1. \end{cases}$$

Then $\operatorname{Var}(g^\#) \leq \operatorname{Var}(g)$, and $h := g - g^\#$ belongs to BV and has countable support, because $g^\#$ differes by g only at the (at most countably many) points of discontinuity of g. Thus, by Theorem 4.1.12, $M_h : BV \to BV$ is compact. Consequently,

$$\|M_g\|_e = \|M_{g-h+h}\|_e = \left\|M_{g^\#} + M_h\right\|_e \leq \left\|M_{g^\#}\right\|_e + \|M_h\|_e = \left\|M_{g^\#}\right\|_e.$$

This and interchanging the roles of g and $g^\#$ show

$$\|M_g\|_e = \left\|M_{g^\#}\right\|_e.$$

We claim that the essential norm of M_g satisfies the upper estimate

$$\|M_g\|_e \leq \left\|g^\#\right\|_{BV} \tag{4.1.10}$$

and, if g is bounded away from zero, also the lower estimate

$$\|M_g\|_e \geq \left\|1/g^\#\right\|_{BV}^{-1}. \tag{4.1.11}$$

In order to show (4.1.10) we obtain similarly as above and with the help of Corollary 4.1.8,

$$\begin{aligned} \|M_g\|_e = \left\|M_{g^\#}\right\|_e \leq \left\|M_{g^\#}\right\|_{BV \to BV} &= \|M_g - M_h\|_{BV \to BV} = \|M_{g-h}\|_{BV \to BV} \\ = \|g - h\|_{BV} = \left\|g^\#\right\|_{BV}. \end{aligned}$$

For the proof of (4.1.11) we assume that g is bounded away from zero. Then $g^\#$ is also bounded away from zero and hence by Corollary 4.1.6 (d) generates a bijective operator $M_{g^\#} : BV \to BV$ with inverse $M_{g^\#}^{-1} = M_{1/g^\#}$. For any compact operator $K : BV \to BV$ we must have, again with Corollary 4.1.8,

$$\left\|M_{g^\#} - K\right\|_{BV \to BV} \geq \frac{1}{\left\|M_{g^\#}^{-1}\right\|_{BV \to BV}} = \frac{1}{\left\|M_{1/g^\#}\right\|_{BV \to BV}} = \frac{1}{\|1/g^\#\|_{BV}},$$

since a compact operator in an infinite dimensional space cannot be invertible. The estimate (4.1.11) now follows by taking the infimum over all compact operators K.

It is clear that equality holds simultaneously in (4.1.10) and (4.1.11) if and only if $g^\#$ is constant. In this case, the conjecture (4.1.9) is indeed true, since then g is of the form $g(t) = c + h(t)$, where $c \equiv g^\#$ is the constant and $h \in BV$ is a function with countable support. But since $h \in BV$, we must have that $\operatorname{supp}_\delta(h)$ is finite for all $\delta > 0$ and hence the quantity in (4.1.9) is just equal to c which then coincides with the bounds given in (4.1.10) and (4.1.11). We illustrate this in the following example.

Example 4.1.14. Let (r_n) be any sequence of pairwise distinct numbers in $(0,1)$. For fixed $c \geq 0$ define g on $[0,1]$ by

$$g(t) = \begin{cases} c + 1/n^2 & \text{for } t = r_n, \\ c & \text{otherwise.} \end{cases}$$

Then $g \in BV$ with $g^{\#} \equiv c$. If $c = 0$, then g has countable support, and hence M_g is compact by Theorem 4.1.12 with $\|M_g\|_e = 0$. For $c > 0$, the function $g^{\#} \equiv c$ is bounded away from zero, and the quantities (4.1.10) and (4.1.11) are all equal to c, i.e. $\|M_g\|_e = c$. Moreover, for arbitrary $c \geq 0$ the set

$$\operatorname{supp}_\delta(g) = \{r_n \mid 1/n^2 \geq \delta - c, n \in \mathbb{N}\}$$

is finite for each $\delta > c$ and becomes infinite for $\delta \leq c$, and thus the quantity given in (4.1.9) is equal to c, as well. So in this case our conjecture is true. ◇

If, however, g is not "essentially constant", then the bounds in (4.1.10) and (4.1.11) may drift apart the closer g comes to zero.

Example 4.1.15. For fixed $\alpha > 0$, the function $g(t) = t + \alpha$ for $t \in [0,1]$ is continuous, bounded away from zero and of bounded variation with $g = g^{\#}$. From (4.1.10) and (4.1.11) we get

$$\frac{\alpha(\alpha+1)}{2+\alpha} = \frac{1}{\|1/g^{\#}\|_{BV}} \leq \|M_g\|_e \leq \left\|g^{\#}\right\|_{BV} = \alpha + 2.$$

Consequently, since

$$\lim_{\alpha \to 0+} \frac{1}{\|1/g^{\#}\|_{BV}} = 0 \quad \text{and} \quad \lim_{\alpha \to 0+} \left\|g^{\#}\right\|_{BV} = 2,$$

the "gap" between (4.1.10) and (4.1.11) becomes 2 for our functions g as $\alpha \to 0+$. However, we clearly have

$$\inf\{\delta > 0 \mid \operatorname{supp}_\delta(g) \text{ is finite}\} = \alpha + 1,$$

which sits almost in the middle between (4.1.10) and (4.1.11). ◇

4.2 Substitution Operators

In this section we investigate the substitution operator $S_g : X \to Y$, generated by some function $g : [0,1] \to [0,1]$ and defined by

$$S_g x(t) = x(g(t)) \qquad \text{for } 0 \leq t \leq 1,$$

where X and Y are linear spaces of real-valued functions on $[0,1]$. Although S_g is a linear operator, not so much is known about it, even in case $X = Y$, particularly if $X = Y$ is one of the spaces BV, WBV_p, YBV_φ, ΛBV or RBV_p which have been introduced in Chapter 1. But since S_g is not the center piece of this thesis, we only give a brief discussion.

First, $S_g : X \to X$ is well-defined if and only if $x \circ g$ belongs to X whenever x does, and using the notation introduced in (2.3.2) this is equivalent to $g \in \Sigma(X)$. For instance, we have seen at the beginning of Section 2.3 that $\Sigma(B)$ contains any function $g : [0,1] \to [0,1]$, while $\Sigma(C)$ consists of all continuous functions $g : [0,1] \to [0,1]$.

For our BV-type spaces things are more difficult. As we have already mentioned prior to Example 2.3.7, the set $\Sigma(BV)$ is precisely the class of those functions $g : [0,1] \to [0,1]$ which are pseudo-monotone, that is, $g : [0,1] \to [0,1]$ belongs to $\Sigma(BV)$ if and only if there is some $N \in \mathbb{N}$ depending only on g, such that for any compact interval $J \subseteq [0,1]$ the preimage $g^{-1}(J)$ can be written as the union of at most N intervals which may be open, closed, half-open or singletons. Any monotone function is pseudo-monotone (with $N = 1$), and any pseudo-monotone function is of bounded variation. However, Example 2.3.7 has shown that not every function of bounded variation is also pseudo-monotone.

That the operator S_g maps BV into itself if and only if g is pseudo-monotone was shown by Josephy in [74]. In fact, the same had been proven later for BV replaced by $X = YBV_\varphi$ with $\varphi \in \delta_2$ by Galkin [65]. In this case, the estimate

$$\mathrm{Var}_\varphi(x \circ g) \leq 2N \, \mathrm{Var}_\varphi(x) \quad \text{for } x \in YBV_\varphi \tag{4.2.1}$$

holds, and we have $\|S_g x\|_{YBV_\varphi} \leq 2N \, \|x\|_{YBV_\varphi}$ for all $x \in YBV_\varphi$ which shows that the operator S_g is automatically continuous. This, of course, is also true in the Wiener spaces WBV_p.

Unfortunately, we do not know what happens when $X = \Lambda BV$ or $X = RBV_p$. However, if S_g maps RBV_p into itself, then g must belong to RBV_p, and hence pseudo-monotonicity alone is not sufficient as an acting condition between Riesz spaces.

Before we give general results concerning injectivity and surjectivity, we start by taking a closer look at $S_g : C \to C$. As said, this operator is well-defined if and only if $g : [0,1] \to [0,1]$ is continuous, and in this case, calculating its norm is trivial: Since the function $\mathbb{1}$ belongs to C, we immediately obtain

$$\|S_g\|_{C \to C} = 1$$

which is independent of g.

However, the following example shows that injectivity or surjectivity of g does not imply the injectivity respectively surjectivity of S_g.

Example 4.2.1. The function $g : [0,1] \to [0,1]$, $t \mapsto t/2$, is injective, but not surjective. However, S_g is not injective, because the function $x : [0,1] \to \mathbb{R}$, defined by $x(t) = t$ for $t \in [0,1/2]$ and arbitrary for $t \in (1/2,1]$ is mapped into the function $y(t) = t/2$ for $t \in [0,1]$, no matter how exactly x is defined on $(1/2,1]$.

On the other hand, the function $g : [0,1] \to [0,1], t \mapsto 4t(1-t)$, is surjective, but not injective. However, S_g is not surjective, because there is no function $x : [0,1] \to \mathbb{R}$

satisfying $x(g(t)) = t$ for all $t \in [0,1]$. Otherwise, we had $0 = x(g(0)) = x(g(1)) = 1$ which is clearly impossible. $\diamondsuit$

The last example shows that the surjectivity of S_g seems to force g to be injective. That this is indeed true is shown by the following result.

Proposition 4.2.2. *Let $g : [0,1] \to [0,1]$ be continuous. Then the following statements hold.*

(a) The operator $S_g : C \to C$ is surjective if and only if g is injective.

(b) The operator $S_g : C \to C$ is injective if and only if g is surjective. In this case, S_g is even an isometry, i.e.

$$\|S_g x\|_\infty = \|x\|_\infty \quad \text{for } x \in C.$$

Proof. (a) Suppose that $S_g : C \to C$ is surjective and fix $r, s \in [0,1]$ with $g(r) = g(s)$. Since $y(t) = t$ belongs to C, there is some function $x \in C$ with $x(g(t)) = y(t) = t$ for all $t \in [0,1]$. This implies $r = x(g(r)) = x(g(s)) = s$ and hence the injectivity of g. Conversely, assume that g is injective. Then $K := g([0,1]) \subseteq [0,1]$ is a compact interval, and the function $g : [0,1] \to K$ is a homeomorphism. Given $y \in C$, the function $y \circ g^{-1} : K \to \mathbb{R}$ is continuous which can be extended to a continuous function x on $[0,1]$. Thus, $x \in C$ and $x \circ g = y$.

(b) If g is surjective, we have

$$\|S_g x\|_\infty = \sup_{s \in [0,1]} |x(g(s))| = \sup_{t \in [0,1]} |x(t)| = \|x\|_\infty ,$$

showing that S_g is an isometry and in particular injective.

Conversely, assume that S_g is injective. Fix $t \in [0,1]$ and consider the functions $x_n \in C$, defined by

$$x_n(s) = \max\left\{0, 1 - n|s-t|\right\} \quad \text{for all } n \in \mathbb{N}.$$

Then $t \in \text{supp}(x_n) \subseteq [t - 1/n, t + 1/n]$ for each $n \in \mathbb{N}$. Since $x_n \neq \mathbb{0}$ and S_g is injective and linear, none of the functions $S_g x_n$ can be zero everywhere, and so there are $s_n \in [0,1]$ with $S_g x_n(s_n) = x_n(g(s_n)) \neq 0$ for each $n \in \mathbb{N}$. This implies $g(s_n) \in \text{supp}(x_n)$ for each $n \in \mathbb{N}$ and hence $g(s_n) \to t$ as $n \to \infty$. But this shows that $g([0,1])$ is dense in $[0,1]$, and since g is continuous, g is surjective. ∎

Observe that the "crossover" between surjectivity and injectivity in our proposition is perfectly symmetric. And on top of the injectivity of S_g we get the isometry property in (b) for free. Also, this proposition shows that it was not accidental that the first function g in Example 4.2.1 is not surjective, while the second one is not injective.

Our aim is now to see how we may imitate the proof to get a similar result in general spaces X and Y of real-valued functions defined on $[0,1]$.

Proposition 4.2.3. *Let $g : [0,1] \to [0,1]$ be so that S_g maps X into Y. The following statements are true.*

(a) If g is surjective, then S_g is injective.

(b) If X separates points strongly and S_g is injective, then $g([0,1]) = [0,1]$, i.e. g is surjective.

(c) If X separates points uniformly and S_g is injective, then $\overline{g([0,1])} = [0,1]$. If, in addition, g is continuous, then g is even surjective.

(d) If Y contains at least one injective function and S_g is surjective, then g is injective.

Proof. For (a) fix $x \in X$ with $S_g x = \mathbb{0}$. For $t \in [0,1]$ we find some $s \in [0,1]$ with $g(s) = t$. This implies $x(t) = x(g(s)) = S_g x(s) = 0$, and since t was arbitrary, $x = \mathbb{0}$. Since S_g is linear, S_g is injective.

For (b) assume that X separates points strongly. Then, for fixed $t \in [0,1]$, the function $x := \chi_{\{t\}}$ belongs to X. Since S_g is injective and $x \neq \mathbb{0}$, the function $S_g x$ cannot be zero everywhere. Thus, there must be some $s \in [0,1]$ such that $S_g x(s) = x(g(s)) \neq 0$ and hence $g(s) = t$ which proves that g is indeed surjective.

If X separates points uniformly, then for fixed $t \in [0,1]$ and each $n \in \mathbb{N}$ we find $x_n \in X$ such that

$$t \in \operatorname{supp}(x_n) \subseteq [t - 1/n, t + 1/n].$$

Since S_g is injective and $x_n \neq \mathbb{0}$, none of the function $S_g x_n$ is zero everywhere. Thus, there must be $s_n \in [0,1]$ such that $S_g x_n(s_n) = x_n(g(s_n)) \neq 0$ and hence

$$|g(s_n) - t| \leq 1/n \quad \text{for all } n \in \mathbb{N}$$

which shows that $g(s_n) \to t$ as $n \to \infty$. This proves $\overline{g([0,1])} = [0,1]$.
If g is continuous, then we even have $g([0,1]) = [0,1]$ which means that g is surjective. This proves (c).

To prove (d) let $y \in Y$ be an injective function. Since S_g is surjective, there must be some $x \in X$ such that $S_g x = y$. For fixed $s, t \in [0,1]$ with $g(s) = g(t)$ we obtain

$$y(s) = S_g x(s) = x(g(s)) = x(g(t)) = S_g x(t) = y(t),$$

thus $s = t$ as y is injective. But then g is also injective. ∎

Comparing Proposition 4.1.3 and Proposition 4.2.3, one can see that injectivity of M_g is related to the support of g, while injectivity of S_g is related to the image of g. Moreover, in contrast to the multiplication operator, there are surjective substitution operators which map BV into itself and are not injective.

Example 4.2.4. Define $g : [0,1] \to [0,1]$ by $g(t) = t/2$. Then g is strictly increasing and hence injective, but not surjective. In particular, S_g maps BV into itself, but cannot be injective by Proposition 4.2.3 (b). However, S_g is indeed surjective, since for fixed $y \in BV$ the function $x : [0,1] \to \mathbb{R}$, defined by

$$x(t) = \begin{cases} y(2t) & \text{for } 0 \leq t \leq 1/2, \\ y(1) & \text{for } 1/2 < t \leq 1, \end{cases}$$

belongs to BV and satisfies $x \circ g = y$. ◊

Let us have a look at Proposition 4.2.3 again and see what it tells us, apart from B and C, about our BV-type spaces.

Corollary 4.2.5. *Let $g : [0,1] \to [0,1]$ be so that S_g maps X into Y. The following statements are true.*

(a) If g is surjective, then S_g is injective, no matter what X and Y are.

(b) If X is one of the spaces B, BV, WBV_p, YBV_φ or ΛBV and if S_g is injective, then g is surjective.

(c) If X is one of the spaces C or RBV_p and if S_g is injective, then $g([0,1])$ is dense in $[0,1]$. If, in addition, g is continuous, then g is surjective.

(d) If Y is one of the spaces B, C, BV, WBV_p, YBV_φ, ΛBV or RBV_p and if S_g is surjective, then g is injective.

By comparing the parts (b) and (c) in Corollary 4.2.5 one might ask if the stronger statement (b) is also true if X is one of the spaces C and RBV_p without putting further constraints on g. In other words, one might ask why (b) and (c) have to be considered separately. The reason is simple: If X is C or RBV_p and S_g maps X injectively into any function space Y, then we *cannot* expect g to be surjective, as illustrated by the following example.

Example 4.2.6. Let $g : [0,1] \to [0,1]$ be given by

$$g(t) = \begin{cases} t & \text{for } 0 < t \leq 1, \\ 1 & \text{for } t = 0. \end{cases}$$

Then the operator S_g maps the space C into the space B. Moreover, S_g is injective. To see this, fix $x \in C$ and assume $S_g x = \mathbb{0}$. For any $0 < t \leq 1$ we have $g(t) = t$ and hence $0 = S_g x(t) = x(g(t)) = x(t)$, and due to the continuity of x we conclude $x = \mathbb{0}$. This implies that S_g is indeed injective. But g is not surjective, because it has no zeros. Note, however, that $g([0,1]) = (0,1]$ is dense in $[0,1]$, and this is in total accordance with part (c) of Corollary 4.2.5. ◊

Note that the injective and surjective criteria of Proposition 4.2.2 can be found entirely within Corollary 4.2.5 (c) and (d). Also observe that all the spaces mentioned in Corollary 4.2.5 (d) contain the injective function $y(t) = t$. The special case $X = Y = YBV_\varphi$ with $\varphi \in \delta_2$ is included in our next set of consequences.

Corollary 4.2.7. *Let $g : [0, 1] \to [0, 1]$ be pseudo-monotone, and let X be one of the spaces BV, WBV_p or YBV_φ with $\varphi \in \delta_2$. Then the following statements are true.*

(a) The operator $S_g : X \to X$ is injective if and only if g is surjective.

(b) If the operator $S_g : X \to X$ is surjective, then g is injective.

(c) The operator $S_g : X \to X$ is bijective if and only if g is bijective and g^{-1} is pseudo-monotone. In this case, $S_g^{-1} = S_{g^{-1}}$.

Proof. Indeed, part (a) follows immediately from Proposition 4.2.3 (a) and (b), whereas part (b) can be deduced from (d) of Proposition 4.2.3. Note that the identity function $x(t) = t$ belongs to YBV_φ and is injective.

To prove (c) note that if S_g is bijective, then it follows from (a) and (b) that g is bijective. Let T be the inverse operator of S_g. Then

$$TS_gx(t) = (Tx)(g(t)) = x(t) \quad \text{for all } t \in [0, 1]$$

and hence

$$Tx(s) = x(g^{-1}(s)) = S_{g^{-1}}x(s) \quad \text{for all } s \in [0, 1].$$

But then g^{-1} must be pseudo-monotone. Conversely, assume that g is bijective with a pseudo-monotone inverse g^{-1}. Then $S_{g^{-1}}$ maps YBV_φ into itself. Moreover,

$$S_gS_{g^{-1}}x(t) = S_{g^{-1}}x(g(t)) = x(g(g^{-1}(t))) = x(t)$$

and similarly $S_{g^{-1}}S_gx(t) = x(t)$ for all $t \in [0, 1]$ and $x \in YBV_\varphi$. This completes the proof. ∎

It is unclear whether the injectivity of g also implies the surjectivity of S_g for the spaces considered in Corollary 4.2.7.

We will now discuss the last of the analytic properties of our interest, namely compactness. For the multiplication operator we have seen in Theorem 4.1.12 that $M_g : BV \to BV$ is compact if and only if g has countable support. This, however, cannot be true for $S_g : BV \to BV$, and here are three reasons: First, if the support of g is countable but contains infinitely many elements, then the function g cannot be pseudo-monotone. Second, the constant function $g = \mathbb{1}$ which has uncountable support obviously generates a compact operator $S_g : BV \to BV$. And a third reason why the support of g is not the appropriate tool for characterizing compactness may be found in Proposition 4.2.3, because we have seen there that the image set of g regulates mapping properties of S_g. But even if the image of g is countable, the operator S_g still does not need to be compact.

Example 4.2.8. Define $g : [0, 1] \to [0, 1]$ by $g(0) = 0$ and $g(t) = 1/n$ for $t \in (\frac{1}{n+1}, \frac{1}{n}] =: I_n$ for all $n \in \mathbb{N}$. Then g is increasing and hence pseudo-monotone, but S_g which indeed maps BV into itself, is not compact. To see this, consider the functions $x_n := \chi_{\{1/n\}}$ for $n \in \mathbb{N}$ which clearly form a bounded sequence in BV with $\|x_n\|_{BV} \leq 3$ for all

$n \in \mathbb{N}$. However, $S_g x_n(t) = x_n(g(t)) = 1$ for $t \in I_n$ and $S_g x_n(t) = 0$ otherwise which can be rewritten as $S_g x_n = M_{\mathbb{1}} \chi_{I_n}$. By Theorem 4.1.12, S_g is not compact, since $\operatorname{supp}(\mathbb{1}) = [0,1]$ is uncountable, although (χ_{I_n}) is a bounded sequence in BV as $\|\chi_{I_n}\|_{BV} \leq 3$ for all $n \in \mathbb{N}$. ◇

Surprisingly, again in contrast to the multiplication operator, the only compact substitution operators are those with finite dimensional range.

Theorem 4.2.9. *Let X be one of the spaces BV, WBV_p, YBV_φ or ΛBV, and let the function $g : [0,1] \to [0,1]$ be pseudo-monotone and so that S_g maps X into itself. Then the following statements are equivalent.*

(a) The operator S_g has finite dimensional range.

(b) The operator S_g is compact.

(c) The set $g([0,1])$ is finite.

In this case, $\dim \operatorname{Im}(S_g) = \#g([0,1])$.

Proof. The implication "(a)⇒(b)" is obviously true. To prove "(b)⇒(c)", assume that the set $g([0,1])$ is infinite from which we then can extract a sequence (s_n) of pairwise distinct numbers. Since each s_n belongs to the image of g, there are numbers $t_n \in [0,1]$ such that $g(t_n) = s_n$. The functions $x_n := \chi_{\{s_n\}}$ now form a bounded sequence in X, but the sequence $(S_g x_n)$ cannot have a Cauchy subsequence in X, as for $m \neq n$,

$$\begin{aligned}\|S_g x_m - S_g x_n\|_X &\geq \|S_g x_m - S_g x_n\|_\infty \geq |S_g x_m(t_n) - S_g x_n(t_n)| \\ &= |x_m(g(t_n)) - x_n(g(t_n))| = |x_m(s_n) - x_n(s_n)| = 1.\end{aligned}$$

Thus, the operator S_g is not compact.

For the last implication "(c)⇒(a)" assume that $g([0,1])$ is finite and hence can be written as $g([0,1]) = \{s_1, \ldots, s_n\}$ for some numbers $s_1, \ldots, s_n \in [0,1]$ and $n = \#g([0,1])$. Then the points s_j are pairwise distinct which ensures that the corresponding preimages $A_j := g^{-1}(\{s_j\})$ partition $[0,1]$. Since g is pseudo-monotone, each set A_j has only finitely many connected components which ensures that the functions $x_j := \chi_{A_j} = S_g \chi_{\{s_j\}}$ belong to X. Moreover, for any $x \in X$ and $t \in A_j$ we have

$$S_g x(t) = x(g(t)) = x(s_j) = x(s_j)\chi_{A_j}(t),$$

and so for arbitrary $t \in [0,1]$,

$$S_g x(t) = \sum_{j=1}^n x(s_j)\chi_{A_j}(t) = \sum_{j=1}^n x(s_j) x_j(t).$$

Consequently, $S_g x \in \operatorname{Span}(\{x_1, \ldots, x_n\})$ which shows that S_g has finite dimensional range. Since the A_j are pairwise disjoint, the functions x_j are linearly independent and thus form a basis of the range of S_g. This shows $\dim \operatorname{Im}(S_g) = \#g([0,1])$. ■

Let us make two final remarks. First, we do not know if the fact that g is pseudo-monotone is sufficient to guarantee that S_g maps also ΛBV into itself. This is why in Theorem 4.2.9 we had to make that as an explicit assumption. Second, many linear operators in functional analysis are proved to be compact by approximating them by operators with finite dimensional range; for instance, we have done so in the proof of Theorem 4.1.12. Remarkably, Theorem 4.2.9 shows that the operator S_g is compact in many BV-spaces of our interest only if it has finite dimensional range itself.

4.3 Integral Operators

In this section we study the integral operator (4.0.3) in spaces of functions of bounded variation. There is a vast literature on the behavior of this operator in the space of continuous or measurable[1] functions, but considerably less is known in the space BV and its various generalizations.

Arbitrary Kernels

To begin with, we state two conditions on the kernel function $g : [0,1] \times [0,1] \to \mathbb{R}$ which will be used over and over in this section. Here, the symbol $\forall\,'s$ means "for almost all s".

$$\forall t \in [0,1]: \quad g(t,\cdot) \in L_1, \tag{A}$$

$$\exists m \in L_1 \ \forall\,'s \in [0,1]: \quad \operatorname{Var}\big(g(\cdot,s)\big) \leq m(s). \tag{B}$$

Condition (A) is needed for the integral in (4.0.3) to make sense. Condition (B) guarantees, that the operator I_g acts in BV. Indeed, the following was shown in [29].

Theorem 4.3.1. *Under the conditions (A) and (B) the operator I_g maps the space BV into itself and is bounded.*

It turns out that the conditions (A) and (B) are too strong and can be relaxed as follows in order to gain the same result. One can show that (B) and the weaker condition

$$\forall t \in [0,1]: \quad g(t,\cdot) \text{ is measurable and } g(0,\cdot) \in L_1 \tag{A'}$$

together imply (A). On the other hand, if we weaken (B) by

$$\exists m \in L_1 \ \forall\,'s \in [0,1]: \quad \|g(\cdot,s)\|_\infty \leq m(s), \tag{B'}$$

that is, if we replace the majorization of the variation by a pointwise majorization, then Theorem 4.3.1 may fail. Here is a simple example.

[1]Recall that the terms "measurable" and "almost everywhere" in this thesis are always understood with respect to the Lebesgue measure.

Example 4.3.2. Let $g(t,s) = \chi_{\mathbb{Q}}(t)$. Then (A) and hence also (A') holds even with $g(t,\cdot) = \mathbb{1} \in L_\infty$ for $t \in \mathbb{Q} \cap [0,1]$ and $g(t,\cdot) = \mathbb{0} \in L_\infty$ for $t \in [0,1]\backslash\mathbb{Q}$, and (B') holds with $m = \mathbb{1}$, because $\|g(\cdot,s)\|_\infty = 1$ for all $s \in [0,1]$. However, (B) does not hold, because $\mathrm{Var}(g(\cdot,s)) = \mathrm{Var}(\chi_{\mathbb{Q}}) = \infty$ for all $s \in [0,1]$. The operator I_g now maps $x = \mathbb{1} \in BV$ onto $I_g x = \chi_{\mathbb{Q}}$ which is not in BV. ◇

In general, however, Theorem 4.3.1 only gives a sufficient condition on g that guarantees $I_g(BV) \subseteq BV$. This condition is not necessary as the following example shows.

Example 4.3.3. For fixed $s,t \in [0,1]$ we have $s-t \in \mathbb{Q}$ if and only if there is some number $q \in [-t, 1-t] \cap \mathbb{Q}$ such that $s = t+q$. We therefore define

$$\mathbb{Q}(t) := \left\{ t+q \mid q \in \mathbb{Q} \cap [-t, 1-t] \right\} \subseteq [0,1]$$

and have $s-t \in \mathbb{Q}$ if and only if $s \in \mathbb{Q}(t)$. Moreover, the set $\mathbb{Q}(t)$ is countable and dense in $[0,1]$ for each fixed $t \in [0,1]$.

The function $g : [0,1] \times [0,1] \to \mathbb{R}$, defined by $g(t,s) = \chi_{\mathbb{Q}}(s-t) = \chi_{\mathbb{Q}(t)}(s)$, clearly satisfies condition (A) and hence also (A'), because for each fixed $t \in [0,1]$ the set $\mathbb{Q}(t)$ is countable and hence $g(t,\cdot) = \mathbb{0}$ almost everywhere. This also implies that $I_g x = \mathbb{0}$ for any $x \in BV$, and so I_g maps BV into itself and is bounded.

But since $\chi_{\mathbb{Q}}(s-t) = \chi_{\mathbb{Q}}(t-s)$ we also have $g(t,s) = \chi_{\mathbb{Q}(s)}(t)$, and for each fixed $s \in [0,1]$ the function $g(\cdot,s) = \chi_{\mathbb{Q}(s)}$ does not belong to BV. This means that (B) is violated. However, (B') is clearly satisfied with $m = \mathbb{1}$. ◇

In fact, the conditions (A) and (B) together are so strong that they are even sufficient for I_g to map WBV_p as well as L_∞ continuously into BV. This statement was proven for WBV_p in [32]; for L_∞ it is true, because for a partition $0 = t_0 < \ldots < t_n = 1$ of $[0,1]$ and $x \in L_\infty$,

$$\begin{aligned} \sum_{j=1}^{n} \left| \int_0^1 g(t_{j-1},s)x(s)\,\mathrm{d}s - \int_0^1 g(t_j,s)x(s)\,\mathrm{d}s \right| &\le \int_0^1 \sum_{j=1}^{n} \left| g(t_{j-1},s) - g(t_j,s) \right| |x(s)|\,\mathrm{d}s \\ &\le \|x\|_{L_\infty} \int_0^1 m(s)\,\mathrm{d}s, \end{aligned}$$

where $m \in L_1$ is the bound from condition (B). Thus,

$$\mathrm{Var}\left(I_g x\right) \le \|x\|_{L_\infty} \int_0^1 m(s)\,\mathrm{d}s < \infty. \tag{4.3.1}$$

We will see in Theorem 4.3.21 below that a similar result is true in all our BV-spaces.

In order to get a milder condition which is both necessary and sufficient, we introduce another requirement for g:

$$\sup_{\tau \in [0,1]} \mathrm{Var}\left(\int_0^\tau g(\cdot,s)\,\mathrm{d}s \right) < \infty. \tag{C}$$

For instance, the function g in Example 4.3.2 cannot satisfy (C), because for any $\tau \in (0,1]$ we have

$$\mathrm{Var}\left(\int_0^\tau g(\cdot,s)\,\mathrm{d}s \right) = \mathrm{Var}\left(\tau \chi_{\mathbb{Q}} \right) = \infty.$$

Note, however, that (B) implies (C). Indeed, if (B) is satisfied, $0 = t_0 < \ldots < t_n = 1$ is a partition of $[0,1]$ and $\tau \in [0,1]$ is fixed, then, similarly as before,

$$\sum_{j=1}^{n} \left| \int_0^\tau g(t_{j-1}, s)\,\mathrm{d}s - \int_0^\tau g(t_j, s)\,\mathrm{d}s \right| \leq \int_0^1 m(s)\,\mathrm{d}s$$

which implies

$$\sup_{\tau\in[0,1]} \operatorname{Var}\left(\int_0^\tau g(\cdot, s)\,\mathrm{d}s\right) \leq \|m\|_{L_1}\,.$$

If we now combine (A) being the overall assumption on g with (C), we get exactly what we want.

Theorem 4.3.4. *Let g satisfy condition (A). Then the following conditions are equivalent.*

(a) The kernel function g satisfies condition (C).

(b) The operator I_g maps the space BV into itself and is bounded. In this case,

$$\operatorname{Var}(I_g x) \leq 2\,\|x\|_{BV} \sup_{\tau\in[0,1]} \operatorname{Var}\left(\int_0^\tau g(\cdot, s)\,\mathrm{d}s\right). \tag{4.3.2}$$

A proof can be found in the paper [32]. In fact, there was proven the following slightly more general result.

Theorem 4.3.5. *Let g satisfy condition (A). Then the following conditions are equivalent.*

(a) The kernel function g satisfies condition (C) with Var *replaced by* Var_p.

(b) The operator I_g maps the space BV into WBV_p and is bounded.

This, however, is also bad news, because analogous conditions (C) for X being one of the spaces YBV_φ, ΛBV and RBV_p may only yield that I_g maps BV continuously into X and not X into itself. Indeed, for ΛBV, the following had been proven in [30].

Theorem 4.3.6. *Let g satisfy condition (A). Then the following conditions are equivalent.*

(a) The kernel function g satisfies condition (C) with Var *replaced by* $\operatorname{Var}_\Lambda$.

(b) The operator I_g maps the space BV into ΛBV and is bounded.

Consequently, the (C)-type conditions seem to be too weak to be sufficient for I_g to map a BV-space into itself, while the (B)-type conditions seem too strong to be necessary. We will see in Theorem 4.3.21 and Corollary 4.3.23 below how strong the really are. The treasure must be hidden somewhere in between, but we do not know where exactly.

For the Riesz spaces, however, the following result is known [7].

Theorem 4.3.7. *If the kernel function g satisfies the conditions (A), (B) and*

$$\forall\,'s \in [0,1] : \; \partial_1 g(\cdot,s) \in C \qquad \text{and} \qquad \left(s \mapsto \|\partial_1 g(\cdot,s)\|_\infty\right) \in L_p, \tag{D}$$

then I_g maps RBV_p into itself and is bounded.

Condition (D) together with Riesz' Theorem 1.2.25 implies that $g(\cdot,s) \in RBV_p$ for almost all $s \in [0,1]$ and any $p \in (1,\infty)$, because from $\partial_1 g(\cdot,s) \in C$ follows $\partial_1 g(\cdot,s) \in L_\infty$ and hence

$$\mathrm{RVar}_p\left(g(\cdot,s)\right) = \int_0^1 \left|\partial_1 g\left(t,s\right)\right|^p \mathrm{d}t < \infty.$$

Moreover, the second condition in (D) then yields that the Riesz variation satisfies

$$\mathrm{RVar}_p\left(g(\cdot,s)\right) \leq \|\partial_1 g(\cdot,s)\|_\infty^p\,,$$

where the right hand side is then an L_1-bound with respect to s. Thus, condition (D) seems to be an analogue to condition (B) for Riesz spaces and therefore not as restrictive as it may appear at first glance. However, we will see later in Corollary 4.3.23 that the actual (B)-type condition for Riesz spaces is strictly weaker than (D) yet more powerful as it is able to guarantee an even stronger result about compactness of the operator I_g.

Let us come back to BV and the conditions (A) and (B) for a moment. Since for $t \in [0,1]$ and $x \in BV$ we get under the assumptions of Theorem 4.3.4 from (4.3.2) that

$$\begin{aligned}|I_g x(t)| &\leq |I_g x(0)| + |I_g x(t) - I_g x(0)| \leq \|x\|_\infty \|g(0,\cdot)\|_{L_1} + \mathrm{Var}(I_g x) \\ &\leq \|x\|_{BV} \|g(0,\cdot)\|_{L_1} + 2\|x\|_{BV} \sup_{\tau\in[0,1]} \mathrm{Var}\left(\int_0^\tau g(\cdot,s)\,\mathrm{d}s\right).\end{aligned}$$

We obtain for the BV-norm of I_g,

$$\|I_g\|_{BV\to BV} \leq \|g(0,\cdot)\|_{L_1} + 4 \sup_{\tau\in[0,1]} \mathrm{Var}\left(\int_0^\tau g(\cdot,s)\,\mathrm{d}s\right). \tag{4.3.3}$$

The authors of [31] gave an example of a function $g : [0,1]\times[0,1] \to \mathbb{R}$ satisfying (A), (B') and (C), but not (B). We give here a simpler example.

Example 4.3.8. Let $g : [0,1]\times[0,1] \to \mathbb{R}$ be defined by $g(t,s) = \chi_{\mathbb{Q}}(t-s)$ as in Example 4.3.3. We have seen there that g satisfies (A), (A') and (B'), but not (B). Because of

$$\int_0^\tau g(t,s)\,\mathrm{d}s = 0 \quad \text{for all } t,\tau \in [0,1],$$

it satisfies condition (C). ◇

There are two special cases for the integral operator (4.0.3) that are important for applications, namely those which are generated by *separated kernels* and by *Volterra kernels*. We do not investigate Volterra kernels intensively in this thesis and only give some brief overview about some known results at the end of this chapter. Separated kernels, however, that is, kernels $g : [0,1] \times [0,1] \to \mathbb{R}$ given by

$$g(t,s) = g_1(t)g_2(s) \tag{4.3.4}$$

with $g_1, g_2 : [0,1] \to \mathbb{R}$ may often be used to find counterexamples. For $g_2 \in L_1$, the integral operator (4.0.3) then has the form

$$I_g x(t) = \int_0^1 g(t,s)x(s)\,\mathrm{d}s = g_1(t)\int_0^1 g_2(s)x(s)\,\mathrm{d}s, \tag{4.3.5}$$

but we point out that the integral in (4.0.3) may make sense also if only $g_1 g_2 \in L_1$. For instance, if $g_1 = \mathbb{0}$, then g_2 can be any function whatsoever, and still I_g is well-defined and maps any function $x : [0,1] \to \mathbb{R}$ onto the zero function $\mathbb{0}$. Let us see how the conditions considered so far translate to separated kernels.

Proposition 4.3.9. *The kernel function $g : [0,1]\times[0,1] \to \mathbb{R}$ given in separated kernels $g(t,s) = g_1(t)g_2(s)$ for $g_1, g_2 : [0,1] \to \mathbb{R}$ satisfies*

(A) if and only if $g_1 = \mathbb{0}$ or $g_2 \in L_1$.

(A') if and only if $g_1 = \mathbb{0}$, or g_2 is measurable and $g_1(0)g_2 \in L_1$.

(B) if and only if g_1 is constant, or $g_2 = \mathbb{0}$ almost everywhere, or $g_1 \in BV$ and $|g_2| \leq m$ for some $m \in L_1$.

(B') if and only if $g_1 = \mathbb{0}$, or $g_2 = \mathbb{0}$ almost everywhere, or $g_1 \in B$ and $|g_2| \leq m$ for some $m \in L_1$.

Moreover, if $g_2 \in L_1$, then g satisfies

(C) if and only if $g_2 = \mathbb{0}$ almost everywhere or $g_1 \in BV$.

Proof. The first four cases follow immediately from the definitions of the conditions (A), (A'), (B) and (B'). We only want to leave some words on (C). It is clear that if $g_2 = \mathbb{0}$ almost everywhere or $g_1 \in BV$ then g satisfies (C). For the converse assume that $g_2 \in L_1$ is not zero almost everywhere. From Theorem 1.1.17 we obtain that

$$G(t) := \int_0^t g_2(s)\,\mathrm{d}s$$

is absolutely continuous with $G' = g_2$ almost everywhere; in particular, G cannot be constant. Thus, there is a $\tau \in (0,1]$ with $G(\tau) \neq 0$, and from (C) we obtain

$$\infty > \operatorname{Var}\left(\int_0^\tau g(\cdot,s)\,\mathrm{d}s\right) = |G(\tau)|\operatorname{Var}(g_1)$$

and hence $g_1 \in BV$. ■

From Proposition 4.3.9 we immediately get that the kernel g from Example 4.3.2 cannot satisfy condition (C).
Even for separated kernels there is a subtle yet significant difference between (A) and (A') as well as between (B) and (B'). It is clear that (A) implies (A') and that (B) together with $g(0,\cdot) \in L_1$ implies (B'), but none of these inclusions may be inverted.

Example 4.3.10. (a) Let $g(t,s) = g_1(t)g_2(s)$ with $g_1 = \chi_{(0,1]}$ and

$$g_2(s) = \begin{cases} 1/s & \text{for } 0 < s \le 1, \\ 0 & \text{for } s = 0. \end{cases}$$

Then Proposition 4.3.9 tells us that (A') is satisfied, because $g_1 \ne \mathbb{0}$ and g_2 is measurable with $g_1(0)g_2 = \mathbb{0} \in L_1$, but (A) is not satisfied, because $g_2 \notin L_1$. The same proposition also guarantees that both conditions (B) and (B') do not hold for this kernel.
Condition (C) cannot be satisfied, because

$$\int_0^T g(t,s)\,\mathrm{d}s = \infty \quad \text{for } 0 < t \le 1.$$

(b) If $g(t,s) = g_1(t)g_2(s)$ with $g_1 = \chi_{\mathbb{Q}}$ and $g_2 = \mathbb{1}$, then (B') is satisfied, but (B) is not, again by Proposition 4.3.9. For this kernel, the same proposition guarantees that both conditions (A) and (A') do hold, while (C) does not. ◇

Proposition 4.3.9 also tells us that for separated kernels the two conditions (A') and (B') together imply (A). This is why we could not combine the two sample functions in (a) and (b) of Example 4.3.10 into one function. More general, the two conditions (A') and (B') indeed imply (A) also for arbitrary kernels. To see this, note that (B') gives

$$|g(t,s)| \le m(s) \quad \text{for all } t \in [0,1] \text{ and almost all } s \in [0,1],$$

and since $g(t,\cdot)$ is measurable due to (A') the condition (A) holds with $\|g(t,\cdot)\|_{L_1} \le \|m\|_{L_1}$; in particular, we even get that $g(t,\cdot)$ is integrable uniformly in t.
However, if one replaces the supremum norm in (B') with the L_∞-norm, the two conditions (A') and (B') together may no longer imply (A), as is shown by the following example.

Example 4.3.11. Consider the kernel

$$g(t,s) = \begin{cases} 1/s & \text{for } t = 1, 0 < s \le 1, \\ 1 & \text{for } t = 1, s = 0, \\ 1 & \text{for } 0 \le t < 1, 0 \le s \le 1. \end{cases}$$

Then (A) is violated, because $g(1,s) = 1/s$ for $0 < s \le 1$ is not Lebesgue integrable with respect to s on $[0,1]$. In particular,

$$\|g(\cdot,s)\|_\infty = 1/s \quad \text{for } 0 < s \le 1,$$

and so (B') is also violated. Moreover, for any $s \in (0,1]$,

$$\operatorname{Var}\big(g(\cdot,s)\big) = 1/s - 1$$

which cannot be bounded by a function $m \in L_1$ in the sense of (B), and so (B) is also violated.

However, (A') holds as $g(t,\cdot)$ is measurable for each fixed $t \in [0,1]$, and $g(0,\cdot) = \mathbb{1} \in L_1$. Finally, $\|g(\cdot,s)\|_{L_\infty} = 1$ holds for any $s \in [0,1]$ and hence (B') with $\|\cdot\|_\infty$ replaced by $\|\cdot\|_{L_\infty}$ is true with $m = \mathbb{1} \in L_1$.
Note that (C) can also not be true, since

$$\int_0^\tau g(1,s)\,\mathrm{d}s = \infty$$

for any $\tau \in (0,1]$. ◇

Separated kernels (4.3.4) cannot generate an instance as in Example 4.3.8, because for those kernels $g = g_1 g_2$ with $g_2 \in L_1$ condition (C) is equivalent to (B). However, if $g_2 = \mathbb{0}$ almost everywhere, then neither (B) nor (C) implies $g_1 \in BV$.

Example 4.3.12. The kernel function $g(t,s) = g_1(t)g_2(s)$ with $g_1 = g_2 = \chi_{\mathbb{Q}}$ satisfies $g_1, g_2 \in L_1$ with $g_1 = g_2 = \mathbb{0}$ almost everywhere and thus (A), (A'), (B), (B') and (C) by Proposition 4.3.9, but $g_1 = \chi_{\mathbb{Q}}$ is clearly not of bounded variation on $[0,1]$. ◇

If $g_1 \in BV$ and $g_2 \in L_1$, we get an improved version of the estimate (4.3.3). Indeed, with the help of (4.3.5) we have for $x \in BV$,

$$\|I_g x\|_{BV} = \|g_1\|_{BV} \left| \int_0^1 g_2(s)x(s)\,\mathrm{d}s \right| \le \|g_1\|_{BV} \|g_2\|_{L_1} \|x\|_{BV}$$

and thus

$$\|I_g\|_{BV \to BV} \le \|g_1\|_{BV} \|g_2\|_{L_1}.$$

The study of solutions to integral equations, both linear and nonlinear, in BV-spaces is motivated by numerous applications to real world problems; we give two examples of such motivations at the beginning of Chapter 7. Sometimes it is useful or even necessary to look for solutions in the space $BV \cap C$, that is, to add continuity. So there is some interest to find conditions which guarantee or are even equivalent to the inclusion

$$I_g(BV \cap C) \subseteq BV \cap C.$$

To this end, we introduce another condition on the kernel function g.

$$\forall \varepsilon > 0\ \exists \delta > 0\ \forall t_1, t_2, \tau \in [0,1]:$$
$$|t_1 - t_2| \le \delta \ \Rightarrow\ \left| \int_0^\tau g(t_1,s) - g(t_2,s)\,\mathrm{d}s \right| \le \varepsilon. \qquad \text{(E)}$$

This new condition (E) is obviously satisfied if $g : [0,1] \times [0,1] \to \mathbb{R}$ is continuous, because then $g(\cdot,s)$ is continuous for each $s \in [0,1]$ and uniformly with respect to $t \in [0,1]$. If this uniformity is dropped, that is, if $g(\cdot,s)$ is merely continuous for each $s \in [0,1]$, then (E) may no longer be true.

Example 4.3.13. Let $g : [0,1] \times [0,1] \to \mathbb{R}$ be given by

$$g(t,s) = \begin{cases} \dfrac{1}{t+s} & \text{for } 0 < s \leq 1, \\ 0 & \text{for } s = 0. \end{cases}$$

Then $g(\cdot, s)$ is continuous for every fixed $s \in [0,1]$. For $\tau = 1$, $t_1 = t^2$ and $t_2 = t \in (0,1]$ we have

$$\begin{aligned} \int_0^1 \left(g(t^2, s) - g(t,s)\right) \mathrm{d}s &= \int_0^1 \left(\frac{1}{t^2+s} - \frac{1}{t+s}\right) \mathrm{d}s \\ &= \log\left(\frac{1+t^2}{t+t^2}\right) \longrightarrow \infty \quad \text{as } t \to 0+ \end{aligned}$$

showing that g cannot satisfy (E).
Also note that g cannot satisfy any of the other conditions (A)–(D). To see this observe that $g(0,s) = 1/s$ for $0 < s \leq 1$ is not Lebesgue integrable with respect to s on $[0,1]$ showing that g cannot satisfy (A') and hence also not (A). Moreover, for fixed $s \in (0,1]$ we have $\|g(\cdot,s)\|_\infty = 1/s$ which cannot be bounded by a Lebesgue integrable function. Thus, (B') is violated. Similarly, for $s \in (0,1]$,

$$\operatorname{Var}\left(g(\cdot, s)\right) = \frac{1}{s} - \frac{1}{1+s} = \frac{1}{s(1+s)}$$

cannot be bounded by an integrable function, and so (B) is also violated. Moreover, (C) does not hold, because

$$\int_0^\tau g(0,s)\,\mathrm{d}s = \infty$$

for any $\tau \in (0,1]$. Condition (D) is also not satisfied, since

$$s \mapsto \|\partial_1 g(\cdot, s)\|_\infty = \sup_{t \in [0,1]} \left| -\frac{1}{(t+s)^2} \right| = \frac{1}{s^2}$$

does not belong to L_p for any $p \geq 1$. ◇

We remark that besides the kernel in Example 4.3.13 none of the remaining kernels in the examples considered so far does satisfy (D) as none of these kernels is differentiable with respect to t for almost all s, with one exception: The kernel $g(t,s) = \chi_{\mathbb{Q}}(t)\chi_{\mathbb{Q}}(s)$ from Example 4.3.12 is zero for almost all s and hence satisfies (D).

The importance of this new condition (E) is now illustrated by the following result which was proven in [32].

Theorem 4.3.14. *Let $g : [0,1] \times [0,1] \to \mathbb{R}$ satisfy condition (A). Then the following statements are equivalent.*

(a) The kernel function g satisfies the conditions (C) and (E).

(b) The integral operator I_g maps the space $BV \cap C$ into itself and is bounded, and the set $\{I_g x \mid x \in BV \cap C, \|x\|_{BV} \leq R\}$ is equicontinuous for every $R > 0$.

A quite similar but slightly stronger condition than (E) is

$$\forall \varepsilon > 0\ \exists \delta > 0\ \forall t_1, t_2 \in [0,1]: \quad |t_1 - t_2| \leq \delta \ \Rightarrow\ \|g(t_1,\cdot) - g(t_2,\cdot)\|_{L_1} \leq \varepsilon. \tag{F}$$

It is clear that (F) implies (E). The converse, however, is not true, if we allow the integral in (E) to be an improper Riemann integral.

Example 4.3.15. Let $\tau_n := 1 - 1/2^n$ and $\mu_n := 1 - 3/2^{n+2}$ for all $n \in \mathbb{N}_0$. Then $\tau_n < \mu_n < \tau_{n+1}$ and $(\tau_n + \tau_{n+1})/2 = \mu_n$ for all $n \in \mathbb{N}_0$. Therefore, the function $h : [0,1] \to \mathbb{R}$, defined to be piecewise linear and continuous on $[0,1)$ by $h(1) := 0$ and $h(\tau_n) := 0$ and $h(\mu_n) := (-1)^n 2^n/(n+1)$ for all $n \in \mathbb{N}_0$, is Riemann integrable on any interval $[0,a]$ for $a \in (0,1)$. We show that h is improperly Riemann integrable on $[0,1]$. Observe that

$$\int_{\tau_{n-1}}^{\tau_n} h(s)\,\mathrm{d}s = \frac{\tau_n - \tau_{n-1}}{2} h(\mu_{n-1}) = \frac{(-1)^{n+1}}{4n},$$

$$\int_{\tau_{n-1}}^{\tau_n} |h(s)|\,\mathrm{d}s = \frac{\tau_n - \tau_{n-1}}{2} |h(\mu_{n-1})| = \frac{1}{4n} \quad \text{for } n \in \mathbb{N}.$$

Fix $a \in (0,1)$ and pick $n \in \mathbb{N}$ such that $\tau_{n-1} \leq a < \tau_n$. Then

$$\left| \int_0^a h(s)\,\mathrm{d}s - \int_0^{\tau_{n-1}} h(s)\,\mathrm{d}s \right| \leq \int_{\tau_{n-1}}^{\tau_n} |h(s)|\,\mathrm{d}s = \frac{1}{4n}$$

and

$$\int_0^{\tau_{n-1}} h(s)\,\mathrm{d}s = \sum_{j=1}^{n-1} \int_{\tau_{j-1}}^{\tau_j} h(s)\,\mathrm{d}s = \sum_{j=1}^{n-1} \frac{(-1)^{j+1}}{4j}.$$

Letting $a \to 1-$ implies $n \to \infty$ and hence

$$\int_0^1 h(s)\,\mathrm{d}s = \lim_{a \to 1-} \int_0^a h(s)\,\mathrm{d}s = \sum_{j=1}^{\infty} \frac{(-1)^{j+1}}{4j} = \frac{\log 2}{4}.$$

However, h is not Lebesgue integrable on $[0,1]$. Even worse, $h \notin L_p$ for any $p \geq 1$. Indeed, a straightforward but cumbersome calculation shows

$$\int_0^1 |h(s)|^p\,\mathrm{d}s = \sum_{j=1}^{\infty} \int_{\tau_{j-1}}^{\tau_j} |h(s)|^p\,\mathrm{d}s = \sum_{j=1}^{\infty} \frac{2^{jp-j-p}}{(p+1)j^p}$$

which is divergent for any $p \geq 1$.
We now define the function $g : [0,1] \times [0,1] \to \mathbb{R}$ by $g(t,s) := t h(s)$. Then

$$\|g(t_1,\cdot) - g(t_2,\cdot)\|_{L_1} = |t_1 - t_2|\, \|h\|_{L_1} = \infty \quad \text{for all } t_1 \neq t_2 \text{ in } [0,1].$$

Thus, g does not satisfy (F). But it does satisfy (E), because

$$\sup_{\tau \in [0,1]} \left| \int_0^\tau g(t_1,s) - g(t_2,s)\,\mathrm{d}s \right| = |t_1 - t_2| \sup_{\tau \in [0,1]} \left| \int_0^\tau h(s)\,\mathrm{d}s \right|,$$

and the right supremum is finite as h is improperly Riemann integrable on $[0,1]$.

Note that this kernel g does not satisfy the conditions (A), (B) and (B'), while it does satisfy the conditions (A'). This follows from Proposition 4.3.9 as g is given in separated kernels.
Moreover, the variation

$$\operatorname{Var}\left(\int_0^\tau g(\cdot,s)\,\mathrm{d}s\right) = \left|\int_0^\tau h(s)\,\mathrm{d}s\right|,$$

where the integral is meant in the sense of Riemann, is finite for each $\tau \in [0,1]$. In particular, h is Kurzweil-Henstock integrable on $[0,1]$ in the sense of Definition 2.1.1, and therefore the function

$$\tau \mapsto \int_0^\tau h(s)\,\mathrm{d}s$$

is continuous [68] and hence bounded on $[0,1]$. Consequently, condition (C) is satisfied. Finally, (D) is not fulfilled, because the function $s \mapsto \|\partial_1 g(\cdot,s)\|_\infty = |h(s)|$ does not belong to L_p for any $p \geq 1$. ◇

Let us now check which of the kernels in the examples considered so far does satisfy the conditions (E) and (F), respectively. The kernel $g(t,s) = \chi_{\mathbb{Q}}(t)$ from the Examples 4.3.2 and 4.3.10 (b) cannot satisfy (E) and hence also not (F), because for any $t \in [0,1]\backslash\mathbb{Q}$ we have

$$\int_0^1 g(0,s) - g(t,s)\,\mathrm{d}s = 1.$$

The kernel $g(t,s) = \chi_{\mathbb{Q}}(t-s) = \chi_{\mathbb{Q}}(s-t)$ from the Examples 4.3.3 and 4.3.8, however, does satisfy (F) and therefore also (E), because for any $t_1, t_2 \in [0,1]$ we have

$$\int_0^1 \big|g(t_1,s) - g(t_2,s)\big|\,\mathrm{d}s = \int_0^1 \big|\chi_{\mathbb{Q}}(t_1-s) - \chi_{\mathbb{Q}}(t_2-s)\big|\,\mathrm{d}s = 0,$$

as $\chi_{\mathbb{Q}}(t-s) = 0$ for fixed $t \in [0,1]$ and almost all $s \in [0,1]$.
The kernel $g(t,s) = g_1(t)g_2(s)$ from Example 4.3.10 (a) with $g_1 = \chi_{(0,1]}$ and

$$g_2(s) = \begin{cases} 1/s & \text{for } 0 < s \leq 1, \\ 0 & \text{for } s = 0 \end{cases}$$

cannot satisfy (E) and hence also not (F), because for any $t \in (0,1]$ we have

$$\int_0^1 \big(g(t,s) - g(0,s)\big)\,\mathrm{d}s = \int_0^1 \frac{1}{s}\,\mathrm{d}s = \infty.$$

A similar reasoning holds for the kernel

$$g(t,s) = \begin{cases} 1/s & \text{for } t = 1, 0 < s \leq 1, \\ 1 & \text{for } t = 1, s = 0, \\ 1 & \text{for } 0 \leq t < 1, 0 \leq s \leq 1 \end{cases}$$

from Example 4.3.11. Indeed, for $t \in [0,1)$ we obtain

$$\int_0^1 \big(g(1,s) - g(t,s)\big)\,\mathrm{d}s = \int_0^1 \left(\frac{1}{s} - 1\right)\mathrm{d}s = \infty,$$

showing that neither (E) nor (F) can hold.

For the kernel $g(t,s) = \chi_{\mathbb{Q}}(t)\chi_{\mathbb{Q}}(s)$ from Example 4.3.12 we have that $g(t,\cdot) = \mathbb{0}$ almost everywhere for fixed $t \in [0,1]$ and so (F) as well as (E) must be satisfied.

As we have seen in Example 4.3.13, the kernel

$$g(t,s) = \begin{cases} \dfrac{1}{t+s} & \text{for } 0 < s \leq 1, \\ 0 & \text{for } s = 0. \end{cases}$$

does not satisfy (E) and hence also not (F).

In general, condition (F) together with (A) now takes care for I_g to map L_∞ into C. But even more is true, namely, a perfect analogue to Theorem 4.3.14 which reads as follows.

Theorem 4.3.16. *Let $g : [0,1] \times [0,1] \to \mathbb{R}$ satisfy condition (A). Then the following statements are equivalent.*

(a) The kernel function g satisfies the condition (F).

(b) The integral operator I_g maps the space L_∞ into the space C and is bounded, and the set $\{I_g x \mid x \in L_\infty, \|x\|_{L_\infty} \leq R\}$ is equicontinuous for every $R > 0$.

Proof. "(a)⇒(b)": That the operator I_g is well-defined follows from (A), and that it maps the space L_∞ into C follows easily from (F). Indeed, for fixed $\varepsilon > 0$ we find accordingly some $\delta > 0$ such that $|t_1 - t_2| \leq \delta$ implies $\|g(t_1,\cdot) - g(t_2,\cdot)\|_{L_1} \leq \varepsilon$. For such $t_1, t_2 \in [0,1]$ and $x \in L_\infty$ it follows that

$$\begin{aligned} \big|I_g x(t_1) - I_g x(t_2)\big| &\leq \int_0^1 |g(t_1,s) - g(t_2,s)||x(s)|\,\mathrm{d}s \leq \|g(t_1,\cdot) - g(t_2,\cdot)\|_{L_1} \|x\|_{L_\infty} \\ &\leq \varepsilon \|x\|_{L_\infty}, \end{aligned}$$

showing that the function $I_g x$ is continuous and that the set $\{I_g x \mid x \in L_\infty, \|x\|_{L_\infty} \leq R\}$ is equicontinuous for any $R > 0$.

It remains to prove that $I_g : L_\infty \to C$ is bounded. To do so, pick for $\varepsilon = 1$ according to condition (F) some $n \in \mathbb{N}$ such that $|t_1 - t_2| \leq 1/n$ implies $\|g(t_1,\cdot) - g(t_2,\cdot)\|_{L_1} \leq 1$. Define $\tau_j := j/n$ for $j \in \{0,\dots,n\}$ and fix $t \in [0,1]$. Then there is some $j \in \{1,\dots,n\}$ such that $\tau_{j-1} \leq t \leq \tau_j$. This gives for all $s \in [0,1]$,

$$\begin{aligned} &|g(t,s)| \\ &\quad\leq |g(t,s) - g(\tau_{j-1},s)| + |g(\tau_{j-1},s) - g(\tau_{j-2},s)| + \ldots + |g(\tau_1,s) - g(\tau_0,s)| + |g(0,s)| \\ &\quad= |g(t,s) - g(\tau_{j-1},s)| + \sum_{i=1}^{j-1} |g(\tau_{i-1},s) - g(\tau_i,s)| + |g(0,s)| \\ &\quad\leq |g(t,s) - g(\tau_{j-1},s)| + \sum_{i=1}^{n} |g(\tau_{i-1},s) - g(\tau_i,s)| + |g(0,s)| \end{aligned}$$

and consequently,

$$\|g(t,\cdot)\|_{L_1} \le \|g(t,\cdot) - g(\tau_{j-1},\cdot)\|_{L_1} + \sum_{i=1}^{n} \|g(\tau_{i-1},\cdot) - g(\tau_i,\cdot)\|_{L_1} + \|g(0,\cdot)\|_{L_1}.$$

Since $|\tau_{j-1} - \tau_j| = 1/n$ and $|t - \tau_{j-1}| \le 1/n$ we obtain

$$\|g(t,\cdot)\|_{L_1} \le n + 1 + \|g(0,\cdot)\|_{L_1}.$$

This implies for $x \in L_\infty$ the estimate

$$|I_g x(t)| \le \|g(t,\cdot)\|_{L_1} \|x\|_{L_\infty} \le \left(n + 1 + \|g(0,\cdot)\|_{L_1}\right) \|x\|_{L_\infty}.$$

Since this is true for any $t \in [0,1]$, we have shown that $I_g : L_\infty \to C$ is bounded with

$$\|I_g\|_{L_\infty \to C} \le n + 1 + \|g(0,\cdot)\|_{L_1}.$$

"(b)⇒(a)": Assume that the set $M := \{I_g x \mid x \in L_\infty, \|x\|_{L_\infty} \le 1\}$ is equicontinuous. This means that for any fixed $\varepsilon > 0$ we find a $\delta > 0$ such that $|t_1 - t_2| \le \delta$ and $\|x\|_{L_\infty} \le 1$ imply $|I_g x(t_1) - I_g x(t_2)| \le \varepsilon$. We now fix such $t_1, t_2 \in [0,1]$ and consider the function

$$h(s) := g(t_1, s) - g(t_2, s)$$

which is measurable due to condition (A). In particular, the sets

$$H_- := \{s \in [0,1] \mid h(s) < 0\} \quad \text{and} \quad H_+ := \{s \in [0,1] \mid h(s) \ge 0\}$$

are measurable, and thus the function $x := \chi_{H_+} - \chi_{H_-}$ belongs to L_∞ with $\|x\|_{L_\infty} \le 1$ and hence $x \in M$. We then have $hx = |h|$ and also

$$\|g(t_1,\cdot) - g(t_2,\cdot)\|_{L_1} = \int_0^1 |h(s)|\,\mathrm{d}s = \int_0^1 h(s)x(s)\,\mathrm{d}s = I_g x(t_1) - I_g x(t_2) \le \varepsilon.$$

But this is nothing else than condition (F), and the proof is complete. ■

For the reader's ease, let us recall in Table 4.3.1 which of the conditions (A)–(F) are fulfilled by the examples considered so far in this section.

Table 4.3.1: Properties of g in the above examples.

Example	(A)	(A')	(B)	(B')	(C)	(D)	(E)	(F)
4.3.2	yes	yes	no	yes	no	no	no	no
4.3.3	yes	yes	no	yes	yes	no	yes	yes
4.3.8	yes	yes	no	yes	yes	no	yes	yes
4.3.10 (a)	no	yes	no	no	no	no	no	no
4.3.10 (b)	yes	yes	no	yes	no	no	no	no
4.3.11	no	yes	no	no	no	no	no	no
4.3.12	yes	yes	yes	yes	yes	yes	yes	yes
4.3.13	no	no	no	no	no	no	no	no
4.3.15	no	yes	no	no	yes	no	yes	no

In Chapter 7 we will apply some of the theoretical results developed in this section to nonlinear integral equations involving linear integral operators like (4.0.3) and nonlinear composition or superposition operators like (5.0.1) and (5.0.2); we will investigate the latter two in the next chapter. In order to prove the existence of solutions of those integral equations we will use fixed point theory, mostly Banach's Fixed Point Theorem for contractions and Schauder's or Darbo's Fixed Point Theorem for compact maps. Since a bounded linear operator is always Lipschitz continuous, for applying Banach's theorem we have to ensure only a Lipschitz condition for the nonlinear operators; this is a difficult problem, as we will see in the next chapter. On the other hand, for applying Schauder's or Darbo's fixed point theorem we need (at least sufficient) conditions on g guaranteeing that I_g is compact, since the nonlinear part is compact only under quite exceptional assumptions.

The following result was proven in [31] and gives a compactness criterion for the operator I_g acting in the Wiener space WBV_p.

Theorem 4.3.17. *Under the conditions (A) and (B), the integral operator I_g maps WBV_p continuously into WBV_q and is compact for any $p, q \geq 1$.*

The proof is based on Helly's Selection Principle, see Theorem 1.2.28.

In order to achieve similar compactness results for other BV-spaces we introduce the following conditions.

$$\exists \theta > 0\ \exists m \in L_1\ \forall\,'s \in [0,1]: \quad \mathrm{Var}_\varphi\Big(\theta g(\cdot, s)\Big) \leq m(s), \tag{B$_\varphi$}$$

$$\forall \theta > 0\ \exists m_\theta \in L_1\ \forall\,'s \in [0,1]: \quad \mathrm{Var}_\varphi\Big(\theta g(\cdot, s)\Big) \leq m_\theta(s), \tag{B$^*_\varphi$}$$

$$\exists m \in L_1\ \forall\,'s \in [0,1]: \quad \mathrm{Var}_\Lambda\Big(g(\cdot, s)\Big) \leq m(s), \tag{B$_\Lambda$}$$

$$\exists m \in L_1\ \forall\,'s \in [0,1]: \quad \mathrm{RVar}_p\Big(g(\cdot, s)\Big) \leq m(s). \tag{B$_p$}$$

Note that the new conditions (B$_\varphi$) respectively (B$^*_\varphi$), (B$_\Lambda$) and (B$_p$) act in the spaces YBV_φ, ΛBV and RBV_p, respectively, as analogues to condition (B) in the space BV. Since these conditions will play an important role in Chapter 7 and since we do not want to repeat all proofs for each individual BV-space we take a uniform approach and summarize the conditions (B$_\varphi$), (B$^*_\varphi$), (B$_\Lambda$) and (B$_p$) for X being one of the spaces BV, WBV_p, YBV_φ, ΛBV and RBV_p in the following conditions.

$$\exists \theta > 0\ \exists m \in L_1\ \forall\,'s \in [0,1]: \quad \mathrm{Var}_X\Big(\theta g(\cdot, s)\Big) \leq m(s), \tag{B$_X$}$$

$$\forall \theta > 0\ \exists m_\theta \in L_1\ \forall\,'s \in [0,1]: \quad \mathrm{Var}_X\Big(\theta g(\cdot, s)\Big) \leq m_\theta(s), \tag{B$^*_X$}$$

where Var_X denotes the variation of the space X, that is, $\mathrm{Var}_{BV} = \mathrm{Var}$, $\mathrm{Var}_{WBV_p} = \mathrm{Var}_p$, $\mathrm{Var}_{YBV_\varphi} = \mathrm{Var}_\varphi$, $\mathrm{Var}_{\Lambda BV} = \mathrm{Var}_\Lambda$ and $\mathrm{Var}_{RBV_p} = \mathrm{RVar}_p$.
We make two comments on these conditions. It is clear that the (B$^*_X$) condition implies

(B_X) in any BV-space X. For the converse, note that for $x \in X$ and $\theta > 0$ we have

$$\begin{aligned} \operatorname{Var}(\theta x) &= \theta \operatorname{Var}(x), \\ \operatorname{Var}_p(\theta x) &= \theta^p \operatorname{Var}_p(x), \\ \operatorname{Var}_\Lambda(\theta x) &= \theta \operatorname{Var}_\Lambda(x), \\ \operatorname{RVar}_p(\theta x) &= \theta^p \operatorname{RVar}_p(x). \end{aligned}$$

As a consequence, the two conditions (B_X) and (B_X^*) are equivalent for X being one of the spaces BV, WBV_p, ΛBV or RBV_p, and we can assume $\theta = 1$ in (B_X) for these spaces. However, in the space $X = YBV_\varphi$ both of these conclusions are false; we give an example.

Example 4.3.18. Let φ and $\mathfrak{J}_{(\alpha_j)}$ be as in Example 1.2.12 and consider $g(t,s) := \mathfrak{J}_{(\alpha_j)}(t)$. We have seen in that example that $\operatorname{Var}_\varphi(g(\cdot,s)) \le c$ for all $s \in [0,1]$ and some $c > 0$, but $\operatorname{Var}_\varphi(4g(\cdot,s)) = \infty$ for all $s \in [0,1]$. Thus, g satisfies (B_φ) but not (B_φ^*). Note that below Example 1.2.12 we have also seen that φ does not satisfy a δ_2-condition. $\diamondsuit$

If the Young function φ is given by $\varphi(t) = t^p$ and thus $YBV_\varphi = WBV_p$, then condition (B_φ) is equivalent to

$$\exists m \in L_1 \; \forall' s \in [0,1]: \quad \operatorname{Var}_p\big(g(\cdot,s)\big) \le m(s);$$

this is exactly the condition in Theorem 4.3.5. In particular, in this case the conditions (B_φ) and (B_φ^*) are equivalent.

For separated kernels we have the following analogue to the (B)-part of Proposition 4.3.9.

Proposition 4.3.19. *The function $g : [0,1] \times [0,1] \to \mathbb{R}$ given in separated kernels $g(t,s) = g_1(t)g_2(s)$ for $g_1, g_2 : [0,1] \to \mathbb{R}$ satisfies*

(B_Λ) if and only if g_1 is constant, or $g_2 = \mathbb{0}$ almost everywhere, or $g_1 \in \Lambda BV$ and $|g_2| \le m$ for some $m \in L_1$.

(B_p) if and only if g_1 is constant, or $g_2 = \mathbb{0}$ almost everywhere, or $g_1 \in RBV_p$ and $|g_2| \le m$ for some $m \in L_p$.

One might wonder why there is no entry for the Young variation. The reason is that for separated kernels with g_2 not being zero almost everywhere the condition

$$g_1 \in YBV_\varphi \text{ and } |g_2| \le m \text{ for some } m \in L_1$$

is simply not equivalent to (B_φ).

Example 4.3.20. Let $g_2 \in L_1$ be given by

$$g_2(s) = \begin{cases} 1/\sqrt{s} & \text{for } 0 < s \le 1, \\ 0 & \text{for } s = 0, \end{cases}$$

take φ and $g_1 = \mathfrak{J}_{(\alpha_j)}$ as in Example 1.2.12 and set $g(t,s) = g_1(t)g_2(s)$. There we have seen that $\mathrm{Var}_\varphi(g_1) < \infty$ and hence $g_1 \in YBV_\varphi$. But we have also seen that $\mathrm{Var}_\varphi(4g_1) = \infty$. Since $g(\cdot,s) \in \mathcal{S}_c$ for any $s \in [0,1]$ we have for any fixed $\theta > 0$ and $0 < s \leq \theta^2/16$ that

$$\mathrm{Var}_\varphi\left(\theta g(\cdot,s)\right) = \mathrm{Var}_\varphi\left(g_1\theta/\sqrt{s}\right) \geq \mathrm{Var}_\varphi\left(4g_1\right) = \infty.$$

But then (B_φ) is violated, although $g_2 \in L_1$. ◊

The importance of the (B)-type conditions summarized in (B_X) become now apparent in one of our main theorems of this section. It generalizes the results of Theorem 4.3.1 to our other BV-spaces.

Theorem 4.3.21. *Let X be one of the spaces BV, WBV_p, YBV_φ, ΛBV or RBV_p. Under the conditions (A) and (B_X) the integral operator I_g maps the space L_∞ into the space X and is bounded with*

$$\|I_g\|_{L_\infty\to X} \leq \|g(0,\cdot)\|_{L_1} + \theta^{-1}\left\{\begin{array}{ll} 2\,\|m\|_{L_1} & \text{for } X = BV, \\ 2\,\|m\|_{L_1}^{1/p} & \text{for } X = WBV_p, \\ \left(\varphi^{-1}(1)+1\right)\max\left\{1, \|m\|_{L_1}\right\} & \text{for } X = YBV_\varphi, \\ \left(1+\lambda_1^{-1}\right)\|m\|_{L_1} & \text{for } X = \Lambda BV, \\ 2\,\|m\|_{L_1}^{1/p} & \text{for } X = RBV_p, \end{array}\right\} \tag{4.3.6}$$

where θ is taken from condition (B_X).

Note that, as pointed out before, in the spaces BV, WBV_p, ΛBV and RBV_p the number θ in (B_X) may be forced to be equal to 1 when m is adjusted properly. This, however, is not true in the space YBV_φ. Therefore, according to Theorem 4.3.21, the operator norm $\|I_g\|_{L_\infty\to X}$ in general depends on both the bound m and the value of θ, both coming from (B_X).

Proof of Theorem 4.3.21. First note that

$$|I_g x(0)| \leq \|g(0,\cdot)\|_{L_1}\|x\|_{L_\infty}. \tag{4.3.7}$$

For $X = WBV_p$ with $1 \leq p < \infty$ assume that g satisfies (B_{WBV_p}) with $\varphi(t) = t^p$ for some $\theta > 0$ and pick $m \in L_1$ so that $\mathrm{Var}_p\left(\theta g(\cdot,s)\right) \leq m(s)$ for almost all $s \in [0,1]$. Let $0 = t_0 < \ldots < t_n = 1$ be a partition of $[0,1]$ and fix $x \in WBV_p$. From Jensen's inequality we obtain

$$\begin{aligned}\sum_{j=1}^n \left|I_g x(t_{j-1}) - I_g x(t_j)\right|^p \mathrm{d}s &\leq \int_0^1 \sum_{j=1}^n \left|g(t_{j-1},s) - g(t_j,s)\right|^p |x(s)|^p \,\mathrm{d}s \\ &\leq \theta^{-p}\int_0^1 m(s)|x(s)|^p \,\mathrm{d}s \leq \theta^{-p}\|x\|_{L_\infty}^p \|m\|_{L_1}.\end{aligned}$$

Consequently $\operatorname{Var}_p(I_g x)^{1/p} \le \theta^{-1} \|x\|_{L_\infty} \|m\|_{L_1}^{1/p}$, and with the help of (4.3.7) and (1.2.11) we obtain

$$\begin{aligned}\|I_g x\|_{WBV_p} &= \|I_g x\|_\infty + \operatorname{Var}_p(I_g x)^{1/p} \le |I_g x(0)| + 2\operatorname{Var}_p(I_g x)^{1/p}\\ &\le \|x\|_{L_\infty} \Big(\|g(0,\cdot)\|_{L_1} + 2\theta^{-1} \|m\|_{L_1}^{1/p} \Big)\end{aligned}$$

which means that I_g maps L_∞ into WBV_p, is bounded and satisfies (4.3.6). This also shows the result and the desired estimate for the space BV.

For $X = YBV_\varphi$ assume that g satisfies (B_{YBV_φ}) for some $\theta > 0$, pick accordingly $m \in L_1$ so that $\operatorname{Var}_\varphi\big(\theta g(\cdot, s)\big) \le m(s)$ for almost all $s \in [0,1]$, and let $0 = t_0 < \ldots < t_n = 1$ be a partition of $[0,1]$. For fixed $x \in YBV_\varphi$ (without loss of generality we may assume $\|x\|_{L_\infty} > 0$), $\mu' := \|x\|_{L_\infty} \max\{1, \|m\|_{L_1}\}$ and $\mu := \theta^{-1}\mu'$ we have with the help of Jensen's inequality

$$\begin{aligned}\sum_{j=1}^n \varphi\Big(\tfrac{1}{\mu}|I_g x(t_{j-1}) - I_g x(t_j)|\Big) &\le \sum_{j=1}^n \varphi\left(\frac{1}{\mu}\int_0^1 \Big|g(t_{j-1},s) - g(t_j,s)\Big||x(s)|\,\mathrm{d}s\right)\\ &\le \int_0^1 \sum_{j=1}^n \varphi\left(\frac{\theta}{\mu'}\Big|g(t_{j-1},s) - g(t_j,s)\Big||x(s)|\right)\mathrm{d}s\\ &\le \frac{\|x\|_{L_\infty}}{\mu'}\int_0^1 \sum_{j=1}^n \varphi\Big(\theta\Big|g(t_{j-1},s) - g(t_j,s)\Big|\Big)\,\mathrm{d}s\\ &\le \frac{\|x\|_{L_\infty}}{\mu'}\int_0^1 m(s)\,\mathrm{d}s \le 1.\end{aligned}$$

Consequently $\mathfrak{M}(I_g x) \le \mu = \theta^{-1} \|x\|_{L_\infty} \max\{1, \|m\|_{L_1}\}$, where the symbol $\mathfrak{M}$ denotes the Minkowski functional (1.2.19). With (4.3.7) and (1.2.24) we obtain

$$\begin{aligned}\|I_g x\|_{YBV_\varphi} &= \|I_g x\|_\infty + \mathfrak{M}(I_g x) \le |I_g x(0)| + \big(\varphi^{-1}(1) + 1\big)\mathfrak{M}(I_g x)\\ &\le \|x\|_{L_\infty}\Big(\|g(0,\cdot)\|_{L_1} + \theta^{-1}\big(\varphi^{-1}(1)+1\big)\max\{1, \|m\|_{L_1}\}\Big)\end{aligned}$$

which means that I_g maps L_∞ into YBV_φ, is bounded and satisfies (4.3.6).

For $X = \Lambda BV$ it was shown in [30] that under the assumptions (A) and ($\mathrm{B}_{\Lambda BV}$) the operator I_g maps L_∞ into ΛBV and is bounded with

$$\operatorname{Var}_\Lambda(I_g x) \le \theta^{-1} \|x\|_{L_\infty} \|m\|_{L_1}.$$

With the help of (1.2.46) and (4.3.7) we obtain

$$\begin{aligned}\|I_g x\|_{\Lambda BV} &= \|I_g x\|_\infty + \operatorname{Var}_\Lambda(I_g x) \le |I_g x(0)| + \big(1 + \lambda_1^{-1}\big)\operatorname{Var}_\Lambda(I_g x)\\ &= \|x\|_{L_\infty}\Big(\|g(0,\cdot)\|_{L_1} + \theta^{-1}\big(1+\lambda_1^{-1}\big)\|m\|_{L_1}\Big).\end{aligned}$$

This shows that I_g maps L_∞ into ΛBV and is bounded with (4.3.6).

For $X = RBV_p$ we proceed as in the proof for $X = YBV_\varphi$, but this time, calculations are much easier. Assume that g satisfies (B_{RBV_p}) for some $\theta > 0$, pick accordingly $m \in$

L_1 so that $\mathrm{RVar}_p(\theta g(\cdot, s)) \le m(s)$ for almost all $s \in [0,1]$, and let $0 = t_0 < \ldots < t_n = 1$ be a partition of $[0,1]$. For fixed $x \in RBV_p$ we have with Jensen's inequality

$$\sum_{j=1}^{n} \frac{|I_g x(t_{j-1}) - I_g x(t_j)|^p}{|t_{j-1} - t_j|^{p-1}} \le \int_0^1 \sum_{j=1}^{n} \frac{|g(t_{j-1}, s) - g(t_j, s)|^p}{|t_{j-1} - t_j|^{p-1}} |x(s)|^p \,\mathrm{d}s$$
$$\le \theta^{-p} \int_0^1 m(s)|x(s)|^p \,\mathrm{d}s \le \theta^{-p} \|x\|_{L_\infty}^p \|m\|_{L_1}.$$

Consequently, $\mathrm{RVar}_p(I_g x)^{1/p} \le \theta^{-1} \|x\|_{L_\infty} \|m\|_{L_1}^{1/p}$. With (4.3.7) and (1.2.53) we obtain

$$\|I_g x\|_{RBV_p} = \|I_g x\|_\infty + \mathrm{RVar}_p(I_g x)^{1/p} \le |I_g x(0)| + 2\,\mathrm{RVar}_p(I_g x)^{1/p}$$
$$\le \|x\|_{L_\infty} \left(\|g(0,\cdot)\|_{L_1} + 2\theta^{-1} \|m\|_{L_1}^{1/p} \right)$$

which means that I_g maps L_∞ into YBV_φ, is bounded and fulfills (4.3.6). ■

Although the condition $I_g(L_\infty) \subseteq X$ may seem somewhat annoying (as $I_g(X) \subseteq X$ seems more natural), it has the advantage that the nonlinear part of the fixed point operators that we consider in Chapter 7 has to map only X into L_∞ which in many cases may be easily achieved. We give one sample criterion for the nonlinear superposition operator in Theorem 5.2.34 in the next chapter.

As announced the conditions (A) and (B_X) are too strong to be responsible only for the boundedness of I_g. Indeed, we get compactness for free. Surprisingly, this is true for the spaces BV, WBV_p, ΛBV and RBV_p, but probably not for the space YBV_φ (we do not know if (B_X) is sufficient). For this space, we need the stronger condition (B_X^*) which is equivalent to (B_X) in all the other BV-spaces. We then get an even stronger result that has been proven in [32] for $X = BV$.

Proposition 4.3.22. *Let X be one of the spaces BV, WBV_p, YBV_φ, ΛBV or RBV_p. Assume that (x_n) is a bounded sequence in L_∞ and converges almost everywhere to some $x \in L_\infty$. If $g : [0,1] \times [0,1] \to \mathbb{R}$ satisfies the conditions (A) and (B_X^*), then $(I_g x_n)$ is a sequence in X that converges in X to $I_g x$.*

Proof. By Theorem 4.3.21 the integral operator I_g maps L_∞ into X, and this shows that the functions $I_g x_n$ belong to X for each $n \in \mathbb{N}$. We now show that they converge in X to $I_g x$. Equivalently we show that the functions $I_g y_n$ converge to $\mathbb{0}$ in X, where $y_n := x_n - x$.

We start with the space $X = YBV_\varphi$. Since g satisfies (B_φ^*) we find for each $\theta > 0$ a function $m_\theta \in L_1$ so that $\mathrm{Var}_\varphi\left(\theta g(\cdot, s)\right) \le m_\theta(s)$ for almost all $s \in [0,1]$. Let $0 = t_0 < \ldots < t_k = 1$ be a partition of $[0,1]$ and let $\lambda > 0$ be fixed. Since (y_n) is a bounded sequence in L_∞ we find some $\mu > 0$ such that $\lambda |y_n(s)| \le \mu$ for all $n \in \mathbb{N}$ and almost all $s \in [0,1]$. Moreover, the convergence of the sequence (y_n) to $\mathbb{0}$ almost everywhere implies that the sequence also converges in measure which means that the measures of the sets $A_n := \{s \in [0,1] \mid |y_n(s)| \ge 1/\lambda\}$ tend to 0 as $n \to \infty$. We obtain again with the help of Jensen's inequality

$$\begin{aligned}
\sum_{j=1}^{k}\varphi\Big(\lambda|I_g y_n(t_{j-1})-I_g y_n(t_j)|\Big) &\le \sum_{j=1}^{k}\varphi\left(\lambda\int_0^1|g(t_{j-1},s)-g(t_j,s)||y_n(s)|\,\mathrm{d}s\right)\\
&\le\left\{\int_{A_n}+\int_{[0,1]\setminus A_n}\right\}\sum_{j=1}^{k}\varphi\Big(\lambda|g(t_{j-1},s)-g(t_j,s)||y_n(s)|\Big)\,\mathrm{d}s\\
&\le\int_{A_n}\sum_{j=1}^{k}\varphi\Big(\mu|g(t_{j-1},s)-g(t_j,s)|\Big)\,\mathrm{d}s\\
&\quad+\lambda\int_0^1|y_n(s)|\sum_{j=1}^{k}\varphi\Big(|g(t_{j-1},s)-g(t_j,s)|\Big)\,\mathrm{d}s\\
&\le\int_{A_n}m_\mu(s)\,\mathrm{d}s+\lambda\int_0^1|y_n(s)|m_1(s)\,\mathrm{d}s. \qquad (4.3.8)
\end{aligned}$$

Since $|A_n|\to 0$ as $n\to\infty$, the first integral in (4.3.8) goes to 0 as $n\to\infty$. Since (y_n) is bounded in L_∞ and converges almost everywhere to $\mathbb{0}$, the second integral in (4.3.8) goes also to 0 by the Dominated Convergence Theorem. Thus, we have shown that $\mathrm{Var}_\varphi(\lambda I_g y_n)\to 0$ as $n\to\infty$ for each $\lambda>0$. By Proposition 1.2.15 (a) we conclude $\mathfrak{M}(I_g y_n)\to 0$ as $n\to\infty$.

Moreover, the Dominated Convergence Theorem also implies $|I_g y_n(0)|\to 0$ as $n\to\infty$, and with the help of (1.2.24) we finally obtain $\|I_g y_n\|_{YBV_\varphi}\to 0$ as $n\to\infty$.

Next, we give attention to the space $X=\Lambda BV$. Let $([a_j,b_j])_{1\le j\le k}$ be a collection of nonoverlapping intervals in $[0,1]$ and let $m\in L_1$ be so that $\mathrm{Var}_\Lambda(g(\cdot,s))\le m(s)$ for almost all $s\in[0,1]$ according to (B_Λ^*) with $\theta=1$. We have

$$\begin{aligned}
\sum_{j=1}^{k}\lambda_j|I_g y_n(a_j)-I_g y_n(b_j)| &\le\int_0^1\left(\sum_{j=1}^{k}\lambda_j|g(a_j,s)-g(b_j,s)|\right)|y_n(s)|\,\mathrm{d}s\\
&\le\int_0^1 m(s)|y_n(s)|\,\mathrm{d}s
\end{aligned}$$

and hence $\mathrm{Var}_\Lambda(I_g y_n)\le\|my_n\|_{L_1}$. From the Dominated Convergence Theorem again follows that $\|my_n\|_{L_1}\to 0$ and hence $\mathrm{Var}_\Lambda(I_g y_n)\to 0$ as $n\to\infty$. Since $|I_g y_n(0)|\to 0$ we conclude with (1.2.46) that $\|I_g y_n\|_{\Lambda BV}\to 0$ as $n\to\infty$.

Finally, we deal with $X=RBV_p$. We proceed as in the proof for $X=YBV_\varphi$, but this time, calculations are again easier. Let $0=t_0<\ldots<t_k=1$ be a partition of $[0,1]$ and let $m\in L_1$ be so that $\mathrm{RVar}_p(g(\cdot,s))\le m(s)$ for almost all $s\in[0,1]$ according to (B_p^*) with $\theta=1$. We have with Jensen's inequality

$$\begin{aligned}
\sum_{j=1}^{k}\frac{|I_g y_n(t_{j-1})-I_g y_n(t_j)|^p}{|t_{j-1}-t_j|^{p-1}} &\le\int_0^1\sum_{j=1}^{k}\frac{|g(t_{j-1},s)-g(t_j,s)|^p}{|t_{j-1}-t_j|^{p-1}}|y_n(s)|^p\,\mathrm{d}s\\
&\le\int_0^1 m(s)|y_n(s)|^p\,\mathrm{d}s.
\end{aligned}$$

Consequently, $\mathrm{RVar}_p(I_g y_n)^{1/p}\le\|m|y_n|^p\|_{L_1}^{1/p}$. As above we conclude with the Dominated Convergence Theorem and (1.2.53) that $\|I_g y_n\|_{RBV_p}\to 0$ as $n\to\infty$. ■

We remark that the sequence (x_n) in Proposition 4.3.22 need not to converge in L_∞ to x. Otherwise, the result would have followed immediately from Theorem 4.3.21.

As a consequence of Proposition 4.3.22 we get the promised compactness criterion.

Corollary 4.3.23. *Let X be one of the spaces BV, WBV_p, YBV_φ, ΛBV or RBV_p. Under the conditions (A) and (B_X^*) the operator I_g maps the space X into itself and is compact.*

Proof. From Theorem 4.3.21 we get that the operator I_g maps L_∞ and hence also X into X. We now show that $I_g : X \to X$ is compact. To this end, let (x_n) be a bounded sequence in X. In view of Helly's Selection Principle (Theorem 1.2.28) we find a subsequence $(x_{n_k})_k$ of (x_n) which converges pointwise to some function $x \in X$. Since (x_n) is bounded in X it is also bounded in L_∞. Proposition 4.3.22 now guarantees that the sequence $(I_g x_{n_k})_k$ converges in X to $I_g x$. Consequently, I_g is compact. ∎

For $X = RBV_p$ the result in Corollary 4.3.23 is indeed stronger than Theorem 4.3.7, because we now get compactness. Even better, the condition (B_{RBV_p}) or equivalently (B_p) is weaker than condition (D), as announced earlier. Note that (D) implies, as already mentioned, that $g(\cdot, s) \in RBV_p$ for almost all $s \in [0,1]$ and any $p \in (1, \infty)$ as well as $\mathrm{RVar}_p(g(\cdot, s)) \leq m(s)$ for some $m \in L_1$. Thus, (D) indeed implies (B_p). But condition (B_p) is strictly weaker than (D), as is shown by the following example.

Example 4.3.24. The function $g : [0,1] \times [0,1] \to \mathbb{R}$, given by

$$g(t, s) = \int_0^t \chi_{[0,1/2]}(\tau)\, \mathrm{d}\tau,$$

is for no $s \in [0,1]$ continuously differentiable with respect to t and thus cannot satisfy (D). However, for all $s \in [0,1]$ we have by Riesz' Theorem 1.2.25 that

$$\mathrm{RVar}_p\left(g(\cdot, s)\right) = \int_0^1 \chi_{[0,1/2]}(t)\, \mathrm{d}t = \frac{1}{2}$$

and hence condition (B_p) is met for any $p \in (1, \infty)$. ◇

As we have seen in Theorem 4.3.21, under the hypotheses of Corollary 4.3.23 the operator I_g maps even L_∞ into the respective spaces BV, WBV_p, YBV_φ, ΛBV and RBV_p. However, we cannot expect $I_g : L_\infty \to X$ for X being one of these BV-spaces to be compact. We illustrate this for the space $X = BV$.

Example 4.3.25. Let $g : [0,1] \times [0,1] \to \mathbb{R}$ be defined by $g(t, s) := \chi_{[0,t]}(s)$. Then g clearly satisfies (A) and (B). Moreover, for $n \in \mathbb{N}$ and $j \in \{1, \ldots, 2^n\}$ define the intervals $I_{j,n} := (\frac{j-1}{2^n}, \frac{j}{2^n})$ and let $x_n \in L_\infty$ be given by

$$x_n(t) = \begin{cases} (-1)^{j+1} & \text{for } t \in I_{j,n}, \\ 0 & \text{for } t \in \{0, 1/2^n, 2/2^n, \ldots, 1\}. \end{cases}$$

The functions x_n form a bounded sequence in L_p with $\|x_n\|_{L_p} = 1$ for any $p \in [1, \infty]$. But for the integrals we obtain

$$y_n(t) := I_g x_n(t) = \int_0^1 g(t, s) x_n(s)\, \mathrm{d}s = \int_0^t x_n(s)\, \mathrm{d}s,$$

and thus each $y_n \in AC$ is a piecewise linear "zigzag" function with $\|y_n\|_\infty = 1/2^n$; in particular, (y_n) converges uniformly to $\mathbb{0}$. But it cannot have a subsequence converging in BV to $\mathbb{0}$, because Theorem 1.1.20 dictates $\operatorname{Var}(y_n) = \|y_n'\|_{L_1} = \|x_n\|_{L_1} = 1$ for each $n \in \mathbb{N}$. ◇

Note that the sequence (x_n) in Example 4.3.25 is bounded in L_∞ but not in BV, because $\operatorname{Var}(x_n) = 2^{n+1}$ for each $n \in \mathbb{N}$. Thus, Example 4.3.25 is not contradictory to Corollary 4.3.23.

Since the (B)-type conditions will play the most important role in Chapter 7 we collect them in Table 4.3.2 below to bring some structure into the thicket of formalism.

Table 4.3.2: Conditions (B_X) and (B_X^*) in our BV-spaces.

X	(B_X)	(B_X^*)
BV	$\exists m \in L_1 \ \forall' s \in [0,1]$: $\operatorname{Var}\big(g(\cdot,s)\big) \leq m(s)$	$\exists m \in L_1 \ \forall' s \in [0,1]$: $\operatorname{Var}\big(g(\cdot,s)\big) \leq m(s)$
WBV_p	$\exists m \in L_1 \ \forall' s \in [0,1]$: $\operatorname{Var}_p\big(g(\cdot,s)\big) \leq m(s)$	$\exists m \in L_1 \ \forall' s \in [0,1]$: $\operatorname{Var}_p\big(g(\cdot,s)\big) \leq m(s)$
YBV_φ	$\exists \theta > 0 \ \exists m \in L_1 \ \forall' s \in [0,1]$: $\operatorname{Var}_\varphi\big(\theta g(\cdot,s)\big) \leq m(s)$	$\forall \theta > 0 \ \exists m_\theta \in L_1 \ \forall' s \in [0,1]$: $\operatorname{Var}_\varphi\big(\theta g(\cdot,s)\big) \leq m_\theta(s)$
ΛBV	$\exists m \in L_1 \ \forall' s \in [0,1]$: $\operatorname{Var}_\Lambda\big(g(\cdot,s)\big) \leq m(s)$	$\exists m \in L_1 \ \forall' s \in [0,1]$: $\operatorname{Var}_\Lambda\big(g(\cdot,s)\big) \leq m(s)$
RBV_p	$\exists m \in L_1 \ \forall' s \in [0,1]$: $\operatorname{RVar}_p\big(g(\cdot,s)\big) \leq m(s)$	$\exists m \in L_1 \ \forall' s \in [0,1]$: $\operatorname{RVar}_p\big(g(\cdot,s)\big) \leq m(s)$

We remark that the integral operator $I_g : X \to X$ can never be injective if X is one of the spaces BV, WBV_p, YBV_φ or ΛBV. Indeed, in all of these spaces lies the characteristic function $x = \chi_{\{0\}}$ which, albeit not being $\mathbb{0}$ itself, is mapped by I_g onto the zero function $\mathbb{0}$.

Moreover, an integral operator with separated kernels $g = g_1 g_2$ cannot be surjective either, because from (4.3.5) we see that $I_g x = c g_1$ for a number c depending on x which shows that the range of I_g is one-dimensional.

Volterra Kernels

A particularly interesting case of linear integral operators is when the kernel function g is a *Volterra kernel* which means

$$g(t,s) = 0 \quad \text{for } 0 \le t < s \le 1. \tag{4.3.9}$$

The corresponding operator I_g has then the form

$$V_g x(t) := I_g x(t) = \int_0^t g(t,s)x(s)\,\mathrm{d}s, \tag{4.3.10}$$

and we call such operators *Volterra operators* in the sequel. Of course, all results for integral operators from the first part of this section remain true also for Volterra operators. Therefore, we present here how the conditions (A), (B) and their relatives (B_X) and (B_X^*), (C), (D), (E) and (F) look like for Volterra kernels. Since Volterra kernels can only attain values different from zero on the triangle

$$T = \{(t,s) \mid 0 \le s \le t \le 1\}$$

those conditions must take in account only points $(t,s) \in T$. Therefore, an arbitrary kernel meeting these conditions only on T and hence satisfying the Volterra versions of the conditions (A)–(F) does not have to satisfy the original conditions. For each such situation we give here in this subsection an explicit example.

We start with condition (A) which is the overall basic assumption on the kernel. It now reads

$$\forall t \in [0,1]: \quad g(t,\cdot) \in L_1[0,t]. \tag{VA}$$

Arbitrary kernels satisfying (VA) do not need to also satisfy (A).

Example 4.3.26. Consider the kernel g, defined by

$$g(t,s) = \begin{cases} 1/s & \text{for } t = 0 < s \le 1, \\ 0 & \text{otherwise.} \end{cases}$$

Clearly, $g(0,\cdot) \in L_1[0,0]$. For $t > 0$ we have $g(t,\cdot) = \mathbb{0}$ and hence also $g(t,\cdot) \in L_1[0,t]$ which shows that (VA) is true. However, $g(0,\cdot) \notin L_1[0,1] = L_1$ showing that g does not satisfy condition (A). ♢

The second most important condition is (B) and its relatives (B_X) and (B_X^*) for our various BV-spaces X. Condition (B) now reads for Volterra kernels

$$\exists m \in L_1 \ \forall' s \in [0,1]: \quad |g(s,s)| + \mathrm{Var}\Big(g(\cdot,s),[s,1]\Big) \le m(s). \tag{VB}$$

Note that the kernel g in Example 4.3.26 does satisfy (VB) but not (B). Indeed, for $s \in (0,1]$,

$$\mathrm{Var}\Big(g(\cdot,s)\Big) = 1/s$$

which cannot be bounded by a function m that is Lebesgue integrable on $[0,1]$ with respect to s. However, since $g(t,s)=0$ for all $t\in[s,1]$, we have

$$|g(s,s)|+\operatorname{Var}\big(g(t,s),[s,1]\big)=0,$$

and so (VB) is indeed satisfied.

Observe that if we require the majorant m in condition (VB) to belong not only to L_1 but also to L_p for some $p>1$, then $g(t,\cdot)\in L_p$ for all $t\in[0,1]$. This follows from $g(t,s)=0$ for $0\le t<s\le 1$ and from the estimate

$$|g(t,s)|\le|g(s,s)|+|g(s,s)-g(t,s)|\le|g(s,s)|+\operatorname{Var}\big(g(\cdot,s),[s,1]\big)\le m(s)$$

for all $t\in[s,1]$ and almost all $s\in[0,1]$.

With these two modified conditions (VA) and (VB) Theorem 4.3.1 reads for Volterra operators as follows.

Theorem 4.3.27. *Under the conditions (VA) and (VB) the operator V_g maps the space BV into itself and is bounded.*

As we have seen in Theorem 4.3.4, condition (B) may be relaxed to condition (C) in order to get the same result. For Volterra kernels g condition (C) becomes

$$\sup_{\tau\in[0,1]}\operatorname{Var}\left(\int_0^{\min\{\tau,\cdot\}} g(\cdot,s)\,\mathrm{d}s\right)<\infty. \tag{VC}$$

This means that the Jordan variation of the function $t\mapsto\int_0^{\min\{\tau,t\}}g(t,s)\,\mathrm{d}s$ stays bounded as τ runs through $[0,1]$. The kernel g from Example 4.3.26 cannot satisfy (C) or (VC), because $g(0,s)=1/s$ is not integrable near 0 with respect to s.

In the next example we show that an arbitrary kernel g may satisfy both (A) and (VC) but not condition (C).

Example 4.3.28. Consider the kernel

$$g(t,s)=\begin{cases}\dfrac{1}{s+t} & \text{for } 0<t\le 1,\\ 0 & \text{for } t=0.\end{cases}$$

Clearly, g satisfies (A) and hence also (VA). Since $g(t,0)=1/t$ for $t\in(0,1]$ is unbounded, (B) cannot hold. And (VB) can also not be satisfied, as

$$|g(s,s)|=\frac{1}{2s}\quad\text{for } 0<s\le 1$$

cannot be bounded by a Lebesgue integrable function on $[0,1]$. On the other hand, g satisfies (VC). To see this, define

$$h(t,\tau):=\int_0^{\min\{\tau,t\}}g(t,s)\,\mathrm{d}s\quad\text{for } 0\le t,\tau\le 1.$$

Then we have $h(0,\tau)=0$ for all $\tau\in[0,1]$. For $t>0$ and $\tau\in[0,1]$ we have

$$h(t,\tau)=\int_0^{\min\{\tau,t\}}\frac{1}{s+t}\,\mathrm{d}s=\log\left(1+\frac{\min\{\tau,t\}}{t}\right)=\begin{cases}\log 2 & \text{for } 0<t\leq\tau,\\ \log(1+\tau/t) & \text{for } \tau<t\leq 1\end{cases}$$

which implies

$$\sup_{\tau\in[0,1]}\operatorname{Var}\Big(h(\cdot,\tau)\Big)=\sup_{\tau\in(0,1]}\operatorname{Var}\Big(h(\cdot,\tau)\Big)=\sup_{\tau\in(0,1]}\Big(2\log 2-\log(1+\tau)\Big)=2\log 2.$$

Thus, (VC) is true. However, (C) is not, because for $0<t\leq 1$ we have

$$\int_0^1 g(t,s)\,\mathrm{d}s=\int_0^1\frac{1}{s+t}\,\mathrm{d}s=\log(1+1/t)$$

and this becomes unbounded and hence of unbounded variation with respect to t when t gets close to 0. $\diamondsuit$

The Theorems 4.3.4, 4.3.5 and 4.3.6 can now be reformulated and summarized in the following Volterra version.

Theorem 4.3.29. *Let X be one of the spaces BV, WBV_p or ΛBV, and let the Volterra kernel g satisfy condition (VA). Then the following conditions are equivalent.*

(a) The Volterra kernel g satisfies condition (VC) with Var *replaced by* Var_X.

(b) The operator V_g maps the space BV into X and is bounded.

Here, $\operatorname{Var}_{BV}=\operatorname{Var}$, $\operatorname{Var}_{WBV_p}=\operatorname{Var}_p$ *and* $\operatorname{Var}_{\Lambda BV}=\operatorname{Var}_\Lambda$.

In order to reformulate the Theorems 4.3.14 and 4.3.16 we need to translate the condition (E) and (F) into the Volterra setting. If g is a Volterra kernel, then for fixed $t_1,t_2\in[0,1]$ we have

$$|g(t_1,s)-g(t_2,s)|=\begin{cases}|g(t_1,s)-g(t_2,s)| & \text{for } 0\leq s\leq\min\{t_1,t_2\},\\ \Big|g(\max\{t_1,t_2\},s)\Big| & \text{for } \min\{t_1,t_2\}<s\leq\max\{t_1,t_2\},\\ 0 & \text{for } \max\{t_1,t_2\}<s\leq 1.\end{cases}$$

Consequently, condition (E) reads

$$\begin{aligned}&\forall\varepsilon>0\ \exists\delta>0\ \forall t_1,t_2,\tau\in[0,1]:\quad |t_1-t_2|\leq\delta\\ &\quad\Rightarrow\Bigg|\int_0^{\min\{t_1,t_2\}}\chi_{[0,\tau]}(s)\Big(g(t_1,s)-g(t_2,s)\Big)\,\mathrm{d}s\\ &\qquad\qquad+\operatorname{sign}(t_1-t_2)\int_{\min\{t_1,t_2\}}^{\max\{t_1,t_2\}}\chi_{[0,\tau]}(s)g\Big(\max\{t_1,t_2\},s\Big)\,\mathrm{d}s\Bigg|\leq\varepsilon.\qquad\text{(VE)}\end{aligned}$$

For instance, the kernel g from Example 4.3.28 does not satisfy this condition and hence also not condition (E). Indeed, for $\tau=1$, $t_1=0$ and $t_2=t\in(0,1]$, we have

$$\begin{aligned}&\left|\int_0^{\min\{0,t\}}\chi_{[0,\tau]}(s)\Big(g(0,s)-g(t,s)\Big)\,\mathrm{d}s+\operatorname{sign}(-t)\int_{\min\{0,t\}}^{\max\{0,t\}}\chi_{[0,\tau]}(s)g\Big(\max\{0,t\},s\Big)\,\mathrm{d}s\right|\\ &\quad=\int_0^t\frac{1}{s+t}\,\mathrm{d}s=\log 2\end{aligned}$$

which cannot be pushed arbitrarily close to 0 when t approaches 0.

Taking into account the special structure of an arbitrary Volterra kernel g condition (VE) can be shortened to the condition

$$\forall \varepsilon > 0\ \exists \delta > 0\ \forall t_1, t_2, \tau \in [0,1]: \\ |t_1 - t_2| \le \delta \ \Rightarrow\ \left| \int_0^{\max\{t_1,t_2,\tau\}} g(t_1,s) - g(t_2,s)\,\mathrm{d}s \right| \le \varepsilon.$$

Again, condition (VE) is weaker than (E) for arbitrary kernels. For instance, the kernel g from Example 4.3.26 does satisfy (VE), because $g(t,s) = 0$ whenever $t \in (0,1]$. But it cannot satisfy (E), since $g(0,s) = 1/s$ is not integrable near 0 with respect to s. However, even if an arbitrary kernel satisfies (A) and (VE), condition (E) does still not have to be fulfilled. This is shown by the next example.

Example 4.3.30. The kernel

$$g(t,s) = \begin{cases} \dfrac{1}{t+s} & \text{for } 0 < t < s \le 1, \\ 0 & \text{for } 0 = t < s \le 1, \\ 0 & \text{for } 0 \le s \le t \le 1 \end{cases}$$

has the property that $g(t,s) = 0$ whenever $0 \le s \le t \le 1$. In particular, all the Volterra conditions (VA), (VB), (VC) and (VE) are satisfied. Moreover, (A) is fulfilled, because $g(0,s) = 0$ and $0 \le g(t,s) \le 1/(t+s)$ for $t \in (0,1]$ and all $s \in [0,1]$. However, (B) is not satisfied, because for $s \in (0,1]$, we have $\mathrm{Var}\big(g(\cdot,s)\big) = 2/s$ which cannot be bounded by an L_1-function. Similarly, (C) is not satisfied, because for $\tau = 1$ and $t \in (0,1)$ we have

$$\int_0^1 g(t,s)\,\mathrm{d}s = \int_t^1 \frac{1}{t+s}\,\mathrm{d}s = \log\frac{t+1}{2t} \longrightarrow \infty \quad \text{as } t \to 0+.$$

Finally, for $0 < t < 1/2 < 1 = \tau$,

$$\begin{aligned} \int_0^1 g(t,s) - g(2t,s)\,\mathrm{d}s &= \int_t^1 \frac{1}{t+s}\,\mathrm{d}s - \int_{2t}^1 \frac{1}{2t+s}\,\mathrm{d}s = \log\frac{t+1}{2t} - \log\frac{2t+1}{4t} \\ &= \log\frac{2t+2}{2t+1} \longrightarrow \log 2 \quad \text{as } t \to 0+. \end{aligned}$$

Thus, (E) is also not true. ◊

Theorem 4.3.14 now reads as follows.

Theorem 4.3.31. *Let $g : [0,1] \times [0,1] \to \mathbb{R}$ be a Volterra kernel satisfying (VA). Then the following statements are equivalent.*

(a) The Volterra kernel g satisfies the conditions (VC) and (VE).

(b) The Volterra operator V_g maps the space $BV \cap C$ into itself and is bounded, and the set $\{V_g x \mid x \in BV \cap C, \|x\|_{BV} \le R\}$ is equicontinuous for every $R > 0$.

Similarly, condition (F) can be translated into

$$\forall \varepsilon > 0\ \exists \delta > 0\ \forall t_1, t_2 \in [0,1]: \quad |t_1 - t_2| \le \delta$$
$$\Rightarrow \int_0^{\min\{t_1,t_2\}} \Big| g(t_1, s) - g(t_2, s) \Big| \, \mathrm{d}s + \int_{\min\{t_1,t_2\}}^{\max\{t_1,t_2\}} \Big| g\Big(\max\{t_1, t_2\}, s\Big) \Big| \, \mathrm{d}s \le \varepsilon. \quad \text{(VF)}$$

Its shorter version for Volterra kernels g is

$$\forall \varepsilon > 0\ \exists \delta > 0\ \forall t_1, t_2 \in [0,1]: \quad |t_1 - t_2| \le \delta \Rightarrow \int_0^{\max\{t_1,t_2\}} \Big| g(t_1, s) - g(t_2, s) \, \mathrm{d}s \Big| \le \varepsilon.$$

The same kernel function as in Example 4.3.30 shows that (VF) does not imply (F) for arbitrary kernels. Now, Theorem 4.3.16 in its Volterra version reads as follows.

Theorem 4.3.32. *Let $g : [0,1] \times [0,1] \to \mathbb{R}$ be a Volterra kernel satisfying (VA). Then the following statements are equivalent.*

(a) The Volterra kernel g satisfies the conditions (VF).

(b) The Volterra operator V_g maps the space L_∞ into the space C and is bounded, and the set $\{V_g x \mid x \in L_\infty, \|x\|_{L_\infty} \le R\}$ is equicontinuous for every $R > 0$.

As an analogue to Table 4.3.1 let us summarize our last three examples and which properties they fulfill in Table 4.3.3 below.

Table 4.3.3: Properties of g in the above examples.

Example	(A)	(VA)	(B)	(VB)	(C)	(VC)	(E)	(VE)
4.3.26	no	yes	no	yes	no	no	no	yes
4.3.28	yes	yes	no	no	no	yes	no	no
4.3.30	yes	yes	no	yes	no	yes	no	yes

We now turn to the most important conditions, namely the (B)-type conditions for one of our BV-spaces X. For a Volterra kernel g in either of the spaces BV or ΛBV both conditions (B_X) and (B_X^*) are equivalent and read

$$\exists m \in L_1\ \forall' s \in [0,1]: \quad |g(s,s)| + \mathrm{Var}_X\Big(g(\cdot, s), [s,1]\Big) \le m(s), \qquad (\mathrm{VB}_X)$$

where again Var_X denotes the variation of the space X, that is, $\mathrm{Var}_{BV} = \mathrm{Var}$ and $\mathrm{Var}_{\Lambda BV} = \mathrm{Var}_\Lambda$. In the general space $X = YBV_\varphi$, however, the situation is more complex. Condition (B_{YBV_φ})=(B_φ) is now

$$\exists \theta > 0\ \exists m \in L_1\ \forall' s \in [0,1]: \quad \varphi\Big(\theta|g(s,s)|\Big) + \mathrm{Var}_\varphi\Big(\theta g(\cdot, s), [s,1]\Big) \le m(s), \quad (\mathrm{VB}_\varphi)$$

and condition ($\mathrm{B}_{YBV_\varphi}^*$)=($\mathrm{B}_\varphi^*$) becomes

$$\forall \theta > 0\ \exists m_\theta \in L_1\ \forall' s \in [0,1]:$$
$$\varphi\Big(\theta|g(s,s)|\Big) + \mathrm{Var}_\varphi\Big(\theta g(\cdot, s), [s,1]\Big) \le m_\theta(s). \qquad (\mathrm{VB}_\varphi^*)$$

In particular, for the Wiener space WBV_p we get that both conditions are equivalent and read

$$\exists m \in L_1 \ \forall' s \in [0,1]: \quad |g(s,s)|^p + \operatorname{Var}_p\Big(g(\cdot,s),[s,1]\Big) \leq m(s). \tag{4.3.11}$$

Since the Young variation often exhibits surprising properties, we give here a more detailed argument for the equivalence of the conditions (B_φ^*) and (VB_φ^*) in the case of Volterra kernels. The argument for condition (VB_φ) is similar and will be skipped.
Let $g : [0,1]\times[0,1] \to \mathbb{R}$ be a Volterra kernel, assume that g satisfies (B_φ^*) and fix $\theta > 0$. Because of (B_φ^*), we find for our θ a function $m \in L_1$ such that $\operatorname{Var}_\varphi\Big(\theta g(\cdot,s)\Big) \leq m(s)$ for almost all $s \in [0,1]$. But this implies

$$\operatorname{Var}_\varphi\Big(\theta g(\cdot,s),[s,1]\Big) \leq \operatorname{Var}_\varphi\Big(\theta g(\cdot,s)\Big) \leq m(s)$$

for almost all $s \in [0,1]$. Moreover, since $g(t,s) = 0$ for $0 \leq t < s$, we have for almost all $s \in (0,1]$ by Proposition 1.2.10 applied on $[0,s]$,

$$\begin{aligned}\varphi\Big(\theta|g(s,s)|\Big) &= \operatorname{Var}\Big(\varphi(\theta|g(\cdot,s)|),[0,s]\Big) \leq \operatorname{Var}_\varphi\Big(\theta g(\cdot,s),[0,s]\Big)\\ &\leq \operatorname{Var}_\varphi\Big(\theta g(\cdot,s)\Big) \leq m(s).\end{aligned}$$

This gives (VB_φ^*) with $m_\theta := 2m$.
Now, assume that the Volterra kernel g satisfies (VB_φ), that is, for any $\alpha > 0$ there is some $m_\alpha \in L_1$ such that

$$\varphi(\alpha|g(s,s)|) + \operatorname{Var}_\varphi(\alpha g(\cdot,s)) \leq m_\alpha(s)$$

for almost all $s \in [0,1]$. Fix $\theta > 0$ and $s \in (0,1]$ and a partition $0 = t_0 < \ldots < t_n = 1$. Then there is some $k \in \{1,\ldots,n\}$ such that $t_{k-1} < s \leq t_k$; in particular, $g(t_j,s) = 0$ for all $j \in \{0,\ldots,k-1\}$. We obtain

$$\begin{aligned}\sum_{j=1}^{n}\varphi\Big(\theta|g(t_{j-1},s)-g(t_j,s)|\Big) &= \varphi\Big(\theta|g(t_k,s)|\Big) + \sum_{j=k+1}^{n}\varphi\Big(\theta|g(t_{j-1},s)-g(t_j,s)|\Big)\\ &\leq \varphi\Big(\theta|g(t_k,s)|\Big) + \operatorname{Var}_\varphi\Big(\theta g(\cdot,s),[s,1]\Big).\end{aligned}$$

In addition, from the convexity and monotonicity of φ we get,

$$\begin{aligned}\varphi\Big(\theta|g(t_k,s)|\Big) &\leq 2^{-1}\Big[\varphi\Big(2\theta|g(s,s)-g(t_k,s)|\Big) + \varphi\Big(2\theta|g(s,s)|\Big)\Big]\\ &\leq 2^{-1}\Big[\operatorname{Var}_\varphi\Big(2\theta|g(\cdot,s)|,[s,1]\Big) + \varphi\Big(2\theta|g(s,s)|\Big)\Big].\end{aligned}$$

In total, we obtain for almost all $s \in [0,1]$,

$$\begin{aligned}\operatorname{Var}_\varphi\Big(\theta|g(\cdot,s)|\Big) &\leq 2^{-1}\varphi\Big(2\theta|g(s,s)|\Big) + 2^{-1}\operatorname{Var}_\varphi\Big(2\theta g(\cdot,s),[s,1]\Big) + \operatorname{Var}_\varphi\Big(\theta g(\cdot,s),[s,1]\Big)\\ &\leq 2^{-1}m_{2\theta}(s) + m_\theta(s)\end{aligned}$$

and thus condition (B_φ) holds with $m = 2^{-1}m_{2\theta} + m_\theta$.

Finally, condition (B_p) for the Riesz space translates to the following condition.

$$\exists m \in L_1 \ \forall' s \in [0,1]: \quad g(s,s) = 0 \text{ and } \mathrm{RVar}_p\Big(g(\cdot,s),[s,1]\Big) \le m(s). \tag{VB$_p$}$$

Here, the strong degeneracy $g(s,s) = 0$ almost everywhere along the line $t = s$ comes from the fact that each function in the Riesz space RBV_p for $p > 1$ is continuous.

To overcome annoying case distinctions in what follows we will extend the symbol (VB_X) to the spaces $X = YBV_\varphi$ and $X = RBV_p$ by setting (VB_{YBV_φ})=(VB_φ) and (VB_{RBV_p})=(VB_p). Analogously, we extend the symbol (VB^*_X) by ($VB^*_{YBV_\varphi}$)=(VB^*_φ) for the space $X = YBV_\varphi$ and by ($VB^*_{RBV_p}$)=(VB_p) for the space $X = RBV_p$. Using this uniform approach we are now in position to restate Theorem 4.3.21, Proposition 4.3.22 and Corollary 4.3.23 also for the Volterra operator. These three results now read as follows.

Theorem 4.3.33. *Let X be one of the spaces BV, WBV_p, YBV_φ, ΛBV or RBV_p, and let $g : [0,1] \times [0,1] \to \mathbb{R}$ be a Volterra kernel. Under the conditions (VA) and (VB_X) the Volterra operator V_g maps the space L_∞ into the space X and is bounded with*

$$\|V_g\|_{L_\infty \to X} \le \theta^{-1} \begin{Bmatrix} 2\,\|m\|_{L_1} & \text{for } X = BV, \\ 2\,\|m\|_{L_1}^{1/p} & \text{for } X = WBV_p, \\ \big(\varphi^{-1}(1)+1\big)\max\big\{1, \|m\|_{L_1}\big\} & \text{for } X = YBV_\varphi, \\ \big(1+\lambda_1^{-1}\big)\,\|m\|_{L_1} & \text{for } X = \Lambda BV, \\ 2\,\|m\|_{L_1}^{1/p} & \text{for } X = RBV_p, \end{Bmatrix} \tag{4.3.12}$$

where θ is taken from condition (VB_X).

Note that since any Volterra kernel $g(t,s)$ vanishes for $s > t$, the norm $\|g(0,\cdot)\|_{L_1}$ in (4.3.6) does not appear here. Furthermore, the condition (VB_p) for the Riesz space is rather strong, because of the almost everywhere degeneracy $g(s,s) = 0$ occurring therein. This condition is far from being necessary for the operator V_g to map L_∞ continuously into RBV_p.

Example 4.3.34. Consider the Volterra kernel g, defined by $g(t,s) = 1$ for $0 \le s \le t \le 1$ and $g(t,s) = 0$ for $0 \le t < s \le 1$. The corresponding Volterra operator

$$V_g x(t) = \int_0^t x(s)\,\mathrm{d}s \quad \text{for } 0 \le t \le 1$$

maps even L_p into RBV_p and is bounded for $1 < p < \infty$, because for $x \in L_p$,

$$\|V_g x\|_{RBV_p} = \|V_g x\|_\infty + \mathrm{RVar}_p(V_g x)^{1/p} \le \|x\|_{L_1} + \left(\int_0^1 |(V_g x)'(s)|^p\,\mathrm{d}s\right)^{1/p} \le 2\,\|x\|_{L_p}.$$

However, $g(s,s) = 1$ for all $s \in [0,1]$. ◇

Proposition (4.3.22) becomes

Proposition 4.3.35. *Let X be one of the spaces BV, WBV_p, YBV_φ, ΛBV or RBV_p, and let $g : [0,1] \times [0,1] \to \mathbb{R}$ be a Volterra kernel. Assume that (x_n) is a bounded sequence in L_∞ and converges almost everywhere to some $x \in L_\infty$. If g satisfies the conditions (VA) and (VB$_X^*$), then $(V_g x_n)$ is a sequence in X that converges in X to $V_g x$.*

Its compactness criterion, namely Corollary 4.3.23 turns into

Corollary 4.3.36. *Let X be one of the spaces BV, WBV_p, YBV_φ, ΛBV or RBV_p, and let $g : [0,1] \times [0,1] \to \mathbb{R}$ be a Volterra kernel. Under the conditions (VA) and (VB$_X^*$) the operator V_g maps the space X into itself and is compact.*

Before we end this chapter with some additional remarks let us summarize the (B)-type conditions in Table 4.3.4 below which is an analogue to Table 4.3.2.

Table 4.3.4: Conditions (VB$_X$) and (VB$_X^*$) in our BV-spaces.

X	(VB$_X$)	(VB$_X^*$)
BV	$\exists m \in L_1 \ \forall' s \in [0,1]$: $\lvert g(s,s)\rvert + \mathrm{Var}(g(\cdot,s),[s,1]) \leq m(s)$	$\exists m \in L_1 \ \forall' s \in [0,1]$: $\lvert g(s,s)\rvert + \mathrm{Var}(g(\cdot,s),[s,1]) \leq m(s)$
WBV_p	$\exists m \in L_1 \ \forall' s \in [0,1]$: $\lvert g(s,s)\rvert^p + \mathrm{Var}_p(g(\cdot,s),[s,1]) \leq m(s)$	$\exists m \in L_1 \ \forall' s \in [0,1]$: $\lvert g(s,s)\rvert^p + \mathrm{Var}_p(g(\cdot,s),[s,1]) \leq m(s)$
YBV_φ	$\exists \theta > 0 \ \exists m \in L_1 \ \forall' s \in [0,1]$: $\varphi(\theta\lvert g(s,s)\rvert) + \mathrm{Var}_\varphi(\theta g(\cdot,s),[s,1]) \leq m(s)$	$\forall \theta > 0 \ \exists m_\theta \in L_1 \ \forall' s \in [0,1]$: $\varphi(\theta\lvert g(s,s)\rvert) + \mathrm{Var}_\varphi(\theta g(\cdot,s),[s,1]) \leq m_\theta(s)$
ΛBV	$\exists m \in L_1 \ \forall' s \in [0,1]$: $\lvert g(s,s)\rvert + \mathrm{Var}_\Lambda(g(\cdot,s),[s,1]) \leq m(s)$	$\exists m \in L_1 \ \forall' s \in [0,1]$: $\lvert g(s,s)\rvert + \mathrm{Var}_\Lambda(g(\cdot,s),[s,1]) \leq m(s)$
RBV_p	$\exists m \in L_1 \ \forall' s \in [0,1]$: $g(s,s) = 0$ and $\mathrm{RVar}_p(g(\cdot,s),[s,1]) \leq m(s)$	$\exists m \in L_1 \ \forall' s \in [0,1]$: $g(s,s) = 0$ and $\mathrm{RVar}_p(g(\cdot,s),[s,1]) \leq m(s)$

We remark that Volterra operators V_g have in general much nicer properties than just ordinary integral operators I_g. For example, the operator V_g has often spectral radius zero, for instance, if one of the iterated kernels is bounded, and this is useful in the search for invariant balls for nonlinear operators of Volterra type. This is not true for the general operator I_g and neither for an arbitrary Volterra operator V_g. We give two examples to show this.

Example 4.3.37. Consider the kernel function $g(t,s) = 1$ for all $t, s \in [0,1]$ which even satisfies (A) and (B). By Theorem 4.3.1, $I_g : BV \to BV$ is well-defined and bounded. The function $x = \mathbb{1}$ belongs to BV and is mapped by I_g into itself. Thus, $\mathbb{1}$ is an eigenvector with corresponding eigenvalue 1. ◇

Example 4.3.38. Consider the Volterra kernel g, defined by

$$g(t,s) = \begin{cases} \dfrac{1}{s+t} & \text{for } 0 \leq s \leq t \leq 1, \\ 0 & \text{for } 0 \leq t < s \leq 1. \end{cases}$$

Since this kernel coincides on the triangle $T := \{(t,s) \mid 0 \le s \le t \le 1\}$ with the kernel investigated in Example 4.3.28 it satisfies (VC) as we have seen therein. By Theorem 4.3.29 the Volterra operator V_g maps the space BV into itself and is bounded. As such, the function $x(t) = t$ belongs to BV and is mapped by V_g into the function $(\log 2)x$. Thus, x is an eigenvector of V_g with corresponding eigenvalue $\log 2$. ◊

The following theorem shows that, due to the special structure (4.3.9) of the kernel function g, a Volterra operator maps a quite large space continuously into BV [31].

Theorem 4.3.39. *Suppose that g satisfies condition (A) and (VB), where $m \in L_p$ for some $p \in [1,\infty)$. Then the operator V_g maps the space $L_{p/(p-1)}$ into BV and is bounded.*

So Theorem 4.3.39 shows that, the milder the condition on the majorant m in (VB) with L_1 replaced by L_p (i.e. the smaller p), the smaller we may choose the space of departure $L_{p/(p-1)}$ for V_g. It also implies that if (x_n) is a sequence that converges in $L_{p/(p-1)}$ to $\mathbb{0}$ for some $p \in [1,\infty)$, then the sequence $(V_g(x_n))$ converges to $\mathbb{0}$ in BV. One may show that this is also true if (x_n) converges merely almost everywhere to $\mathbb{0}$ and not necessarily in $L_{p/(p-1)}$; see Proposition 4.3.22 and 4.3.35 for $p = 1$. However, this is no longer true for $p = \infty$, that is, for $L_{p/(p-1)} = L_1$, as the following example shows [31]:

Example 4.3.40. For the kernel function take $g(t,s) = \chi_{[0,t]}(s)$ which satisfies (A) and (VB). The sequence $x_n := n\chi_{[0,1/n]}$ is a bounded sequence in L_1 with $\|x_n\|_{L_1} = 1$ and converges almost everywhere but not in L_1 to $\mathbb{0}$. But

$$V_g x_n(t) = \int_0^t x_n(s)\,\mathrm{d}s = n\min\{t, 1/n\} \quad \text{for } 0 \le t \le 1,$$

which shows $\|V_g x_n\|_{BV} = 2$ for all $n \in \mathbb{N}$. ◊

For Waterman spaces ΛBV an exact analogue of Theorem 4.3.39 holds where condition (VB) is replaced by ($\mathrm{VB}_{\Lambda BV}$) [30].

Theorem 4.3.41. *Suppose that g satisfies condition (A) and ($VB_{\Lambda BV}$), where $m \in L_p$ for some $p \in [1,\infty)$. Then the operator V_g maps the space $L_{p/(p-1)}$ into ΛBV and is bounded.*

Note that Example 4.3.25 may serve to show that in the Theorems 4.3.39 and 4.3.41 we cannot expect the operator $V_g : L_{p/(p-1)} \to BV$ respectively $V_g : L_{p/(p-1)} \to \Lambda BV$ to be compact for any $p \ge 1$.

Chapter 5

Nonlinear Operators between BV-Spaces

The purpose of this chapter is to study two nonlinear operators mainly in BV-spaces X and Y, where the symbols X and Y represent one of the spaces BV, WBV_p, YBV_φ, ΛBV or RBV_p introduced in Chapter 1. In detail we will consider

- the *composition operator* $C_g : X \to Y$, generated by $g : \mathbb{R} \to \mathbb{R}$ and defined by
$$C_g x(t) = g\big(x(t)\big), \tag{5.0.1}$$
- the *superposition operator* $N_g : X \to Y$, generated by $g : [0,1] \times \mathbb{R} \to \mathbb{R}$ and defined by
$$N_g x(t) = g\big(t, x(t)\big). \tag{5.0.2}$$

The superposition operator N_g is often called *Nemytskij operator*, especially in Russian literature. This and the fact that the character S is reserved for the substitution operator that we have studied briefly in Section 4.2 is the reason why we use the symbol N for the superposition operator.

For both the composition and the superposition operator we are particularly interested in analytic properties like acting conditions, continuity, boundedness and compactness. But we will also investigate set-theoretic properties like injectivity, surjectivity and bijectivity. Although the composition operator C_g is similarly defined as the substitution operator S_g that we have studied in Section 4.2, it exhibits due to its nonlinearity a completely different behavior than its linear counterpart. However, almost all its analytic properties may be fully characterized in terms of the generating function g; we will do this in Section 5.1 and give a summary at the end in Table 5.1.2.

The superposition operator N_g is only a slight generalization of the composition operator C_g, but the dependence on t leads to quite unexpected properties and will make both the investigations and the formulation of results much more complicated. Not all analytic properties are fully understood; we give a summary especially referring to the disparities of the composition operator C_g and the superposition operator N_g at the end of Section 5.2.

5.1 Composition Operators

In this section we are going to investigate the composition operator $C_g : X \to Y$, generated by some function $g : \mathbb{R} \to \mathbb{R}$, which is defined by

$$C_g x(t) = g\big(x(t)\big) \quad \text{for } 0 \leq t \leq 1,$$

where X and Y are one of the spaces BV, WBV_p, YBV_φ, ΛBV or RBV_p which have been introduced in Chapter 1. However, we primarily focus on the case when X and Y are the same BV-space and give only a few results for $X \neq Y$. For $X = Y$, the composition operator is well-defined if and only if $g \circ x$ belongs to X whenever x does; in short: $C_g(X) \subseteq X$. For instance, it is easy to show that this is the case for $X = B$ and $X = C$ if and only if g is locally bounded respectively continuous, and we have given these and other examples in Section 2.3. Observe that, in contrast to the substitution operator S_g, the operator C_g is nonlinear, and so boundedness and continuity are independent here.

For our BV-type spaces, things are a little more difficult. Recall that Josephy proved in [75] that C_g maps BV into itself if and only if g is locally Lipschitz continuous which we denote by $g \in Lip_{loc}(\mathbb{R})$, and the same had been proven later on for $X = RBV_p$ by Marcus and Mizel [101]. Moreover, Ciemnoczolowski and Orlicz proved in [42] that C_g maps WBV_p into itself if and only if $g \in Lip_{loc}(\mathbb{R})$. More generally, they proved that $g \in Lip_{loc}(\mathbb{R})$ is necessary and sufficient also in YBV_φ, where for necessity the assumption $\varphi, \varphi^{-1} \in \delta_2$ is needed. We do not know what happens when φ or φ^{-1} do not satisfy a δ_2-condition. Finally, Pierce and Waterman proved $C_g(\Lambda BV) \subseteq \Lambda BV$ if and only if $g \in Lip_{loc}(\mathbb{R})$ in [130]. We summarize these results in the following

Proposition 5.1.1. *Let $g : \mathbb{R} \to \mathbb{R}$. The following statements are true.*

(a) If X is one of the spaces BV, WBV_p, ΛBV or RBV_p, then C_g maps X into itself if and only if $g \in Lip_{loc}(\mathbb{R})$.

(b) If $g \in Lip_{loc}(\mathbb{R})$, then C_g maps the space YBV_φ into itself. If both φ and φ^{-1} satisfy a δ_2-condition, the converse is also true.

Note that (a) for $X = WBV_p$ follows indeed from (b), since in this case $\varphi(t) = t^p$ has the property that both φ and $\varphi^{-1}(t) = t^{1/p}$ satisfy a δ_2-condition.

We also remark that acting conditions for $C_g : X \to Y$ are sometimes also known if X and Y are distinct spaces. For instance, $C_g(WBV_p) \subseteq WBV_q$ holds for $p \leq q$ if and only if $g \in Lip_{loc}^{p/q}(\mathbb{R})$, where $Lip_{loc}^{\alpha}(\mathbb{R})$ denotes the space of all locally Hölder continuous functions with exponent $\alpha \leq 1$.

If X is one of the spaces BV, WBV_p, YBV_φ, ΛBV or RBV_p and $\|\cdot\|_X$ is its respective norm, then the inequality

$$\|g \circ x\|_X \leq |g(0)| + \operatorname{lip}\left(g, [-R, R]\right) \|x\|_X$$

holds for $g \in Lip[-R, R]$ and $x \in X$ with $\|x\|_\infty \leq R$, where

$$\operatorname{lip}\left(g, [-R, R]\right) = \sup\left\{\frac{|g(u) - g(v)|}{|u - v|} \mid u, v \in [-R, R], u \neq v\right\}$$

denotes the optimal Lipschitz constant of g on $[-R, R]$. This means that C_g is (locally) bounded in this case. Of course, the situation in Proposition 5.1.1 for YBV_φ is unsatisfactory because we have to deal with an additional δ_2-condition. However, if we additionally assume boundedness of C_g, then we can show that C_g maps any of our spaces X into itself and is bounded if and only if $g \in Lip_{loc}(\mathbb{R})$. We will reformulate and prove this statement later in Theorem 5.1.19, when we talk about analytic properties of C_g like compactness and continuity. Since C_g, in contrast to the multiplication and substitution operator, is not linear, conditions on g characterizing pointwise, locally uniform and locally Lipschitz continuity, may differ. We present a full characterization of uniform and Lipschitz continuity on bounded sets and on the entire space. However, pointwise continuity is a much harder problem and - as far as we known - only solved in BV and RBV_p; we will give some sample results at the end of this section. In Theorem 6.2.8 in Section 6.2 we will present a new proof for the pointwise continuity of $C_g : BV \to BV$. But for the moment we will focus on more basic properties.

The following statements are almost immediate consequences of the definitions and the fact that constant functions belong to either of the mentioned BV-spaces.

Proposition 5.1.2. *Let X be any of the spaces BV, WBV_p, YBV_φ with $\varphi, \varphi^{-1} \in \delta_2$, ΛBV or RBV_p, and let $g \in Lip_{loc}(\mathbb{R})$. Then the following statements are true.*

(a) The operator C_g is injective if and only if g is injective.

(b) If the operator C_g is surjective, then g is surjective.

(c) The operator C_g is bijective if and only if g is bijective and $g^{-1} \in Lip_{loc}(\mathbb{R})$. In this case, $C_g^{-1} = C_{g^{-1}}$.

Proof. (a) Assume that $C_g : X \to X$ is injective and fix $u, v \in \mathbb{R}$ with $g(u) = g(v)$. The constant functions $x \equiv u$ and $y \equiv v$ belong to X and satisfy $C_g x = g \circ x = g \circ y = C_g y$, and since C_g is injective, we conclude $x = y$ and hence $u = v$. This shows that g is also injective.
Conversely, assume that g is injective, and fix $x, y \in X$ with $C_g x = C_g y$. For fixed $t \in [0, 1]$ we then have $g(x(t)) = C_g x(t) = C_g y(t) = g(y(t))$, and since g is injective, we obtain $x(t) = y(t)$. Since this is true for each $t \in [0, 1]$, it follows that $x = y$ and hence the injectivity of C_g.

(b) Assume that C_g is surjective and fix $v \in \mathbb{R}$. The constant function $y \equiv v$ belongs to X, and since C_g is surjective, there is a function $x \in X$ with $C_g x = y$. In particular, $v = y(0) = g(x(0)) = g(u)$ with $u := x(0)$ which shows that g is surjective, as well.

(c) If C_g is bijective, then from (a) and (b) follows that g must be bijective. If $y \in X$ is fixed, we find some $x \in X$ such that $C_g x = y$. Consequently, $x = g^{-1} \circ y = C_{g^{-1}} y$,

and since y was arbitrary we conclude that the operator $C_{g^{-1}}$ maps X into itself. By Proposition 5.1.1, $g^{-1} \in Lip_{loc}(\mathbb{R})$, and since $C_g C_{g^{-1}} x = x$ for all $x \in X$, we obtain $C_g^{-1} = C_{g^{-1}}$.
If, conversely, g is bijective and satisfies $g^{-1} \in Lip_{loc}(\mathbb{R})$, then the operator $C_{g^{-1}}$ maps X into itself. As $C_g C_{g^{-1}} x = x = C_{g^{-1}} C_g x$ for all $x \in X$, the operator C_g must be bijective with $C_g^{-1} = C_{g^{-1}}$. ∎

We remark that the assumption $\varphi, \varphi^{-1} \in \delta_2$ is only needed for the "only if"-part (c) of Proposition 5.1.2. The parts (a) and (c) fully characterize injectivity and bijectivity of the operator $C_g : X \to X$ in all the spaces BV, WBV_p, YBV_φ, ΛBV or RBV_p. Condition (b), however, provides only a necessary condition on g that makes C_g surjective, namely that g is surjective itself.
In Table 5.1.1 we summarize basic properties concerning injectivity and surjectivity of the multiplication operator M_g from (4.0.1), the substitution operator S_g from (4.0.2) and the composition operator C_g from (5.0.1) in the space BV.

Table 5.1.1: Some mapping properties of some operators in BV.

M_g is injective	$\iff$	$\operatorname{supp}(g) = [0,1]$
M_g is surjective	$\iff$	$\operatorname{supp}_\delta(g) = [0,1]$ for some $\delta > 0$
S_g is injective	$\iff$	g is surjective
S_g is surjective	$\implies$	g is injective
C_g is injective	$\iff$	g is injective
C_g is surjective	$\implies$	g is surjective

We have seen that if the operator C_g is surjective, then g must be surjective, as well. But does a surjective g generate a surjective operator C_g? The answer for all our BV-type spaces is negative, and we illustrate this by the composition operator $C_g : X \to X$, generated by $g(u) = u^3$, and the functions $\mathfrak{J}_{(\alpha_j)}$, defined in (1.2.1). The first example lives in $X = BV$.

Example 5.1.3. The function $g : \mathbb{R} \to \mathbb{R}$, $u \mapsto u^3$, is surjective (even bijective), but C_g is not surjective. Take, for instance $y = \mathfrak{J}_{(1/j^3)}$, that is, $y(1/(2j)) = 1/j^3$ for $j \in \mathbb{N}$ and $y(t) = 0$ elsewhere on $[0,1]$. Then $y \in BV$ by Corollary 1.2.1.
But any x satisfying $C_g x = y$ also satisfies $x(t) = \sqrt[3]{y(t)}$ for all $t \in [0,1]$ which gives $x(1/(2j)) = 1/j$ for $j \in \mathbb{N}$ and $x(t) = 0$ elsewhere on $[0,1]$, that is, $x = \mathfrak{J}_{(1/j)}$ which does not belong to BV, again by Corollary 1.2.1. ◇

The next example treats the case $X = YBV_\varphi$ and therefore generalizes the idea of the previous example.

Example 5.1.4. Let $g(u) = u^3$ be as in Example 5.1.3. Define $\Phi : (0,\infty) \times [0,\infty) \to [0,\infty)$ by

$$\Phi(\lambda, t) = \varphi\left(\lambda \sqrt[3]{\varphi^{-1}(t)}\right).$$

Then $\Phi(\cdot,t)$ is increasing for each $t \in [0,\infty)$. Since φ is a homeomorphism of $[0,\infty)$, we have for all $\lambda > 0$,

$$\limsup_{t\to 0+} \frac{\Phi(\lambda,t)}{t} = \limsup_{s\to 0+} \frac{\varphi(\lambda s)}{\varphi(s^3)} \geq \limsup_{s\to 0+} \frac{\varphi(s^2)}{s\varphi(s^2)} = \infty.$$

By Lemma 1.2.16 we find a sequence (u_j) of positive numbers such that

$$\sum_{j=1}^{\infty} u_j < \infty \quad \text{and} \quad \sum_{j=1}^{\infty} \Phi(\lambda, u_j) = \infty \quad \text{for all } \lambda > 0,$$

and the substitution $v_j := \sqrt[3]{\varphi^{-1}(u_j)}$ yields

$$\sum_{j=1}^{\infty} \varphi\left(v_j^3\right) < \infty \quad \text{and} \quad \sum_{j=1}^{\infty} \varphi(\lambda v_j) = \infty \quad \text{for all } \lambda > 0.$$

But this implies that the function $y = \mathfrak{I}_{(v_j^3)}$ belongs to YBV_φ by Corollary 1.2.11, while the only function x with $C_g x = y$ is $x = \mathfrak{I}_{(v_j)}$ which does not belong to YBV_φ, again by Corollary 1.2.11. Thus, $C_g : YBV_\varphi \to YBV_\varphi$ is not surjective. ♢

The third example in this series shows an analogous behavior of C_g in ΛBV.

Example 5.1.5. Let $g(u) = u^3$ as in Example 5.1.3, and let (λ_j) denote the Waterman sequence Λ. If we define $\Lambda_n := \lambda_1 + \ldots + \lambda_n$, then a theorem of Abel and Dini [77] says

$$\sum_{j=1}^{\infty} \frac{\lambda_j}{\Lambda_j^3} < \infty \quad \text{and} \quad \sum_{j=1}^{\infty} \frac{\lambda_j}{\Lambda_j} = \infty.$$

This implies for any permutation σ of $\mathbb{N}$ that

$$\sum_{j=1}^{\infty} \lambda_{\sigma(j)} \frac{1}{\Lambda_j^3} \leq \sum_{j=1}^{\infty} \frac{\lambda_j}{\Lambda_j^3} < \infty,$$

as both sequences (λ_j) and $(1/\Lambda_j^3)$ are decreasing. Thus, the function $y = \mathfrak{I}_{(1/\Lambda_j^3)}$ belongs to ΛBV by Corollary 1.2.21, while the only function x with $C_g x = y$ is $x = \mathfrak{I}_{(1/\Lambda_j)}$ which cannot belong to ΛBV, again by Corollary 1.2.21. Thus, again, $C_g : \Lambda BV \to \Lambda BV$ is not surjective. ♢

The last example handles the case $X = RBV_p$.

Example 5.1.6. Let $g(u) = u^3$ as in Example 5.1.3, and let $p > 1$ and $q := 2(p-1)/p$. The function $y : [0,1] \to \mathbb{R}$, defined by $y(t) = t^q$, belongs to RBV_p, since $y'(t) = qt^{q-1}$ belongs to L_p as $p(q-1) = 2(p-1) - p = p - 2 > -1$. Since g is bijective, the only preimage of y is the function $x(t) = t^{q/3}$. However, x does not belong to RBV_p, since $y'(t) = \frac{q}{3} t^{q/3-1}$ and $(q/3 - 1)p = 2/3(p-1) - p = -p/3 - 2/3 < -1$. Consequently, $C_g : RBV_p \to RBV_p$ cannot be surjective. ♢

The function $g(u) = u^3$ used in the last four examples defined injective composition operators by Proposition 5.1.2. These examples therefore also show that injectivity of C_g does not imply surjectivity of C_g. However, in contrast to the multiplication operator, where surjectivity implies injectivity, there are composition operators that are surjective but not injective. We will give an example in BV below (see Example 5.1.8) for which we need some technical result in advance.

Lemma 5.1.7. *Let y be a member of one of the classes BV, WBV_p, YBV_φ or ΛBV, and let $a, b, c \in \mathbb{R}$ with $a < b < c$. Then there is a set $A \subseteq [0,1]$ with finitely many connected components such that*

$$y^{-1}\big([a,b]\big) \subseteq A \subseteq y^{-1}\big([a,c)\big).$$

Proof. For better readability we write $A_b := y^{-1}([a,b])$ and $A_c := y^{-1}([a,c))$. Let $\mathcal{I}$ be the system of the connected components of A_c. Then the sets in $\mathcal{I}$ are pairwise disjoint intervals. We show that

$$A := \bigcup\big\{I \in \mathcal{I} \mid I \cap A_b \neq \emptyset\big\}$$

has the desired properties. First, A is clearly a subset of A_c. Moreover, A_b is a subset of A. To see this fix $t \in A_b$. Since $b < c$, we have $A_b \subseteq A_c$ and hence $t \in A_c$; in particular, there is some $I \in \mathcal{I}$ such that $t \in I$. But then $I \cap A_b \neq \emptyset$ and hence $I \subseteq A$ which implies $t \in A$. This shows $A_b \subseteq A \subseteq A_c$.

We now show that A has only finitely many connected components. Assume the opposite, that is, A has infinitely many connected components. Then we can extract from them pairwise disjoint intervals $I_1, I_2, I_3, \ldots \in \mathcal{I}$. By construction, $I_j \cap A_b \neq \emptyset$ and this ensures that we can pick $t_j \in I_j \cap A_b$ for all $j \in \mathbb{N}$; in particular, $a \le y(t_j) \le b$ for all $j \in \mathbb{N}$. Since the sequence (t_j) is bounded, we can assume (after passing to a suitable subsequence if necessary) that (t_j) is strictly monotone, and without loss of generality we may assume that (t_j) is strictly increasing; the other case is similar.

We claim that for each $j \in \mathbb{N}$ there is some $s_j \in (t_j, t_{j+1})$ with $y(s_j) \ge c$. If not, we have $y(s) < c$ for all $s \in (t_j, t_{j+1})$, and since $y(t_j) \le b < c$ and $y(t_{j+1}) \le b < c$ we even have $y(s) < c$ for all $s \in [t_j, t_{j+1}]$ and consequently $[t_j, t_{j+1}] \subseteq A_c$. Since $t_j \in I_j$ and $[t_j, t_{j+1}]$ is connected, we must have $[t_j, t_{j+1}] \subseteq I_j$. But the same argument also shows $[t_j, t_{j+1}] \subseteq I_{j+1}$ and hence $[t_j, t_{j+1}] \subseteq I_j \cap I_{j+1} = \emptyset$ which is clearly impossible. Thus, we indeed find $s_j \in (t_j, t_{j+1})$ with $y(s_j) \ge c$ for all $j \in \mathbb{N}$.

Now, if $X = YBV_\varphi$, then for all $\lambda > 0$,

$$\operatorname{Var}_\varphi(\lambda y) \ge \sum_{j=1}^{\infty} \varphi\big(\lambda|y(s_j) - y(t_j)|\big) \ge \sum_{j=1}^{\infty} \varphi\big(\lambda(c-b)\big) = \infty$$

and hence $y \notin YBV_\varphi$, a contradiction. If $X = \Lambda BV$, then

$$\operatorname{Var}_\Lambda(y) \ge \sum_{j=1}^{\infty} \lambda_j |y(s_j) - y(t_j)| \ge (b-c) \sum_{j=1}^{\infty} \lambda_j = \infty,$$

and so again $y \notin \Lambda BV$. ■

We remark that Lemma 5.1.7 also holds for $y \in RBV_p$, but this will not be needed in the sequel.

Now, here comes the promised example of a composition operator $C_g : BV \to BV$ which is surjective, but not injective.

Example 5.1.8. Define the function $g : \mathbb{R} \to \mathbb{R}$ by $g(u) := \min\{u+2, |u|\}$ which is shown in Figure 5.1.1.

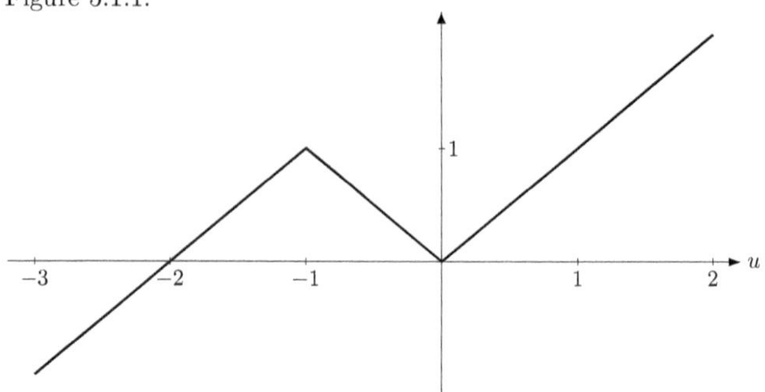

Figure 5.1.1: The function g on $[-3, 2]$.

Then for fixed $v \in \mathbb{R}$ we have

$$g^{-1}(\{v\}) = \begin{cases} \{v-2\} & \text{for } v < 0, \\ \{v-2, -v, v\} & \text{for } 0 \leq v \leq 1, \\ \{v\} & \text{for } v > 1. \end{cases}$$

In particular, g is surjective but not injective, and so $C_g : BV \to BV$ is not injective by Proposition 5.1.2 (a). However, we claim that C_g is surjective. To see this, fix $y \in BV$. Then $y([0,1]) \subseteq [a,b]$ for some $a, b \in \mathbb{R}$ with $a < 1/3 \leq b$. By Lemma 5.1.7 there is a set $A \subseteq [0,1]$ with only finitely many connected components such that

$$y^{-1}([a, 1/3]) \subseteq A \subseteq y^{-1}([a, 2/3)).$$

Define $x : [0,1] \to \mathbb{R}$ by

$$x(t) = \left\{ \begin{array}{ll} y(t) - 2 & \text{for } t \in A, \\ y(t) & \text{for } t \in [0,1] \backslash A \end{array} \right\} = \big(y(t) - 2\big)\chi_A(t) + y(t)\chi_{[0,1]\backslash A}(t).$$

Since A and hence $[0,1]\backslash A$ have only finitely many connected components, both functions χ_A and $\chi_{[0,1]\backslash A}$ belong to BV, and so does x, as BV is an algebra. Moreover, by construction, $g(x(t)) = y(t)$ for all $t \in [0,1]$. ◊

The same example also works if the space BV is replaced by the other spaces WBV_p, YBV_φ or ΛBV. However, we will see later on that it will not work in the Riesz space RBV_p.

The fact that for each $v \in \mathbb{R}$ there is some $u \in \mathbb{R}$ such that $g(u) = v$ (and hence pure surjectivity) is only necessary but not sufficient for C_g to be surjective, as was shown by the Examples 5.1.3, 5.1.4, 5.1.5 and 5.1.6. We therefore need to impose more on those u to ensure surjectivity of C_g.

Definition 5.1.9. We call a function $g : \mathbb{R} \to \mathbb{R}$ *nonflat at* $u \in \mathbb{R}$ if there are compact intervals $I, J \subseteq \mathbb{R}$ such that $u \in I^\circ$, $g|_I : I \to J$ is bijective and $(g|_I)^{-1} \in Lip(J)$.

Example 5.1.10. The function $g_1(u) = u^3$ from the Examples 5.1.3, 5.1.4, 5.1.5 and 5.1.6 which generates a surjective operator C_{g_1} in neither of the spaces BV, WBV_p, ΛBV and RBV_p is nonflat at all $u \in \mathbb{R}\backslash\{0\}$, but not at $u = 0$ which is the only preimage for $v = 0$.
The function $g_2(u) = \min\{u-2, |u|\}$ from Example 5.1.8, however, which did generate a surjective operator $C_{g_2} : X \to X$ in all the spaces BV, WBV_p, YBV_φ and ΛBV, is nonflat at all $u \in \mathbb{R}\backslash\{-1, 0\}$, where $g_2(-1) = 1$ and $g_2(0) = 0$. But for $v = 0$ we can choose $u = -2$ as a preimage at which g_2 is nonflat, while for $v = 1$ we can choose $u = 1$ where g_2 is nonflat. Consequently, this function has the property that for each $v \in \mathbb{R}$ there is always some $u \in \mathbb{R}$ that $g_2(u) = v$ *and* that g_2 is nonflat at u. ♢

The last example suggests that the missing property for surjectivity of C_g is that for each real number the generating function g has at least one preimage at which g is nonflat. This is indeed true and content of the following result which gives at least a sufficient condition for the surjectivity of C_g.

Theorem 5.1.11. *Let X be one of the spaces BV, WBV_p, YBV_φ or ΛBV, and let $g \in Lip_{loc}(\mathbb{R})$. Assume that for each $v \in \mathbb{R}$ there is some $u \in \mathbb{R}$ at which g is nonflat such that $g(u) = v$. Then $C_g : X \to X$ is surjective.*

Proof. Fix $y \in X$ and let $[c, d] := y([0, 1])$. By hypothesis, for each $v \in [c, d]$ there exists some $u \in \mathbb{R}$ at which g is nonflat such that $g(u) = v$. This means that for each $v \in \mathbb{R}$ there is some $u \in \mathbb{R}$ and compact intervals I_v, J_v such that $u \in I_v^\circ$, $g(u) = v$, $g|_{I_v} : I_v \to J_v$ is bijective and $(g|_{I_v})^{-1} \in Lip(J_v)$. In particular, $g|_{I_v}$ is strictly monotone on I_v which implies $v = g(u) \in g(I_v^\circ) = J_v^\circ$. Hence, the system

$$\left\{ J_v^\circ \mid v \in [c, d] \right\}$$

forms an open cover of $[c, d]$, and since the interval $[c, d]$ is compact, we need only finitely many of these intervals to cover it. Let $v_1, \ldots, v_n \in [c, d]$ be points generating these covering intervals, and let $I_j := I_{v_j}$ and $J_j := J_{v_j} = [c_j, d_j]$ with $c_j < d_j$ be so that $g|_{I_j} : I_j \to J_j$ is bijective, $(g|_{I_j})^{-1} \in Lip(J_j)$ and $[c, d] \subseteq \bigcup_{j=1}^n J_j^\circ$; in particular,

$$[0, 1] = y^{-1}\Big([c, d]\Big) = \bigcup_{j=1}^{n} y^{-1}\big(J_j^\circ\big) = \bigcup_{j=1}^{n} y^{-1}\Big([c_j, d_j)\Big). \tag{5.1.1}$$

Without loss of generality we can assume that

$$\begin{aligned} &c_j < d_{j-1} < c_{j+1} < d_j \qquad \text{for } j \in \{2, \ldots, n-1\}, \\ &c_1 \leq c \quad \text{and} \quad c_n \leq d \leq d_n. \end{aligned} \tag{5.1.2}$$

Taking (5.1.2) into account we find by Lemma 5.1.7 sets $A_j \subseteq [0, 1]$ with only finitely many connected components such that

$$y^{-1}\Big([c_j, c_{j+1}]\Big) \subseteq A_j \subseteq y^{-1}\Big([c_j, d_j)\Big) \qquad \text{for } j \in \{1, \ldots, n-1\}.$$

Writing $A_n := [0,1]$, the sets B_j, recursively defined by

$$B_j := \left([0,1]\setminus\bigcup_{i=1}^{j-1} B_i\right)\cap A_j \qquad \text{for } j\in\{1,\dots,n\},$$

also have only finitely many connected components and are pairwise disjoint.

We now show

$$[0,1]=\bigcup_{j=1}^{n} B_j. \tag{5.1.3}$$

To see this first note that each B_j is a subset of $[0,1]$, and thus we only need to show that $[0,1]\subseteq\bigcup_{j=1}^n B_j$. But this is clear, since if $t_0\in[0,1]\setminus\bigcup_{j=1}^{n-1}B_j$, then

$$t_0\in[0,1]\setminus\bigcup_{j=1}^{n-1}B_j=\left([0,1]\setminus\bigcup_{i=1}^{n-1}B_i\right)\cap A_n=B_n$$

which shows $t_0\in\bigcup_{j=1}^n B_j$ and hence (5.1.3). Consequently, the sets B_j form a partition of the interval $[0,1]$.

Writing $h_j=(g|_{I_j})^{-1}:J_j\to I_j$ for $j\in\{1,\dots,n\}$, each function $\varphi_j:[c,d]\to\mathbb{R}$, defined by

$$\varphi_j(v):=\begin{cases} h_j(v) & \text{for } v\in J_j,\\ h_j(c_j) & \text{for } v\in[c,c_j),\\ h_j(d_j) & \text{for } v\in(d_j,d],\end{cases}$$

belongs to $Lip[c,d]$ for $j\in\{1,\dots,n\}$. Let us now define the function $x:[0,1]\to\mathbb{R}$ by

$$x(t)=\sum_{j=1}^{n}\varphi_j\big(y(t)\big)\chi_{B_j}(t).$$

Since each φ_j is Lipschitz continuous, we have $\varphi_j\circ y\in X$ for each $j\in\{1,\dots,n\}$, and since each B_j has only finitely many connected components, we also have $\chi_{B_j}\in X$ for each $j\in\{1,\dots,n\}$. Finally, since X is an algebra, $x\in X$.

It remains to show that $g(x(t))=y(t)$ holds for each $t\in[0,1]$. To this end, fix $t_0\in[0,1]$. Then there is exactly one $j\in\{1,\dots,n\}$ such that $t_0\in B_j$. If $j<n$, then $t_0\in A_j$ and hence $y(t_0)\in[c_j,d_j)\subseteq J_j$ which implies

$$g\big(x(t_0)\big)=g\left(\sum_{k=1}^{n}\varphi_k\big(y(t)\big)\chi_{B_k}(t_0)\right)=g\Big(\varphi_j\big(y(t_0)\big)\Big)=y(t_0).$$

If $j=n$, then $t_0\notin A_k$ for all $k\in\{1,\dots,n-1\}$, since otherwise we had $t_0\in A_k$ for some $k\in\{1,\dots,n-1\}$ and hence $t_0\in\left([0,1]\setminus\bigcup_{i=1}^{k-1}B_i\right)\cap A_k=B_k$ which is not possible. But this implies $y(t_0)\notin[c_i,c_{i+1}]$ for all $i\in\{1,\dots,n-1\}$ and hence

$$y(t_0)\in[c,d]\setminus\bigcup_{i=1}^{n-1}[c_i,c_{i+1}]=[c,d]\setminus[c_1,c_n]=(c_n,d]\subseteq[c_n,d_n]=J_n.$$

Again we obtain

$$g\big(x(t_0)\big) = g\left(\sum_{k=1}^{n} \varphi_k\big(y(t)\big)\chi_{B_k}(t_0)\right) = g\Big(\varphi_n\big(y(t_0)\big)\Big) = y(t_0)$$

and this shows indeed $C_g x = y$ and hence the surjectivity of C_g. ■

It is, however, not clear if the condition given in Theorem 5.1.11 is also necessary. In order to find out if weakening this condition slightly still remains sufficient, we need to have a closer look at the term "nonflat".
Recall that local Lipschitz continuity geometrically means that the slope of the function is locally bounded. For a function to be nonflat at a point it is therefore reasonable to ensure that the slope in a neighborhood of that point needs to be bounded away from zero. This is indeed true and the content of the following

Proposition 5.1.12. *A continuous function $g : \mathbb{R} \to \mathbb{R}$ is nonflat at $u \in \mathbb{R}$ if and only if there are numbers $\delta, m > 0$ such that $|g(u_1) - g(u_2)| \geq m|u_1 - u_2|$ for $u_1, u_2 \in [u - \delta, u + \delta]$.*

Proof. Let g be nonflat at u. Then there exist compact intervals I, J such that $u \in I^\circ$, $g|_I : I \to J$ is bijective and $h := (g|_I)^{-1} \in Lip(J)$. Then there is some $\delta > 0$ such that $[u - \delta, u + \delta] \subseteq I$, and by replacing I with $[u - \delta, u + \delta]$ and J with $g(I)$ we can assume that $I = [u - \delta, u + \delta]$ and $J = g(I)$. Note that due to the continuity of g, the set $g(I)$ is again a compact interval. Since $h \in Lip(J)$ there is some $L > 0$ such that $|h(v_1) - h(v_2)| \leq L|v_1 - v_2|$ for all $v_1, v_2 \in J$. Substituting $u_1 = h(v_1)$ and $u_2 = h(v_2)$ we obtain $|u_1 - u_2| \leq L|g(u_1) - g(u_2)|$ and hence $|g(u_1) - g(u_2)| \geq m|u_1 - u_2|$ for $m := 1/L$ and all $u_1, u_2 \in I = [u - \delta, u + \delta]$, as claimed.
Now assume that there are constants $\delta, m > 0$ such that $|g(u_1) - g(u_2)| \geq m|u_1 - u_2|$ for $u_1, u_2 \in [u - \delta, u + \delta]$. Then g is injective on $I := [u - \delta, u + \delta]$. Since g is continuous, the set $J := g(I)$ is also a compact interval, and $g|_I : I \to J$ is bijective. Let h be its inverse, i.e. $h = (g|_I)^{-1} : J \to I$. Then for $v_1, v_2 \in J$ we set $u_1 := h(v_1)$ and $u_2 := h(v_2)$. By writing $L := 1/m$ we obtain $|h(v_1) - h(v_2)| = |u_1 - u_2| \leq L|g(u_1) - g(u_2)| = L|g(h(v_1)) - g(h(v_2))| = L|v_1 - v_2|$, and this shows $(g|_I)^{-1} = h \in Lip(J)$. Consequently, g is nonflat at u. ■

Thus, g being nonflat at a point u_0 means that the slope of g *in a neighborhood* of u_0 is bounded away from zero. In particular,

$$\liminf_{u \to u_0} \frac{|g(u) - g(u_0)|}{|u - u_0|} > 0. \tag{5.1.4}$$

It is now reasonable to ask whether the condition being nonflat at $u_0 = u$ in Theorem 5.1.11 may be replaced by the weaker condition (5.1.4). Surprisingly, the answer is negative, even if this weaker condition is satisfied only at one single point while the function remains nonflat at all other points. Even this case may cause the corresponding operator C_g to be not surjective anymore. We illustrate this in the following example, where we will construct a function $g : \mathbb{R} \to \mathbb{R}$ with $g(0) = 0$ which is bijective, globally

Lipschitz continuous and nonflat at every point of $\mathbb{R}$ except at $u_0 = 0$, where it at least satisfies (5.1.4), that is,

$$\liminf_{u\to 0} \left|\frac{g(u)}{u}\right| > 0. \tag{5.1.5}$$

Example 5.1.13. Consider $g : \mathbb{R} \to \mathbb{R}$, defined by

$$g(u) = \begin{cases} \dfrac{u}{n} + 2^{-n}\dfrac{n-1}{n} & \text{for } 2^{-n} \le u < 2^{-n}\frac{3n-1}{2n-1} \text{ and } n \in \mathbb{N}, \\ 2u - 2^{-n+1} & \text{for } 2^{-n}\frac{3n-1}{2n-1} \le u < 2^{-n+1} \text{ and } n \in \mathbb{N}, \\ u & \text{for } u \in (-\infty, 0] \cup [1, \infty). \end{cases}$$

Figure 5.1.2 shows the relevant part of the graph of g.

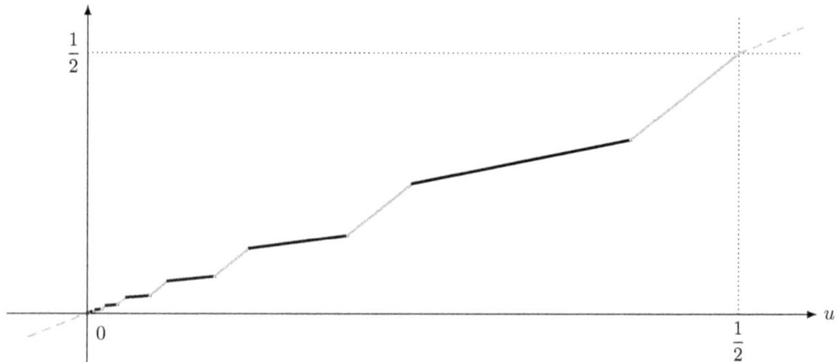

Figure 5.1.2: Graph of g on $[0, 1/2]$.

Then g is globally Lipschitz continuous with $\mathrm{lip}(g) = 2$,

$$g(2^{-n}) = 2^{-n} \quad \text{and} \quad g\left(2^{-n}\frac{3n-1}{2n-1}\right) = \frac{2^{-n+1}n}{2n-1} \quad \text{for } n \in \mathbb{N}.$$

Moreover, g is strictly increasing and bijective with inverse

$$g^{-1}(v) = \begin{cases} nv - 2^{-n}(n-1) & \text{for } 2^{-n} \le v < \frac{2^{-n+1}n}{2n-1} \text{ and } n \in \mathbb{N}, \\ \frac{v}{2} + 2^{-n} & \text{for } \frac{2^{-n+1}n}{2n-1} \le v < 2^{-n+1} \text{ and } n \in \mathbb{N}, \\ v & \text{for } v \in (-\infty, 0] \cup [1, \infty). \end{cases}$$

By Proposition 5.1.2 (a), the operator $C_g : BV \to BV$ is injective. If it was surjective and so bijective, then by Proposition 5.1.2 (c) the function g had a locally Lipschitz continuous inverse. However, g^{-1} is not locally Lipschitz continuous at $v = 0$.

We also can see directly that C_g cannot be surjective. To this end, define $T_n \subseteq [2^{-n}, 2^{-n+1})$, $y_n > 0$ and $m_n \in \mathbb{N}$ so that

$$y_n := \frac{2^{-n-1}}{n} \qquad \text{and} \qquad m_n \leq \frac{1}{2y_n n^2} < 2m_n$$

and

$$T_n := \left\{2^{-n} + \frac{2^{-n}}{m_n} \cdot j \mid j = 0, \ldots, m_n - 1\right\},$$

and set $y : [0,1] \to \mathbb{R}$ by $y(0) = 0$, $y(1) = 1/2$ and

$$y(t) = \begin{cases} 2^{-n} + y_n & \text{for } t \in T_n, \\ 2^{-n} & \text{for } [2^{-n}, 2^{-n+1}) \backslash T_n. \end{cases}$$

Then $y \in BV$ with

$$\begin{aligned} \operatorname{Var}(y) &= \sum_{n=1}^{\infty} \operatorname{Var}(y, [2^{-n}, 2^{-n+1}]) = \sum_{n=1}^{\infty} (2y_n m_n + 2^{-n+1} - 2^{-n}) \\ &\leq \sum_{n=1}^{\infty} \frac{1}{n^2} + 1 = \frac{\pi^2}{6} + 1 < \infty. \end{aligned}$$

Moreover, $y([2^{-n}, 2^{-n+1}]) \subseteq [2^{-n}, \frac{2^{-n+1}n}{2n-1})$, and hence the only function $x : [0,1] \to \mathbb{R}$ satisfying $C_g x = y$ is given by $x(0) = 0, x(1) = 1/2$ and

$$x(t) = \begin{cases} 2^{-n} + ny_n & \text{for } t \in T_n, \\ 2^{-n} & \text{for } [2^{-n}, 2^{-n+1}) \backslash T_n. \end{cases}$$

However, $x \notin BV$, since

$$\operatorname{Var}(x) = \sum_{n=1}^{\infty} \operatorname{Var}(x, [2^{-n}, 2^{-n+1}]) = \sum_{n=1}^{\infty} (2ny_n m_n + 2^{-n+1} - 2^{-n}) \geq \sum_{n=1}^{\infty} \frac{1}{2n} + 1 = \infty.$$

Consequently, $C_g : BV \to BV$ cannot be surjective.
Finally, g is obviously nonflat at any $u \neq 0$ but cannot be nonflat at $u = 0$, as the gray line segments near the origin (see Figure 5.1.2) have slopes that get arbitrarily close to 0. However, g satisfies (5.1.5), because for $u < 0$ we have $g(u) = u$, and for $2^{-n} \leq u < 2^{-n+1}$ with $n \in \mathbb{N}$ we have due to the monotonicity of g,

$$g(u) \geq g\big(2^{-n}\big) = 2^{-n} \geq u/2.$$

Consequently,

$$\liminf_{u \to 0} \frac{g(u)}{u} \geq \frac{1}{2}$$

and so (5.1.5) is indeed true. ◊

The last example shows that weakening the condition given in Theorem 5.1.11 only slightly turns down the surjectivity of C_g. We therefore conjecture that this condition is also necessary, but we were not able to prove it.

The previous results referred to BV-spaces which contain characteristic functions and therefore excluded our last space RBV_p. Here, the situation is again different, but apparently as complicated as in BV. However, the condition given in Theorem 5.1.11 is not sufficient to guarantee that C_g from RBV_p into itself is surjective.

Example 5.1.14. Define $g(u) := \min\{u+2, |u|\}$ as in Example 5.1.8. There we have seen that C_g maps the space BV into itself and is surjective. However, since g is Lipschitz continuous, C_g maps also the space RBV_p into itself. But this time, $C_g : RBV_p \to RBV_p$ is not surjective. Recall that for fixed $v \in \mathbb{R}$,

$$g^{-1}(\{v\}) = \begin{cases} \{v-2\} & \text{for } v < 0, \\ \{v-2, -v, v\} & \text{for } 0 \leq v \leq 1, \\ \{v\} & \text{for } v > 1. \end{cases}$$

The function $y(t) = 3t - 1$ maps the interval $[0,1]$ bijectively onto $[-1,2]$. If $x \in RBV_p$ is a function satisfying $C_g x = y$, then x must be injective and continuous. But since $y(0) = -1 < 0$ and $y(1) = 2 > 1$, we must have $x(t) = y(t) - 2$ for $t \in [0, 2/3]$ but simultaneously $x(t) = y(t)$ for $t \in [1/3, 1]$ which is not possible. ◊

The following result is a sufficient condition on g to ensure that $C_g : RBV_p \to RBV_p$ is surjective. It says basically that those "zigzag" patterns like in Example 5.1.14 must be compensated somewhere else.

Proposition 5.1.15. *Let $g \in Lip_{loc}(\mathbb{R})$. Then the composition operator $C_g : RBV_p \to RBV_p$ is surjective if for any compact interval $J \subseteq \mathbb{R}$ we find a compact interval $I \subseteq \mathbb{R}$ such that $g|_I : I \to J$ is bijective and $(g|_I)^{-1} \in Lip(J)$.*

Proof. Assume that g possesses the property that for any compact interval $J \subseteq \mathbb{R}$ we find a compact interval $I \subseteq \mathbb{R}$ such that $g|_I : I \to J$ is bijective and $(g|_I)^{-1} \in Lip(J)$. Fix $y \in RBV_p$. Then y is bounded and continuous, and by the Intermediate Value Theorem, $J := y([0,1])$ is a compact interval. By assumption, there is some compact interval $I \subseteq \mathbb{R}$ such that $g|_I : I \to J$ is bijective and $g|_I^{-1} \in Lip(J)$. Therefore, the function $x := g|_I^{-1} \circ y$ is well-defined, belongs to RBV_p and satisfies $g \circ x = y$, that is, $C_g x = y$. Since y was arbitrary, $C_g : RBV_p \to RBV_p$ is surjective. ■

The function g from the Examples 5.1.8 and 5.1.14 does not have the property assumed in Proposition 5.1.15. Indeed, if one takes $J = [-1,3]$, then g maps no interval bijectively onto J. This explains why g in Example 5.1.14 did not generate a surjective operator $C_g : RBV_p \to RBV_p$.

One could argue that the condition given in Proposition 5.1.15 is equivalent to g being nonflat at any $u \in \mathbb{R}$. This is not true. In fact, neither of these two conditions implies the other. The next example shows that being nonflat at any $u \in \mathbb{R}$ does not imply the condition given in Proposition 5.1.15.

Example 5.1.16. The function $g(u) = e^u$ is nonflat at any $u \in \mathbb{R}$. However, for the compact interval $J = [-1,0]$ there is no compact interval $I \subseteq \mathbb{R}$ such that $g|_I : I \to J$ is bijective. ◊

The reason for this is that the condition in Proposition 5.1.15 implies the surjectivity of g whereas being nonflat at each $u \in \mathbb{R}$ does not. However, if one requires g to be nonflat at any $u \in \mathbb{R}$ while being surjective, then g must be a homeomorphism of $\mathbb{R}$ with locally Lipschitz continuous inverse. On the other hand, there are functions satisfying the condition of Proposition 5.1.15 which are not bijective on $\mathbb{R}$.

Example 5.1.17. The function $g : \mathbb{R} \to \mathbb{R}$, defined by $g(u) := u^2 \cos(u)$, is locally Lipschitz continuous, surjective but not injective. Therefore, it cannot be nonflat at any $u \in \mathbb{R}$. Indeed, g is differentiable with $g'(0) = 0$, and so (5.1.4) is violated at $u = 0$ (and, similarly, at the infinitely many locally extremal points of g).
However, g does meet the requirements of Proposition 5.1.15. Indeed, if $J \subseteq \mathbb{R}$ is a compact interval, then one can choose $k \in \mathbb{N}$ so large that

$$J \subseteq \Big(g\big(k\pi\big), g\big((k+1)\pi\big)\Big) = g\Big(\big(k\pi, (k+1)\pi\big)\Big).$$

But since g is injective on $[k\pi, (k+1)\pi]$, there is a compact interval $I \subseteq [k\pi, (k+1)\pi]$ such that g maps I bijectively onto J with $g^{-1} \in Lip(J)$. ◊

We now come back to analytic properties of C_g and formulate and prove the promised result concerning the boundedness of C_g. For this and the rest of this section we need the following technical auxiliary result.

Lemma 5.1.18. *Let X be one of the spaces BV, WBV_p, YBV_φ, ΛBV or RBV_p, and let $g : \mathbb{R} \to \mathbb{R}$ be so that C_g maps X into itself. For fixed $R \geq 1$ there is $\varrho > 0$ such that for $u, v \in \mathbb{R}$, $\alpha \in (0, \varrho]$, $\beta \in [0, \alpha]$ and $\gamma \in (0, 1]$ there are functions $x, y \in X$ which satisfy*

(i) $\|x\|_X \leq \max\{|u|, |u+\alpha|\} + 2R/\gamma$ *and* $\|y\|_X \leq \max\{|v|, |v+\beta|\} + 2R/\gamma$,

(ii) $\|x - y\|_X = |u - v|$ *for* $\alpha = \beta$,

(iii) $$\left|\frac{g(u+\alpha) - g(u)}{\alpha} - \frac{g(v+\beta) - g(v)}{\alpha}\right| \leq 16\gamma \, \|C_g x - C_g y\|_X.$$

The functions x and y can be chosen of the form $x(t) = u + \alpha h(t)$ and $y(t) = v + \beta h(t)$, where $h \in X$ satisfies $\|h\|_\infty = 1$.

The most important inequality is of course the one given in (iii). It basically says that the difference of the slopes of g at two prescribed points u and v can be estimated from above by the norm of $C_g x - C_g y$, where x and y are suitably chosen functions that are close to u and v, respectively. In this sense the growth of g may be estimated by the mapping behavior of C_g; in particular, if $C_g x$ stays close to $C_g y$, then g cannot grow too rapidly. This is of course a very vague interpretation of Lemma 5.1.18, but it will be of great use later on, where it will be applied in detail.

Proof of Lemma 5.1.18. Fix $u, v \in \mathbb{R}$ and $R \geq 1$. We handle the three cases $X = YBV_\varphi$, $X = \Lambda BV$ and $X = RBV_p$ separately and start with $X = YBV_\varphi$.
First, pick $\varrho > 0$ so that $0 < \varphi(\varrho) \leq R$. This is possible since $\varphi(r) \to 0$ as $r \to 0+$. Fix $\alpha \in (0, \varrho]$, $\beta \in [0, \alpha]$ and $\gamma \in (0, 1]$. Then we have $R/\varphi(\alpha\gamma) \geq R/\varphi(\varrho) \geq 1$, and this is why we find some $n \in \mathbb{N}$ so large that

$$n \leq \frac{R}{\varphi(\alpha\gamma)} \leq 2n. \tag{5.1.6}$$

Define the functions $x, y : [0, 1] \to \mathbb{R}$ by

$$x := u + \mathfrak{J}_{(\alpha_j)} \qquad \text{with} \qquad \alpha_j := \begin{cases} \alpha & \text{for } 1 \leq j \leq n, \\ 0 & \text{for } j > n, \end{cases} \tag{5.1.7}$$

$$y := v + \mathfrak{J}_{(\beta_j)} \qquad \text{with} \qquad \beta_j := \begin{cases} \beta & \text{for } 1 \leq j \leq n, \\ 0 & \text{for } j > n, \end{cases} \tag{5.1.8}$$

where we use the functions defined in (1.2.1). Then $x, y \in YBV_\varphi$ with

$$\|x\|_\infty = \max\{|u|, |u + \alpha|\} \quad \text{and} \quad \|y\|_\infty = \max\{|v|, |v + \beta|\}. \tag{5.1.9}$$

Moreover, since $R \geq 1$, we have due to the convexity of φ the estimate $\varphi(\alpha\gamma/R) \leq \varphi(\alpha\gamma)/R$ and hence by Corollary 1.2.11 and (5.1.6),

$$\begin{aligned} \mathrm{Var}_\varphi\left(\frac{x}{2R/\gamma}\right) &= \mathrm{Var}_\varphi\left(\frac{u + \mathfrak{J}_{(\alpha_j)}}{2R/\gamma}\right) = \mathrm{Var}_\varphi\left(\frac{\mathfrak{J}_{(\alpha_j)}}{2R/\gamma}\right) \leq n\varphi\left(\frac{\alpha\gamma}{R}\right) \leq n\frac{\varphi(\alpha\gamma)}{R} \leq 1, \\ \mathrm{Var}_\varphi\left(\frac{y}{2R/\gamma}\right) &= \mathrm{Var}_\varphi\left(\frac{v + \mathfrak{J}_{(\beta_j)}}{2R/\gamma}\right) = \mathrm{Var}_\varphi\left(\frac{\mathfrak{J}_{(\beta_j)}}{2R/\gamma}\right) \leq n\varphi\left(\frac{\beta\gamma}{R}\right) \leq n\varphi\left(\frac{\alpha\gamma}{R}\right) \leq 1. \end{aligned}$$

This implies $\mathfrak{M}(x) \leq 2R/\gamma$ and $\mathfrak{M}(y) \leq 2R/\gamma$, and therefore

$$\|x\|_{YBV_\varphi} \leq \max\{|u|, |u + \alpha|\} + 2R/\gamma, \qquad \|y\|_{YBV_\varphi} \leq \max\{|v|, |v + \beta|\} + 2R/\gamma,$$

which is property (i).
If $\alpha = \beta$, then $x - y \equiv u - v$, and hence $\mathfrak{M}(x - y) = 0$. In this case, $\|x - y\|_{YBV_\varphi} = \|x - y\|_\infty = |u - v|$ which is property (ii).
To show (iii), let $\lambda > 0$ be so that

$$\mathrm{Var}_\varphi\left(\frac{C_g x - C_g y}{\lambda}\right) \leq 1.$$

In order to estimate the Young variation of $C_g x - C_g y$, note that $C_g x - C_g y - g(u) + g(v) = \mathfrak{J}_{(\eta_j)}$, where $\eta_j = g(u + \alpha) - g(u) - g(v + \beta) + g(v)$ for $1 \leq j \leq n$ and $\eta_j = 0$ for $j > n$. Thus, from Corollary 1.2.11 we obtain by using (5.1.6) and $R \geq 1$,

$$\begin{aligned} 1 \geq \mathrm{Var}_\varphi\left(\frac{g \circ x - g \circ y}{\lambda}\right) &\geq 2n\varphi\left(\frac{|g(u + \alpha) - g(u) - g(v + \beta) + g(v)|}{\lambda}\right) \\ &\geq \frac{1}{\varphi(\alpha\gamma)}\varphi\left(\frac{|g(u + \alpha) - g(u) - g(v + \beta) + g(v)|}{\lambda}\right). \end{aligned}$$

Multiplying by $\varphi(\alpha\gamma)$, applying φ^{-1}, dividing by $\alpha\gamma$ and multiplying by λ on both sides gives

$$\frac{1}{\gamma}\left|\frac{g(u+\alpha)-g(u)}{\alpha}-\frac{g(v+\beta)-g(v)}{\alpha}\right| \le \lambda.$$

If we now take the infimum over all such λ, we get

$$\frac{1}{\gamma}\left|\frac{g(u+\alpha)-g(u)}{\alpha}-\frac{g(v+\beta)-g(v)}{\alpha}\right| \le \mathfrak{M}(C_g x - C_g y) \le \|C_g x - C_g y\|_{YBV_\varphi},$$

and this proves (iii).

We now prove the statement for $X = \Lambda BV$, where $\Lambda = (\lambda_j)$ is the Waterman sequence for ΛBV, and $\Lambda_n := \lambda_1 + \lambda_2 + \ldots + \lambda_n$ its partial sums. This time, choose ϱ so that $R/\varrho = \lambda_1$. Fix $\alpha \in (0,\varrho]$, $\beta \in [0,\alpha]$ and $\gamma \in (0,1]$. Then we have $R/(\alpha\gamma) \ge R/\varrho = \lambda_1 = \Lambda_1$, and this is why we find some $n \in \mathbb{N}$ so large that

$$\Lambda_n \le \frac{R}{\alpha\gamma} \le 2\Lambda_n. \tag{5.1.10}$$

Note that since (λ_j) is decreasing we have $\Lambda_{n+1} \le 2\Lambda_n$ for all $n \in \mathbb{N}$, and this implies that the intervals $([\Lambda_n, 2\Lambda_n])_{n\in\mathbb{N}}$ cover $[\Lambda_1, \infty)$, because $\Lambda_n \to \infty$ as $n \to \infty$.
Define the functions x and y as in (5.1.7) and (5.1.8). Then $x, y \in \Lambda BV$ with (5.1.9). Moreover, by (5.1.10) and Corollary 1.2.21,

$$\operatorname{Var}_\Lambda(x) = \operatorname{Var}_\Lambda(\mathfrak{J}_{(\alpha_j)}) \le 2\sum_{j=1}^{n} \lambda_j \alpha = 2\alpha\Lambda_n \le \frac{2R}{\gamma},$$
$$\operatorname{Var}_\Lambda(y) = \operatorname{Var}_\Lambda(\mathfrak{J}_{(\beta_j)}) \le 2\sum_{j=1}^{n} \lambda_j \beta = 2\beta\Lambda_n \le 2\alpha\Lambda_n \le \frac{2R}{\gamma}.$$

These estimates imply property (i).
Property (ii) is again fulfilled as for $\alpha = \beta$ we have $\operatorname{Var}_\Lambda(x-y) = 0$.
To show (iii) we argue similarly as for $X = YBV_\varphi$, but this time we use Corollary 1.2.21 and (5.1.10) together with $R \ge 1$. Accordingly, we obtain

$$\begin{aligned}\|C_g x - C_g y\|_{\Lambda BV} &\ge \operatorname{Var}_\Lambda(g\circ x - g\circ y) \ge \Lambda_n\big|g(u+\alpha)-g(u)-g(v+\beta)+g(v)\big| \\ &\ge \frac{1}{2\gamma}\left|\frac{g(u+\alpha)-g(u)}{\alpha}-\frac{g(v+\beta)-g(v)}{\alpha}\right|,\end{aligned}$$

and so (iii) is established. Note that in the last two cases the functions x and y have indeed the form $x = u + \alpha h$ and $y = v + \beta h$, where $h = \mathfrak{J}_{(\zeta_j)}$ with $\zeta_j = 1$ for $1 \le j \le n$ and $\zeta_j = 0$ for $j > n$ satisfies $\|h\|_\infty = 1$.

Finally, we show the statement for $X = RBV_p$, but for that we need to proceed a little different than before. First, we choose ϱ so that $R/(8\varrho) = 1$, because then we have for $\alpha \in (0,\varrho]$, $\beta \in [0,\alpha]$ and $\gamma \in (0,1]$ that $R/(8\alpha\gamma) \ge 1$. For such fixed α and γ we therefore find an $n \in \mathbb{N}$ so large that

$$n^{2-1/p} \le \frac{R}{8\alpha\gamma} \le (n+1)^{2-1/p}. \tag{5.1.11}$$

The function $h : [0,1] \to \mathbb{R}$, defined by

$$h(t) = \begin{cases} 0 & \text{for } 0 \le t \le \frac{1}{2n+1}, \\ \left[\cos(\frac{\pi}{2t})\right]^2 & \text{for } \frac{1}{2n+1} < t \le 1, \end{cases}$$

belongs to RBV_p, because $h \in AC$, and by Theorem 1.2.25,

$$\begin{aligned} \mathrm{RVar}_p(h)^{1/p} &= \left(\int_{1/(2n+1)}^1 |h'(t)|^p \, \mathrm{d}t\right)^{1/p} = \frac{\pi}{2}\left(\int_{1/(2n+1)}^1 \frac{|\sin(\pi/t)|^p}{t^{2p}} \, \mathrm{d}t\right)^{1/p} \\ &\le \frac{\pi}{2}\left(\int_{1/(2n+1)}^1 \frac{1}{t^{2p}} \, \mathrm{d}t\right)^{1/p} = \frac{\pi}{2}\left(\frac{(2n+1)^{2p} - 2n - 1}{(2n+1)(2p-1)}\right)^{1/p} \le 16n^{2-1/p}. \end{aligned}$$

Moreover, $0 \le h(t) \le 1$ for all $t \in [0,1]$ with

$$h\left(\frac{1}{2j}\right) = 1 \quad \text{and} \quad h\left(\frac{1}{2j-1}\right) = 0 \quad \text{for } j \in \{1, \dots, n\};$$

in particular, $\|h\|_\infty = 1$. The functions $x := u + \alpha h$ and $y := v + \beta h$ then also belong to RBV_p and satisfy

$$\begin{aligned} \mathrm{RVar}_p(x)^{1/p} &= \mathrm{RVar}_p(\alpha h) \le 16\alpha n^{2-1/p} \le 2R/\gamma, \\ \mathrm{RVar}_p(y)^{1/p} &= \mathrm{RVar}_p(\beta h) \le 16\beta n^{2-1/p} \le 16\alpha n^{2-1/p} \le 2R/\gamma, \end{aligned}$$

by (5.1.11). In addition, $\|x\|_\infty \le \max\{|u|, |u+\alpha|\}$ and $\|y\|_\infty \le \max\{|v|, |v+\beta|\}$, and this shows part (i).

Again, (ii) is satisfied for $\alpha = \beta$ as then $\mathrm{RVar}_p(x - y) = 0$.

For (iii) note that

$$\begin{aligned} \|C_g x - C_g y\|_{RBV_p} &\ge \left(\sum_{j=1}^n \frac{\left|(g \circ x - g \circ y)\left(\frac{1}{2j-1}\right) - (g \circ x - g \circ y)\left(\frac{1}{2j}\right)\right|^p}{\left|\frac{1}{2j-1} - \frac{1}{2j}\right|^{p-1}}\right)^{1/p} \\ &= \Big|g(u) - g(u+\alpha) - g(v) + g(v+\beta)\Big| \left(\sum_{j=1}^n \left(4j^2 - 2j\right)^{p-1}\right)^{1/p} \\ &\ge \frac{(n+1)^{2-1/p}}{2} \Big|g(u) - g(u+\alpha) - g(v) + g(v+\beta)\Big|, \end{aligned} \tag{5.1.12}$$

where we have used the identity

$$\sum_{j=1}^n \left(4j^2 - 2j\right)^{p-1} \ge \frac{(n+1)^{2p-1}}{2^p} \tag{5.1.13}$$

which we are going to prove now by induction. For $n = 1$ the estimate (5.1.13) is clearly true (even with equality). Assume that (5.1.13) has already been established for some fixed $n \in \mathbb{N}$. Then we obtain

$$\sum_{j=1}^{n+1} \left(4j^2 - 2j\right)^{p-1} \ge \frac{(n+1)^{2p-1}}{2^p} + 2^{p-1}\left(2n^2 + 3n + 1\right)^{p-1}.$$

We show that the inequality

$$\frac{(n+1)^{2p-1}}{2^p} + 2^{p-1}\left(2n^2+3n+1\right)^{p-1} \geq \frac{(n+2)^{2p-1}}{2^p} \tag{5.1.14}$$

holds, because then (5.1.13) is proven for n replaced by $n+1$. Note that (5.1.14) is equivalent to the estimate

$$\psi(n+2) - \psi(n+1) \leq 2^{p-1}\left(2n^2+3n+1\right)^{p-1}, \tag{5.1.15}$$

where $\psi(t) := t^{2p-1}/2^p$. From the Mean Value Theorem we obtain $\psi(n+2)-\psi(n+1) = \psi'(\xi)$ for some $\xi \in [n+1, n+2]$, and this yields

$$\psi(n+2) - \psi(n+1) \leq \frac{2p-1}{2^p}\left(n^2+4n+4\right)^{p-1}. \tag{5.1.16}$$

However,

$$(2p-1)\left(n^2+4n+4\right)^{p-1} \leq 2^{2p-1}\left(2n^2+3n+1\right)^{p-1} \tag{5.1.17}$$

is true for our n. Indeed, if $n=1$, then the inequality (5.1.17) reduces to $2p-1 \leq 2^{2p-1}(2/3)^{p-1}$ which is true for all $p \geq 1$, and if $n=2$, then the estimate (5.1.17) reads $2p-1 \leq 2^{2p-1}(15/16)^{p-1}$ which again is true for all $p \geq 1$. If $n \geq 3$, then (5.1.17) is also true as $2p-1 \leq 2^{2p-1}$ and $n^2+4n+4 \leq 2n^2+3n+1$. Consequently, (5.1.17) holds for our n. Combining (5.1.17) and (5.1.16) we see that (5.1.15) and hence also (5.1.14) is true. Finally, (5.1.13) is indeed established.
Using (5.1.11) and $R \geq 1$, we obtain from (5.1.12),

$$\|C_g x - C_g y\|_{RBV_p} \geq \frac{1}{16\gamma}\left|\frac{g(u)-g(u+\alpha)}{\alpha} - \frac{g(v)-g(v+\beta)}{\alpha}\right|,$$

which again shows (iii) and finally completes the proof of the Lemma. ■

We are now in a position to formulate our result about the acting conditions and boundedness of C_g. Recall that

$$\mathbb{B}_R(X) = \left\{x \in X \mid \|x\|_X \leq R\right\}$$

denotes the closed ball around 0 with radius R in a normed vector space $(X, \|\cdot\|_X)$.

Theorem 5.1.19. *Let $g : \mathbb{R} \to \mathbb{R}$, and let X be one of the spaces BV, WBV_p, YBV_φ, ΛBV or RBV_p. Then C_g maps X into itself and is bounded if and only if $g \in Lip_{loc}(\mathbb{R})$.*

Proof. We have already seen in the comments around Proposition 5.1.1 that if $g \in Lip_{loc}(\mathbb{R})$, then C_g maps X into itself and is bounded. We now prove the converse which, again by Proposition 5.1.1, is only necessary for $X = YBV_\varphi$. To this end, fix $R \geq 1$ and consider

$$A := \mathbb{B}_{34R}(YBV_\varphi) = \{x \in YBV_\varphi \mid \|x\|_{YBV_\varphi} \leq 34R\}$$

which is a bounded subset of YBV_φ. Since C_g is assumed to be bounded, there is some $L > 0$ such that $\|C_g x\|_{YBV_\varphi} \leq L/2$ for all $x \in A$.
By Lemma 5.1.18 there is some $\varrho \in (0, R]$ such that for $u \in [-R, R]$, $v = 0$, $\alpha \in (0, \varrho]$, $\beta = 0$ and $\gamma = 1/16$ there are functions $x, y \in YBV_\varphi$ of the form $x = u + \alpha h$ and $y = v + \beta h = \mathbb{O}$ with $h \in YBV_\varphi$, $\|x\|_{YBV_\varphi} \leq \max\{|u|, |u + \alpha|\} + 2R/\gamma \leq 34R$, $\|y\|_{YBV_\varphi} = 0$ and

$$\left|\frac{g(u+\alpha) - g(u)}{\alpha}\right| \leq \|C_g x - C_g y\|_{YBV_\varphi} \leq \|C_g x\|_{YBV_\varphi} + \|C_g \mathbb{O}\|_{YBV_\varphi} \leq L;$$

note that $x, y \in A$. We get $|g(u + \alpha) - g(u)| \leq L\alpha$ or equivalently $|g(u') - g(u)| \leq L(u' - u)$ for $u, u' \in [-R, R]$, $0 < u' - u \leq \varrho$. This implies that g is Lipschitz continuous on $[-R, R]$, and since R was arbitrary, it is locally Lipschitz continuous on all of $\mathbb{R}$. ■

Lemma 5.1.18 can also be used to characterize compactness. Recall that an arbitrary operator $T : X \to X$ from a normed space X into itself is compact if and only if every bounded sequence is mapped by T into a sequence from which a convergent subsequence may be extracted. It turns out that, in contrast to the multiplication operator, compactness of the composition operator C_g leads to a very strong degeneracy of g. Here and in the upcoming results we will frequently combine Lemma 5.1.18 with Lemma 1.1.27.

Theorem 5.1.20. *Let X be any of the spaces BV, WBV_p, YBV_φ, ΛBV or RBV_p, and let $g \in Lip_{loc}(\mathbb{R})$. Then $C_g : X \to X$ is compact if and only if g is constant.*

Proof. It is clear that $C_g : X \to X$ is compact if g is constant, and so it remains to prove the converse. To this end, let $D \subseteq \mathbb{R}$ be the set of points of differentiability of g which has full measure as $g \in Lip_{loc}(\mathbb{R})$. According to part (d) of Lemma 1.1.27 in order to show that g is constant it suffices to show that $g'|_D$ is zero. So let us fix $u \in D$. By Lemma 5.1.18 (with $v = u$) we find for $R = 1$ some $\varrho > 0$ and for $\gamma = 1/16$, $\beta = 0$ and $\alpha_n = 1/n, n \in \mathbb{N}, n \geq 1/\varrho$ functions x_n and y_n with $\|x_n\|_X \leq \max\{|u|, |u + \alpha_n|\} + 32R \leq |u| + 1 + 32R$ and

$$\left|\frac{g(u+\alpha_n) - g(u)}{\alpha_n}\right| \leq \|C_g x_n - C_g y_n\|_X, \tag{5.1.18}$$

where $x_n = u + \alpha_n h_n$ for proper $h_n \in X$ with $\|h_n\|_\infty = 1$, and $y_n \equiv u$. In particular, the x_n form a bounded sequence in X.
Since $\alpha_n \to 0$ as $n \to \infty$, the functions x_n converge pointwise to $y \equiv u$. Moreover, since g is continuous, the compositions $C_g x_n$ converge pointwise to $C_g y \equiv g(u)$; in particular, the limit function of any subsequence of $(C_g x_n)$ which converges in X must be $C_g y \equiv g(u)$.
But since C_g is compact, there must be a convergent subsequence $(C_g x_{n_k})_k$ of $(C_g x_n)$ in X, and since g is differentiable at u, we obtain from (5.1.18),

$$\begin{aligned}|g'(u)| &= \lim_{k\to\infty}\left|\frac{g(u+\alpha_{n_k}) - g(u)}{\alpha_{n_k}}\right| \leq \lim_{k\to\infty}\|C_g x_{n_k} - C_g y_{n_k}\|_X \\ &= \lim_{k\to\infty}\|C_g x_{n_k} - C_g y\|_X = 0.\end{aligned}$$

Thus, $g'(u) = 0$, and as $u \in D$ has been chosen arbitrarily, the result follows indeed from part (d) of Lemma 1.1.27. ∎

We now turn to more delicate questions, a characterization of Lipschitz, uniform and pointwise continuity of $C_g : X \to X$ for X being one of our BV-type spaces. We start with Lipschitz continuity, both on the entire space and on bounded sets.

By a local Lipschitz continuity of $C_g : X \to X$ we mean a condition of the type

$$\|C_g x - C_g y\|_X \leq L(R) \|x - y\|_X \quad \text{for } \|x\|_X, \|y\|_X \leq R, \tag{5.1.19}$$

that is, Lipschitz continuity on closed balls in X. It turns out that C_g satisfies an estimate of the form (5.1.19) if and only if g is continuously differentiable with locally Lipschitz continuous derivative. Denoting the space of such functions by $Lip^1_{loc}(\mathbb{R})$ the following result is known.

Theorem 5.1.21. *Let $g \in Lip_{loc}(\mathbb{R})$, and let X be one of the spaces BV, WBV_p, YBV_φ, ΛBV or RBV_p. Then $C_g : X \to X$ is locally Lipschitz continuous in the sense of (5.1.19) if and only if $g \in Lip^1_{loc}(\mathbb{R})$.*

Proofs of this theorem can be found in [9] for BV, WBV_p, YBV_φ and RBV_p, and even for other BV-type spaces like, for instance, the space of functions of bounded variation in the sense of Korenblum [117]. For our BV-spaces, the "if"-part can alternatively be proved using Lemma 5.48 of [6] and Lemma 1.2.26; we will show how to do that in a moment. However, the proofs for the "only if"-part in the aforementioned literature do not cover all our variations at once or are quite complicated, because they are mostly based on Helly's Selection Principle. Therefore, we show that our all-round Lemma 5.1.18 also provides a quite short proof for all our BV-spaces X at once without any additional ingredients.

Proof of Theorem 5.1.21. First assume that $g \in Lip^1_{loc}(\mathbb{R})$, fix $R > 0$ and $x, y \in X$ with $\|x\|_X, \|y\|_X \leq R$; in particular, $x(t), y(t) \in [-R, R]$ for all $t \in [0, 1]$. By Lemma 5.48 of [6] there are $L_1(R), L_2(R) > 0$ only depending on R such that

$$\begin{aligned}
&\Big|g\big(x(s)\big) - g\big(y(s)\big) - g\big(x(t)\big) + g\big(y(t)\big)\Big| \\
&\quad \leq L_1(R)\Big(\big|x(s) - x(t)\big| + \big|y(s) - y(t)\big|\Big)\Big(\big|x(s) - y(s)\big| + \big|x(t) - y(t)\big|\Big) \\
&\qquad + L_2(R)\big|x(s) - y(s) - x(t) + y(t)\big| \\
&\quad \leq 2L_1(R)\Big(\big|x(s) - x(t)\big| + \big|y(s) - y(t)\big|\Big)\, \|x - y\|_\infty \\
&\qquad + L_2(R)\big|x(s) - y(s) - x(t) + y(t)\big| \quad \text{for all } s, t \in [0, 1].
\end{aligned}$$

From Lemma 1.2.26 we obtain

$$\begin{aligned}
\Phi_X(g \circ x - g \circ y) &\leq 2L_1(R)\big(\Phi_X(x) + \Phi_X(y)\big)\, \|x - y\|_\infty + L_2(R)\Phi_X(x - y) \\
&\leq 4RL_1(R)\, \|x - y\|_\infty + L_2(R)\Phi_X(x - y),
\end{aligned}$$

where Φ_X denotes the seminorm part of $\|\cdot\|_X$ as in Table 1.2.1.
Moreover, since $g \in C^1(\mathbb{R})$ and hence $g \in Lip_{loc}(\mathbb{R})$ there is some $L_3(R) > 0$ only depending on R such that

$$\|g \circ x - g \circ y\|_\infty = \sup_{t\in[0,1]} \left|g\big(x(t)\big) - g\big(y(t)\big)\right|$$
$$\leq L_3(R) \sup_{t\in[0,1]} \left|x(t) - y(t)\right| = L_3(R) \|x - y\|_\infty .$$

In total, we obtain

$$\|C_g x - C_g y\|_X = \|g \circ x - g \circ y\|_\infty + \Phi_X(g \circ x - g \circ y) \leq L(R) \|x - y\|_X ,$$

where

$$L(R) := \max\big\{L_3(R) + 4RL_1(R), L_2(R)\big\}.$$

Since R was arbitrary, C_g is locally Lipschitz continuous.

We now show the converse and assume that $C_g : X \to X$ is locally Lipschitz continuous in the sense of (5.1.19). In order to show that g is differentiable with locally Lipschitz continuous derivative, by Lemma 1.1.27 (b) it suffices to show that $g'|_D$ is Lipschitz continuous on $[-R, R]$ for each $R \geq 1$, where D is the set of points at which g is differentiable. So let $R \geq 1$ be fixed. Since the set

$$A := \mathbb{B}_{34R}(X) = \{x \in X \mid \|x\|_X \leq 34R\}$$

is bounded in X, we find some $L > 0$ such that $\|C_g x - C_g y\|_X \leq L \|x - y\|_X$ for all $x, y \in A$. By Lemma 5.1.18 we find some $\varrho \in (0, R]$ and for $u, v \in [-R, R] \cap D$, $\alpha = \beta \in (0, \varrho]$ and $\gamma = 1/16$ functions $x, y \in A$ with $\|x - y\|_X = |u - v|$ and

$$\left|\frac{g(u + \alpha) - g(u)}{\alpha} - \frac{g(v + \alpha) - g(v)}{\alpha}\right| \leq \|C_g x - C_g y\|_X \leq L \|x - y\|_X = L|u - v|.$$

But letting $\alpha \to 0+$ immediately yields $|g'(u) - g'(v)| \leq L|u - v|$, and the claim is proven. ■

Thus, the local Lipschitz condition (5.1.19) imposed on C_g leads to a stronger regularity condition of g, as expected. However, the natural question arises what happens when C_g is supposed to be even globally Lipschitz continuous. This is possible only for highly degenerate functions g, namely only if g is affine. This phenomenon has been discussed, even in the nonautonomous case, for many function spaces and especially for spaces of functions of bounded variation of various types. A detailed survey can be found in [6]. In many cases, especially in all our BV-spaces, this degeneracy occurs even if "globally Lipschitz continuous" is replaced by "globally uniformly continuous"; we give a short proof again with the help of Lemma 5.1.18.

Theorem 5.1.22. *Let $g \in Lip_{loc}(\mathbb{R})$, and let X be one of the spaces BV, WBV_p, YBV_φ, ΛBV or RBV_p. Then $C_g : X \to X$ is globally uniformly continuous if and only if g is affine.*

Proof. If $g(u) = au + b$ is affine, then the uniform continuity is immediate. Indeed, for all $x, y \in X$ we have

$$C_g x - C_g y = M_h(x - y),$$

where $h(t) = a$ for all $t \in [0,1]$. By Corollary 4.1.8, $\|C_g x - C_g y\|_X \le \|h\|_X \|x - y\|_X = |a| \|x - y\|_X$. Thus, we even have globally Lipschitz continuity in this case.

For the converse, assume that $C_g : X \to X$ is uniformly continuous on all of X. In order to show that g is affine, it suffices to show that $g'|_D$ is constant, where D is the set of points at which g is differentiable, according to Lemma 1.1.27 (c). Due to the globally uniform continuity of C_g there is some $\delta > 0$ such that for all $x, y \in X$ with $\|x - y\|_X \le \delta$ we have $\|C_g x - C_g y\|_X \le 1$. Now, fix $u, v \in D$ with $|u - v| \le \delta$. By Lemma 5.1.18 (with $R = 1$) there is some $\varrho \in (0, 1]$ such that for $u, v \in D$ with $|u - v| \le \delta$, $\alpha = \beta \in (0, \varrho]$ and $\gamma \in (0, 1]$ there are functions $x, y \in X$ with $\|x - y\|_X = |u - v|$ and

$$\left| \frac{g(u+\alpha) - g(u)}{\alpha} - \frac{g(v+\alpha) - g(v)}{\alpha} \right| \le 16\gamma \|C_g x - C_g y\|_X .$$

But since $|u - v| \le \delta$ we have $\|x - y\|_X \le \delta$, so $\|C_g x - C_g y\|_X \le 1$ and hence

$$\left| \frac{g(u+\alpha) - g(u)}{\alpha} - \frac{g(v+\alpha) - g(v)}{\alpha} \right| \le 16\gamma.$$

Now, if we let first $\alpha \to 0+$ and afterwards $\gamma \to 0+$, then we get $g'(u) = g'(v)$. And since the points $u, v \in D$ have been chosen arbitrarily, $g'|_D$ is constant. ∎

The last question to answer in the framework of uniform continuity is what conditions have to be imposed on g to guarantee that $C_g : X \to X$ is locally uniformly continuous, that is, uniformly continuous on bounded subsets of X, where X is one of our BV-spaces introduced in Chapter 1. The authors of [31] and [32] proved for $X = BV$ and $X = \Lambda BV$ that $g \in C^1(\mathbb{R})$ generates a continuous operator $C_g : X \to X$, and they did that by approximating the operator C_g by locally Lipschitz continuous composition operators uniformly converging on bounded subsets of X to C_g. We will come back to this principle later in Section 6.2. However, they actually proved (but did not mention) that this operator C_g is then even *locally uniformly continuous*. Another proof for this was given earlier in 1969 for $X = BV$ in [55]. We give here a more elementary proof for this fact and show even more: $g \in C^1(\mathbb{R})$ is in fact equivalent to the locally uniform continuity of C_g in all our BV-spaces.

Theorem 5.1.23. *Let $g \in Lip_{loc}(\mathbb{R})$, and let X be any of the spaces BV, WBV_p, YBV_φ, ΛBV or RBV_p. Then $C_g : X \to X$ is uniformly continuous on bounded subsets of X if and only if g is continuously differentiable on $\mathbb{R}$.*

Proof. We first assume that C_g is uniformly continuous on bounded subsets of X. In order to show that g is differentiable and has a continuous derivative, we show that $g'|_D$ is uniformly continuous on $[-R, R]$ for each $R \ge 1$, where D is the set of points at which g is differentiable, in accordance with Lemma 1.1.27 (a). To this end, let $R \geq 1$ and $\varepsilon > 0$ be fixed, and consider the ball

$$A := \mathbb{B}_{34R}(X) = \{x \in X \mid \|x\|_X \leq 34R\}$$

which is bounded in X. Since C_g is uniformly continuous on A, there is some $\delta \in (0, R)$ such that for all $x, y \in A$ with $\|x - y\|_X \leq \delta$, we have $\|C_g x - C_g y\|_X \leq \varepsilon$. Now, fix $u, v \in D$ with $|u - v| \leq \delta$. By Lemma 5.1.18 (with $\gamma = 1/16$) there is some $\varrho \in (0, R]$ such that for all $\alpha = \beta \in (0, \varrho]$ we get $x, y \in A$ with $\|x - y\|_X = |u - v|$ and

$$\left| \frac{g(u + \alpha) - g(u)}{\alpha} - \frac{g(v + \alpha) - g(v)}{\alpha} \right| \leq \|C_g x - C_g y\|_X \leq \varepsilon.$$

Letting $\alpha \to 0+$ yields $|g'(u) - g'(v)| \leq \varepsilon$, and this proves the claim.

For the converse, assume that g is continuously differentiable in all of $\mathbb{R}$, and fix $\varepsilon > 0$ and a bounded set $A \subseteq X$. Then there is some $R > 0$ such that $\|x\|_X \leq R$ for all $x \in A$. Because of the continuity of g', the function $G : \mathbb{R}^2 \to \mathbb{R}$, defined by

$$G(u, v) := \begin{cases} \dfrac{g(u) - g(v)}{u - v} & \text{for } u \neq v, \\ g'(u) & \text{for } u = v, \end{cases}$$

is continuous with respect to the norm $\|(u, v)\|_{\mathrm{m}} := \max\{|u|, |v|\}$ for $\mathbb{R}^2$ and the Euclidean norm for $\mathbb{R}$, and in particular uniformly continuous on compact subsets of $\mathbb{R}^2$. Thus, we find a $\delta \in (0, \varepsilon]$ such that

$$|G(u_1, v_1) - G(u_2, v_2)| \leq \varepsilon$$

for all $(u_1, v_1), (u_2, v_2) \in [-R, R]^2$ with $\|(u_1, v_1) - (u_2, v_2)\|_{\mathrm{m}} \leq \delta$, and why G is bounded on $[-R, R]^2$, i.e. $|G(u, v)| \leq M$ for some $M > 0$ and all $u, v \in [-R, R]$. Moreover, since g itself is uniformly continuous on $[-R, R]$, we can assume that δ is so small that $|g(u) - g(v)| \leq \varepsilon$ holds for all $u, v \in [-R, R]$ with $|u - v| \leq \delta$.

We now fix $x, y \in A$ with $\|x - y\|_X \leq \delta$; in particular, we have $\|x\|_\infty, \|y\|_\infty \leq R$ and $\|x - y\|_\infty \leq \delta$ which implies $x(t), y(t) \in [-R, R]$ and $|x(t) - y(t)| \leq \delta$ for all $t \in [0, 1]$. For fixed $s, t \in [0, 1]$ we write for abbreviation $x_s := x(s), y_s := y(s), x_t := x(t), y_t := y(t)$. Then $\|(x_s, x_t) - (y_s, y_t)\|_{\mathrm{m}} \leq \|x - y\|_\infty \leq \delta$, and thus

$$\Big|G(x_s, x_t) - G(y_s, y_t)\Big| \leq \varepsilon.$$

This implies

$$\begin{aligned} \Big|(C_g x - C_g y)(s) - (C_g x - C_g y)(t)\Big| &= \Big|g(x_s) - g(y_s) - g(x_t) + g(y_t)\Big| \\ &= \Big|G(x_s, x_t)(x_s - x_t) - G(y_s, y_t)(y_s - y_t)\Big| \\ &\leq \Big|G(x_s, x_t) - G(y_s, y_t)\Big||x_s - x_t| + \Big|G(y_s, y_t)\Big||x_s - x_t - y_s + y_t| \\ &\leq \varepsilon|x(s) - x(t)| + M\Big|x(s) - y(s) - x(t) + y(t)\Big|. \end{aligned} \tag{5.1.20}$$

Let Φ be the seminorm part of $\|\cdot\|_X$ as in Table 1.2.1. Then by (5.1.20) and Lemma 1.2.26,

$$\Phi(C_g x - C_g y) \leq \varepsilon\,\Phi(x) + M\,\Phi(x-y). \tag{5.1.21}$$

Finally, by our choice of δ,

$$\|C_g x - C_g y\|_\infty = \sup_{t\in[0,1]} \big|g\big(x(t)\big) - g\big(y(t)\big)\big| \leq \varepsilon. \tag{5.1.22}$$

Combining (5.1.21) with (5.1.22) while taking $\delta \leq \varepsilon$ into account gives

$$\|C_g x - C_g y\|_X \leq \varepsilon + \varepsilon\,\|x\|_X + M\,\|x-y\|_X \leq \varepsilon(1+R+M),$$

and this completes the proof. ∎

As a last result, we discuss the most difficult problem which is a characterization of pointwise continuity of $C_g : X \to X$ if X is one of our BV spaces. This problem is far from being fully understood and seems to be extremely complicated. The first one who discussed this problem was Morse [118] who proved 1937 that $C_g : BV \to BV$ is pointwise continuous if and only if $g \in Lip_{loc}(\mathbb{R})$. In other words, as soon as the composition operator $C_g : BV \to BV$ is well-defined (that is, $g \in Lip_{loc}(\mathbb{R})$), we get boundedness and pointwise continuity of C_g for free. However, Morse's proof including all auxiliary results is around 30 pages long which shows that the continuity problem seems to be highly nontrivial indeed. In the recent paper [96], the author gives a more straightforward and elegant proof. In Section 6.2 we will discuss this problem in more detail and give a third proof in Theorem 6.2.8. We will follow another idea and give some insight into the convergence behavior of sequences of composition operators in the space BV. However, it seems that all the aforementioned ideas cannot be generalized to ΛBV or YBV_φ, not even to WBV_p.

In the Riesz space RBV_p, however, we may use a trick to get continuity. As was shown by Marcus and Mizel [101], the composition operator C_g for $g \in Lip_{loc}(\mathbb{R})$ maps the Sobolev space $W^{1,p}$ for $p > 1$ into itself and is continuous. Since RBV_p contains the continuous representatives of $W^{1,p}$, and since their RBV_p-norms agree with their $W^{1,p}$-norms, we conclude that $C_g : RBV_p \to RBV_p$ is also continuous. Summarizing these observations, we obtain

Theorem 5.1.24. *Let $g \in Lip_{loc}(\mathbb{R})$, and let X be BV or RBV_p. Then $C_g : X \to X$ is continuous.*

As said, we do not know if the same result is true for X being one of the spaces WBV_p, YBV_φ or ΛBV. However, similar continuity results are known. For instance, the authors of [31] proved that for $g \in Lip_{loc}(\mathbb{R})$ the operator C_g maps BV into WBV_p and is continuous for any $p > 1$. The same authors achieved in [32] a particularly noteworthy result: If $g : \mathbb{R} \to \mathbb{R}$ is *continuous* and Λ is a given Waterman sequence, then one can construct another Waterman sequence Γ such that C_g maps ΛBV into

ΓBV and is continuous. It is, however, not clear under which circumstances $\Lambda = \Gamma$ holds.
Finally, note that many of the mapping properties of the composition operator discussed here in both BV-spaces and other spaces may also be found in the survey paper [8].

Table 5.1.2 summarizes what we know about the composition operator $C_g : X \to X$ for X being one of our BV spaces BV, WBV_p, YBV_φ with $\varphi \in \delta_2$, ΛBV or RBV_p:

Table 5.1.2: Mapping properties of C_g reflected by those of g.

$C_g : X \to X$ is	if and only if
well-defined	$g \in Lip_{loc}(\mathbb{R})$
bounded	$g \in Lip_{loc}(\mathbb{R})$
continuous (known only for $X = BV$ and $X = RBV_p$)	$g \in Lip_{loc}(\mathbb{R})$
locally uniformly continuous	$g \in C^1(\mathbb{R})$
locally Lipschitz continuous	$g \in Lip^1_{loc}(\mathbb{R})$
globally uniformly continuous	g is affine
globally Lipschitz continuous	g is affine
compact	g is constant

Although boundedness and continuity are in general independent for a nonlinear operator, as mentioned before, the table shows that for the operator C_g they are in fact equivalent, at least in the spaces BV and RBV_p.

5.2 Superposition Operators

In this section we investigate the superposition operator $N_g : X \to Y$ defined by (5.0.2) between two function spaces X and Y of real-valued functions on $[0, 1]$, that is,

$$N_g x(t) = g\big(t, x(t)\big) \quad \text{for } 0 \leq t \leq 1,$$

where $g : [0, 1] \times \mathbb{R} \to \mathbb{R}$ is a given function. The behavior of N_g is much more complex than that of its little brother C_g which we have studied in the previous section. This section is dedicated to investigate the superposition operator (5.0.2) in our BV-spaces with respect to analytic properties like continuity and compactness. Unfortunately, not so much is known in the spaces WBV_p, YBV_φ and RBV_p, and the superposition operator in contrast to the composition operator C_g reveals very often quite weird and unexpected properties. We therefore focus ourselves mainly on the space BV and give comments for the other BV-spaces.

For making the presentation more coherent and for not overburdening the formulation of the upcoming results, we collect here right from the beginning seven technical conditions (A)–(G) on the generating function $g : [0,1] \times \mathbb{R} \to \mathbb{R}$.

$$\exists L > 0\ \forall u, v \in \mathbb{R} : \quad \|g(\cdot, u) - g(\cdot, v)\|_\infty \le L|u - v|, \tag{A}$$

$$\forall R > 0\ \exists L_R > 0\ \forall u, v \in [-R, R] : \quad \|g(\cdot, u) - g(\cdot, v)\|_\infty \le L_R|u - v|, \tag{B}$$

$$\exists M > 0\ \forall u \in \mathbb{R} : \quad \operatorname{Var}\big(g(\cdot, u)\big) \le M, \tag{C}$$

$$\forall R > 0\ \exists M_R > 0\ \forall u \in [-R, R] : \quad \operatorname{Var}\big(g(\cdot, u)\big) \le M_R, \tag{D}$$

$$\begin{aligned}&\exists M > 0\ \forall 0 = t_0 < \ldots < t_n = 1\ \forall u_0, \ldots, u_{n-1} \in \mathbb{R} :\\ &\quad \sum_{j=1}^{n} \big|g(t_{j-1}, u_{j-1}) - g(t_j, u_{j-1})\big| \le M,\end{aligned} \tag{E}$$

$$\begin{aligned}&\forall R > 0\ \exists M_R > 0\ \forall 0 = t_0 < \ldots < t_n = 1\ \forall u_0, \ldots, u_{n-1} \in [-R, R] :\\ &\quad \sum_{j=1}^{n-1} |u_{j-1} - u_j| \le R \implies \sum_{j=1}^{n} \big|g(t_{j-1}, u_{j-1}) - g(t_j, u_{j-1})\big| \le M_R,\end{aligned} \tag{F}$$

$$\begin{aligned}&\forall R > 0\ \exists M_R > 0\ \forall 0 = t_0 < \ldots < t_n = 1\ \forall u_0, \ldots, u_n \in [-R, R] :\\ &\quad \sum_{j=1}^{n} |u_{j-1} - u_j| \le R\\ &\quad \implies \quad \sum_{j=1}^{n} \big|g(t_{j-1}, u_j) - g(t_j, u_j)\big| \le M_R\\ &\quad\quad \text{and} \quad \sum_{j=1}^{n} \big|g(t_{j-1}, u_{j-1}) - g(t_{j-1}, u_j)\big| \le M_R.\end{aligned} \tag{G}$$

One could call (A) a *Lipschitz condition* for $g(t, \cdot)$, uniformly in t, (B) a *local Lipschitz condition* for $g(t, \cdot)$, uniformly in t, (C) a *variation condition* for $g(\cdot, u)$, uniformly in u, (D) a *variation condition* for $g(t, \cdot)$, locally uniformly in u, (E) a *mixed condition* for g, (F) a *local mixed condition* for g and (G) a *local crossed mixed condition* for g.

Note that for functions g not depending on its first argument, that is, $g(t, u) = h(u)$ for all $t \in [0, 1], u \in \mathbb{R}$ and some function $h : \mathbb{R} \to \mathbb{R}$, condition (A) reduces to $h \in Lip(\mathbb{R})$, (B) is the same as $h \in Lip_{loc}(\mathbb{R})$, and the conditions (C)–(F) are always satisfied. Condition (G) is equivalent to $h \in SBV[-R, R]$, where SBV is the class of function of super bounded variation introduced in Section 1.2. But in Theorem 1.1.22 we have seen that $SBV[-R, R] = Lip[-R, R]$ and so (G) is equivalent to $h \in Lip_{loc}(\mathbb{R})$.

Let us come back to the general case when g depends on both arguments. There are some obvious interconnections between the conditions (A)–(G) which we collect in Figure 5.2.1. Here, (X)$\longrightarrow$(Y) means that (X) implies (Y).

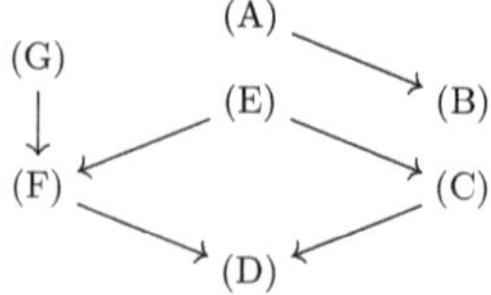

Figure 5.2.1: Relations between the conditions (A)–(G).

None of these implications can be inverted, and we will show this in the sequel by a series of examples. In order to give such examples, we will frequently consider generating functions g that have the form

$$g(t,u) = \begin{cases} \varphi_j(u) & \text{for } t = \frac{1}{2j},\, j \in \mathbb{N}, \\ 0 & \text{otherwise}, \end{cases} \tag{5.2.1}$$

where $\varphi_j : \mathbb{R} \to \mathbb{R}$ for $j \in \mathbb{N}$ are arbitrary functions. In the following result we give precise criteria on the sequence (φ_j) under which the function g in (5.2.1) satisfies the conditions (A)–(G). Note that for fixed $u \in \mathbb{R}$ we have for g in (5.2.1) the identity

$$g(t,u) = \mathfrak{J}_{\left(\varphi_j(u)\right)}(t), \tag{5.2.2}$$

where $\mathfrak{J}$ denotes the functions defined in (1.2.1).

Proposition 5.2.1. *Let $\varphi_j : \mathbb{R} \to \mathbb{R}$ for $j \in \mathbb{N}$ be arbitrary functions. Then g in (5.2.1) satisfies condition*

(A) if and only if there is an $L > 0$ such that $|\varphi_j(u) - \varphi_j(v)| \leq L|u - v|$ for all $j \in \mathbb{N}$ and all $u, v \in \mathbb{R}$,

(B) if and only if for each $R > 0$ there is an $L_R > 0$ such that $|\varphi_j(u) - \varphi_j(v)| \leq L_R|u - v|$ for all $j \in \mathbb{N}$ and all $u, v \in [-R, R]$,

(C) if and only if there is an $M > 0$ such that $\sum_{j=1}^{\infty} \left|\varphi_j(u)\right| \leq M$ for all $u \in \mathbb{R}$,

(D) if and only if for each $R > 0$ there is an $M_R > 0$ such that $\sum_{j=1}^{\infty} \left|\varphi_j(u)\right| \leq M_R$ for all $u \in [-R, R]$,

(E) if and only if there is an $M > 0$ such that $\sum_{j=1}^{\infty} \left|\varphi_j(u_j)\right| \leq M$ for all sequences (u_j) in $\mathbb{R}$,

(F) if and only if for each $R > 0$ there is an $M_R > 0$ such that $\sum_{j=1}^{\infty} \left|\varphi_j(u_j)\right| \leq M_R$ for all sequences (u_j) in $[-R, R]$ satisfying $\sum_{j=1}^{\infty} |u_{j-1} - u_j| \leq R$,

(G) if and only if it satisfies (F).

Proof. Let the conditions mentioned here in Proposition 5.2.1 be labeled by (A*)–(G*). We need to prove that (A) $\Leftrightarrow$ (A*), (B) $\Leftrightarrow$ (B*) and so on. First note that the equivalences (A) $\Leftrightarrow$ (A*), (B) $\Leftrightarrow$ (B*), (C) $\Leftrightarrow$ (C*) and (D) $\Leftrightarrow$ (D*) are clear, where for the latter two we use (5.2.2) and (1.2.2). For the remaining proof let $\tau_j := 1/(2j)$ for $j \in \mathbb{N}$.

For "(E)$\Rightarrow$(E*)" assume that g in (5.2.1) satisfies condition (E) with $M > 0$. Let (v_j) be a sequence in $\mathbb{R}$ and $n \in \mathbb{N}$ be fixed. For $j \in \{0, \ldots, n-1\}$ set $t_{2j+1} := \tau_{n-j}$, $t_0 := 0$, $t_{2n} := 1$ and then pick $t_{2j} \in (t_{2j-1}, t_{2j+1})$ arbitrarily for $j \in \{1, \ldots, n-1\}$. Moreover, set $u_{2j} := u_{2j+1} := v_{n-j}$ for $j \in \{0, \ldots, n-1\}$. Condition (E) yields

$$M \geq \sum_{j=1}^{2n} \left| g(t_{j-1}, u_{j-1}) - g(t_j, u_{j-1}) \right| = 2 \sum_{j=1}^{n} \left| \varphi_j(v_j) \right|$$

and letting $n \to \infty$ gives (E*).

For the converse assume that g satisfies (E*) with $M > 0$ and fix a partition $0 = t_0 < \ldots < t_n = 1$ of $[0,1]$ and numbers $u_0, \ldots, u_{n-1} \in \mathbb{R}$. If we set $u_j := 0$ for $j \geq n$, we get by (E*),

$$\begin{aligned} \sum_{j=1}^{n} \left| g(t_{j-1}, u_{j-1}) - g(t_j, u_{j-1}) \right| &\leq \sum_{j=1}^{n} \left| g(t_{j-1}, u_{j-1}) \right| + \sum_{j=1}^{n} \left| g(t_j, u_{j-1}) \right| \\ &\leq \sum_{j=2}^{\infty} \left| \varphi_{j-1}(u_{j-1}) \right| + \sum_{j=1}^{\infty} \left| \varphi_j(u_{j-1}) \right| \leq 2M \end{aligned}$$

which establishes (E).

For "(F)$\Rightarrow$(F*)" assume that g in (5.2.1) satisfies condition (F) with $M_R > 0$ for $R > 0$. Let (v_j) be a sequence in $[-R, R]$ with $\sum_{j=1}^{\infty} |v_{j-1} - v_j| \leq R$, and let $n \in \mathbb{N}$ be fixed. For $j \in \{0, \ldots, n-1\}$ set $t_{2j+1} := \tau_{n-j}$, $t_0 := 0$, $t_{2n} := 1$ and then pick $t_{2j} \in (t_{2j-1}, t_{2j+1})$ arbitrarily for $j \in \{1, \ldots, n-1\}$. Moreover, set $u_{2j} := u_{2j+1} := v_{n-j}$ for $j \in \{0, \ldots, n-1\}$. Then $u_0, \ldots, u_{2n+1} \in [-R, R]$ with

$$\sum_{j=1}^{2n-1} |u_{j-1} - u_j| = \sum_{j=2}^{n} |v_{j-1} - v_j| \leq \sum_{j=1}^{\infty} |v_{j-1} - v_j| \leq R.$$

Consequently, condition (F) yields

$$M_R \geq \sum_{j=1}^{2n} \left| g(t_{j-1}, u_{j-1}) - g(t_j, u_{j-1}) \right| = 2 \sum_{j=1}^{n} \left| \varphi_j(v_j) \right|$$

and letting $n \to \infty$ gives (F*).

For the converse assume that g satisfies (F*) with M_R for fixed $R > 0$, and fix a partition $0 = t_0 < \ldots < t_n = 1$ of $[0,1]$ and numbers $u_0, \ldots, u_{n-1} \in [-R, R]$ with $\sum_{j=1}^{n-1} |u_{j-1} - u_j| \leq R$. Let $T := \{\tau_j, \mid j \in \mathbb{N}\}$, where the τ_j have been defined at the beginning of this proof. From the numbers $t_0, \ldots, t_n$ we extract those which belong to T and relabel them $t_{j_0}, \ldots, t_{j_m} = \tau_{l_0}, \ldots, \tau_{l_m}$; note that $t_n \notin T$ and hence $m < n$. The numbers u_{j_k} as well as the numbers $u_{j_k - 1}$ then satisfy

$$\sum_{k=1}^{m} |u_{j_{k-1}} - u_{j_k}|, \sum_{k=1}^{m} |u_{j_{k-1}-1} - u_{j_k - 1}| \leq \sum_{j=1}^{n-1} |u_{j-1} - u_j| \leq R.$$

Consequently, we can apply (F*) and obtain

$$\sum_{j=1}^{n} \left|g(t_{j-1}, u_{j-1}) - g(t_j, u_{j-1})\right| \le \sum_{j=1}^{n} \left|g(t_{j-1}, u_{j-1})\right| + \sum_{j=1}^{n} \left|g(t_j, u_{j-1})\right|$$
$$= \sum_{k=0}^{m} \left|g(t_{j_k}, u_{j_k})\right| + \sum_{k=0}^{m} \left|g(t_{j_k}, u_{j_k-1})\right| = \sum_{k=0}^{m} \left|\varphi_{l_k}(u_{j_k})\right| + \sum_{k=0}^{m} \left|\varphi_{l_k}(u_{j_k-1})\right| \le 2M_R$$

which establishes (F).

Since in general (G) implies (F) we only have to prove that (F) also implies (G) for the function (5.2.1). Thus, assume that g satisfies (F). We already know that then it satisfies also (F*). Because of

$$\sum_{j=1}^{n} \left|g(t_{j-1}, u_j) - g(t_j, u_j)\right| \le \sum_{j=1}^{n} \left|g(t_{j-1}, u_j)\right| + \sum_{j=1}^{n} \left|g(t_j, u_j)\right|$$

and

$$\sum_{j=1}^{n} \left|g(t_{j-1}, u_{j-1}) - g(t_{j-1}, u_j)\right| \le \sum_{j=1}^{n} \left|g(t_{j-1}, u_{j-1})\right| + \sum_{j=1}^{n} \left|g(t_{j-1}, u_j)\right|$$

we can use the same argument as in the implication "(F*)⇒(F)" to prove that g also satisfies the condition (G). ■

We can now use Proposition 5.2.1 to construct examples showing that none of the implications given in Figure 5.2.1 can be inverted, with one exception: Since for the function g in (5.2.1) the conditions (F) and (G) are equivalent, in order to show that (F) does not imply (G) we need an example that has not the form (5.2.1). Here is one:

Example 5.2.2. It suffices to consider a function g which is independent of t. Indeed, take any function $\varphi : \mathbb{R} \to \mathbb{R}$ and set $g(t, u) := \varphi(u)$. Then (C), (D), (E) and (F) are clearly satisfied, because $g(s, u) - g(t, u) = 0$ for all $s, t \in [0, 1]$ and $u \in \mathbb{R}$. Condition (G) now translates to the following condition:

$$\forall R > 0 \; \exists M_R > 0 \; \forall 0 = t_0 < \ldots < t_n = 1 \; \forall u_0, \ldots, u_n \in [-R, R] :$$
$$\sum_{j=1}^{n} |u_{j-1} - u_j| \le R \quad \Longrightarrow \quad \sum_{j=1}^{n} |\varphi(u_{j-1}) - \varphi(u_j)| \le M_R.$$

By Theorem 1.1.22, this is equivalent to $\varphi \in Lip[-R, R]$ for each $R > 0$. Thus, for any function φ which is not locally Lipschitz continuous the corresponding generator g will satisfy (F) but not (G). For instance, $\varphi(u) = \sqrt{|u|}$ will do the job. Note that this function cannot satisfy (A) or (B), because it is not locally Lipschitz continuous with respect to u. ◇

None of the ostensibly strong conditions (A) and (B) does imply any of the other conditions (C), (D), (E), (F) or (G).

Example 5.2.3. Let $\varphi_j := \mathbb{1}$ for $j \in \mathbb{N}$ and consider g in (5.2.1). Then clearly (A) and (B) are met, but neither (C), (D), (E), (F) nor (G) hold, because none of the series in Proposition 5.2.1 can converge. ◇

The next example shows that neither (C), (D) nor (F) implies (E).

Example 5.2.4. Let $\varphi_j := \chi_{(j,j+1)}$ for $j \in \mathbb{N}$ and consider g in (5.2.1). Any fixed $u \in \mathbb{R}$ belongs to at most one interval $(j, j+1)$. Thus, $\varphi_j(u) = 0$ for all but at most one $j \in \mathbb{N}$; in particular, g satisfies (C) and (D) by Proposition 5.2.1 with $M = 1$ and $M_R = 1$, respectively.
Moreover, for fixed $R > 0$ and any sequence (u_j) in $[-R, R]$ we have that $\varphi_n(u_j) = 0$ for all $j \in \mathbb{N}$ and $n \geq R$; in particular, g satisfies (F) and hence (G) by Proposition 5.2.1 with $M_R = R$. But by the same Proposition, g cannot satisfy (E), because for the sequence $u_j := j + 1/2$ we have that $\varphi_j(u_j) = 1$ for all $j \in \mathbb{N}$. Note that g can also not fulfill (A) or (B), because none of the functions φ_j is locally Lipschitz continuous. ◇

A slight modification of the previous example shows that neither (C) nor (D) implies any of the conditions (E), (F) and (G).

Example 5.2.5. Let $\varphi_j := \chi_{(\frac{1}{j+1}, \frac{1}{j})}$ for $j \in \mathbb{N}$ and consider g in (5.2.1). Any fixed $u \in \mathbb{R}$ belongs to at most one interval $(\frac{1}{j+1}, \frac{1}{j})$. Thus, $\varphi_j(u) = 0$ for all but at most one $j \in \mathbb{N}$; in particular, g satisfies (C) and (D) by Proposition 5.2.1 with $M = 1$ and $M_R = 1$, respectively.
Moreover, g cannot satisfy (E), (F) or (G), again by Proposition 5.2.1, because for $R = 1$ and the sequence $u_j := \frac{1}{2}\left(\frac{1}{j+1} + \frac{1}{j}\right)$ for $j \in \mathbb{N}$ in $[-1, 1] = [-R, R]$ we have on the one hand

$$\sum_{j=2}^{\infty} |u_{j-1} - u_j| = u_1 - \lim_{n \to \infty} u_n = \frac{3}{4} \leq 1 = R$$

and on the other hand $\varphi_j(u_j) = 1$ for all $j \in \mathbb{N}$. Finally, g cannot satisfy (A) and (B), because none of the functions φ_j is locally Lipschitz continuous. ◇

The last example in this series proves that (D) does not imply (C) and that (A) cannot be deduced from (B).

Example 5.2.6. Let $\varphi_j(u) := u^2/j^2$ for $u \in \mathbb{R}$ and $j \in \mathbb{N}$. Then

$$\sum_{j=1}^{\infty} \varphi_j(u_j) = \sum_{j=1}^{\infty} \frac{u_j^2}{j^2} \tag{5.2.3}$$

for any real sequence (u_j). In particular, if $u_j = u$ for some fixed $u \in \mathbb{R}$ and all $j \in \mathbb{N}$, (5.2.3) becomes

$$\sum_{j=1}^{\infty} \varphi_j(u_j) = \frac{\pi^2}{6} u^2,$$

and so g in (5.2.1) satisfies (D) but not (C), as well as (B) but not (A) by Proposition 5.2.1. Moreover, (E) is not satisfied, because for the sequence $u_j := j$ the series in (5.2.3) diverges. Condition (F) and hence also (G), however, are satisfied, because if (u_j) is bounded by $R > 0$, then the series in (5.2.3) is majorized by $R^2\pi^2/6$. ◇

We remark that even if (A) and (B) alone do not imply any of the other conditions, (B) together with (F) implies (G). This explains why the function φ in Example 5.2.2 had to be chosen so that it was not locally Lipschitz continuous.

We summarize in the following table which of the previous five example satisfies the conditions (A)–(G).

Table 5.2.1: Conditions (A)–(G) in the above examples.

Example	(A)	(B)	(C)	(D)	(E)	(F)	(G)
5.2.2	no	no	yes	yes	yes	yes	no
5.2.3	yes	yes	no	no	no	no	no
5.2.4	no	no	yes	yes	no	yes	yes
5.2.5	no	no	yes	yes	no	no	no
5.2.6	no	yes	no	yes	no	yes	yes

From this table we see immediately that none of the implications in Figure 5.2.1 can be inverted. We therefore present the same diagram again here in Figure 5.2.2, but this time, the numbers labeling the arrows refer to the examples which show that the corresponding implications cannot be inverted.

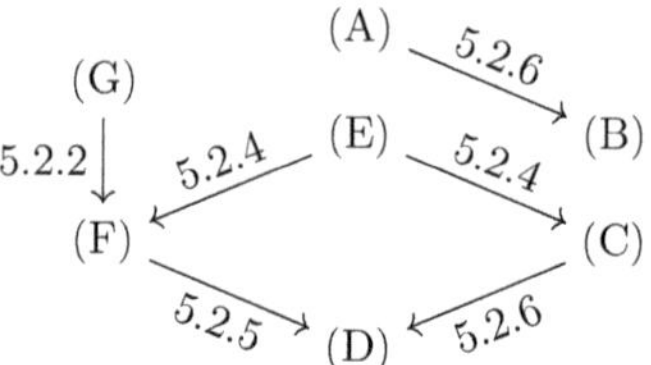

Figure 5.2.2: Relations between the conditions (A)–(G).

We now come back to the general superposition operator $N_g : BV \to BV$. Let us check the sufficiency (or necessity) of the conditions (A)–(G) for the acting condition $N_g(BV) \subseteq BV$ and the analytic properties of N_g. To begin with, we remark that Lyamin [94] claimed that the conditions (B) and (D) together imply $N_g(BV) \subseteq BV$. However, Maćkowiak showed in [95] by means of a sophisticated example that this is in fact false, even if (A) and (D) are assumed to be true. With the help of our special functions (5.2.1) and Proposition 5.2.1 we can now give a much simpler example.

Example 5.2.7. For $j \in \mathbb{N}$ we define $\varphi_j : \mathbb{R} \to \mathbb{R}$ by

$$\varphi_j(u) = \max\Big\{0, 1/j - \big||u| - 1/j\big|\Big\}.$$

Figure 5.2.3 shows the relevant part of φ_j for fixed $j \in \mathbb{N}$.

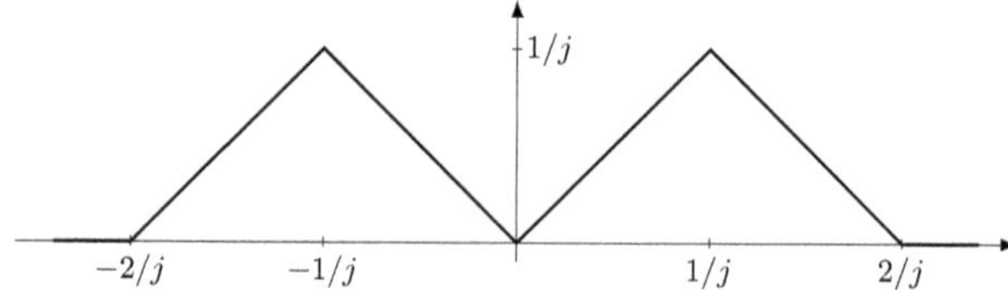

Figure 5.2.3: The function φ_j for fixed $j \in \mathbb{N}$.

Then the φ_j are globally Lipschitz continuous with $\text{lip}(\varphi_j) = 1$ for all $j \in \mathbb{N}$; in particular, the corresponding function g in (5.2.1) satisfies (A) by Proposition 5.2.1. Moreover, $\varphi_j(0) = 0$ for all $j \in \mathbb{N}$, and for fixed $u \in \mathbb{R}\backslash\{0\}$ we have $\varphi_j(u) = 0$ if $|u| \geq 2/j$. Since $0 \leq \varphi_j(u) \leq |u|$ for all $j \in \mathbb{N}$ and $u \in \mathbb{R}$ we obtain in this case

$$\sum_{j=1}^{\infty} \varphi_j(u) \leq \sum_{j \leq 2/|u|} |u| \leq 2$$

which shows that g satisfies (C) and hence (D), again by Proposition 5.2.1. But N_g does not map BV into itself. For instance, the function $x(t) := t$ clearly belongs to BV, but

$$N_g\big(t, x(t)\big) = \begin{cases} \varphi_j(\frac{1}{2j}) & \text{for } t = \frac{1}{2j}, \\ 0 & \text{otherwise}, \end{cases} = \begin{cases} \dfrac{1}{2j} & \text{for } t = \dfrac{1}{2j}, \\ 0 & \text{otherwise}, \end{cases}$$

does not.

The same idea leads to the sequence $u_j := \frac{1}{2j}$ for $j \in \mathbb{N}$ and $u_0 := 1$ with $\sum_{j=1}^{\infty} |u_{j-1} - u_j| = 1$ and

$$\sum_{j=1}^{\infty} \varphi_j(u_j) = \sum_{j=1}^{\infty} \varphi_j\left(\frac{1}{2j}\right) = \sum_{j=1}^{\infty} \frac{1}{2j} = \infty.$$

Consequently, neither of the conditions (E), (F) and (G) can be satisfied, again by our Proposition 5.2.1. $\diamondsuit$

The first correct sufficient conditions for the inclusion $N_g(BV) \subseteq BV$ have been obtained by Bugajewska in 2010 [25] and Bugajewska et al. in 2016 [26] and read as follows.

Theorem 5.2.8. *If the generating function $g : [0,1] \times \mathbb{R} \to \mathbb{R}$ satisfies (A) and (E), then N_g maps BV into itself.*

Theorem 5.2.9. *If the generating function $g : [0,1] \times \mathbb{R} \to \mathbb{R}$ satisfies (B) and (F), then N_g maps BV into itself and is bounded.*

Since (A)$\Rightarrow$(B) and (E)$\Rightarrow$(F), but neither (B)$\Rightarrow$(A) nor (F)$\Rightarrow$(E), Theorem 5.2.9 is actually stronger than Theorem 5.2.8. Moreover, the Theorems 5.2.8 and 5.2.9 explain why g in Example 5.2.7 could not satisfy (E) and (F), because $N_g(BV) \not\subseteq BV$ in this case.

However, the conditions given in the Theorems 5.2.8 and 5.2.9 are only sufficient for the boundedness of the operator N_g. We illustrate this for the condition (B) by the following simple

Example 5.2.10. Define $\varphi_j : \mathbb{R} \to \mathbb{R}$ by $\varphi_1(u) = \min\{\sqrt{|u|}, 1\}$ and $\varphi_j = \mathbb{0}$ for $j \geq 2$. Then by Proposition 5.2.1 the corresponding function g in (5.2.1) satisfies (C), (D), (E), (F) and (G), but neither (A) nor (B). Nonetheless, N_g maps BV into itself, because for $x \in BV$ we have

$$N_g x(t) = g\big(t, x(t)\big) = \begin{cases} \min\left\{\sqrt{|x(1/2)|}, 1\right\} & \text{for } t = 1/2, \\ 0 & \text{for } t \in [0,1]\backslash\{1/2\}, \end{cases}$$

and hence

$$\|N_g x\|_{BV} = 3 \min\left\{\sqrt{|x(1/2)|}, 1\right\} \leq 3\sqrt{\|x\|_{BV}}$$

which also shows that $N_g : BV \to BV$ is bounded. ♢

However, the condition (B) and (D) are "almost" necessary for the boundedness of N_g. The precise formulation is as follows [55].

Theorem 5.2.11. *Let $g : [0,1] \times \mathbb{R} \to \mathbb{R}$ be so that N_g maps BV into itself and is bounded. Then g can be written as*

$$g(t,u) = g_1(t,u) + g_2(t,u),$$

where the functions $g_1, g_2 : [0,1] \times \mathbb{R} \to \mathbb{R}$ have the following properties.

(a) The function g_1 satisfies the conditions (B) and (D).

(b) The function g_2 vanishes on $([0,1]\backslash\mathcal{C}) \times \mathbb{R}$, where $\mathcal{C} \subseteq [0,1]$ is some countable set.

It turns out that condition (G) alone is exactly what we need to characterize bounded superposition operators in BV [26]:

Theorem 5.2.12. *The superposition operator N_g maps BV into itself and is bounded if and only if the generating function $g : [0,1] \times \mathbb{R} \to \mathbb{R}$ satisfies condition (G).*

This explains why the function g in Example 5.2.10 had to satisfy condition (G), in contrast to the function g in Example 5.2.7.

The fact that boundedness is included in Theorem 5.2.12 is somewhat unsatisfactory: One could ask whether or not condition (G) is also necessary for the mere inclusion $N_g(BV) \subseteq BV$ without the boundedness requirement on N_g. This is not true, because in contrast to the composition operator C_g which we have studied in the previous section the superposition operator N_g need neither be bounded nor continuous if it maps BV into itself. This is illustrated by the following example which is a slight modification of Example 5.2.10.

Example 5.2.13. Define $\varphi_j : \mathbb{R} \to \mathbb{R}$ by

$$\varphi_1(u) = \begin{cases} 1/u & \text{for } u \neq 0, \\ 0 & \text{for } u = 0, \end{cases}$$

and $\varphi_j = \mathbb{0}$ for $j \geq 2$. Then by Proposition 5.2.1 the corresponding function g in (5.2.1) satisfies none of the conditions (A)–(G). Nonetheless, N_g maps BV into itself, because for $x \in BV$ with $x(1/2) \neq 0$ we have

$$N_g x(t) = g\big(t, x(t)\big) = \begin{cases} 1/x(1/2) & \text{for } t = 1/2, \\ 0 & \text{for } t \in [0,1]\backslash\{1/2\}, \end{cases}$$

and $N_g x = \mathbb{0}$ if $x(1/2) = 0$. This implies $\|N_g x\|_{BV} = 3/|x(1/2)|$ for functions $x \in BV$ with $x(1/2) \neq 0$ and becomes infinitely large the closer $x(1/2)$ gets to 0. Consequently, in this case the operator N_g maps BV into itself but cannot be bounded. ♢

The previous example shows that neither of the condition (A)–(G) is necessary for the acting condition $N_g(BV) \subseteq BV$ which again illustrates impressively the weird behavior of the operator N_g in contrast to its quite well behaving little brother C_g.
As far as we know, a practical condition both necessary and sufficient for the acting condition $N_g(BV) \subseteq BV$ is not known.[1] Such a criterion should be weaker than (G), but include the function g from Example 5.2.13.

Concerning boundedness of N_g, the following result seems to be of independent interest. It shows that the boundedness of g is reflected in the boundedness of N_g [26].

Theorem 5.2.14. *The following statements are true.*

(a) Under the condition (B) the operator N_g is bounded in BV if and only if g is locally bounded.

(b) If $N_g(BV) \subseteq BV$, then the set

$$T_R := \left\{ t \in [0,1] \mid \sup_{|u| \leq r} |g(t,u)| = \infty \right\}$$

is finite for each $R > 0$.

Part (b) of Theorem 5.2.14 says, roughly speaking, that the points t for which $g(t,\cdot)$ is unbounded on $[-R,R]$ must be isolated. That T_R can be nonempty for all $R > 0$ was shown in Example 5.2.13: For the function g therein we have $T_R = \{1/2\}$ for all $R > 0$. For the functions g in all other examples considered in this section so far the set T_R is empty for any $R > 0$, as these functions are locally bounded with respect to $u \in \mathbb{R}$ for each fixed $t \in [0,1]$. However, even $T_R = \emptyset$ for all $R > 0$ is not sufficient for $N_g(BV) \subseteq BV$, as Example 5.2.7 shows.

For Waterman spaces ΛBV an analogue of Theorem 5.2.9 is true where condition (F) has to be replaced by the following condition.

$$\begin{aligned}
&\forall R > 0\ \exists M_R > 0\ \forall 0 = t_0 < \ldots < t_n = 1\ \forall u_0, \ldots, u_n \in [-R,R]: \\
&\quad \sup_\sigma \sum_{j=1}^n \lambda_{\sigma(j)} |u_{j-1} - u_j| \leq R \implies \sup_\sigma \sum_{j=1}^n \lambda_{\sigma(j)} |g(t_{j-1}, u_{j-1}) - g(t_j, u_{j-1})| \leq M_R,
\end{aligned} \tag{F_Λ}$$

and the suprema have to be taken over all permutations σ of $\mathbb{N}$. In [30] the following result was shown.

Theorem 5.2.15. *If the generating function $g : [0,1] \times \mathbb{R} \to \mathbb{R}$ satisfies (B) and (F_Λ), then N_g maps ΛBV into itself and is bounded.*

[1]Actually, there is a condition both necessary and sufficient for $N_g(BV) \subseteq BV$ given in [31], but it is far from being practical and extremely technical.

Again, also for the Waterman space we have a perfect analogue to Theorem 5.2.12, but condition (G) has now to be adjusted in the following way.

$$\begin{aligned}&\forall R>0\ \exists M_R>0\ \forall 0=t_0<\ldots<t_n=1\ \forall u_0,\ldots,u_n\in[-R,R]:\\ &\quad\sup_\sigma\sum_{j=1}^n\lambda_{\sigma(j)}|u_{j-1}-u_j|\le R\\ &\qquad\Longrightarrow\begin{cases}\sup_\sigma\sum_{j=1}^n\lambda_{\sigma(j)}|g(t_{j-1},u_j)-g(t_j,u_j)|\le M_R & \text{and}\\ \sup_\sigma\sum_{j=1}^n\lambda_{\sigma(j)}|g(t_{j-1},u_{j-1})-g(t_{j-1},u_j)|\le M_R,\end{cases}\end{aligned} \tag{G_Λ}$$

where the suprema are taken over all permutations σ of $\mathbb{N}$. In [30] the authors proved

Theorem 5.2.16. *The superposition operator N_g maps ΛBV into itself and is bounded if and only if g satisfies condition (G_Λ).*

We do not know if there are any conditions similar to (G) and (G_Λ) for the spaces YBV_φ and RBV_p. However, we prove a necessary condition for an operator N_g that maps RBV_p into itself and is bounded which will be needed in the sequel and might be of its own interest.

Proposition 5.2.17. *Let $g:[0,1]\times\mathbb{R}\to\mathbb{R}$ be so that N_g maps RBV_p into itself and is bounded. Then $g(t,\cdot)$ is continuous in $\mathbb{R}$ for each fixed $t\in[0,1]$.*

Proof. Fix $u\in\mathbb{R}$, $s\in[0,1]$ and $\varepsilon>0$. We assume $s\in[0,1)$; the proof for $s=1$ is similar. Since N_g maps the space RBV_p into itself the function $g(\cdot,u)$ is continuous at s. This is why we find some $\delta>0$ such that

$$|t-s|\le\delta\quad\Rightarrow\quad|g(s,u)-g(t,u)|\le\varepsilon/2. \tag{5.2.4}$$

Moreover, since N_g is bounded there is some $M>0$ such that

$$\|x\|_{RBV_p}\le|u|+2\quad\Rightarrow\quad\mathrm{RVar}_p(N_gx)^{1/p}\le M/2. \tag{5.2.5}$$

Fix $v\in[u-1,u+1]\backslash\{u\}$ so that

$$|u-v|\le\min\left\{(1-s)^{(p-1)/p},\delta^{(p-1)/p},\frac{\varepsilon}{M}\right\}, \tag{5.2.6}$$

and define $t:=s+|u-v|^{p/(p-1)}$ which implies $0\le s<t\le1$ and $0<t-s=|u-v|^{p/(p-1)}\le\delta$. From (5.2.4) we get

$$|g(s,u)-g(t,u)|\le\varepsilon/2. \tag{5.2.7}$$

The function $x:[0,1]\to\mathbb{R}$, defined to be piecewise linear and continuous by $x(0)=v=x(s)$ and $x(t)=u=x(1)$, has norm

$$\|x\|_{RBV_p}=\max\{|u|,|v|\}+\frac{|u-v|}{(t-s)^{(p-1)/p}}\le|u|+2.$$

Therefore, by (5.2.5), $\mathrm{RVar}_p(N_g x)^{1/p} \leq M/2$ which implies

$$\frac{M}{2} \geq \frac{\left|g\big(s,x(s)\big) - g\big(t,x(t)\big)\right|}{(t-s)^{(p-1)/p}} = \frac{\left|g\big(s,v\big) - g\big(t,u\big)\right|}{|u-v|}.$$

From this, (5.2.6) and (5.2.7) we obtain

$$|g(s,v) - g(s,u)| \leq |g(s,v) - g(t,u)| + |g(t,u) - g(s,u)| \leq \frac{M}{2}|u-v| + \frac{\varepsilon}{2} \leq \varepsilon,$$

and this finishes the proof. ■

There are three questions naturally arising when we look at Proposition 5.2.17. The first is whether the statement follows from the inclusion $RBV_p \subseteq C$ for $p > 1$. However, this inclusion gives us that if N_g maps RBV_p into itself, then $g(\cdot, u)$ is continuous on $[0,1]$ for each fixed $u \in \mathbb{R}$. Proposition 5.2.17, on the other hand, guarantees that under the additional assumption that N_g is bounded in RBV_p, also $g(t,\cdot)$ is continuous on $\mathbb{R}$ for each fixed $t \in [0,1]$. Thus, g is continuous with respect to both of its arguments separately.

Therefore, the second question is, whether under the hypotheses of Proposition 5.2.17 the function g as a function of two variables is continuous on $[0,1] \times \mathbb{R}$. Unfortunately, we do not know the answer.

However, the third question, namely whether Proposition 5.2.17 is true in other BV-spaces like BV itself, is easily answered by a counterexample.

Example 5.2.18. The function $g : [0,1] \times \mathbb{R} \to \mathbb{R}$, defined by

$$g(t,u) = \begin{cases} 1 & \text{for } (t,u) = (0,0), \\ 0 & \text{for } (t,u) \neq (0,0), \end{cases}$$

generates a bounded operator $N_g : X \to BV$ on any function space X, since

$$N_g x = \begin{cases} \chi_{\{0\}} & \text{for } x(0) = 0, \\ \mathbb{0} & \text{for } x(0) \neq 0, \end{cases}$$

and thus $\|N_g x\|_{BV} \in \{0, 2\}$. However, $g(0,\cdot) = \chi_{\{0\}}$ is discontinuous at $u = 0$. ◇

The previous example represents a more general fact about superposition operators. Indeed, Proposition 5.2.17 is wrong in other "non-regular" BV-spaces, especially in BV itself. The authors of [26] have shown that if N_g maps BV into itself, then *nothing* can be said about the function $u \mapsto g(t,u)$ for fixed $t \in [0,1]$.

We now turn to continuity properties of the superposition operator N_g defined in (5.0.2). As we have seen in the previous section, boundedness and continuity as well as the pure acting condition are equivalent for the composition operator $C_g : BV \to BV$ and also equivalent to a certain regularity on g, namely a local Lipschitz continuity in $\mathbb{R}$. Moreover, locally uniform continuity of $C_g : X \to X$ was characterized in Theorem 5.1.23 for all our BV-spaces. Accordingly, it is equivalent to g being continuously differentiable in $\mathbb{R}$. Consequently, there is a nice symmetry in the regularity of g and its operator C_g: The more regular g is, the more regular C_g is, as well, and we have illustrated and summarized this in Table 5.1.2.

In the case of a superposition operator, however, things are completely different, not to say much worse. It turns out that the regularity of the generating function seems to have not so much to do with the regularity of the corresponding superposition operator N_g. Maćkowiak showed in [96] that even a *globally Lipschitz continuous* function $g : [0,1] \times \mathbb{R} \to \mathbb{R}$ may generate a *discontinuous* superposition operator $N_g : BV \to BV$. His example is a piecewise linear function. In the following proposition we describe a general technique on how to construct such functions. The idea is that $g(\cdot, u_n) = g_n$ is a Lipschitz function for each $n \in \mathbb{N}$ with a uniform Lipschitz constant but so that its BV-norm is uniformly bounded away from 0, where the number sequence (u_n) decreases sufficiently fast to 0. On the other hand, the functions g_n have to converge uniformly to $\mathbb{0}$. We can then take the sequence (x_n) of constant functions $x_n :\equiv u_n$ that converges in BV to $\mathbb{0}$, but $N_g x_n(t) = g(t, x_n(t)) = g_n(t)$ has, as said, a BV-norm that is bounded away from 0 and hence cannot converge to $\mathbb{0}$. This implies that N_g is discontinuous as an operator from BV to BV. Here come the details.

Proposition 5.2.19. *For every $n \in \mathbb{N}$ choose functions $g_n \in BV$ and numbers $u_n \in (0,1]$ such that the following requirements are all met.*

(i) $g_n \in Lip[0,1]$ *with* $\mathrm{lip}(g_n) \le L$ *for some* $L > 0$ *and all* $n \in \mathbb{N}$,

(ii) $\mathrm{Var}(g_n) \ge 1$,

(iii) The sequence (u_n) strictly decreases to 0 as $n \to \infty$,

(iv) There is some $M > 0$ such that for all $n \in \mathbb{N}$,

$$\frac{\|g_n\|_\infty + \|g_{n+1}\|_\infty}{u_n - u_{n+1}} \le M.$$

Then the function $g : [0,1] \times \mathbb{R} \to \mathbb{R}$, defined by

$$g(\cdot, u) = \begin{cases} g_1 & \text{for } u \ge u_1, \\ \dfrac{u - u_{n+1}}{u_n - u_{n+1}} g_n + \dfrac{u_n - u}{u_n - u_{n+1}} g_{n+1} & \text{for } u_{n+1} \le u < u_n, n \in \mathbb{N}, \\ \mathbb{0} & \text{for } u \le 0, \end{cases} \tag{5.2.8}$$

is globally Lipschitz continuous and generates a superposition operator $N_g : BV \to BV$ that is discontinuous.

Proof. Condition (iii) guarantees that g is well-defined.

We first show that g is Lipschitz continuous on $[0,1] \times \mathbb{R}$. Observe that with the help of (i) we get $\mathrm{lip}(g(\cdot, u)) \le L$ for all fixed $u \in \mathbb{R} \backslash (0, u_1)$, since for $u \ge u_1$ we have $g(\cdot,u) = g_1$, and $g(\cdot,u) = \mathbb{0}$ holds for $u \le 0$. For fixed $u \in (0,u_1)$ we have $u_{n+1} \le u < u_n$ for some $n \in \mathbb{N}$ and hence

$$\left|\partial_1 g(t,u)\right| \le \frac{u-u_{n+1}}{u_n-u_{n+1}}|g_n'(t)| + \frac{u_n-u}{u_n-u_{n+1}}|g_{n+1}'(t)| \le \frac{u-u_{n+1}}{u_n-u_{n+1}}L + \frac{u_n-u}{u_n-u_{n+1}}L = L$$

by (i) for almost all $t \in [0,1]$. This implies

$$\left|g(s,u) - g(t,u)\right| \le L|s-t| \quad \text{for all } s,t \in [0,1], u \in \mathbb{R}.$$

Conversely, for fixed $t \in [0,1]$ the function $g(t,\cdot)$ is constant on $\mathbb{R}\backslash(0,1)$ and hence $\partial_2 g(t,\cdot) = 0$ there. On each interval (u_{n+1},u_n) we have

$$|\partial_2 g(t,\cdot)| \le \frac{\|g_n\|_\infty + \|g_{n+1}\|_\infty}{u_n - u_{n+1}} \le M$$

by (iv) and hence

$$\left|g(t,u) - g(t,v)\right| \le M|u-v| \quad \text{for all } t \in [0,1], u,v \in \mathbb{R}.$$

This gives in total

$$\left|g(s,u) - g(t,v)\right| \le L|s-t| + M|u-v| \quad \text{for all } s,t \in [0,1], u,v \in \mathbb{R}$$

showing that $g \in Lip([0,1]\times\mathbb{R})$.
The sequence (x_n) of functions $x_n \in BV$, defined by $x_n(t) := u_n$ for all $t \in [0,1]$, converges in BV to $\mathbb{0}$, due to $\|x_n\|_{BV} = |u_n|$ and (iii). Moreover, $N_g x_n(t) = g(t,u_n) = g_n(t)$ and thus,

$$\|N_g x_n - N_g \mathbb{0}\|_{BV} = \|g_n\|_{BV} \ge \operatorname{Var}(g_n) \ge 1,$$

by (ii). But this means nothing than that the operator N_g is not continuous at $\mathbb{0}$ with respect to the BV-norm. ■

We now give a practical example of such a construction.

Example 5.2.20. For $n \in \mathbb{N}$ the functions

$$g_n(t) := \frac{\sin(2^n\pi t)}{2^{n+1}} \quad \text{together with the numbers} \quad u_n := \frac{1}{2^n} \tag{5.2.9}$$

satisfy all four conditions (i)–(iv) of Proposition 5.2.19. Indeed,

$$|g_n'(t)| = \pi|\cos(2^n\pi t)|/2 \le \pi/2 \quad \text{for } t \in [0,1]$$

and hence (i) is fulfilled with $L = \pi/2$. Moreover, by Theorem 1.1.20 we have

$$\operatorname{Var}(g_n) = \int_0^1 \left|g_n'(t)\right| \mathrm{d}t = \frac{\pi}{2}\int_0^1 \left|\cos(2^n\pi t)\right| \mathrm{d}t = 1$$

for all $n \in \mathbb{N}$ which proves (ii). The sequence (u_n) clearly decreases to 0, and finally

$$\frac{\|g_n\|_\infty + \|g_{n+1}\|_\infty}{u_n - u_{n+1}} \le \frac{2^{-n-1} + 2^{-n-2}}{2^{-n} - 2^{-n-1}} = \frac{3}{2}.$$

This shows that also (iii) and (iv) are met with $M = 3/2$.

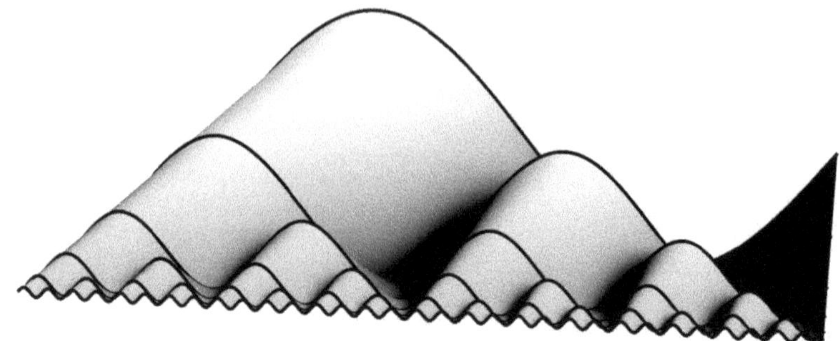

Figure 5.2.4: Lipschitz continuous g generating a discontinuous operator N_g in BV.

Figure 5.2.4 shows g of (5.2.8) generated by the g_n defined in (5.2.9) on $[0,1] \times [1/64, 1/2]$. The thick black waves represent the functions $g_1, \dots, g_6$, where g_1 is in the back and g_6 is in the front.

Let us now check which of the conditions (A)–(G) are satisfied by g. Note that since g is globally Lipschitz continuous we find some $K > 0$ such that

$$|g(s,u) - g(t,v)| \le K\big(|s-t| + |u-v|\big) \qquad \text{for } s,t \in [0,1],\, u,v \in \mathbb{R}. \tag{5.2.10}$$

In particular, $\|g(\cdot,u) - g(\cdot,v)\|_\infty \le K|u-v|$ for all $u,v \in \mathbb{R}$ showing that (A) and hence (B) are satisfied. Moreover, for a partition $0 = t_0 < \ldots < t_n = 1$ of $[0,1]$ and arbitrary numbers $u_0, \dots, u_{n-1} \in \mathbb{R}$ we have

$$\sum_{j=1}^{n} \big|g(t_{j-1}, u_{j-1}) - g(t_j, u_{j-1})\big| \le K \sum_{j=1}^{n} |t_{j-1} - t_j| = K$$

and thus condition (E) with $M = K$. But this implies that also the conditions (C), (D) and (F) are fulfilled, and that the operator N_g maps BV into itself and is bounded by Theorem 5.2.8. Furthermore, by Theorem 5.2.12, condition (G) must also be satisfied. Indeed, if $R > 0$, a partition $0 = t_0 < \ldots < t_n = 1$ of $[0,1]$ and numbers $u_0, \dots, u_n \in [-R, R]$ with $\sum_{j=1}^{n} |u_{j-1} - u_j| \le R$ are given, then we obtain from (5.2.10) that

$$\sum_{j=1}^{n} \big|g(t_{j-1}, u_j) - g(t_j, u_j)\big| \le K \sum_{j=1}^{n} |t_{j-1} - t_j| = K,$$

$$\sum_{j=1}^{n} \big|g(t_{j-1}, u_{j-1}) - g(t_{j-1}, u_j)\big| \le K \sum_{j=1}^{n} |u_{j-1} - u_j| \le KR,$$

and this proves that g also satisfies (G) with $M_R = \max\{K, KR\}$. ◇

We have now a method to construct functions $g : [0,1] \times \mathbb{R} \to \mathbb{R}$ that are globally Lipschitz continuous yet generate a discontinuous operator $N_g : BV \to BV$. Conversely, we now give an example of a *discontinuous* function $g : [0,1] \times \mathbb{R} \to \mathbb{R}$ that generates a *constant* (and therefore an utmost smooth) operator N_g.

Example 5.2.21. Let $\varphi_j : \mathbb{R} \to \mathbb{R}$ for $j \in \mathbb{N}$ be defined by $\varphi_1 = \mathbb{1}$ and $\varphi_j = \mathbb{0}$ for $j \geq 2$. Then the function g, defined by (5.2.1), has the form $g(t,u) = \chi_{\{1/2\}}(t)$ and is discontinuous at each point $(1/2, u) \in [0,1] \times \mathbb{R}$. For an arbitrary function $x : [0,1] \to \mathbb{R}$ we have $N_g x = \chi_{\{1/2\}}$ and so N_g is constant. In particular, this operator N_g maps any function space X whatsoever into any of the spaces BV, WBV_p, YBV_φ and ΛBV and, if X is normed, is globally Lipschitz continuous. ◇

Interestingly, a function $g \in C^1$ does not only generate a locally uniformly continuous composition operator (see Theorem 5.1.23), but also a locally uniformly continuous superposition operator [96].

Theorem 5.2.22. *If $g : [0,1] \times \mathbb{R} \to \mathbb{R}$ is continuously differentiable, then N_g maps BV into itself and is uniformly continuous on bounded subsets of BV.*

As we have seen in Example 5.2.21, the converse of Theorem 5.2.22 is far from being true. This is again in contrast to the composition operator $C_g : BV \to BV$, where $g : \mathbb{R} \to \mathbb{R}$ being continuously differentiable is equivalent to C_g being uniformly continuous on bounded subsets of BV (see Theorem 5.1.23).

Sometimes one is not interested in global continuity on the entire space, but rather in continuity at a particular point. For fixed $x \in BV$ we impose the following condition.

$$\begin{aligned}
&\forall \varepsilon > 0\ \exists \delta > 0\ \forall 0 = t_0 < \ldots < t_n = 1\ \forall u_0, \ldots, u_n \in [-\delta, \delta] : \quad \sum_{j=1}^{n} |u_{j-1} - u_j| \leq \delta \\
&\quad \Longrightarrow \sum_{j=1}^{n} \Big| \Big[g\big(t_{j-1}, u_j + x(t_{j-1})\big) - g\big(t_j, u_j + x(t_j)\big) \Big] \\
&\qquad\qquad - \Big[g\big(t_{j-1}, x(t_{j-1})\big) - g\big(t_j, x(t_j)\big) \Big] \Big| \leq \varepsilon \\
&\qquad \text{and} \quad \sum_{j=1}^{n} \Big| g\big(t_{j-1}, u_{j-1} + x(t_{j-1})\big) - g\big(t_{j-1}, u_j + x(t_{j-1})\big) \Big| \leq \varepsilon. \qquad (\mathrm{H}(x))
\end{aligned}$$

For the special case that $N_g \mathbb{0} = \mathbb{0}$, condition $(\mathrm{H}(\mathbb{0}))$ reduces to the following condition.

$$\begin{aligned}
&\forall \varepsilon > 0\ \exists \delta > 0\ \forall 0 = t_0 < \ldots < t_n = 1\ \forall u_0, \ldots, u_n \in [-\delta, \delta] : \quad \sum_{j=1}^{n} |u_{j-1} - u_j| \leq \delta \\
&\quad \Longrightarrow \sum_{j=1}^{n} \Big| g(t_{j-1}, u_j) - g(t_j, u_j) \Big| \leq \varepsilon \quad \text{and} \quad \sum_{j=1}^{n} \Big| g(t_{j-1}, u_{j-1}) - g(t_{j-1}, u_j) \Big| \leq \varepsilon. \quad (\mathrm{H}_0)
\end{aligned}$$

The conditions (H_0) and (G) look very similar yet neither of them implies the other.

Example 5.2.23. Let $g : [0,1] \times \mathbb{R} \to \mathbb{R}$ be defined by $g(t,u) = h(u)$ with

$$h(u) = \begin{cases} u & \text{for } |u| \leq 1, \\ 0 & \text{for } |u| > 1. \end{cases}$$

In particular, $N_g = C_h$, that is, the superposition operator is in fact a composition operator with generator h and satisfies $N_g \mathbb{0} = \mathbb{0}$.

We now show that g satisfies (H$_0$) but not (G). For $\varepsilon > 0$ pick $\delta := \min\{\varepsilon, 1\}$. Then for any partition $0 = t_0 < \ldots < t_n = 1$ and any collection $u_0, \ldots, u_n \in [-\delta, \delta]$ of real numbers with

$$\sum_{j=1}^{n} |u_{j-1} - u_j| \leq \delta$$

we have

$$\sum_{j=1}^{n} \left|g(t_{j-1}, u_j) - g(t_j, u_j)\right| = \sum_{j=1}^{n} \left|h(u_j) - h(u_j)\right| = 0 \leq \varepsilon$$

and, as $|u_j| \leq \delta \leq 1$ for all $j \in \{0, \ldots, n\}$,

$$\sum_{j=1}^{n} \left|g(t_{j-1}, u_{j-1}) - g(t_{j-1}, u_j)\right| = \sum_{j=1}^{n} \left|h(u_{j-1}) - h(u_j)\right| = \sum_{j=1}^{n} \left|u_{j-1} - u_j\right| \leq \delta \leq \varepsilon.$$

Thus, g satisfies (H$_0$). However, for $R := 3$ and $v_j := 1 + (-1)^j 2^{-j}$ for $j \in \mathbb{N}_0$ we have

$$\sum_{j=1}^{2n} |v_{j-1} - v_j| = 3 - \frac{3}{4^n} \leq 3 = R \quad \text{for } n \in \mathbb{N},$$

but, as $v_j > 1$ for even j and $0 < v_j < 1$ for odd j,

$$\sum_{j=1}^{2n} \left|g(t_{j-1}, v_{j-1}) - g(t_{j-1}, v_j)\right| = \sum_{j=1}^{2n} \left|h(v_{j-1}) - h(v_j)\right| = 2\sum_{j=1}^{n} \left(1 - 2^{-2j+1}\right) \geq n$$

which becomes unbounded as n increases. Consequently, g cannot satisfy condition (G). This is also clear by the fact that for composition operators C_h condition (G) is equivalent to $h \in Lip_{loc}(\mathbb{R})$. But our h here is not even continuous. ◇

In a moment we will see that also (G) does not imply (H$_0$). But first we mention the following result which states that (H(x)) provides a pointwise continuity criterion for N_g [96].

Theorem 5.2.24. *Let $g : [0,1] \times \mathbb{R} \to \mathbb{R}$ be so that N_g maps BV into itself, and let $x \in BV$ be fixed. Then N_g is continuous at $x \in BV$ if and only if for each $t \in [0,1]$ the function $g(t, \cdot)$ is continuous at $u = x(t)$ and ($H(x)$) holds.*

Let us have a quick look back again at the function g of Example 5.2.20 which has been constructed in such a way that N_g is discontinuous at $x = \mathbb{0}$. As we have seen there, g satisfies all the conditions (A)–(G) and hence generates a superposition operator N_g that maps BV into itself and is bounded. Moreover, since g is globally Lipschitz continuous, the function $g(t, \cdot)$ is continuous at $u = 0$ for each fixed $t \in [0,1]$. However, by Theorem 5.2.24, condition (H(x))=(H$_0$) cannot be satisfied at $x := \mathbb{0}$. Indeed, let g_n be the functions defined in (5.2.9), and set $\varepsilon := 1/2$. Pick $\delta > 0$ arbitrarily and choose $k \in \mathbb{N}$ so large that $v := 1/2^k \leq \delta$. Since $\mathrm{Var}(g_k) = 1$ we find a partition $0 = t_0 < \ldots < t_m = 1$ of $[0,1]$ such that

$$\sum_{j=1}^{m} |g_k(t_{j-1}) - g_k(t_j)| > \frac{1}{2}.$$

Now, consider $u_j := v = 1/2^k$ for $j \in \{0,\dots,m\}$. Then $u_j \in [-\delta,\delta]$ for all $j \in \{0,\dots,m\}$ and $\sum_{j=1}^m |u_{j-1} - u_j| = 0$, but

$$\sum_{j=1}^m \big|g(t_{j-1},u_j) - g(t_j,u_j)\big| = \sum_{j=1}^m \big|g(t_{j-1},v) - g(t_j,v)\big| = \sum_{j=1}^m \big|g_k(t_{j-1}) - g_k(t_j)\big| > \frac{1}{2} = \varepsilon.$$

Thus, (H_0) cannot hold. In particular, none of the conditions (A)–(G) implies (H_0).

We now consider the global Lipschitz condition of N_g, i.e. a condition of the form

$$\|N_g x - N_g y\|_{BV} \le L\,\|x-y\|_{BV} \qquad \text{for } x,y \in BV, \tag{5.2.11}$$

where $L > 0$ is a constant independent of x and y. The following result is similar to Theorem 5.1.22 and was proven in [108].

Theorem 5.2.25. *Let N_g map the space BV into itself.*

(a) If N_g is Lipschitz continuous in the sense of (5.2.11), then

$$|g(t,u) - g(t,v)| \le L|u-v| \qquad \textit{for } t \in [0,1], u,v \in \mathbb{R}, \tag{5.2.12}$$

and the right regularization $g^\#$, defined by

$$g^\#(t,u) = \begin{cases} \lim_{s\to t+} g(s,u) & \textit{for } 0 \le t < 1, \\ g(1,u) & \textit{for } t = 1, \end{cases} \tag{5.2.13}$$

is affine, that is, there are two functions $a, b \in BV$ such that

$$g^\#(t,u) = a(t)u + b(t) \qquad \textit{for } t \in [0,1], u \in \mathbb{R}. \tag{5.2.14}$$

(b) Conversely, if g coincides with $g^\#$ defined by (5.2.13) and is of the form (5.2.14) for two functions $a, b \in BV$, then N_g is Lipschitz continuous in the sense of (5.2.11).

Let us remark that, unfortunately, there is a tiny gap between the statements (a) and (b) of Theorem 5.2.25. Part (a) says that from (5.2.11) it follows that $g^\#$ is affine with respect to u. However, part (b) states that if $g^\#$ is affine with respect to u *and* $g = g^\#$, then N_g satisfies (5.2.11). In particular, the function g from Example 5.2.21 generating a constant and hence globally Lipschitz continuous operator $N_g : BV \to BV$ does not contradict Theorem 5.2.25, because although being not affine itself, $g^\# \equiv 0$ is. Thus, $g^\#$ being affine is only necessary for the global Lipschitz condition for N_g. We show in the following example, that it is not sufficient, and that (5.2.12) cannot be dropped.

Example 5.2.26. The function $g : [0,1] \times \mathbb{R} \to \mathbb{R}$, defined by

$$g(t,u) = \begin{cases} u^2 & \text{for } t = 0, \\ 0 & \text{for } 0 < t \le 1, \end{cases}$$

does not satisfy (5.2.12) and hence generates an operator N_g that cannot be globally Lipschitz continuous in the sense of (5.2.11). Moreover, $g^\#(t,u) = 0$ for all $t \in [0,1]$ and $u \in \mathbb{R}$, and so $g^\#$ has the form (5.2.14) with $a = b = \mathbb{0}$. ◇

It is clear that in the autonomous case $g(t,\cdot) = \tilde{g}$ for all $t \in [0,1]$ and some function $\tilde{g} : \mathbb{R} \to \mathbb{R}$ there is no difference between $g^{\#}$ and $\tilde{g}$ since the regularization refers only to the variable t. In this case, Theorem 5.2.25 reduces to Theorem 5.1.22.

One might ask whether under the hypothesis of part (a) of Theorem 5.2.25 the function g itself must be affine with respect to u and not only its regularization $g^{\#}$. We give here an example which shows that the answer is negative. Our example is a slight simplification of the example given in [108].

Example 5.2.27. Let g be as in (5.2.1) with $\varphi_j(u) := 2^{-j}\sin(u)$ for all $j \in \mathbb{N}$. By Proposition 5.2.1 the function g satisfies the conditions (B) and (F) (even (A) and (E)) and therefore generates an operator N_g that maps BV into itself and is bounded by Theorem 5.2.9. For any partition $0 = t_0 < \ldots < t_n = 1$ and $x, y \in BV$ we have

$$\begin{aligned}
\sum_{j=1}^{n}\Big|N_g x(t_{j-1}) - N_g y(t_{j-1}) - N_g x(t_j) + N_g y(t_j)\Big| &\leq 2\sum_{j=1}^{n}\Big|N_g x(t_j) - N_g y(t_j)\Big| \\
\leq 2\sum_{j=1}^{\infty}\left|g\left(\frac{1}{2j}, x\left(\frac{1}{2j}\right)\right) - g\left(\frac{1}{2j}, y\left(\frac{1}{2j}\right)\right)\right| &= 2\sum_{j=1}^{\infty}2^{-j}\left|\sin x\left(\frac{1}{2j}\right) - \sin y\left(\frac{1}{2j}\right)\right| \\
\leq 2\sum_{j=1}^{\infty}2^{-j}\left|x\left(\frac{1}{2j}\right) - y\left(\frac{1}{2j}\right)\right| \leq 2\,\|x-y\|_\infty \sum_{j=1}^{\infty}2^{-j} &\leq 2\,\|x-y\|_{BV}\,.
\end{aligned}$$

Moreover,

$$\begin{aligned}
\|N_g x - N_g y\|_\infty &= \sup_{t\in[0,1]}\Big|g\big(t, x(t)\big) - g\big(t, y(t)\big)\Big| \leq \sup_{t\in[0,1]}\Big|\sin x(t) - \sin y(t)\Big| \leq \|x-y\|_\infty \\
&\leq \|x-y\|_{BV}
\end{aligned}$$

which gives in total $\|N_g x - N_g y\|_{BV} \leq 3\,\|x-y\|_{BV}$. This proves that N_g is globally Lipschitz continuous in BV. However, $g(\frac{1}{2j}, u)$ is clearly not affine with respect to u for any $j \in \mathbb{N}$. ◇

Note that $g^{\#}(t,u) = 0$ for all $t \in [0,1]$ and $u \in \mathbb{R}$ for g in the last example, in accordance with Theorem 5.2.25 (a).

We point out that a stronger degeneracy phenomenon has been proved for many other normed function spaces X, namely, if N_g maps X into itself and is globally Lipschitz continuous, then g is affine with respect to u, that is, $g(t,u) = a(t)u + b(t)$ for functions $a, b \in X$. For instance, this has been proved for $X = C^n$ in [104, 105], for the Sobolev space $X = W^{1,p}$ in [106] and for the space $X = WBV_p^2$ of functions of bounded $(p,2)$-variation in [107]. Likewise, an analogous result was shown in [103] for the space $X = Lip_\alpha$ of Hölder continuous functions with exponent $\alpha \leq 1$ and in [93] for the space $X = C^{n,\alpha}$ of functions with Hölder continuous n-th derivative.

As we have seen in Theorem 5.1.22, already uniform continuity of C_g leads to a strong degeneracy of g. In fact, a similar result is true for the superposition operator in all our BV-spaces. The degeneracy of g is then expressed in terms of its right regularization as in Theorem 5.2.25.

Theorem 5.2.28. *Let X be any of the spaces BV, WBV_p, YBV_φ or ΛBV. If N_g maps X into itself and is uniformly continuous, then $g^\#$ is of the form (5.2.14) for $a, b \in X$.*

In fact, Theorem 5.2.28 remains true if the operator N_g is merely uniformly bounded [6].
In [1] the authors have shown that for the Riesz space an even stronger degeneracy occurs. To be more precise, they proved the following.

Theorem 5.2.29. *If N_g maps RBV_p into itself and is uniformly continuous, then g is of the form $g(t,u) = a(t)u + b(t)$ for some functions $a, b \in RBV_p$.*

It is clear that the converse of Theorem 5.2.29 is also true, even for all our BV-spaces X. Indeed, if g is of the form $g(t,u) = a(t)u + b(t)$ for $a, b \in X$, then for $x, y \in X$ we have

$$N_g x(t) - N_g y(t) = g\big(t, x(t)\big) - g\big(t, y(t)\big) = a(t)\big(x(t) - y(t)\big) \quad \text{for } t \in [0,1]$$

and hence

$$\|N_g x - N_g y\|_\infty \leq \|a\|_\infty \|x - y\|_\infty .$$

Moreover,

$$\begin{aligned} \Big|N_g x(s) - N_g y(s) - N_g x(t) + N_g y(t)\Big| &= \Big|a(s)\big(x(s) - y(s)\big) - a(t)\big(x(t) - y(t)\big)\Big| \\ &= \Big|a(s)\big[x(s) - y(s) - x(t) + y(t)\big] + \big[a(s) - a(t)\big]\big[x(t) - y(t)\big]\Big| \\ &\leq \|a\|_\infty \Big|x(s) - y(s) - x(t) + y(t)\Big| + \Big|a(s) - a(t)\Big| \, \|x - y\|_\infty \end{aligned}$$

for any $s, t \in [0,1]$. Using the symbol Φ_X for the seminorm part of our BV-norms as summarized in Table 1.2.1 we obtain by Lemma 1.2.26,

$$\Phi_X\big(N_g x - N_g y\big) \leq \|a\|_\infty \Phi_X(x - y) + \Phi_X(a) \|x - y\|_\infty$$

and thus in total $\|N_g x - N_g y\|_X \leq \|a\|_X \|x - y\|_X$. This shows that $N_g : X \to X$ is even globally Lipschitz continuous.
The following is an extension of Example 5.2.26 and shows that the converse of Theorem 5.2.28 is not true in general.

Example 5.2.30. Let g be as in Example 5.2.26, and let X be any of the spaces BV, WBV_p, YBV_φ or ΛBV. Then N_g maps X into itself and is bounded, because for any $x \in X$ we have

$$N_g x(t) = \chi_{\{0\}} x(0)^2.$$

Moreover, $g^\#(t,u) = 0$ for all $t \in [0,1]$ and $u \in \mathbb{R}$, as we have seen in Example 5.2.26. But N_g cannot be uniformly continuous. To see this, pick any $\delta, u > 0$. The constant functions $x \equiv u$ and $y \equiv u + \delta$ belong to X with $\|x - y\|_X = \delta$, but

$$\|N_g x - N_g y\|_X \geq \|N_g x - N_g y\|_\infty = (u + \delta)^2 - u^2 = 2u\delta + \delta^2$$

which gets infinitely large as $u \to \infty$. 

Uniform continuity of N_g implies that $g(t,\cdot)$ is uniformly continuous for each fixed $t \in [0,1]$ (which is not true for the function g in the previous example), and so this and probably something else is missing in Theorem 5.2.28 to gain a necessary *and sufficient* condition for the uniform continuity of N_g in X. But we do not know what exactly.

Later in Chapter 7 we apply some of the theoretical results presented here to integral equations which we will solve with fixed point theorems. For most applications the Banach-Caccioppoli Fixed Point Theorem will do the job. However, applying it to the entire space would require the underlying superposition operators to be globally Lipschitz continuous which in most cases is too restrictive, as we have seen before. Therefore, we will use it only locally in order to gain solutions that are at least locally unique. The advantage is that then the corresponding superposition operators need to be only locally Lipschitz continuous. As in (5.1.19) we mean by that a condition of the form

$$\|N_g x - N_g y\|_X \leq L_R \|x - y\|_X \quad \text{for } \|x\|_X, \|y\|_X \leq R, \tag{5.2.15}$$

where X is one of our BV-spaces. As far as we know apart from trivial sufficient conditions there are no conditions known for g to make N_g satisfy (5.2.15). We now give here a sufficient condition which is very similar to Theorem 5.1.21. Therein we have seen that the composition operator C_h maps any of our BV-spaces locally Lipschitz continuously into itself if and only if $h \in Lip^1_{loc}(\mathbb{R})$. A similar result is also true for the superposition operator.

Theorem 5.2.31. *Let X be any of the spaces BV, WBV_p, YBV_φ, ΛBV or RBV_p. Assume that $g : [0,1] \times \mathbb{R} \to \mathbb{R}$ satisfies the following conditions.*

(i) $g(\cdot, 0) \in X$,

(ii) $g(t,\cdot) \in C^1(\mathbb{R})$ *for each fixed* $t \in [0,1]$,

(iii) $\partial_2 g(\cdot, 0) \in B$,

(iv) For each $R > 0$ there is some $A_R > 0$ such that

$$|\partial_2 g(t,u) - \partial_2 g(t,v)| \leq A_R |u - v|$$

whenever $t \in [0,1]$, $u, v \in [-R, R]$.

(v) For each $R > 0$ there is some $B_R > 0$ and a function $z_R \in X$ such that

$$|g(s,u) - g(s,v) - g(t,u) + g(t,v)| \leq B_R |z_R(s) - z_R(t)||u - v|$$

whenever $s, t \in [0,1]$ and $u, v \in [-R, R]$.

Then N_g maps X into itself and is locally Lipschitz continuous in the sense of (5.2.15).

Proof. First note that once (5.2.15) has been established, it follows from (i) that N_g maps X into itself, because

$$\|N_g x\|_X \leq \|N_g x - N_g \mathbb{0}\|_X + \|N_g \mathbb{0}\|_X \leq L_R \|x\|_X + \|N_g \mathbb{0}\|_X < \infty.$$

We thus have to focus only on (5.2.15). For that fix $R > 0$. We first show that g satisfies for fixed $u_1, u_2, v_1, v_2 \in [-R, R]$ and $s, t \in [0,1]$ the estimate

$$\begin{aligned}
&|g(s,u_1) - g(s,v_1) - g(t,u_2) + g(t,v_2)| \\
&\qquad \leq A_R\Big(|u_1 - u_2| + |v_1 - v_2|\Big)\Big(|u_1 - v_1| + |u_2 - v_2|\Big) \\
&\qquad\quad + M_R|u_1 - v_1 - u_2 + v_2| + B_R|z_R(s) - z_R(t)||u_2 - v_2|, \qquad (5.2.16)
\end{aligned}$$

where M_R is given by

$$M_R := RA_R + \|\partial_2 g(\cdot, 0)\|_\infty ; \qquad (5.2.17)$$

note that (iii) guarantees that M_R is finite. To this end, first note that for $u_1, u_2, v_1, v_2 \in [-R, R]$ and $s, t \in [0,1]$,

$$\begin{aligned}
&|g(s,u_1) - g(s,v_1) - g(t,u_2) + g(t,v_2)| \\
&\qquad \leq |g(s,u_1) - g(s,v_1) - g(s,u_2) + g(s,v_2)| \\
&\qquad\quad + |g(s,u_2) - g(s,v_2) - g(t,u_2) + g(t,v_2)| \\
&\qquad \leq |g(s,u_1) - g(s,v_1) - g(s,u_2) + g(s,v_2)| + B_R|z_R(s) - z_R(t)||u_2 - v_2|
\end{aligned}$$

by (v). To estimate the remaining term $|g(s,u_1) - g(s,v_1) - g(s,u_2) + g(s,v_2)|$ consider the function $h(u) := g(s,u)$ for fixed $s \in [0,1]$. This function is continuously differentiable in $\mathbb{R}$ by (ii) and satisfies $|h'(u) - h'(v)| \leq A_R|u - v|$ by (iv). By [6, Lemma 5.48] we obtain

$$\begin{aligned}
&|h(u_1) - h(v_1) - h(u_2) + h(v_2)| \\
&\qquad \leq A_R\Big(|u_1 - u_2| + |v_1 - v_2|\Big)\Big(|u_1 - v_1| + |u_2 - v_2|\Big) \\
&\qquad\quad + \tilde{M}_R(s)|u_1 - v_1 - u_2 + v_2|, \qquad (5.2.18)
\end{aligned}$$

where

$$\tilde{M}_R(s) := \|h'\|_{[-R,R]} = \sup_{|u| \leq R} |\partial_2 g(s,u)|.$$

Since $|\partial_2 g(s,u)| \leq |\partial_2 g(s,u) - \partial_2 g(s,0)| + |\partial_2 g(s,0)| \leq A_R|u| + \|\partial_2 g(\cdot,0)\|_\infty \leq M_R$ for $|u| \leq R$ by (iv), the estimate (5.2.16) follows.

For functions $x, y : [0,1] \to \mathbb{R}$ and $s, t \in [0,1]$ we obtain from (5.2.16),

$$\begin{aligned}
&\Big|N_g x(s) - N_g y(s) - N_g x(t) + N_g y(t)\Big| \\
&\quad = \Big|g\big(s, x(s)\big) - g\big(s, y(s)\big) - g\big(t, x(t)\big) + g\big(t, y(t)\big)\Big| \\
&\quad \leq A_R\Big(|x(s) - x(t)| + |y(s) - y(t)|\Big)\Big(|x(s) - y(s)| + |x(t) - y(t)|\Big) \\
&\qquad + M_R\Big|x(s) - y(s) - x(t) + y(t)\Big| + B_R\Big|z_R(s) - z_R(t)\Big|\Big|x(t) - y(t)\Big|
\end{aligned}$$

and thus,

$$
\begin{aligned}
&|N_g x(s) - N_g y(s) - N_g x(t) + N_g y(t)| \\
&\quad \le 2A_R \|x - y\|_\infty \Big(|x(s) - x(t)| + |y(s) - y(t)|\Big) \\
&\qquad + M_R|x(s) - y(s) - x(t) + y(t)| + B_R \|x - y\|_\infty |z_R(s) - z_R(t)|. \qquad (5.2.19)
\end{aligned}
$$

By Lemma 1.2.26 we get for $x, y \in \mathbb{B}_R(X)$,

$$
\begin{aligned}
\Phi_X(N_g x - N_g y) &\le 2A_R \|x - y\|_\infty \big(\Phi_X(x) + \Phi_X(y)\big) \\
&\qquad + M_R \Phi_X(x - y) + B_R \|x - y\|_\infty \Phi_X(z_R) \\
&\le \big(4RA_R + B_R \Phi_X(z_R)\big) \|x - y\|_\infty + M_R \Phi_X(x - y), \qquad (5.2.20)
\end{aligned}
$$

where Φ_X is as in Table 1.2.1. Moreover, for $s = t \in [0, 1]$ we obtain from (5.2.16) with $u_1 = x(t)$, $v_1 = y(t)$, $u_2 = v_2 = 0$,

$$
\begin{aligned}
\big|N_g x(t) - N_g y(t)\big| &= \big|g\big(t, x(t)\big) - g\big(t, y(t)\big)\big| \\
&\le A_R\Big(|x(t)| + |y(t)|\Big)|x(t) - y(t)| + M_R|x(t) - y(t)| \\
&\le \big(2RA_R + M_R\big) \|x - y\|_\infty . \qquad (5.2.21)
\end{aligned}
$$

Combining (5.2.20) and (5.2.21) yields

$$
\begin{aligned}
\|N_g x - N_g y\|_X &\le \big(6RA_R + B_R \Phi_X(z_R) + M_R\big) \|x - y\|_\infty + M_R \Phi_X(x - y) \\
&\le \big(6RA_R + B_R \Phi_X(z_R) + M_R\big) \|x - y\|_X .
\end{aligned}
$$

Finally, taking (5.2.17) into account, we obtain for any of our BV-spaces X,

$$
\|N_g x - N_g y\|_X \le \Big(7RA_R + B_R \Phi_X(z_R) + \|\partial_2 g(\cdot, 0)\|_\infty\Big) \|x - y\|_X \qquad (5.2.22)
$$

which is the desired estimate (5.2.15) with

$$
L_R = 7RA_R + B_R \Phi_X(z_R) + \|\partial_2 g(\cdot, 0)\|_\infty .
$$

The proof is complete. ■

As we have seen, under the assumptions of Theorem 5.2.31 the superposition operator N_g satisfies the estimate

$$
\|N_g x - N_g y\|_X \le \Big(7RA_R + B_R \Phi_X(z_R) + \|\partial_2 g(\cdot, 0)\|_\infty\Big) \|x - y\|_X \qquad (5.2.23)
$$

for all $x, y \in X$ with $\|x\|_X , \|y\|_X \le R$, where Φ_X is as in Table 1.2.1.

We remark that condition (iv) in Theorem 5.2.31 says that $\partial_2 g$ satisfies condition (B). Moreover, any $g \in C^2([0, 1] \times \mathbb{R})$ satisfies all the hypotheses of Theorem 5.2.31. Indeed, for such g the conditions (i)–(iv) are clearly fulfilled. Moreover, the function

$$
G(t, u, v) := \begin{cases} \dfrac{g(t, u) - g(t, v)}{u - v} & \text{for } u \ne v, \\ \partial_2 g(t, u) & \text{for } u = v, \end{cases}
$$

belongs to $C^1([0,1] \times \mathbb{R} \times \mathbb{R})$; in particular, for each $R > 0$ there is some $B_R > 0$ such that $|G(s,u,v) - G(t,u,v)| \leq B_R|s-t|$ for all $s,t \in [0,1], u,v \in [-R,R]$, and this implies (v) with $z(t) = t$. Thus, the hypotheses of Theorem 5.2.31 are not as artificial as they may appear at first glance.

However, Theorem 5.2.31 may also be applied to discontinuous functions g satisfying the conditions (i)–(v). For instance, the function g in Example 5.2.21 is discontinuous yet meets all the conditions of Theorem 5.2.31 for $X = BV$. Indeed, we have seen that this g generates a constant operator $N_g : BV \to BV$. Here is another more general example that deals with separated variables in one of our BV-spaces X.

Example 5.2.32. If $g : [0,1] \times \mathbb{R} \to \mathbb{R}$ is given by $g(t,u) = g_1(t)g_2(u)$ for $g_1 \in X$ and $g_2 \in Lip^1_{loc}(\mathbb{R})$, then the conditions (i)–(iv) of Theorem 5.2.31 are satisfied. Indeed, $g(\cdot,0) = g_1g_2(0)$ and $\partial_2 g(\cdot,0) = g_1g_2'(0)$ are in X and bounded, because g_1 is, and so (i) and (iii) are fulfilled with

$$\|\partial_2 g(\cdot,0)\|_\infty = |g_2'(0)|\,\|g_1\|_\infty .$$

Moreover, $g(t,\cdot) = g_1(t)g_2$ belongs to $C^1(\mathbb{R})$ for each $t \in [0,1]$ which is (ii). The partial derivatives with respect to the second argument fulfill

$$|\partial_2 g(t,u) - \partial_2 g(t,v)| = |g_1(t)||g_2'(u) - g_2'(v)| \leq \|g_1\|_\infty \operatorname{lip}\left(g_2', [-R,R]\right)|u-v|$$

for $u,v \in [-R,R]$ and thus (iv) holds with

$$A_R = \|g_1\|_\infty \operatorname{lip}\left(g_2', [-R,R]\right).$$

Finally,

$$|g(s,u) - g(s,v) - g(t,u) + g(t,v)| = |g_1(s) - g_1(t)||g_2(u) - g_2(v)|$$

for $s,t \in [0,1], u,v \in [-R,R]$, and because of $g_1 \in X$ and $g_2 \in Lip_{loc}(\mathbb{R})$ condition (v) is satisfied with

$$z_R = g_1 \quad \text{and} \quad B_R = \operatorname{lip}\left(g_2, [-R,R]\right).$$

In particular, $N_g : X \to X$ is locally Lipschitz continuous by Theorem 5.2.31, but g_1 and hence also g can be discontinuous. ◇

As a special case of Example 5.2.32, let $g_1 \in X$ be arbitrary and $g_2(u) = u$ for all $u \in \mathbb{R}$. Then $N_g = M_{g_1}$ with

$$\|M_{g_1}x - M_{g_1}y\|_X \leq \|g_1\|_X\,\|x-y\|_X \quad \text{for } x,y \in X,$$

and this shows that Theorem 5.2.31 covers Corollary 4.1.8.
If we take $g_1 = \mathbb{1}$ and $g_2 \in Lip^1_{loc}(\mathbb{R})$ in Example 5.2.32 instead, then $N_g = C_{g_2}$ with

$$\|C_{g_2}x - C_{g_2}y\|_X \leq \left(7R\operatorname{lip}\left(g_2', [-R,R]\right) + |g_2'(0)|\right)\|x-y\|_X \quad \text{for } x,y \in \mathbb{B}_R(X),$$

and this shows that Theorem 5.2.31 also covers Theorem 5.1.21. However, in contrast to Theorem 5.1.21 the hypotheses of Theorem 5.2.31 are *not* necessary.

Example 5.2.33. Consider the function $g : [0,1] \times \mathbb{R} \to \mathbb{R}$, defined by $g(t,u) = \chi_{\{0\}}(t)|u|$. Then $g(0,u) = |u|$ is not differentiable with respect to u, and so g does not meet (ii) in Theorem 5.2.31. However, g generates an even globally Lipschitz continuous superposition operator $N_g : BV \to BV$. To see this, note that for any $x, y \in BV$,

$$N_g x(t) - N_g y(t) = g\big(t, x(t)\big) - g\big(t, y(t)\big) = \begin{cases} \big|x(0)\big| - \big|y(0)\big| & \text{for } t = 0, \\ 0 & \text{for } 0 < t \leq 1; \end{cases}$$

in particular, $\|N_g x\|_{BV} = 2|x(0)|$, and so N_g maps BV into itself. Moreover,

$$\|N_g x - N_g y\|_{BV} = 2\Big||x(0)| - |y(0)|\Big| \leq 2\,\|x - y\|_\infty \leq 2\,\|x - y\|_{BV}$$

for any $x, y \in BV$, and thus N_g is indeed globally Lipschitz continuous in BV. ◊

Note that Example 5.2.33 is not contradictory to Theorem 5.2.25, because for the generator $g(t,u) = \chi_{\{0\}}(t)|u|$, any $t \in [0,1]$ and $u \in \mathbb{R}$ we have $g^{\#}(t,u) = 0$ which is affine in the sense of (5.2.14).

Instead of considering operators N_g from a BV-space X into itself it will be of great use later on in Chapter 7 to also consider N_g as an operator from a BV-space X into L_∞. In order to find conditions for the local Lipschitz continuity of such operators one needs to find for each $R > 0$ a number $L_R > 0$ such that

$$\|N_g x - N_g y\|_{L_\infty} \leq L_R \|x - y\|_X \quad \text{for } \|x\|_X, \|y\|_X \leq R. \tag{5.2.24}$$

To establish that one can give much milder conditions on g than those given in Theorem 5.2.31. We end up with

Theorem 5.2.34. *Let X be any of the spaces BV, WBV_p, YBV_φ, ΛBV or RBV_p, and let $g : [0,1] \times \mathbb{R} \to \mathbb{R}$ be given. Then the operator N_g maps the space X into L_∞ and is locally Lipschitz continuous in the sense of (5.2.24) if g satisfies the following two conditions.*

(i) $g(\cdot, u)$ is measurable for each $u \in \mathbb{R}$, and $g(\cdot, 0) \in L_\infty$.

(ii) For each $R > 0$ there is some $a_R \in L_\infty$ such that $\big|g(t,u) - g(t,v)\big| \leq a_R(t)|u - v|$ for all $t \in [0,1]$ and all $u, v \in [-R, R]$.

Proof. Assume that g satisfies (i) and (ii) and fix $R > 0$. Since $a_R \in L_\infty$, there is some constant $L_R > 0$ such that $a_R(t) \leq L_R$ for almost all $t \in [0,1]$. For $x, y \in X$ with $\|x\|_X, \|y\|_X \leq R$ we then have for almost all $t \in [0,1]$ by (ii),

$$\Big|g\big(t, x(t)\big) - g\big(t, y(t)\big)\Big| \leq L_R |x(t) - y(t)|.$$

Thus, since $\|z\|_{L_\infty} \leq \|z\|_\infty \leq \|z\|_X$ for all $z \in X$,

$$\|N_g x - N_g y\|_{L_\infty} \leq L_R \|x - y\|_{L_\infty} \leq L_R \|x - y\|_X .$$

This shows that (5.2.24) is satisfied. Moreover, from this and (i) we obtain

$$\|N_g x\|_{L_\infty} \le \|N_g x - N_g \mathbb{0}\|_{L_\infty} + \|N_g \mathbb{0}\|_{L_\infty} \le RL_R + \|g(\cdot,0)\|_{L_\infty},$$

and thus N_g maps the space X into L_∞. ∎

Clearly, Theorem 5.2.34 remains true if one replaces L_∞ by the space B of bounded functions; one then just has to replace (ii) by condition (B). Moreover, under the hypothesis of Theorem 5.2.34 the operator N_g maps even L_∞ into itself and is locally Lipschitz continuous.

At this point we remark that condition (i) in Theorem 5.2.34 together with condition (ii) implies that $g(t,\cdot)$ is continuous for (almost) all $t \in [0,1]$. This means in particular that g is a so called *Carathéodory function* which guarantees that the superposition operator N_g maps the space of measurable functions into itself [37]. However, being a Carathéodory function is only sufficient but not necessary for this acting condition on the space of measurable functions, and the problem of finding sufficient and necessary conditions is delicate. For a detailed discussion of this and related problems see also [12].

Moreover, we remark that Theorem 5.2.34 only provides a sufficient condition for the local Lipschitz continuity of $N_g : X \to L_\infty$, and we do not know if it is also necessary. The crucial part in condition (ii) is that the function $a_R \in L_\infty$ must be independent of u and v, that is, (ii) is equivalent to the condition that for each $R > 0$ there is some $L_R > 0$ and some null set $N \subseteq [0,1]$ such that

$$|g(t,u) - g(t,v)| \le L_R|u - v| \tag{5.2.25}$$

for all $t \in [0,1]\backslash N$ and all $u, v \in [-R, R]$; in particular, N is independent of u and v. Therefore, one might think that (ii) could be weakened by

$$\forall R > 0 \; \exists L_R > 0 : \quad \|g(\cdot,u) - g(\cdot,v)\|_{L_\infty} \le L_R|u - v|, \tag{5.2.26}$$

because this means that only for each fixed $u, v \in [-R, R]$ there is a null set N depending on u and v such that (5.2.25) holds for all $t \in [0,1]\backslash N$. By considering constant functions it is easy to show that (5.2.26) is necessary for $N_g : X \to L_\infty$ to be locally Lipschitz continuous. However, as the following example shows, it is not sufficient, even when (i) is added.

Example 5.2.35. Let $g : [0,1] \times \mathbb{R} \to \mathbb{R}$ be defined by

$$g(t,u) = \begin{cases} 1/t & \text{for } 0 < t = u \le 1, \\ 0 & \text{otherwise.} \end{cases}$$

Then $g(\cdot,u)$ is measurable for each $u \in \mathbb{R}$, and $g(t,0) = 0$ for any $t \in [0,1]$ showing that (i) in Theorem 5.2.34 is satisfied. Moreover, for fixed $u, v \in \mathbb{R}$ we have $g(t,u) = 0$ for all $t \in [0,1]\backslash\{u,v\}$ and hence $\|g(\cdot,u) - g(\cdot,v)\|_{L_\infty} = 0$. Thus, (5.2.26) is also satisfied. However, the function $x(t) = t$ clearly belongs to any of our BV-spaces X, but $g(t, x(t)) = 1/t$ for $0 < t \leq 1$ does not belong to L_∞. Thus, N_g does not map X into L_∞. Note that g does also not satisfy (ii) of Theorem 5.2.34, because otherwise we had $|g(t,t) - g(t,0)| \leq a_1(t)t$ for some function $a_1 \in L_\infty$ and all $t \in (0,1]$ which would imply $a_1(t) \geq 1/t^2$ for such t. But this is impossible. ♢

The following example shows that we cannot drop assumption (i) in Theorem 5.2.34.

Example 5.2.36. Let $g : [0,1] \times \mathbb{R} \to \mathbb{R}$ be defined by

$$g(t,u) = u + \begin{cases} 1/t & \text{for } 0 < t \leq 1, \\ 0 & \text{for } t = 0. \end{cases}$$

Then $|g(t,u) - g(t,v)| = |u - v|$ for all $u, v \in \mathbb{R}$ and $t \in [0,1]$ and hence (ii) of Theorem 5.2.34 (even globally) is satisfied.
However, condition (i) fails for g, because $g(t,0) = 1/t$ for $0 < t \leq 1$ is not essentially bounded; in particular, $N_g \mathbb{0} \notin L_\infty$, although $\mathbb{0} \in X$. This also shows that even if g is globally Lipschitz continuous N_g does not have to map *any* linear function space into the space L_∞. ♢

Let us now take a closer look at compactness. As Theorem 5.1.20 shows, the composition operator C_g is compact only if the generating function g degenerates to a constant function. The situation is of course different for superposition operators. To see this, let us again have a look back at the linear multiplication operator that we have exhaustively studied in Section 4.1. This is because a multiplication operator M_h can be considered as a superposition operator N_g with $g(t,u) = h(t)u$, and we have already done so in Example 5.2.32 and the special cases thereafter. According to Theorem 4.1.12 such operators are compact in BV if and only if $\operatorname{supp}(h)$ is countable. In particular, there are many compact superposition operators N_g where g is not constant. Here is an example.

Example 5.2.37. Our function g in (5.2.1) with $\varphi_j(u) = u$ for all $j \in \mathbb{N}$ generates a superposition operator which is in fact a multiplication operator $N_g = M_h$ with generating function $h = \chi_A$ and support $A := \{1/(2j) \mid j \in \mathbb{N}\}$. By Theorem 4.1.12, N_g is compact. ♢

One could conjecture that the result for multiplication operators carries over to superposition operators, requiring that $\operatorname{supp} g(\cdot, u)$ is countable for each $u \in \mathbb{R}$. However, the following two examples show that this condition is not necessary for the compactness of N_g, and it even does not guarantee the acting condition $N_g(BV) \subseteq BV$.

Example 5.2.38. The function $g(t,u) \equiv 1$ generates a constant and hence compact operator $N_g : BV \to BV$, but $\operatorname{supp} g(\cdot, u) = [0,1]$ for all $u \in \mathbb{R}$. ♢

Example 5.2.39. The support $\operatorname{supp} g(\cdot, u)$, where g is as in (5.2.1) with $\varphi_j = \chi_{1/(2j)}$ for all $j \in \mathbb{N}$, has either only one element or is empty. But N_g does not map BV into itself, since the identity function $x(t) = t$ which belongs to BV is mapped to the function $g(t,t) = \chi_A(t)$ with $A := \{1/(2j) \mid j \in \mathbb{N}\}$ which does not belong to BV. ♢

The last two examples have shown that we cannot expect $\operatorname{supp} g(\cdot, u)$ to be countable for every $u \in \mathbb{R}$ if N_g is compact. However, if we shift $g(\cdot, u)$ into $g(\cdot, u) - g(\cdot, 0)$, then we obtain the following result.

Theorem 5.2.40. *Let X be any of the spaces BV, WBV_p, YBV_φ and ΛBV, and let the generator $g : [0,1] \times \mathbb{R} \to \mathbb{R}$ be so that N_g maps X into itself and is compact. Then*

$$\operatorname{supp}\Big(g(\cdot, u) - g(\cdot, 0)\Big) \tag{5.2.27}$$

is countable for each $u \in \mathbb{R}$.

Proof. Fix $t \in [0,1]$ and $u \in \mathbb{R}$ and let (τ_n) be a sequence in $(0,1)\backslash\{t\}$ that converges to t. The functions $x_n = u\chi_{\{\tau_n\}}$ form a bounded sequence (x_n) in X. Indeed, we have $\|x_n\|_\infty = |u|$ for all $n \in \mathbb{N}$. For $X = YBV_\varphi$ we have by Proposition 1.2.10 for $\lambda = 2|u|/\varphi^{-1}(1)$,

$$\operatorname{Var}_\varphi\left(\frac{x_n}{\lambda}\right) \leq \varphi\left(\frac{2|u|}{\lambda}\right) = 1.$$

and hence $\mathfrak{M}(x_n) \leq 2|u|/\varphi^{-1}(1)$. In total, this gives $\|x_n\|_{YBV_\varphi} \leq \Big(1 + 2/\varphi^{-1}(1)\Big)|u|$ for all $n \in \mathbb{N}$ in this case.
For $X = \Lambda BV$ we get from Proposition 1.2.20 that $\operatorname{Var}_\Lambda(x_n) \leq 2\lambda_1|u|$ and hence $\|x_n\|_{\Lambda BV} \leq \Big(1 + 2\lambda_1\Big)|u|$ for all $n \in \mathbb{N}$.
In any case, the sequence (x_n) is mapped by N_g into the sequence

$$y_n(s) = g\Big(s, x_n(s)\Big) = \begin{cases} g(s,u) & \text{for } s = \tau_n, \\ g(s,0) & \text{for } s \neq \tau_n. \end{cases}$$

Since N_g is compact in X, the sequence (y_n) has a subsequence $(y_{n_k})_k$ that converges in X and hence also pointwise to some function $y \in X$. For fixed $s \in [0,1]\backslash\{t\}$ we have $s \neq \tau_n$ for sufficiently large $n \in \mathbb{N}$ as the τ_n converge to t, and hence $y_{n_k}(s) = g(s,0)$ for sufficiently large $k \in \mathbb{N}$. For $s = t$ we have $s \neq \tau_n$ even for all $n \in \mathbb{N}$ and thus again $y_{n_k}(s) = g(s,0)$ for all $k \in \mathbb{N}$. Consequently, $y(s) = g(s,0)$ for all $s \in [0,1]$.
The convergence of (y_{n_k}) to y in $X = YBV_\varphi$ implies that $\mathfrak{M}(y_{n_k} - y) \to 0$ as $k \to \infty$. Therefore, there exists a sequence (μ_k) of positive real numbers tending to 0 such that

$$\begin{aligned} 1 \geq \operatorname{Var}_\varphi\left(\frac{y_{n_k} - y}{\mu_k}\right) &\geq \varphi\left(\frac{\Big|y_{n_k}(t) - y(t) - y_{n_k}(\tau_{n_k}) + y(\tau_{n_k})\Big|}{\mu_k}\right) \\ &= \varphi\left(\frac{\Big|g(\tau_{n_k}, u) - g(\tau_{n_k}, 0)\Big|}{\mu_k}\right) = \varphi\left(\frac{|h(\tau_{n_k})|}{\mu_k}\right) \end{aligned}$$

and thus

$$|h(\tau_{n_k})| \leq \mu_k \varphi^{-1}(1)$$

for all $k \in \mathbb{N}$, where $h(s) := g(s,u) - g(s,0)$. For $k \to \infty$ the right hand side goes to zero, and thus taking the limit inferior on both sides with respect to k yields

$$\liminf_{s \to t} |h(s)| = 0. \tag{5.2.28}$$

For $X = \Lambda BV$ we obtain

$$\begin{aligned}\mathrm{Var}_\Lambda(y_{n_k} - y) &\geq \lambda_1 \big| y_{n_k}(t) - y(t) - y_{n_k}(\tau_{n_k}) + y(\tau_{n_k}) \big| = \lambda_1 \big| g(\tau_{n_k}, u) - g(\tau_{n_k}, 0) \big| \\ &= \lambda_1 |h(\tau_{n_k})|\end{aligned}$$

for all $k \in \mathbb{N}$. Again, for $k \to \infty$ the left hand side goes to zero, and thus taking the limit inferior on both sides with respect to k yields also in this case (5.2.28).
Since t was picked arbitrarily, (5.2.28) holds in fact for all $t \in [0,1]$. By assumption, N_g maps X into itself which implies that h belongs to X. Thus, h has at most countably many points of discontinuity. But at every point $t \in [0,1]$ of continuity, (5.2.28) implies that $h(t) = 0$. Consequently, h has countable support, and this was exactly what we had to establish. ■

First, let us quickly discuss the weird looking condition that (5.2.27) is countable for each $u \in \mathbb{R}$ in Theorem 5.2.40. If N_g maps BV into BV, the function $g(\cdot, u)$ belongs to BV and possesses one-sided limits at each point in $[0,1]$ for every fixed $u \in \mathbb{R}$. If, in addition, these limits coincide with the values of the function at the corresponding points, that is,

$$g(t,u) = \lim_{s \to t+} g(s,u) \quad \text{for } 0 \leq t < 1 \qquad \text{or} \qquad g(t,u) = \lim_{s \to t-} g(s,u) \quad \text{for } 0 < t \leq 1$$

and

$$\lim_{s \to 0+} g(s,u) = g(0,u) \qquad \text{and} \qquad \lim_{s \to 1-} g(s,u) = g(1,u)$$

holds true for each $u \in \mathbb{R}$, then the countability of (5.2.27) actually implies $g(t,u) = g(t,0)$ for all $t \in [0,1]$ and $u \in \mathbb{R}$ which means that g does not depend on u whatsoever! In particular, Theorem 5.1.20 is a special case of Theorem 5.2.40 which in turn covers the necessity in Theorem 4.1.12. Indeed, if N_g is a multiplication operator in BV, that is, $g(t,u) = h(t)u$ for some $h \in BV$, then the countability of (5.2.27) implies that $\mathrm{supp}(h)$ is countable. Observe that Example 5.2.38 is now also covered by Theorem 5.2.40.

Theorem 5.2.40 does not cover the Riesz spaces which we will consider now. In Theorem 4.1.13 we have seen that the multiplication operator $M_g : RBV_p \to RBV_p$ is compact if and only if $g \equiv 0$. In Theorem 5.1.20 we have shown that the composition operator $C_g : RBV_p \to RBV_p$ is compact if and only if g is constant. Thus, it is reasonable to suspect that a compact superposition operator $N_g : RBV_p \to RBV_p$ which enshrines the properties of both the multiplication and the composition operator should behave in a similar way. Indeed, our suspicion is correct.

Theorem 5.2.41. *Let $g : [0,1] \times \mathbb{R} \to \mathbb{R}$ be so that N_g maps RBV_p into itself. Then N_g is compact if and only if there is some $h \in RBV_p$ such that $g(\cdot, u) = h$ for all $u \in \mathbb{R}$.*

Proof. It is clear that if $g(\cdot, u) = h$ for some $h \in RBV_p$ and all $u \in \mathbb{R}$, then N_g is compact since $N_g x = h$ for all $x \in RBV_p$.

For the converse let $N_g : RBV_p \to RBV_p$ be well-defined and compact. We show that g does not depend on its second argument and hence has the predicted form. In order to do so fix $\tau \in [0,1]$ and $u \in \mathbb{R}$ and let

$$S := \limsup_{v\to 0} \left| \frac{g(\tau, u+v) - g(\tau, u)}{v} \right| .$$

Then there is a sequence (v_n) in $[-1,1]\backslash\{0\}$ converging to 0 such that

$$\lim_{n\to\infty} \left| \frac{g(\tau, u+v_n) - g(\tau, u)}{v_n} \right| = S.$$

Consider the functions $z_n : \mathbb{R} \to \mathbb{R}$, defined to be continuous and piecewise linear by $z_n(t) = u$ for $|\tau - t| \geq \eta_n$ and $z_n(\tau) = u + v_n$, where $\eta_n := |v_n|^{p/(p-1)}$; in particular, $\eta_n \to 0$ as $n \to \infty$. The functions $x_n := z_n|_{[0,1]}$ then form a bounded sequence in RBV_p, since

$$\int_0^1 |x_n'(t)|^p \, \mathrm{d}t \leq \int_{\tau-\eta_n}^{\tau+\eta_n} \left| \frac{v_n}{\eta_n} \right|^p \mathrm{d}t = 2 \frac{|v_n|^p}{|\eta_n|^{p-1}} = 2$$

which implies

$$\|x_n\|_{RBV_p} = \|x_n\|_\infty + \mathrm{RVar}_p(x_n)^{1/p} \leq |u| + |v_n| + 2^{1/p} \leq |u| + 1 + 2^{1/p} \quad \text{for } n \in \mathbb{N}.$$

Since the operator N_g is compact it is also bounded which implies that the function $g(\tau, \cdot)$ is continuous in $\mathbb{R}$ by Proposition 5.2.17. Moreover, the compactness of N_g also tells us that the functions $y_n := N_g x_n$ must have a subsequence (y_{n_k}) that converges in RBV_p to some $y \in RBV_p$. For fixed $t \in [0,1]\backslash\{\tau\}$ we have $y_n(t) = g(t, x_n(t)) = g(t, u)$ for sufficiently large n and hence $y(t) = g(t, u)$. At τ we have

$$y(\tau) = \lim_{k\to\infty} g\big(\tau, x_{n_k}(\tau)\big) = \lim_{k\to\infty} g(\tau, u + v_{n_k}) = g(\tau, u),$$

and here we used the continuity of $g(\tau, \cdot)$ and the fact that $v_{n_k} \to 0$ as $k \to \infty$. Thus, $y(t) = g(t, u)$ for all $t \in [0,1]$.

We now obtain

$$\begin{aligned}
\mathrm{RVar}_p(y_{n_k} - y)^{1/p} &\geq \frac{\left| y_{n_k}(\tau + \eta_{n_k}) - y(\tau + \eta_{n_k}) - y_{n_k}(\tau) + y(\tau) \right|}{\eta_{n_k}^{(p-1)/p}} \\
&= \frac{\left| g\big(\tau + \eta_{n_k}, x_{n_k}(\tau + \eta_{n_k})\big) - g\big(\tau + \eta_{n_k}, u\big) - g\big(\tau, x_{n_k}(\tau)\big) + g(\tau, u) \right|}{\eta_{n_k}^{(p-1)/p}} \\
&= \frac{\left| g\big(\tau, u + v_{n_k}\big) - g\big(\tau, u\big) \right|}{|v_{n_k}|}
\end{aligned}$$

for sufficiently larke $k \in \mathbb{N}$; if $\tau = 1$ we have to replace $\tau + \eta_{n_k}$ by $\tau - \eta_{n_k}$.
Since $\mathrm{RVar}_p(y_{n_k} - y)^{1/p}$ goes to 0 as $k \to \infty$ we obtain

$$0 = \lim_{k\to\infty} \left| \frac{g\big(\tau, u + v_{n_k}\big) - g\big(\tau, u\big)}{v_{n_k}} \right| = S.$$

This shows $\partial_2 g(\tau,u)=0$ for all $u\in\mathbb{R}$ and hence $g(\tau,u)=:h(\tau)$ for all $u\in\mathbb{R}$. Since τ was arbitrary and N_g maps RBV_p into itself we conclude $g(t,u)=h(t)$ for all $t\in[0,1]$ and $u\in\mathbb{R}$ with $h\in RBV_p$. ■

Let us now come back to the space BV. The natural question is now if the countability of (5.2.27) for each $u\in\mathbb{R}$ is in fact equivalent to compactness of N_g. Unfortunately, the answer is negative, and we give three examples to illustrate this. The first shows that the countability of (5.2.27) alone (without the requirement $N_g(BV)\subseteq BV$) is not sufficient to guarantee that we can always extract from $(N_g x_n)$ a subsequence converging in BV even if both (x_n) and $(N_g x_n)$ are bounded sequences in BV.

Example 5.2.42. Let $g(t,u):=\chi_{\{0\}}(t-u)$, that is,

$$g(t,u)=\begin{cases}1 & \text{for } t=u,\\ 0 & \text{for } t\neq u.\end{cases}$$

In particular, for $u=0$ we have $g(t,u)-g(t,0)=0$ for all $t\in[0,1]$, and for $u\neq 0$ we get

$$g(t,u)-g(t,0)=\begin{cases}-1 & \text{for } t=0,\\ 1 & \text{for } t=u,\\ 0 & \text{for } t\in[0,1]\backslash\{0,u\},\end{cases}$$

and thus $\operatorname{supp}\big(g(\cdot,u)-g(\cdot,0)\big)\in\big\{\emptyset,\{0\},\{0,u\}\big\}$ which shows that (5.2.27) is even finite for each $u\in\mathbb{R}$. The functions $x_n(t):=t_n\chi_{\{t_n\}}(t)$ with $t_n:=1/(2n)$ for all $n\in\mathbb{N}$ (or any other bounded sequence of pairwise distinct numbers $t_n\in(0,1)$) form a bounded sequence in BV, since $\|x_n\|_{BV}=3t_n$ for all $n\in\mathbb{N}$. Moreover, since $x_n(t)=t$ if and only if $t\in\{0,t_n\}$ we have

$$N_g x_n(t)=\begin{cases}1 & \text{for } t\in\{0,t_n\},\\ 0 & \text{for } t\in(0,1]\backslash\{t_n\},\end{cases}$$

and so $(N_g x_n)$ is also a bounded sequence in BV with $\|N_g x_n\|_{BV}=4$ for all $n\in\mathbb{N}$. However, $(N_g x_n)$ cannot have a subsequence converging in BV, since it cannot have a Cauchy subsequence in BV. This is, because for $m\neq n$ we have

$$N_g x_m(t)-N_g x_n(t)=\begin{cases}1 & \text{for } t=t_m,\\ -1 & \text{for } t=t_n,\\ 0 & \text{for } t\in[0,1]\backslash\{t_m,t_n\},\end{cases}$$

and thus $\|N_g x_m-N_g x_n\|_{BV}=5$ for all $m\neq n$.
Finally, N_g does not map BV into BV. To see this, consider the function $x(t)=\varphi_{3,0,1}(t)+t$ which was defined in (1.1.1). It is easy to verify that x is continuously differentiable on $[0,1]$ and hence belongs to BV. Moreover, $x(t)=t$ if and only if $t\in\{0\}\cup\{1/(\pi n)\mid n\in\mathbb{N}\}=:A$. But x is mapped by N_g into the function χ_A which is not an element of BV. ◇

The next example shows that even an operator N_g which maps BV into BV with countable set (5.2.27) for each $u \in \mathbb{R}$ may not be compact.

Example 5.2.43. Define $g : [0,1] \times \mathbb{R} \to \mathbb{R}$ by

$$g(t,u) = \begin{cases} n & \text{for } t = 0, u = \frac{1}{n}, n \in \mathbb{N}, \\ 0 & \text{otherwise.} \end{cases}$$

For any function $x : [0,1] \to \mathbb{R}$ we then have $g(\cdot, x) = n\chi_{\{0\}}$ if $x(0) = 1/n$ for some $n \in \mathbb{N}$ and $g(\cdot, x) = \mathbb{0}$ if $x(0) \neq 1/n$ for all $n \in \mathbb{N}$. Clearly, the sets (5.2.27) contain at most $t = 0$ for each $u \in \mathbb{R}$, and N_g maps any space of real-valued functions on $[0,1]$ into BV. But N_g as an operator from BV into itself is not compact, because the constant functions $x_n \equiv 1/n$ form a bounded (even convergent) sequence in BV, but the functions $N_g x_n = n\chi_{\{0\}}$ cannot have a convergent subsequence in BV. Note that N_g is also discontinuous and unbounded. ♢

The last example of this series is more complicated but presents a bounded but not compact operator $N_g : BV \to BV$ generated by a function g that has finite sets (5.2.27) for each $u \in \mathbb{R}$.

Example 5.2.44. For $n \in \mathbb{N}$ put $A_n := \{k/2^n \mid k \in \{1, \dots, 2^n - 1\}\}$ and define the functions $g_n := 2^{-n}\chi_{A_n}$. Consider $g : [0,1] \times \mathbb{R} \to \mathbb{R}$, defined by

$$g(t,u) = \begin{cases} g_n(t) & \text{for } u = 2^{-n}, n \in \mathbb{N}, \\ 0 & \text{otherwise.} \end{cases}$$

Figure 5.2.5 shows those points $(t,u) \in [0,1] \times [0,1/2]$ at which g is not zero. At these points the value of g is u.

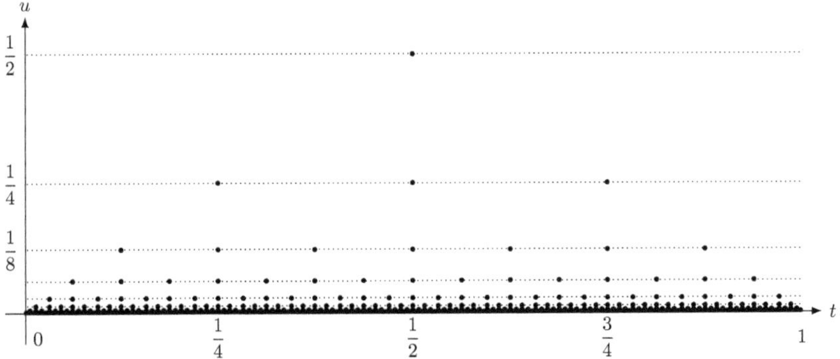

Figure 5.2.5: The points at which g is not zero.

We first show that g generates a bounded operator $N_g : BV \to BV$ which is the hardest part. In order to do so fix $x \in BV$ and set $y(t) := g(t, x(t))$. Then $\|y\|_\infty \leq 1/2$, and

$\mathrm{Var}(y)$ being finite or infinite is given by

$$\mathrm{Var}(y) = 2 \sum_{t \in \mathrm{supp}(y)} y(t)$$

according to Proposition 1.1.8, because g has countable support, and so has y. If $\mathrm{supp}(y)$ is finite, $y \in BV$ and we are done. We therefore assume that $\mathrm{supp}(y)$ is infinite. Note that then for each such $t \in \mathrm{supp}(y)$ we find $m \in \mathbb{N}$ and $k \in \{1, \ldots, 2^m - 1\}$ such that

$$t = \frac{k}{2^m} \quad \text{and} \quad x(t) = \frac{1}{2^m}.$$

We now pick $n \in \mathbb{N}$ and pairwise distinct $t_0, \ldots, t_n \in \mathrm{supp}(y)$; without loss of generality we may assume $0 \leq t_0 < \cdots < t_n \leq 1$. Defining $u_j := x(t_j)$ and $J := \{j \in \{1, \ldots, n\} \mid u_{j-1} \neq u_j\}$ gives

$$\sum_{j=1}^{n} y(t_j) = \sum_{j=1}^{n} g(t_j, u_j) = \sum_{j \in J} g(t_j, u_j) + \sum_{j \notin J} g(t_j, u_j). \tag{5.2.29}$$

To estimate the last sum in (5.2.29) we fix $j \in \{1, \ldots, n\} \backslash J$. Then $u_{j-1} = u_j$, and we find numbers $m, k_{j-1}, k_j \in \mathbb{N}$, $k_{j-1} < k_j$, such that $t_{j-1} = k_{j-1}/2^m$, $t_j = k_j/2^m$ and $u_{j-1} = u_j = 1/2^m$. Then

$$g(t_j, u_j) = \frac{1}{2^m} \leq \frac{k_j - k_{j-1}}{2^m} = t_j - t_{j-1}$$

and consequently

$$\sum_{j \notin J} g(t_j, u_j) \leq \sum_{j \notin J} t_j - t_{j-1} \leq 1. \tag{5.2.30}$$

We now take care of the second to last sum in (5.2.29) and fix $j \in J$. Since $u_{j-1} \neq u_j$, either $u_{j-1} \geq 2u_j$ or $u_{j-1} \leq u_j/2$. In the first case we get $u_{j-1} - u_j \geq u_j$ and in the latter we obtain $u_j - u_{j-1} \geq u_j/2$. But in any case we have

$$u_j \leq 2|u_{j-1} - u_j|,$$

and this leads to

$$\sum_{j \in J} g(t_j, u_j) \leq 2 \sum_{j \in J} |u_{j-1} - u_j| \leq 2\,\mathrm{Var}(x). \tag{5.2.31}$$

Thus, putting (5.2.29), (5.2.30) and (5.2.31) together yields

$$\sum_{j=1}^{n} y(t_j) \leq 2\,\mathrm{Var}(x) + 1,$$

and since n has been picked arbitrarily, this is true for any $n \in \mathbb{N}$. Consequently,

$$\mathrm{Var}(y) \leq 4\,\mathrm{Var}(x) + 2.$$

Finally,

$$\|N_g x\|_{BV} = \|y\|_{BV} = \|y\|_\infty + \mathrm{Var}(y) \leq 4\,\|x\|_{BV} + 5/2$$

which shows that N_g is indeed a bounded operator from BV into itself. However, N_g is not compact: The sequence (x_n) of constant functions $x_n \equiv 2^{-n}$ is mapped into the sequence (g_n) with $\|g_n\|_\infty = 2^{-n} \to 0$ as $n \in \mathbb{N}$. Therefore, any subsequence of (g_n) must (uniformly) converge to $\mathbb{0}$. However, such a subsequence cannot converge with respect to the BV-norm, since

$$\mathrm{Var}(g_n) = 2^{1-n}\#A_n = \frac{2^n - 1}{2^{n-1}} = 2 - \frac{1}{2^{n-1}} \geq 1 \quad \text{for } n \in \mathbb{N}.$$

Consequently, $N_g : BV \to BV$ cannot be compact. ♢

We now give some examples of compact superposition operators in BV. First, let us mention that if N_g has the form

$$N_g = C_f \circ M_\mu \circ C_h, \tag{5.2.32}$$

that is,

$$N_g x(t) = f\Big(\mu(t)h\big(x(t)\big)\Big) \quad \text{for } 0 \leq t \leq 1,$$

where C_f and C_h are composition operators generated by functions $f, h : \mathbb{R} \to \mathbb{R}$, respectively, and M_μ is a multiplication operator generated by a function $\mu : [0,1] \to \mathbb{R}$, then the operator $N_g : BV \to BV$ is compact if $f, h \in Lip_{loc}(\mathbb{R})$ and $\mu \in BV \cap \mathcal{S}_c$. Indeed, if $h \in Lip_{loc}(\mathbb{R})$, then C_h is bounded by Theorem 5.1.19, and $\mu \in BV \cap \mathcal{S}_c$ implies that M_μ is compact by Theorem 4.1.12. Consequently, $M_\mu \circ C_h$ is compact in BV. Again, $f \in Lip_{loc}(\mathbb{R})$ yields together with Theorem 5.1.24 that C_f is continuous in BV and so in total N_g is compact as an operator from BV into itself.

Based on these general observations we give now two examples of compact superposition operators $N_g : BV \to BV$. The first one is generated by a function g in separated variables which cannot be written as a multiplication operator. The second one is generated by a function g which cannot be written in separated variables.

Example 5.2.45. Let $g(t,u) = \chi_{\mathbb{Q}\cap[0,1]}(t)u^2$ be given in separated variables. Writing $f(u) = u$, $\mu(t) = \chi_{\mathbb{Q}\cap[0,1]}(t)$ and $h(u) = u^2$ we have $f, h \in Lip_{loc}(\mathbb{R})$ and $\mu \in BV \cap \mathcal{S}_c$, and we conclude by what we have observed above that $N_g = C_f \circ M_\mu \circ C_h$ maps BV into itself and is compact. However, since h is not linear, the operator N_g is not a multiplication operator. ♢

Example 5.2.46. The function $g(t,u) = \sin(\chi_{\mathbb{Q}\cap[0,1]}(t)u)$ cannot be written in separated variables. Setting $f(u) = \sin(u)$, $\mu(t) = \chi_{\mathbb{Q}\cap[0,1]}(t)$ and $h(u) = u$, we again have $f, h \in Lip_{loc}(\mathbb{R})$ and $\mu \in BV \cap \mathcal{S}_c$ and conclude that $N_g = C_f \circ M_\mu \circ C_h$ maps BV into itself and is compact. ♢

It is now time to summarize what we know about the disparities between the composition operator C_g and the superposition operator N_g in the space BV.

- Whenever the composition operator C_g maps BV into itself, it is automatically bounded; this is not true for the superposition operator N_g, see Example 5.2.13.
- Whenever the operator C_g maps BV into itself, it is automatically continuous; this is not true for the superposition operator N_g, see Example 5.2.20.
- The acting condition $C_g(BV) \subseteq BV$ holds precisely for locally Lipschitz continuous functions $g : \mathbb{R} \to \mathbb{R}$; the acting condition $N_g(BV) \subseteq BV$ may hold even for discontinuous functions $g : [0,1] \times \mathbb{R} \to \mathbb{R}$, see Example 5.2.18.
- The local Lipschitz continuity of $g : \mathbb{R} \to \mathbb{R}$ guarantees the continuity of C_g in BV; not even the global Lipschitz continuity of $g : [0,1] \times \mathbb{R} \to \mathbb{R}$ guarantees the continuity of N_g in BV, see Example 5.2.20.
- Only affine functions $g : \mathbb{R} \to \mathbb{R}$ generate globally Lipschitz continuous operators C_g in BV; this is not true for the superposition operator N_g for $g : [0,1] \times \mathbb{R} \to \mathbb{R}$, see Example 5.2.21.
- Only constant functions $g : \mathbb{R} \to \mathbb{R}$ generate compact operators C_g in BV; this is not true for the superposition operator N_g for $g : [0,1] \times \mathbb{R} \to \mathbb{R}$, see Example 5.2.37.

On the last pages of this section we make some remarks concerning so called *locally defined operators* which are defined as follows.

Definition 5.2.47. Let X and Y be spaces of real-valued functions on $[0,1]$. A (linear or nonlinear) operator $T : X \to Y$ is called *locally defined* if for each open interval $I \subseteq \mathbb{R}$ the implication

$$x|_{I\cap[0,1]} = y|_{I\cap[0,1]} \qquad \Longrightarrow \qquad (Tx)|_{I\cap[0,1]} = (Ty)|_{I\cap[0,1]}$$

holds for all functions $x, y \in X$.

Locally defined operators are also called *operators with memory* in the literature.

Clearly, any multiplication, composition and superposition operator is locally defined as long as it is well-defined. Indeed, from $x(t) = y(t)$ for $t \in I \cap [0,1]$ it follows that $M_g x(t) = g(t)x(t) = g(t)y(t) = M_g y(t)$ and $C_g x(t) = g(x(t)) = g(y(t)) = C_g y(t)$ and $N_g x(t) = g(t, x(t)) = g(t, y(t)) = N_g y(t)$ for all $t \in I \cap [0,1]$, respectively.

Here are three examples of operators from the space BV into itself which are not locally defined.

Example 5.2.48. The substitution operator $S_g : BV \to BV$, defined in (4.0.2) and generated by $g : [0,1] \to [0,1]$, $t \mapsto 1$, is not locally defined. The functions $x := \chi_{[1/2,1]}$ and $y := 2\chi_{[1/2,1]}$ both belong to BV and agree on the open interval $I := (0, 1/2)$, but $S_g x(t) = x(g(t)) = x(1) = 1 \neq 2 = y(1) = y(g(t)) = S_g y(t)$ for any $t \in [0,1]$. $\diamond$

Example 5.2.49. The integral operator $I_g : BV \to BV$, defined in (4.0.3) and generated by $g : [0,1] \times [0,1] \to \mathbb{R}$, $(t,s) \mapsto 1$, is not locally defined. The functions $x := \chi_{[0,1/2]}$ and $y := 2\chi_{[0,1/2]}$ both belong to BV and agree on the open interval $I := (1/2, 1)$, but

$$\begin{aligned} I_g x(t) &= \int_0^1 g(t,s)x(s)\,\mathrm{d}s = \int_0^{1/2} 1\,\mathrm{d}s = \frac{1}{2} \\ &\neq 1 = \int_0^{1/2} 2\,\mathrm{d}s = \int_0^1 g(t,s)y(s)\,\mathrm{d}s = I_g y(t) \end{aligned}$$

for any $t \in [0,1]$. ◇

Example 5.2.50. The Volterra operator $V_g : BV \to BV$, defined in (4.3.10) and generated by $g : [0,1] \times [0,1] \to \mathbb{R}$, given by

$$g(t,s) = \begin{cases} 1 & \text{for } 0 \leq s \leq t \leq 1, \\ 0 & \text{for } 0 \leq t < s \leq 1, \end{cases}$$

is not locally defined. The functions $x := \chi_{[0,1/2]}$ and $y := 2\chi_{[0,1/2]}$ both belong to BV and agree on the open interval $I := (1/2, 1)$, but

$$\begin{aligned} V_g x(t) &= \int_0^t g(t,s)x(s)\,\mathrm{d}s = \int_0^{1/2} 1\,\mathrm{d}s = \frac{1}{2} \\ &\neq 1 = \int_0^{1/2} 2\,\mathrm{d}s = \int_0^t g(t,s)y(s)\,\mathrm{d}s = V_g y(t) \end{aligned}$$

for any $t \in I$. ◇

The point is now that if X and Y are certain spaces of continuous functions, then *any* locally defined operator $T : X \to Y$ is in fact a superposition operator. For instance, the following has been established in [89].

Theorem 5.2.51. *For each locally defined operator $T : C \to C$ there is exactly one continuous function $g : [0,1] \times \mathbb{R} \to \mathbb{R}$ such that $T = N_g$.*

In [89] it was also shown that the same result remains true if C is replaced by C^1. Surprisingly, in this case, the function g may not be continuous anymore.

Similar representation results have been achieved for other function spaces. For instance, if $T : X \to C$ is locally defined on $X = C^n$, the space of n-times continuously differentiable functions, it was shown in [89] that then T has the form $Tx(t) = g(t, x(t), x'(t), \ldots, x^{(n)}(t))$ for some function $g \in [0,1] \times \mathbb{R}^{n+1} \to \mathbb{R}$. Other results in this direction are also known for $X = Lip_\alpha$, the space of Hölder continuous functions [157], and for X being the space of Whitney differentiable functions [109, 110, 156].

It turns out that we have a similar result if X is one of our BV-spaces intersected with the space C.

Theorem 5.2.52. *Let X be any of the spaces Lip, BV, WBV_p, YBV_φ, ΛBV or RBV_p. Then for any locally defined operator $T : X \cap C \to C$ there is exactly one function $g : [0,1] \times \mathbb{R} \to \mathbb{R}$ such that $T = N_g$.*

Theorem 5.2.52 was proven in [158] for $X = YBV_\varphi$ and hence also for $X = WBV_p$ and $X = BV$ and in [18] for $X = RBV_p$. Theorem 5.2.52 remains true even for other BV-type spaces like $X = WBV_{p(\cdot)}$, the space of functions of bounded Wiener variation with variable exponent $p : [0,1] \to (1,\infty)$ [69] . The proofs for all these results are very similar. We present here the proof for $X = Lip$ and for $X = \Lambda BV$ for which we could not find a reference.

Proof of Theorem 5.2.52 for $X = \Lambda BV$ and $X = Lip$. We begin by showing *Claim 1: The following implication holds.*

$$\forall x, y \in X \cap C\ \forall \tau \in (0,1): \quad x(\tau) = y(\tau) \quad \Longrightarrow \quad Tx(\tau) = Ty(\tau). \tag{5.2.33}$$

To this end, fix $x, y \in X \cap C$ and $\tau \in (0,1)$ with $x(\tau) = y(\tau)$. Then the function $z : [0,1] \to \mathbb{R}$, defined by

$$z(t) = \begin{cases} x(t) & \text{for } 0 \le t \le \tau, \\ y(t) & \text{for } \tau < t \le 1, \end{cases}$$

belongs to $X \cap C$. Moreover, z coincides on $[0,\tau]$ with x and on $[\tau, 1]$ with y. Therefore, there is some $\eta > 0$ such that the open intervals $I_\varepsilon := (\tau - 2\varepsilon, \tau)$ and $J_\varepsilon := (\tau, \tau + 2\varepsilon)$ are contained in $[0,1]$ and satisfy $x|_{I_\varepsilon} = z|_{I_\varepsilon}$ and $y|_{J_\varepsilon} = z|_{J_\varepsilon}$ for all $\varepsilon \in (0,\eta)$. Since T is locally defined on $X \cap C$ it follows that

$$(Tx)|_{I_\varepsilon} = (Tz)|_{I_\varepsilon} \quad \text{and} \quad (Ty)|_{J_\varepsilon} = (Tz)|_{J_\varepsilon} \qquad \text{for all } \varepsilon \in (0,\eta).$$

But this implies

$$Tx(\tau - \varepsilon) = Tz(\tau - \varepsilon) \quad \text{and} \quad Ty(\tau + \varepsilon) = Tz(\tau + \varepsilon) \qquad \text{for all } \varepsilon \in (0,\eta),$$

and since Tx, Ty and Tz are continuous at τ we conclude by letting $\varepsilon \to 0$ that $Tx(\tau) = Tz(\tau) = Ty(\tau)$. Thus, (5.2.33) is true, indeed.

Claim 2: For $x, y \in X \cap C$ with $x(0) = y(0)$ there are sequences (s_n) and (t_n) in $(0,1]$ both decreasing to 0 and a function $z \in X \cap C$ such that $z(s_n) = x(s_n)$ and $z(t_n) = y(t_n)$ for all $n \in \mathbb{N}$.

Claim 2 follows for $X = \Lambda BV$ from [158]. Therein it was shown that z can be constructed in such a way that $z \in BV \cap C$, even if x and y are merely continuous without having bounded Waterman variation.

We now show Claim 2 for $X = Lip$. To this end, let $s_n := 2^{2-2n}$ and $t_n := 2^{1-2n}$ for $n \in \mathbb{N}$. Then both (s_n) and (t_n) are sequences in $(0,1]$ that converge to 0 so that $s_{n+1} < t_n < s_n$ for all $n \in \mathbb{N}$. Consider the function $z : [0,1] \to \mathbb{R}$, defined by $z(0) = x(0) = y(0)$ as well as by $z(s_n) = x(s_n)$, $z(t_n) = y(t_n)$ and linear between s_{n+1} and t_n and between t_n and s_n for all $n \in \mathbb{N}$; Figure 5.2.6 shows how the function z (thick black lines) "zigzags" between x and y.

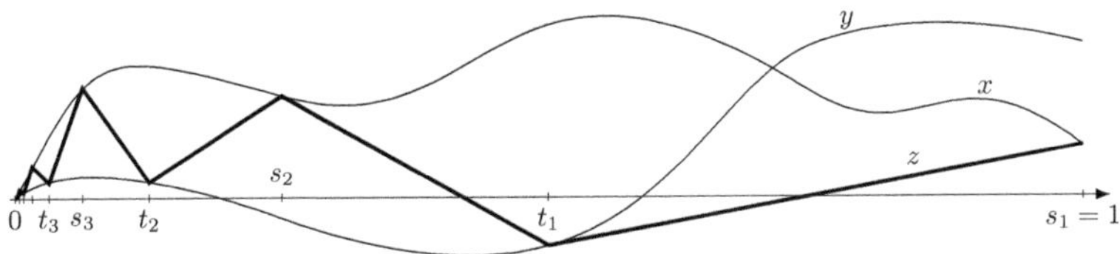

Figure 5.2.6: The function z "zigzagging" between x and y.

Since $x, y \in Lip$ we also have $x - y \in Lip$, and since $x(0) - y(0) = 0$,

$$|x(s) - y(s)| \leq \Big(\operatorname{lip}(x) + \operatorname{lip}(y)\Big)s \quad \text{for } s \in [0,1].$$

Thus, for all $n \in \mathbb{N}$ we have

$$\begin{aligned}\frac{|z(s_n) - z(t_n)|}{s_n - t_n} &= \frac{|x(s_n) - y(t_n)|}{s_n - t_n} \leq \frac{|x(s_n) - x(t_n)|}{s_n - t_n} + \frac{|x(t_n) - y(t_n)|}{s_n - t_n} \\ &\leq \operatorname{lip}(x) + \Big(\operatorname{lip}(x) + \operatorname{lip}(y)\Big)\frac{t_n}{s_n - t_n} = 2\operatorname{lip}(x) + \operatorname{lip}(y)\end{aligned}$$

and

$$\begin{aligned}\frac{|z(t_n) - z(s_{n+1})|}{t_n - s_{n+1}} &= \frac{|y(t_n) - x(s_{n+1})|}{t_n - s_{n+1}} \leq \frac{|y(t_n) - x(t_n)|}{t_n - s_{n+1}} + \frac{|x(t_n) - x(s_{n+1})|}{t_n - s_{n+1}} \\ &\leq \Big(\operatorname{lip}(x) + \operatorname{lip}(y)\Big)\frac{t_n}{t_n - s_{n+1}} + \operatorname{lip}(x) = 3\operatorname{lip}(x) + 2\operatorname{lip}(y).\end{aligned}$$

Consequently, $z \in X \cap C$ with $\operatorname{lip}(z) \leq 3\operatorname{lip}(x) + 2\operatorname{lip}(y)$, and Claim 2 is established.

Claim 3: The implication 5.2.33 holds also for $\tau \in \{0, 1\}$.
Let $\tau = 0$ (the argument for $\tau = 1$) is similar. Fix $x, y \in X \cap C$ with $x(0) = y(0)$. By Claim 2 there are sequences (s_n) and (t_n) in $(0, 1]$ both decreasing to 0 and a function $z \in X \cap C$ such that

$$z(s_n) = x(s_n) \quad \text{and} \quad z(t_n) = y(t_n) \qquad \text{for all } n \in \mathbb{N}.$$

Without loss of generality we may assume that $0 < s_n, t_n < 1$ for all $n \in \mathbb{N}$. Then by Claim 1 it follows that

$$Tx(s_n) = Tz(s_n) \quad \text{and} \quad Ty(t_n) = Tz(t_n) \qquad \text{for all } n \in \mathbb{N}.$$

Finally, the continuity of Tx, Ty and Tz yields $Tx(0) = Tz(0) = Ty(0)$.

Claim 4: There is exactly one function $g : [0,1] \times \mathbb{R} \to \mathbb{R}$ such that $T = N_g$.
Define $g : [0,1] \times \mathbb{R} \to \mathbb{R}$ which generates the desired superposition operator by

$$g(t, u) := T(u\mathbb{1})(t).$$

For any function $x \in X \cap C$ and fixed $\tau \in [0,1]$ we have $x(\tau) = \big(x(\tau)\mathbb{1}\big)(\tau)$, where both x and $x(\tau)\mathbb{1}$ belong to $X \cap C$. We get with the help (5.2.33) and Claim 3,

$$g\big(\tau, x(\tau)\big) = T\big(x(\tau)\mathbb{1}\big)(\tau) = Tx(\tau).$$

This shows the existence of g. If $h : [0,1] \times \mathbb{R} \to \mathbb{R}$ is another function with $T = N_h$, then for any $t \in [0,1]$ and $u \in \mathbb{R}$,

$$g(t,u) = T(u\mathbb{1})(t) = h\Big(t, (u\mathbb{1})(t)\Big) = h(t,u).$$

This shows the uniqueness of g and completes the proof. ■

We end this section with two examples of locally defined operators which cannot be represented as superposition operators.

Example 5.2.53. Define $T : BV \to R$ by $Tx := x^{\#}$, where $x^{\#}$ denotes the right regularization of x defined in (1.1.18). To show that T is locally defined, fix $x, y \in BV$ and an open interval $I \subseteq \mathbb{R}$ such that $x|_{I\cap[0,1]} = y|_{I\cap[0,1]}$. Without loss of generality we may assume that $I \cap [0,1] \neq \emptyset$. Fix $\tau \in I \cap [0,1]$. Since I is open and x and y coincide on $I \cap [0,1]$ they also coincide in a neighborhood of τ. In particular, $x^{\#} = y^{\#}$ in that neighborhood. Since τ was chosen arbitrarily in $I \cap [0,1]$ we conclude $Tx = x^{\#} = y^{\#} = Ty$ on $I \cap [0,1]$. Consequently, T is locally defined.
However, T is not a superposition operator. To see this assume the opposite, that is, assume that there is a function $g : [0,1] \times \mathbb{R} \to \mathbb{R}$ such that $x^{\#}(t) = Tx(t) = g(t, x(t))$ for all $x \in BV$ and $t \in [0,1]$. If we take $x = \chi_{\{0\}} \in BV$ then $g(0,1) = g(0, x(0)) = x^{\#}(0) = 0$, but on the other hand $g(0,1) = g(0, \mathbb{1}(0)) = \mathbb{1}^{\#}(0) = 1$, a contradiction. ◇

This example also shows that we cannot drop the requirement in Theorem 5.2.52 that the domain and the target space of T are subsets of C. The same is true if $X \cap C$ in Theorem 5.2.52 is replaced by a space that contains all functions that agree with a continuous function almost everywhere, even if the operator T maps this space into C. This is illustrated by our last example in this chapter.

Example 5.2.54. Let $X := \{y + z \mid y \in C, z = \mathbb{0} \text{ almost everywhere}\}$, and define the operator $T : X \to C$ by

$$Tx(t) = \lim_{\delta \to 0} \frac{1}{\delta} \int_t^{t+\delta} x(s)\, \mathrm{d}s. \tag{5.2.34}$$

Then T is well-defined, because $X \subseteq L_1$, and if $x \in X$, then x can be written as $x = y + z$ for some $y \in C$ and $z : [0,1] \to \mathbb{R}$ with $z = \mathbb{0}$ almost everywhere. Such x is then mapped by T into the function y, because

$$Tx(t) = \lim_{\delta \to 0} \frac{1}{\delta} \int_t^{t+\delta} \Big(y(s) + z(s)\Big)\, \mathrm{d}s = \lim_{\delta \to 0} \frac{1}{\delta} \int_t^{t+\delta} y(s)\, \mathrm{d}s = y(t)$$

holds for all $t \in [0,1]$, where the last equality follows from Theorem 2.1.3 (b).
Moreover, it is clear that T is locally defined, because the integral in (5.2.34) needs to know only how x looks like in a neighborhood of t. But T is not a superposition operator. Otherwise we had $Tx(t) = g(t, x(t))$ for some function $g : [0,1] \times \mathbb{R} \to \mathbb{R}$ and all $t \in [0,1]$. For the two functions $\mathbb{1}$ and $\chi_{\{0\}}$ both belonging to X this would imply on the one hand $g(0,1) = g(0, \mathbb{1}(0)) = T\mathbb{1}(0) = \mathbb{1}(0) = 1$, but on the other hand $g(0,1) = g(0, \chi_{\{0\}}(0)) = T\chi_{\{0\}}(0) = \mathbb{0}(0) = 0$, a contradiction. ◇

We remark that the fixed points of the operator T in Example 5.2.54 are precisely all continuous functions.
If therein X is replaced by the space KH of all Kurzweil-Henstock integrable functions, then T is no longer well-defined, for two reasons: First, the limit in (5.2.34) may not exist at some $t \in [0,1]$. Second, the resulting function Tx does not have to be continuous. However, T can be replaced by the operator $\hat{T} : KH \to KH$, defined by

$$\hat{T}x(t) = \begin{cases} \lim\limits_{\delta \to 0} \dfrac{1}{\delta} \displaystyle\int_t^{t+\delta} x(s)\,\mathrm{d}s & \text{if the limit exists and is finite,} \\ 0 & \text{otherwise,} \end{cases}$$

which can be considered as an extension of T onto KH that is now well and still locally defined. It can be shown that $\hat{T}x = x$ almost everywhere [84]. In this setting the fixed points of $\hat{T}$ in the subspace

$$KH_0 := \left\{ x \in KH \mid \forall t \in [0,1] : \lim_{\delta \to 0} \frac{1}{\delta} \int_t^{t+\delta} x(s)\,\mathrm{d}s \ \text{ exists and is finite} \right\}$$

of those KH-integrable functions x for which the limit (5.2.34) exists and is finite are precisely those functions which have a primitive, and this is exactly the assertion of Theorem 2.1.3 (d). The fact that T and hence also $\hat{T}$ cannot be represented as superposition operators might be one tiny of the many reasons why the class Δ of functions with primitive is still entangled by some mysteries.

Chapter 6

Types of Convergence which Preserve Continuity

It is a well-known fact that a locally uniformly convergent sequence of continuous functions always has a continuous limit function. That is, the continuity is preserved under locally uniform convergence. In probably every Analysis course it is also pointed out by means of the example $f_n(x) = x^n$ for $x \in [0,1]$ that the locally uniform convergence cannot be replaced by pointwise convergence. However, continuity of the limit function is not equivalent to locally uniform convergence.

Example 6.0.1. Consider the continuous "hump"-functions $f_n(x) = n^2x^2 \exp\left(-n^2x^2\right)$ for $n \in \mathbb{N}$ and $x \in \mathbb{R}$. For $x \neq 0$ we have $f_n(x) \to 0$ as $n \to \infty$, and $f_n(0) = 0$ and $f_n(1/n) = 1/e$ for each $n \in \mathbb{N}$. Consequently, (f_n) converges pointwise but not locally uniformly on $\mathbb{R}$ to the continuous function $\mathbb{0}$. ◇

Note that these "hump"-functions form a bounded sequence with respect to the supremum norm that has no locally uniformly convergent subsequence.

This raises the question what additional assumption on top of pointwise convergence can be made in order to guarantee that the limit function of a sequence of continuous functions is always continuous. The first one who solved this problem was Arzelà in 1883, who introduced in [14] and [15] the term *piecewise uniform convergence*[1] and investigated series of continuous functions on compact real intervals which was also discussed in [155]. His term was later renamed into *quasi uniform convergence* and generalized and modified in many directions, for instance, for a discussion of the same question in topological spaces [2]. A detailed survey including Arzelà's original definition and the historical development of this type of convergence can be found in [38]. Nowadays, the term *quasi uniform convergence* seems to be used not consistently; we will give the definition of which we will make use at the beginning of Section 6.1 below.

[1]His original Italian notion is *convergenza uniforme a tratti*. In English literature, one sometimes finds the translation *uniform convergence by segments*.

In this chapter which will also be published as [134] we investigate - in addition to pointwise and locally uniform convergence - three further types of convergence in metric spaces which are, besides a variant of the classical *quasi uniform convergence*, also *semi uniform* and *continuously uniform* convergence. We give criteria under which a sequence converges in one of these types and keep our eyes on those which preserve continuity. In addition, we give some conditions on sequences of functions and their underlying spaces under which convergent subsequences can be extracted and recall that several types of convergence considered herein can be used to characterize compactness of the domains the functions under consideration live in. Eventually, we compare all five types of convergence with each other and give a summary of the relations among them at the end of Section 6.1. However, we will focus our attention on *semi uniform convergence* which will be of particular importance in Section 6.2.
Therein we are going to apply some of the results from Section 6.1 to the (autonomous) composition operators in the space BV which we have discussed in detail in Section 5.1. As we have seen in Proposition 5.1.1 there are known criteria under which such operators map the space BV into itself. Moreover, we gave criteria under which the composition operator is locally Lipschitz continuous (Theorem 5.1.21), globally uniformly continuous (Theorem 5.1.22) and uniformly continuous on bounded subsets of BV (Theorem 5.1.23). But as we have already mentioned at the end of Section 5.1 the question of whether the composition operator is automatically pointwise continuous in the space BV has an interesting history. Its positive answer has two quite technical proofs, the first of which given in [118] is almost 30 pages long, and the second was given quite recently in [96]. In Section 6.2 we will apply some of the theoretical results developed in Section 6.1 and present a new and short proof for this fact. We also give criteria under which sequences of composition operators converge locally uniformly and semi uniformly in the space BV.

6.1 Five Types of Convergence in Comparison

In what follows, let (X, d_X) and (Y, d_Y) be metric spaces. Just for the sake of completeness, let us start with recalling the definition of locally uniform convergence.

Definition 6.1.1. Let $f_n, f : X \to Y$ for $n \in \mathbb{N}$ be functions, and let $x \in X$ be fixed. We say that the sequence (f_n) converges *locally uniformly* to f at x if there is some $\delta > 0$ such that for each $\varepsilon > 0$ there is $N \in \mathbb{N}$ such that for all $n \geq N$ and $y \in X$ with $d_X(x, y) \leq \delta$ we have $d_Y\big(f_n(y), f(y)\big) \leq \varepsilon$.
Using quantifiers, this reads

$$\exists \delta > 0\ \forall \varepsilon > 0\ \exists N \in \mathbb{N}\ \forall n \geq N\ \forall y \in X : \quad d_X(x, y) \leq \delta\ \Rightarrow\ d_Y\big(f_n(y), f(y)\big) \leq \varepsilon.$$

Moreover, we say that (f_n) converges locally uniformly to f in X if (f_n) converges locally uniformly to f at each $x \in X$.

It is clear from this definition that locally uniform convergence implies pointwise convergence.

As written at the beginning of this chapter, a necessary and sufficient condition on top of pointwise convergence of a sequence (f_n) of continuous functions to guarantee that f is continuous is that the convergence is not only pointwise but also *quasi uniform*. Since there are many different definitions for this term, we state the definition we will work with here in detail. It is a pointwise version of Arzelà's original definition[2] and was also used in [54, Definition 1] under the name "almost uniform convergence"; for metric spaces it reads as follows.

Definition 6.1.2. Let $f_n, f : X \to Y$ for $n \in \mathbb{N}$ be functions, and let $x \in X$ be fixed. We say that the sequence (f_n) converges *quasi uniformly* to f at x if for each $\varepsilon > 0$ and each $N \in \mathbb{N}$ there are $\delta > 0$ and $n \geq N$ such that for all $y \in X$ with $d_X(x, y) \leq \delta$ we have $d_Y\big(f_n(y), f(y)\big) \leq \varepsilon$.
Using quantifiers, this reads

$$\forall \varepsilon > 0 \; \forall N \in \mathbb{N} \; \exists \delta > 0 \; \exists n \geq N \; \forall y \in X : \quad d_X(x, y) \leq \delta \;\Rightarrow\; d_Y\big(f_n(y), f(y)\big) \leq \varepsilon.$$

Moreover, we say that (f_n) converges quasi uniformly to f in X if (f_n) converges quasi uniformly to f at each $x \in X$.

Let us make five comments on this definition. First, locally uniform convergence clearly implies quasi uniform convergence, but quasi uniform convergence does not imply pointwise and hence also not locally uniform convergence:

Example 6.1.3. The functions $f_n \equiv (-1)^n$ for $n \in \mathbb{N}$ converge quasi uniformly in $\mathbb{R}$ to the constant function $\mathbb{1}$, but not pointwise and thus also not locally uniformly. ◇

Second, and more surprising, Example 6.1.3 also shows that the limit function of a quasi uniformly convergent sequence may not be unique! Indeed, $\tilde{f} = -\mathbb{1}$ is another quasi uniform limit (in the above Definition 6.1.2 just take even n for f and odd n for $\tilde{f}$); in particular, the quasi uniform convergence cannot be induced by a metric. Third, the functions $f_n \equiv ((-1)^n + 1)n$ for $n \in \mathbb{N}$ converge quasi uniformly to the zero function $\mathbb{0}$ and show that a quasi uniformly convergent sequence may neither be bounded nor a Cauchy sequence, not even pointwise. Fourth, the same example shows that an arbitrary subsequence of a quasi uniformly convergent sequence may not be quasi uniformly convergent anymore (consider the subsequence $f_{2n} \equiv 2n$). Lastly, a sequence which has a quasi uniformly convergent subsequence must be quasi uniformly convergent itself. Therefore, quasi uniform convergence behaves completely different than ordinary types of convergence and always requires caution when used.

Let us come back to continuity. With Definition 6.1.2 at hand one can show

Theorem 6.1.4. *Let (f_n) be a sequence of continuous functions $f_n : X \to Y$, let $f : X \to Y$ be a function and let $x \in X$ be fixed. Then the following statements are equivalent.*

(a) The sequence (f_n) converges quasi uniformly to f at x.

(b) The function f is continuous at x and $f(x)$ is a limit point of $(f_n(x))$.

[2] See [14] and [15] and also [38] for his original definition.

Proof. "(a)⇒(b)": Fix $\varepsilon > 0$. Since (f_n) converges quasi uniformly to f at x, we find for each $N \in \mathbb{N}$ some $\delta > 0$ and $m \geq N$ such that

$$d_X(x,y) \leq \delta \quad \Rightarrow \quad d_Y\big(f_m(y), f(y)\big) \leq \frac{\varepsilon}{3}.$$

In particular, for $y = x$, we obtain $d_Y(f_m(x), f(x)) \leq \frac{\varepsilon}{3}$, and this shows that $f(x)$ is a limit point of $(f_n(x))$.
Now, for $N = 1$ we pick δ and m accordingly and keep it fixed. Since f_m is continuous at x, we find some $\eta > 0$ such that

$$d_X(x,y) \leq \eta \quad \Rightarrow \quad d_Y\big(f_m(x), f_m(y)\big) \leq \frac{\varepsilon}{3}.$$

Thus, we obtain for $y \in X$ with $d_X(x,y) \leq \min\{\delta, \eta\}$ that

$$d_Y\big(f(x), f(y)\big) \leq d_Y\big(f(x), f_m(x)\big) + d_Y\big(f_m(x), f_m(y)\big) + d_Y\big(f_m(y), f(y)\big) \leq \varepsilon,$$

which shows the continuity of f at x.
"(b)⇒(a)": To show that (f_n) converges quasi uniformly to f at x, fix $\varepsilon > 0$ and $N \in \mathbb{N}$. Since $f(x)$ is a limit point of $(f_n(x))$, we find some $m \geq N$ such that

$$d_Y\big(f_m(x), f(x)\big) \leq \frac{\varepsilon}{3}.$$

Since f_m is continuous at x we find some $\eta > 0$ such that

$$d_X(x,y) \leq \eta \quad \Rightarrow \quad d_Y\big(f_m(x), f_m(y)\big) \leq \frac{\varepsilon}{3},$$

and since f is continuous at x we find some $\delta > 0$ such that

$$d_X(x,y) \leq \delta \quad \Rightarrow \quad d_Y\big(f(x), f(y)\big) \leq \frac{\varepsilon}{3}.$$

Thus, we obtain for $y \in X$ with $d_X(x,y) \leq \min\{\delta, \eta\}$ that

$$d_Y\big(f_m(y), f(y)\big) \leq d_Y\big(f_m(y), f_m(x)\big) + d_Y\big(f_m(x), f(x)\big) + d_Y\big(f(x), f(y)\big) \leq \varepsilon,$$

which shows the quasi uniform convergence of (f_n) at x. ■

Note that the additional requirement in (b) of Theorem 6.1.4, namely, that $f(x)$ is a limit point of $(f_n(x))$, is a natural requirement for the implication "(b)⇒(a)", and interconnects the limit function with the sequence.

Theorem 6.1.4 implies that the limit function of a *pointwise* convergent sequence of continuous functions is continuous if and only if the convergence is quasi uniform, and this is just Arzelà's original result mentioned at the beginning; in particular, the sequence (f_n) given in Example 6.0.1 converges quasi uniformly.
Moreover, Theorem 6.1.4 also says that the convergence cannot be quasi uniform if the pointwise limit function of a sequence of continuous functions is discontinuous. We illustrate this explicitly in the following

Example 6.1.5. The continuous functions $f_n : [0,1] \to \mathbb{R}$, defined by $f_n(x) = x^n$, converge pointwise to the discontinuous characteristic function $f := \chi_{\{1\}}$ on $[0,1]$. Theorem 6.1.4 now implies that the convergence cannot be quasi uniform at $x = 1$. This can be seen also directly: Note that

$$\left|f_n(y) - f(y)\right| = \begin{cases} y^n & \text{for } 0 \le y < 1, \\ 0 & \text{for } y = 1. \end{cases}$$

For $\varepsilon = \frac{1}{2}$, $N = 1$ and arbitrary $\delta > 0$ and $n \in \mathbb{N}$ we can pick $y := \max\{1-\delta/2, 2^{-1/n}\} \in (0,1)$. Then $|y-1| = 1-y \le \delta/2 < \delta$ and $|f_n(y) - f(y)| = y^n \ge \frac{1}{2} = \varepsilon$, and this shows that (f_n) does not converge quasi uniformly to f at $x = 1$. ◇

However, even if a sequence of everywhere discontinuous functions converges pointwise to a *continuous* function the convergence may not be quasi uniform:

Example 6.1.6. Denote by p_n the n-the prime number, and let

$$A_n := \left\{ \frac{m}{p_n^k} \mid m \in \mathbb{Z}, k \in \mathbb{N}, \gcd(m, p_n) = 1 \right\} \quad \text{for } n \in \mathbb{N}.$$

Then the sets A_n are all dense in $\mathbb{R}$. To see this, fix $n \in \mathbb{N}$, $x \ge 0$ and $\varepsilon > 0$. Choose $k \in \mathbb{N}$ so large that $p_n^k(x+\varepsilon) - p_n^k x = p_n^k \varepsilon \ge 2$. Then the interval $\left[p_n^k x,\, p_n^k(x+\varepsilon)\right]$ contains at least two consecutive integers, of which at least one is coprime to p_n. If we call this integer m, we obtain $p_n^k(x+\varepsilon) \ge m \ge p_n^k x$ and hence $m/p_n^k \in [x, x+\varepsilon] \cap A_n$, as desired. The same reasoning works for $x < 0$.
To prove that the A_n are pairwise disjoint, assume that $x \in A_{n_1} \cap A_{n_2}$ for some $n_1 \ne n_2$. Then there are $m_1, m_2 \in \mathbb{Z}, k_1, k_2 \in \mathbb{N}$ such that

$$\frac{m_1}{p_{n_1}^{k_1}} = x = \frac{m_2}{p_{n_2}^{k_2}} \quad \text{with} \quad \gcd(m_1, p_{n_1}) = 1 = \gcd(m_2, p_{n_2}).$$

This implies $m_1 p_{n_2}^{k_2} = m_2 p_{n_1}^{k_1}$, and thus m_1 must be dividable by p_{n_1}, as p_{n_1} and p_{n_2} are distinct prime numbers, contradicting $\gcd(m_1, p_{n_1}) = 1$.
We now consider the everywhere discontinuous functions $f_n := \chi_{A_n}$ on $\mathbb{R}$; these satisfy $f_n(0) = 0$ for all $n \in \mathbb{N}$, as $0 \notin A_n$ for all $n \in \mathbb{N}$. Moreover, each fixed $x \in \mathbb{R}\backslash\{0\}$ belongs to at most one A_n as the sets A_n are pairwise disjoint. Thus, $f_n(x) \ne 0$ for that x and at most one $n \in \mathbb{N}$. This shows that (f_n) converges pointwise to the everywhere continuous function $\mathbb{0}$. However, the convergence to $\mathbb{0}$ can nowhere be quasi uniform, since for each $x \in \mathbb{R}$, $n \in \mathbb{N}$ and $\delta > 0$ we have

$$\sup_{|x-y| \le \delta} |f_n(y)| = \sup_{|x-y| \le \delta} |\chi_{A_n}(y)| = 1,$$

as each A_n is dense in $\mathbb{R}$. ◇

Also observe that Example 6.1.5 shows that only the second assertion in Theorem 6.1.4 (b), that is, the limit point part, alone does not suffice to imply (a), while Example 6.1.6 shows that only assertion (b) in Theorem 6.1.4 without the overall assumption that the sequence consists of continuous functions does also not suffice to imply (a).

A tiny modification of Definition 6.1.2 leads to another type of convergence which in contrast to quasi uniform convergence does imply pointwise convergence.

Definition 6.1.7. Let $f_n, f : X \to Y$ for $n \in \mathbb{N}$ be functions, and let $x \in X$ be fixed. We say that the sequence (f_n) converges *semi uniformly* to f at $x \in X$ if for each $\varepsilon > 0$ there exist $N \in \mathbb{N}$ and $\delta > 0$ such that for all $n \geq N$ and $y \in X$ with $d_X(x, y) \leq \delta$ we have $d_Y\big(f_n(y), f(y)\big) \leq \varepsilon$.
Using quantifiers, this reads

$$\forall \varepsilon > 0\ \exists N \in \mathbb{N}\ \exists \delta > 0\ \forall n \geq N\ \forall y \in X : \quad d_X(x, y) \leq \delta \ \Rightarrow\ d_Y\big(f_n(y), f(y)\big) \leq \varepsilon.$$

We say that (f_n) converges *semi uniformly* to f in X if (f_n) converges semi uniformly to f at each $x \in X$.

This new definition has some benefits. First, the major difference between Definition 6.1.1 and Definition 6.1.7 is that the δ in the definition of semi uniform convergence may depend on ε. Second, it is clear from that definition that semi uniform convergence implies both pointwise and quasi uniform convergence and hence always ensures uniqueness of the limit function. In particular, the two sequences constructed in the Examples 6.1.5 and 6.1.6 can neither converge semi uniformly nor locally uniformly, since they do not converge quasi uniformly. Moreover, it is also clear from the definition that any subsequence of a semi uniformly convergent sequence is again semi uniformly convergent. Thus, the definition of semi uniform convergence recovers most of the familiar properties a "nice" type of convergence should have. But we pay a price for this. Semi uniform convergence is indeed stronger than quasi uniform convergence which is again shown by the functions $f_n \equiv (-1)^n$ of Example 6.1.3. They converge quasi uniformly to either $\mathbb{1}$ or $-\mathbb{1}$, but they cannot converge semi uniformly, because they do not even converge pointwise. Thus, semi uniform convergence requires more restrictive assumptions on the sequence. However, semi uniform convergence is weaker than locally uniform convergence, and we give three examples for this, the first one being a sequence of functions which are everywhere discontinuous on $\mathbb{R}$.

Example 6.1.8. Let the sets A_n be defined for all $n \in \mathbb{N}$ as in Example 6.1.6 and consider the modifications $g_n : \mathbb{R} \to \mathbb{R}$ of the functions in the same example, defined by

$$g_n(x) = \begin{cases} x\chi_{A_n}(x) & \text{for } x \neq 0, \\ \frac{1}{n} & \text{for } x = 0. \end{cases}$$

Since the χ_{A_n} converge pointwise to $\mathbb{0}$, as we have seen in Example 6.1.6, and $g_n(0) = \frac{1}{n} \to 0$ as $n \to \infty$, also (g_n) converges pointwise to $\mathbb{0}$. Moreover, since each A_n is dense in $\mathbb{R}$, for fixed $n \in \mathbb{N}$ and $\delta > 0$ we have

$$\sup_{|y| \leq \delta} |g_n(y)| = \max\left\{\frac{1}{n}, \sup_{|y| \leq \delta} |y\chi_{A_n}(y)|\right\} = \max\left\{\frac{1}{n}, \delta\right\},$$

and (g_n) fails to converge locally uniformly at $x = 0$.

If, however, for fixed $\varepsilon > 0$ we choose $\delta = \varepsilon$, then we have for $|y| \leq \delta$ and $n \geq \frac{1}{\delta}$,

$$|g_n(y)| \leq \max\left\{\frac{1}{n}, \delta\right\} = \varepsilon,$$

and this shows that (g_n) converges semi uniformly to $\mathbb{0}$ at $x = 0$. But for the same reason as in Example 6.1.6 at every other point $x \neq 0$ the convergence again cannot be semi uniform. ◊

The next example consists of functions which are everywhere continuous and, exactly as in Example 6.1.8, converge semi uniformly but not locally uniformly to $\mathbb{0}$ at $x = 0$.

Example 6.1.9. For $n \in \mathbb{N}$ define $g_n : \mathbb{R} \to \mathbb{R}$ by

$$g_n(x) = \begin{cases} 2^{n+2}x(1-2^n x) & \text{for } 0 \leq x \leq 2^{-n}, \\ 0 & \text{otherwise.} \end{cases}$$

Then each g_n is continuous everywhere on $\mathbb{R}$ and attains values different from zero only for $x \in (0, 2^{-n})$ and its global maximum 1 at $x = 2^{-n-1}$. Moreover, the sequence (g_n) converges pointwise on $\mathbb{R}$ to $\mathbb{0}$. We now consider the sequence (f_n) of functions $f_n : \mathbb{R} \to \mathbb{R}$, defined by

$$f_n(x) = \begin{cases} x g_n\left(2^k x - 1\right) & \text{for } 2^{-k} < x \leq 2^{-k+1} \text{ and } k \in \mathbb{N}, \\ 0 & \text{for } x \in \mathbb{R}\backslash(0,1]. \end{cases}$$

Then the functions f_n while being continuous on $\mathbb{R}$ for each $n \in \mathbb{N}$ converge pointwise to 0.

For fixed $\delta > 0$ we find $k \in \mathbb{N}$ such that $2^{-k+1} \leq \delta$ and obtain for all $n \in \mathbb{N}$,

$$\sup_{|y|\leq\delta} |f_n(y)| \geq \sup_{2^{-k}<y\leq 2^{-k+1}} y g_n\left(2^k y - 1\right) \geq 2^{-k},$$

which shows that (f_n) cannot converge locally uniformly to $\mathbb{0}$ at $x = 0$.

However, the convergence is semi uniform at $x = 0$. To see this fix $\varepsilon > 0$ and choose $\delta = \varepsilon$. Then we obtain

$$\sup_{|y|\leq\delta} |f_n(y)| \leq \sup_{|y|\leq\delta} |y| = \delta = \varepsilon$$

which holds for all $n \in \mathbb{N}$. ◊

Since semi uniform convergence implies quasi uniform convergence, the two sequences constructed in the Examples 6.1.8 and 6.1.9 converge quasi uniformly, as well.

We will give a third example showing that semi uniform convergence is indeed weaker than locally uniform convergence in Example 6.2.9 in the next section.

As we have seen, semi uniform convergence is situated between locally uniform and pointwise convergence. The following result gives a comprehensive and pointwise criterion for semi uniform convergence of continuous functions.

Theorem 6.1.10. *Let (f_n) be a sequence of continuous functions $f_n : X \to Y$, let $f : X \to Y$ be a function and let $x \in X$ be fixed. Then the following statements are equivalent.*

(a) The sequence (f_n) converges semi uniformly to f at x.

(b) The sequence (f_n) is equicontinuous at x, the function f is continuous at x and $(f_n(x))$ converges to $f(x)$.

Proof. "(a)⇒(b)": Since semi uniform convergence implies quasi uniform convergence, we obtain from Theorem 6.1.4 that f is continuous at x. To show that (f_n) is equicontinuous at x, fix $\varepsilon > 0$ and pick according to the continuity of f at x some $\delta_\infty > 0$ such that

$$d_X(x,y) \le \delta_\infty \quad \Rightarrow \quad d_Y\big(f(x), f(y)\big) \le \frac{\varepsilon}{3}. \tag{6.1.1}$$

Since (f_n) converges semi uniformly to f at x, we find $\delta > 0$ and $N \in \mathbb{N}$ such that

$$\forall n \ge N: \quad d_X(x,y) \le \delta \quad \Rightarrow \quad d_Y\big(f_n(y), f(y)\big) \le \frac{\varepsilon}{3}. \tag{6.1.2}$$

Finally, since each function f_n is continuous at x, we find for each $n < N$ some $\delta_n > 0$ such that

$$d_X(x,y) \le \delta_n \quad \Rightarrow \quad d_Y\big(f_n(x), f_n(y)\big) \le \varepsilon.$$

Letting $\eta := \min\{\delta_1, \delta_2, \dots, \delta_{N-1}, \delta, \delta_\infty\}$ implies for $n < N$ and $y \in X$ with $d_X(x,y) \le \eta$ that $d_Y(f_n(x), f_n(y)) \le \varepsilon$. For $n \ge N$ we obtain for those y

$$d_Y\big(f_n(x), f_n(y)\big) \le d_Y\big(f_n(x), f(x)\big) + d_Y\big(f(x), f(y)\big) + d_Y\big(f(y), f_n(y)\big) \le \varepsilon$$

and hence the equicontinuity of (f_n) at x. Here we have estimated the first and last term by (6.1.2) and the middle term by (6.1.1). That $(f_n(x))$ converges to $f(x)$ follows also from (6.1.2). Consequently, (b) is proven.

"(b)⇒(a)": To show that (f_n) converges semi uniformly to f at x, fix $\varepsilon > 0$. Since $(f_n(x))$ converges to $f(x)$, we find some $N \in \mathbb{N}$ such that

$$\forall n \ge N: \quad d_Y\big(f_n(x), f(x)\big) \le \frac{\varepsilon}{3}. \tag{6.1.3}$$

Since (f_n) is equicontinuous at x we find some $\eta > 0$ such that

$$\forall n \in \mathbb{N}: \quad d_X(x,y) \le \eta \quad \Rightarrow \quad d_Y\big(f_n(x), f_n(y)\big) \le \frac{\varepsilon}{3}, \tag{6.1.4}$$

and since f is continuous at x we find some $\delta > 0$ such that

$$d_X(x,y) \le \delta \quad \Rightarrow \quad d_Y\big(f(x), f(y)\big) \le \frac{\varepsilon}{3}. \tag{6.1.5}$$

Thus, we obtain for $y \in X$ with $d_X(x,y) \le \min\{\delta, \eta\}$ and $n \ge N$ that

$$d_Y\big(f_n(y), f(y)\big) \le d_Y\big(f_n(y), f_n(x)\big) + d_Y\big(f_n(x), f(x)\big) + d_Y\big(f(x), f(y)\big) \le \varepsilon.$$

Here we have used (6.1.4) for the first, (6.1.3) for the second and (6.1.5) for the last term. This shows (a) and completes the proof. ■

Let us compare the requirements (b) of Theorem 6.1.4 and of Theorem 6.1.10. Quasi uniform convergence at a point x implies that the limit function is continuous at x, and that the limit function can be approximated at x at least by a subsequence. From semi uniform convergence at x, however, we get in addition that the limit function can be approximated by the entire sequence at x. But more important is that the sequence itself must be equicontinuous at x. Thus, in contrast to quasi uniform convergence, the limit function of a pointwise convergent sequence of continuous functions is continuous if - but not necessarily only if - the convergence is semi uniform.
Note that since the pointwise limit of an everywhere equicontinuous sequence of functions is continuous, the "global" version of Theorem 6.1.10 reads as follows.

Corollary 6.1.11. *Let (f_n) be a sequence of continuous functions $f_n : X \to Y$ and let $f : X \to Y$ be a function. Then the following statements are equivalent.*

(a) The sequence (f_n) converges semi uniformly to f.

(b) The sequence (f_n) is equicontinuous and converges pointwise to f.

Of course, we cannot expect the implication "(b)⇒(a)" in Corollary 6.1.11 to remain true if we drop the pointwise convergence in part (b); the functions $f_n \equiv n$ may serve as a counterexample. Moreover, the same example shows that we even cannot extract a semi uniformly convergent subsequence if we only assume (f_n) to be equicontinuous and not necessarily pointwise convergent. However, if we make sure that the spaces X and Y are sufficiently "small", then we indeed find a semi uniformly convergent subsequence. Note that the following result is very similar to the famous and well-known theorem of Arzelà and Ascoli.

Corollary 6.1.12. *Let (f_n) be an equicontinuous sequence of functions $f_n : X \to Y$, where X is separable and Y is compact. Then (f_n) has a semi uniformly convergent subsequence with a continuous limit function.*

Proof. Since X is separable, there is a set $D := \{x_1, x_2, x_3, \dots\}$ which is countable and dense in X. Since Y is compact, we find a subsequence $(f_{n,1})$ of (f_n) such that $(f_{n,1}(x_1))$ converges. Again, since Y is compact, we find a subsequence $(f_{n,2})$ of $(f_{n,1})$ such that $(f_{n,2}(x_2))$ converges. Continuing this process, we obtain for each $k \in \mathbb{N}$ and $k > 1$ a subsequence $(f_{n,k})$ of $(f_{n,k-1})$ such that $(f_{n,k}(x_k))$ converges. By a diagonal argument, the sequence (g_n), defined by $g_n := f_{n,n}$, converges at each x_k.
Fix $x \in X$ and $\varepsilon > 0$. First, since (f_n) and hence (g_n) is equicontinuous at x, we find a $\delta > 0$ so that

$$\forall n \in \mathbb{N} : \quad d_X(x, y) \le \delta \quad \Rightarrow \quad d_Y\big(g_n(x), g_n(y)\big) \le \frac{\varepsilon}{3}.$$

Second, since D is dense in X, we find some $k \in \mathbb{N}$ such that $d_X(x_k, x) \le \delta$. Lastly, since $(g_n(x_k))$ converges, it is also a Cauchy sequence, and we find some $N \in \mathbb{N}$ such that

$$\forall m, n \ge N : \quad d_Y\big(g_m(x_k), g_n(x_k)\big) \le \frac{\varepsilon}{3}.$$

By combining all the three arguments we obtain for $m, n \geq N$,

$$d_Y\Big(g_m(x), g_n(x)\Big) \leq d_Y\Big(g_m(x), g_m(x_k)\Big) + d_Y\Big(g_m(x_k), g_n(x_k)\Big) + d_Y\Big(g_n(x_k), g_n(x)\Big) \leq \varepsilon,$$

showing that $(g_n(x))$ is a Cauchy sequence in Y. Since x was arbitrary and Y is compact and hence complete, (g_n) converges pointwise to some function $f : X \to Y$. Finally, f must be continuous as a pointwise limit of an equicontinuous sequence. By Corollary 6.1.11, (g_n) converges semi uniformly to f. ∎

The following two examples show that Corollary 6.1.12 turns wrong if X is not separable or Y is not compact. The first example is of particular interest, since it provides another sequence of functions that converges quasi uniformly but not semi uniformly.

Example 6.1.13. Let X be the metric space of all functions on $\mathbb{R}$ with values in $[-1, 1]$, equipped with the supremum norm $\|\cdot\|_\infty$, and let $Y = [-1, 1]$ be equipped with the Euclidean norm. Then X is not separable, but Y is compact. The functions $f_n : X \to [-1, 1]$, $x \mapsto x(n)$, form an (even uniformly) equicontinuous sequence, since

$$|f_n(x) - f_n(y)| = |x(n) - y(n)| \leq \|x - y\|_\infty \quad \text{for } x, y \in X, n \in \mathbb{N},$$

but cannot have a semi uniformly convergent subsequence. Even worse, they cannot have a pointwise convergent subsequence. Indeed, assume that the subsequence $(f_{n_k})_k$ converges pointwise, where $n_1 < n_2 < n_3 < \ldots$. The function $\xi : \mathbb{R} \to [-1, 1]$, defined by

$$\xi(t) = \begin{cases} (-1)^k & \text{for } t = n_k, k \in \mathbb{N}, \\ 0 & \text{otherwise,} \end{cases}$$

belongs to X. But then $(f_{n_k}(\xi))_k$ is divergent, since

$$f_{n_k}(\xi) = \xi(n_k) = (-1)^k$$

diverges as $k \to \infty$. In particular, (f_n) converges neither locally uniformly nor semi uniformly nor pointwise.

We now show that (f_n) converges quasi uniformly to the function

$$f : X \to [-1, 1], \; x \mapsto \limsup_{n\to\infty} x(n).$$

Note that f is well-defined since for $x \in X$ the sequence $(x(n))_{n\in\mathbb{N}}$ taking values only in $[-1, 1]$ has a finite limit superior in $[-1, 1]$. Observe that if $x \in X$ is fixed, then there is a subsequence $(x(n_k))_k$ of $(x(n))_n$ such that the numbers $x(n_k)$ converge to

$$\limsup_{n\to\infty} x(n) = f(x).$$

This means that $f(x)$ is a limit point of $(f_n(x))$. If $\varepsilon > 0$ is fixed and $y \in X$ so that $\|x - y\|_\infty \leq \varepsilon$, then

$$x(n) - \varepsilon \leq y(n) \leq x(n) + \varepsilon$$

for large $n \in \mathbb{N}$, and taking the limit superior on both sides yields

$$\left|f(x) - f(y)\right| = \left|\limsup_{n\to\infty} x(n) - \limsup_{n\to\infty} y(n)\right| \leq \varepsilon.$$

This shows that f is continuous at x. By Theorem 6.1.4, (f_n) converges quasi uniformly to f at x. As x had been chosen arbitrarily, the quasi uniform convergence on X is established. ◇

We remark that the same idea as presented in Example 6.1.13 shows that (f_n) also converges quasi uniformly to the function

$$\tilde{f} : X \to [-1, 1], \ x \mapsto \liminf_{n\to\infty} x(n).$$

This function is different from f in Example 6.1.13, as the function $x(t) = \cos(\pi t)$ that belongs to X shows, because we have

$$\begin{aligned} f(x) &= \limsup_{n\to\infty} \cos(n\pi) = \limsup_{n\to\infty} (-1)^n = 1 \\ &\neq -1 = \liminf_{n\to\infty} (-1)^n = \liminf_{n\to\infty} \cos(n\pi) = \tilde{f}(x). \end{aligned}$$

This again illustrates that a quasi uniform limit may not be unique.

Example 6.1.14. Let $X = \mathbb{R}$ and $Y = (0, 1]$ be equipped with the Euclidean norm. Then X is separable, but Y is not compact. The functions $f_n : \mathbb{R} \to (0, 1], \ x \mapsto \frac{1}{n}$, form an equicontinuous sequence, since $|f_n(x) - f_n(y)| = 0$ for all $x, y \in \mathbb{R}$ and $n \in \mathbb{N}$, but cannot have a semi uniformly convergent subsequence. Even worse, they cannot have a pointwise convergent subsequence, since any subsequence $(f_{n_k})_k$ of (f_n) would pointwise (even uniformly) converge to 0 which does not belong to $Y = (0, 1]$.
The same reasoning works to show that (f_n) neither converges quasi uniformly nor has a quasi uniformly convergent subsequence. ◇

We add another notion of convergence which appears sometimes in books even for beginners[3]. This definition is also only a tiny modification of Definition 6.1.7, but exhibits some weird properties.

Definition 6.1.15. Let $f_n, f : X \to Y$ for $n \in \mathbb{N}$ be functions, and let $x \in X$ be fixed. We say that the sequence (f_n) converges *continuously uniformly* to f at $x \in X$ if for each $\varepsilon > 0$ there exist $N \in \mathbb{N}$ and $\delta > 0$ such that for all $n \geq N$ and $y \in X$ with $d_X(x, y) \leq \delta$ we have $d_Y(f_n(y), f(x)) \leq \varepsilon$.
Using quantifiers, this reads

$$\forall \varepsilon > 0 \ \exists N \in \mathbb{N} \ \exists \delta > 0 \ \forall n \geq N \ \forall y \in X : \quad d_X(x, y) \leq \delta \ \Rightarrow \ d_Y\big(f_n(y), f(x)\big) \leq \varepsilon.$$

We say that (f_n) converges *continuously uniformly* to f in X if (f_n) converges continuously uniformly to f at each $x \in X$.

[3] See, for instance, [91].

Note that continuously uniform convergence is in the literature often just called "continuous convergence" or "α-convergence", and similar to the definition of pointwise continuity, defined via sequences [73, 91]. But both definitions are equivalent, as the following result shows.

Proposition 6.1.16. *Let $f_n, f : X \to Y$ for $n \in \mathbb{N}$ be arbitrary functions, and let $x \in X$ be fixed. Then the following statements are equivalent.*

(a) The sequence (f_n) converges continuously uniformly to f at x.

(b) For each sequence (x_n) in X converging to x the sequence $(f_n(x_n))$ converges to $f(x)$.

Proof. "(a)⇒(b)": Let (f_n) be continuously uniformly converging to f at x, fix a sequence (x_n) in X which converges to x, and pick $\varepsilon > 0$. Due to the continuous convergence of (f_n) to f at x we find some $N \in \mathbb{N}$ and some $\delta > 0$ such that

$$\forall n \geq N : \quad d_X(x, y) \leq \delta \quad \Rightarrow \quad d_Y\big(f_n(y), f(x)\big) \leq \varepsilon.$$

Since (x_n) converges to x, we find some $M \in \mathbb{N}$ such that

$$\forall n \geq M : \quad d_X(x, x_n) \leq \delta.$$

For $n \geq \max\{M, N\}$ we therefore obtain $d_Y\big(f_n(x_n), f(x)\big) \leq \varepsilon$ which proves (b).
"(b)⇒(a)": Assume that (f_n) does not converge continuously uniformly to f at x. Then there exists some $\varepsilon > 0$ such that for all $m \in \mathbb{N}$ and $\delta = 1/m$ we find $n_m \geq m$ and $y_m \in X$ with

$$d_X(y_m, x) \leq 1/m \quad \text{and} \quad d_Y\big(f_{n_m}(y_m), f(x)\big) > \varepsilon.$$

The sequence (x_k), defined by

$$x_k = \begin{cases} y_m & \text{for } k = n_m, m \in \mathbb{N}, \\ x & \text{for } k \notin \{n_1, n_2, n_3, \ldots\}, \end{cases}$$

converges to x, since (y_m) does, but $d_Y\big(f_{n_m}(x_{n_m}), f(x)\big) > \varepsilon$, and consequently $(f_n(x_n))$ cannot converge to $f(x)$. ∎

Proposition 6.1.16 (b) can be used to show that the sequence (f_n) in Example 6.0.1 does not converge continuously uniformly. Indeed, if we choose $x_n = 1/n$ which converges to 0, we have $f_n(x_n) = 1/e$ which cannot converge to $0 = f(0)$.

The only but very subtle difference between the Definitions 6.1.7 and 6.1.15 which differ only by one letter is that continuously uniform convergence directly measures the distance from $f_n(y)$ to $f(x)$ and not to $f(y)$. In particular, continuously uniform convergence implies pointwise convergence. This has a surprising consequence: The limit function of a continuously convergent sequence is always continuous, no matter if the functions forming the sequence are continuous. Even more is true: The following result shows how continuously and semi uniform convergence are related.

Theorem 6.1.17. *Let $f_n, f : X \to Y$ for $n \in \mathbb{N}$ be arbitrary functions, and let $x \in X$ be fixed. Then the following statements are equivalent.*

(a) The sequence (f_n) converges continuously uniformly to f at x.

(b) The sequence (f_n) converges semi uniformly to f at x, and f is continuous at x.

Proof. "(a)⇒(b)": To show that f is continuous at x, fix $\varepsilon > 0$. Since (f_n) converges continuously uniformly to f at x we find some $N \in \mathbb{N}$ and $\delta > 0$ such that

$$\forall n \geq N: \quad d_X(x, y) \leq \delta \quad \Rightarrow \quad d_Y\big(f_n(y), f(x)\big) \leq \frac{\varepsilon}{3}. \tag{6.1.6}$$

For fixed $y \in X$ with $d_X(x, y) \leq \delta$, the sequence $(f_n(y))$ converges to $f(y)$, and that is why we can find some $n \geq N$ such that $d_Y(f_n(y), f(y)) \leq \varepsilon/3$. Using (6.1.6) we obtain

$$d_Y\big(f(x), f(y)\big) \leq d_Y\big(f(x), f_n(y)\big) + d_Y\big(f_n(y), f(y)\big) \leq \frac{2}{3}\varepsilon \leq \varepsilon,$$

that is, f is continuous at x. Moreover, for all $n \geq N$, we get again with (6.1.6),

$$d_Y\big(f_n(y), f(y)\big) \leq d_Y\big(f_n(y), f(x)\big) + d_Y\big(f(x), f(y)\big) \leq \varepsilon,$$

and this shows that (f_n) converges also semi uniformly to f at x and proves (b).

"(b)⇒(a)": Fix $\varepsilon > 0$. Since (f_n) converges semi uniformly to f at x, we find some $N \in \mathbb{N}$ and $\delta > 0$ such that

$$\forall n \geq N: \quad d_X(x, y) \leq \delta \quad \Rightarrow \quad d_Y\big(f_n(y), f(y)\big) \leq \frac{\varepsilon}{2}. \tag{6.1.7}$$

Moreover, since f is continuous at x, we find some $\eta > 0$ such that

$$d_X(x, y) \leq \eta \quad \Rightarrow \quad d_Y\big(f(y), f(x)\big) \leq \frac{\varepsilon}{2}. \tag{6.1.8}$$

For $y \in X$ with $d_X(x, y) \leq \min\{\delta, \eta\}$ and $n \geq N$ we therefore obtain from (6.1.7) and (6.1.8) that

$$d_Y\big(f_n(y), f(x)\big) \leq d_Y\big(f_n(y), f(y)\big) + d_Y\big(f(y), f(x)\big) \leq \varepsilon$$

holds which proves (a). ■

Theorem 6.1.17 shows that continuously uniform convergence always implies semi uniform convergence; in particular, the sequences constructed in the Examples 6.1.3, 6.1.5, 6.1.6, 6.1.13 and 6.1.14 cannot converge continuously uniformly as they do not converge semi uniformly. However, the implication "(b)⇒(a)" of Theorem 6.1.17 applied to the Examples 6.1.8 and 6.1.9 shows that the two sequences given therein converge not only semi uniformly but also continuously uniformly at $x = 0$. In addition, since the sequence (f_n) of Example 6.0.1 does not converge continuously uniformly but pointwise to a continuous limit function, we get from Theorem 6.1.17 that it does also not converge semi uniformly.

Furthermore, Theorem 6.1.17 also shows that continuously uniform convergence of (f_n) always implies the continuity of the limit function, even if every function f_n is discontinuous everywhere. For example, the everywhere discontinuous functions $f_n := \frac{1}{n}\chi_{\mathbb{Q}}$ converge (continuously) uniformly on $\mathbb{R}$ to the everywhere continuous function $\mathbb{0}$. One could also naively argue that with the help of Theorem 6.1.17 every function f can be shown to be continuous everywhere by just taking $f_n = f$ for all $n \in \mathbb{N}$. However, such a sequence converges semi uniformly (even uniformly) to f, but only continuously uniformly if f is continuous. This leads to another interesting fact: constant sequences of functions being all equal to a function f do converge pointwise, (locally) uniformly, semi uniformly and quasi uniformly on the entire space, but they may not converge continuously uniformly, namely if the limit function f is discontinuous. Such a sequence is given in our last example of this section.

Example 6.1.18. The functions $f_n = \chi_{\mathbb{Q}}$ do not converge continuously uniformly on $\mathbb{R}$ by Theorem 6.1.17, since the pointwise limit $f = \chi_{\mathbb{Q}}$ is discontinuous everywhere. However, as a constant sequence, it clearly converges locally, semi and quasi uniformly as well as pointwise. ◊

After the proof of Theorem 2.1.8 we raised the question whether the set Δ of derivatives is closed under not only uniform convergence but also under one of the weaker types of convergence considered in this section. Since the interval $[0, 1]$ on which the functions in Δ live is compact, we first check how the types of convergence of this section behave on compact metric spaces in general.

For instance, it is well known that locally uniform convergence coincides with uniform convergence if we impose compactness on the metric space X. Surprisingly, in this case continuously and semi uniform convergence not only do behave calmly, they can even be used to *characterize* compactness.

Theorem 6.1.19. *The following statements are equivalent.*

(a) The space X is compact.

(b) Each sequence (f_n) of arbitrary functions $f_n : X \to Y$ which converges semi uniformly to a function $f : X \to Y$ also converges uniformly to f.

(c) Each sequence (f_n) of arbitrary functions $f_n : X \to Y$ which converges continuously uniformly to a function $f : X \to Y$ also converges uniformly to f.

Proof. To show "(a)⇒(b)", fix $\varepsilon > 0$ and let (f_n) be a sequence of arbitrary functions $f_n : X \to Y$ which converges semi uniformly to a function $f : X \to Y$. Then for each $x \in X$ there is some $N(x) \in \mathbb{N}$ and some $\delta(x) > 0$ such that

$$\forall n \geq N(x)\ \forall y \in X : \quad d_X(x, y) \leq \delta(x) \quad \Rightarrow \quad d_Y\Big(f_n(y), f(y)\Big) \leq \varepsilon.$$

Since the open balls $B(x) := \{z \in X \mid d_X(z, x) < \delta(x)\}$ cover X and X is compact, we find $x_1, \ldots, x_m \in X$ such that the balls $B(x_1), \ldots, B(x_m)$ are sufficient to cover X.

For fixed $y \in X$ we therefore find some $k \in \{1, \ldots, m\}$ such that $y \in B(x_k)$, and hence for $n \geq \max\{N(x_1), \ldots, N(x_m)\}$ we obtain $d_Y\big(f_n(y), f(y)\big) \leq \varepsilon$ which shows that (f_n) indeed converges uniformly to f.

For "(b)⇒(c)", let (f_n) be a sequence of arbitrary functions $f_n : X \to Y$ which converges continuously uniformly to a function $f : X \to Y$. By Theorem 6.1.17, the sequence (f_n) also converges semi uniformly to f, and by (b), (f_n) converges even uniformly to f.

Finally, the implication "(c)⇒(a)" follows from [73, Theorem 3.2] and Proposition 6.1.16. ■

Note that if X is not necessarily compact the implication "(b)⇒(c)" in Theorem 6.1.19 is still true, whereas the implication "(c)⇒(b)" is true if f in part (b) is additionally assumed to be continuous; this follows immediately from Theorem 6.1.17. Thus, for continuous limits f the statements (b) and (c) are in fact equivalent, no matter whether X is compact. This also explains why the sequences constructed in the Examples 6.1.8 and 6.1.9 cannot converge semi respectively continuously uniformly in an entire neighborhood of $x = 0$, since then they would need to converge also uniformly in this neighborhood by Theorem 6.1.19, but they do not converge locally uniformly at $x = 0$.

Caution within non compact spaces: In this case, Theorem 6.1.19 only says, that there exist sequences of functions which converge semi/continuously uniformly but not uniformly. It does not say that any sequence which does not converge uniformly also not converges semi/continuously uniformly. For instance, the functions $f_n(x) = x^n$ on $[0, 1)$ converge pointwise and semi uniformly (even locally uniformly) but not uniformly to $\mathbb{0}$, although $[0, 1)$ is certainly not compact.

In the next section we will investigate sequences of functions which do converge semi uniformly but not locally uniformly. The spaces such sequences live in therefore cannot be compact by Theorem 6.1.19. In our case, this will be balls in the space BV which - as BV is an infinite dimensional vector space - cannot be compact, indeed.

Let us now compare all notions of convergence considered so far at a fixed point $x \in X$. Here is a compact comparison between pointwise convergence

$$\text{(P)}\ \forall \varepsilon > 0\, \exists N \in \mathbb{N}\, \forall n \geq N: \quad d_Y\big(f_n(x), f(x)\big) \leq \varepsilon,$$

as well as locally (L), semi (S), continuously (C) and quasi uniform convergence (Q) at x:

$$\text{(L)}\ \exists \delta > 0\, \forall \varepsilon > 0\, \exists N \in \mathbb{N}\, \forall n \geq N\, \forall y \in X: \quad d_X(x, y) \leq \delta \Rightarrow d_Y\big(f_n(y), f(y)\big) \leq \varepsilon,$$

$$\text{(S)}\ \forall \varepsilon > 0\, \exists N \in \mathbb{N}\, \exists \delta > 0\, \forall n \geq N\, \forall y \in X: \quad d_X(x, y) \leq \delta \Rightarrow d_Y\big(f_n(y), f(y)\big) \leq \varepsilon,$$

$$\text{(C)}\ \forall \varepsilon > 0\, \exists N \in \mathbb{N}\, \exists \delta > 0\, \forall n \geq N\, \forall y \in X: \quad d_X(x, y) \leq \delta \Rightarrow d_Y\big(f_n(y), f(x)\big) \leq \varepsilon,$$

$$\text{(Q)}\ \forall \varepsilon > 0\, \forall N \in \mathbb{N}\, \exists \delta > 0\, \exists n \geq N\, \forall y \in X: \quad d_X(x, y) \leq \delta \Rightarrow d_Y\big(f_n(y), f(y)\big) \leq \varepsilon.$$

In particular, (L)⇒(S)⇒(Q), and (L), (S) and (C) imply (P). Theorem 6.1.17 proved that (C) implies (S) and hence also (Q). We end up with the following diagram.

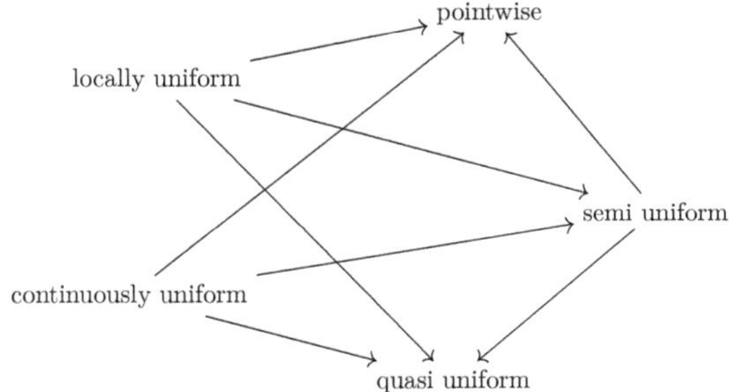

Figure 6.1.1: Relations between types of convergence.

Other implications than those in Figure 6.1.1 do not hold, as was shown by our examples. To make this a little more visible, let us collect the convergence properties of the sequences constructed in all the examples in Table 6.1.1 below. A "yes" means that the sequence converges at least at a particular point and not necessarily globally. Similarly, a "no" means that the sequence diverges at a particular point under consideration.

Table 6.1.1: Convergence properties of example sequences.

Example	locally uniform	continuously uniform	semi uniform	quasi uniform	pointwise
6.0.1	no	no	no	yes	yes
6.1.3	no	no	no	yes	no
6.1.5	no	no	no	no	yes
6.1.6	no	no	no	no	yes
6.1.8	no	yes	yes	yes	yes
6.1.9	no	yes	yes	yes	yes
6.1.13	no	no	no	yes	no
6.1.14	no	no	no	no	no
6.1.18	yes	no	yes	yes	yes

This table can now be used to show that none of the implications in the above diagram may be inverted. For instance, Example 6.0.1 shows that neither (P) nor (Q) implies any of the types (S), (L) and (C). Example 6.1.18 proves that (S) does not imply (C) which in turn is not implied by (L). Conversely, (L) can also not be deduced from (C) or (S), as is shown by the Examples 6.1.8 and 6.1.9. Finally, (P) does not imply (Q) due to the Examples 6.1.5 and 6.1.6, and (Q) does not imply (P), as we have seen in Example 6.1.3.

6.2 Continuity of Composition Operators in BV

In this section we are going to apply Theorem 6.1.10 to composition operators $C_g : BV \to BV$ as defined in (5.0.1) in the space BV of functions of bounded Jordan variation. By Proposition 5.1.1 (a) such operators C_g are well-defined if and only if $g \in Lip_{loc}(\mathbb{R})$.

A key ingredient in the proof for the continuity of C_g will be the following generalization of Definition 1.1.7.

Definition 6.2.1. For a set $U \subseteq \mathbb{R}$, a partition $P : 0 = t_0 < \ldots < t_n = 1$ of $[0,1]$ and some function $x : [0,1] \to \mathbb{R}$ we define the index set

$$J(x, U, P) := \Big\{ j \in \{1, \ldots, n\} \mid [x(t_{j-1}), x(t_j)] \subseteq U \Big\}$$

as well as the short cut

$$\mathrm{rVar}(x, U, P) := \sum_{j \in J(x,U,P)} \Big| x(t_{j-1}) - x(t_j) \Big|.$$

Moreover, we define the *variation of x restricted to U* by

$$\mathrm{rVar}(x, U) := \sup_P \mathrm{rVar}(x, U, P) = \sup_P \sum_{j \in J(x,U,P)} \Big| x(t_{j-1}) - x(t_j) \Big|,$$

where the supremum is taken over all partitions P of $[0,1]$. Here, we agree that the sums are 0 if $J(x, U, P) = \emptyset$.

Observe that Definition 6.2.1 comprises Definition 1.1.7 as we have $\mathrm{Var}(x) = \mathrm{rVar}(x, \mathbb{R})$ for any function $x : [0,1] \to \mathbb{R}$.

The restricted variation measures the variation of those parts of the function x whose values lie in U. It is clear from the definitions that $\mathrm{rVar}(x, \cdot)$ is increasing in the sense that $\mathrm{rVar}(x, U) \le \mathrm{rVar}(x, V)$ whenever $U \subseteq V$.

Example 6.2.2. Each monotone function $x : [0,1] \to \mathbb{R}$ is of bounded variation, and for any Lebesgue measurable set $U \subseteq \mathbb{R}$ we have

$$\mathrm{rVar}(x, U) \le \min\{|U|, \mathrm{Var}(x)\},$$

where $|U|$ denotes the Lebesgue measure of U. However, we cannot expect equality. In fact, for $x(t) = t$ and $U = \mathbb{R}$ we clearly do have equality, but for $U = \mathbb{R} \backslash \mathbb{Q}$ we have $\mathrm{rVar}(x, U) = 0 < 1 = \min\{|U|, \mathrm{Var}(x)\}$. ◇

The last example not only shows that the restricted variation only makes sense for sets U which are not too “thin”, so, for instance, for open sets, because in general we always have $\mathrm{rVar}(x, \mathbb{R} \backslash \mathbb{Q}) = 0$, no matter what x is. Moreover, if the range of x does not lie within U, then the restricted variation cannot reflect the overall behavior of x in a proper way. In particular, $\mathrm{rVar}(x, U) = 0$ does not mean that x is constant, in contrast to the ordinary variation, where $\mathrm{Var}(x) = 0$ implies that x is constant.

In Example 6.2.2 it is shown that we may have $\mathrm{rVar}(x, U, P) = 0$ for every partition P of $[0,1]$ and an appropriate U lying dense in $\mathbb{R}$, although x is not constant. The following dual example is of interest which shows that we may have $\mathrm{rVar}(x, U, P) = 0$ for every U and infinitely many partitions P whose mesh size shrinks down to 0.

Example 6.2.3. Consider the characteristic function $x := \chi_{\mathbb{Q}\cap[0,1]}$ and the partitions $P_n : 0 = t_{0,n} < \ldots < t_{n,n}$, given by $t_{j,n} := j/n$ for $j \in \{0, \ldots, n\}$ and $n \in \mathbb{N}$. Then the mesh size of P_n tends to zero as $n \to \infty$. But

$$\mathrm{rVar}(x, \mathbb{R}, P_n) = \sum_{j=1}^{n} \left| x(t_{j-1,n}) - x(t_{j,n}) \right| = 0$$

for each $n \in \mathbb{N}$, and this gives $\mathrm{rVar}(x, U, P_n) = 0$ for any set $U \subseteq \mathbb{R}$ and all $n \in \mathbb{N}$, although x is of unbounded Jordan variation! $\diamondsuit$

Example 6.2.2 also suggests that the restricted variation of a given function becomes smaller and smaller, the smaller $|U|$ gets, and hence decreases to zero if $|U|$ does so. This also makes perfect sense, since then less and less of the values of x belong to U. However, we can do better. The restricted variation is also continuous with respect to x and U, even when the size of U is measured with the Lebesgue outer measure $|U|^*$.

Lemma 6.2.4. *Let $x \in BV$. Then $\mathrm{rVar}(x, U)$ is continuous with respect to x and U in the following sense: For each $\varepsilon > 0$ there is some $\delta > 0$ such that for all $y \in BV$ and all sets $U \subseteq \mathbb{R}$ we have*

$$\|x - y\|_{BV} \leq \delta \quad \text{and} \quad |U|^* \leq \delta \qquad \Longrightarrow \qquad \mathrm{rVar}(y, U) \leq \varepsilon.$$

Proof. Fix $\varepsilon > 0$ and $x \in BV$. Then there is a partition $0 = s_0 < \ldots < s_p = 1$ of $[0,1]$ such that

$$\sum_{j=1}^{p} |x(s_{j-1}) - x(s_j)| \geq \mathrm{Var}(x) - \frac{\varepsilon}{4}. \tag{6.2.1}$$

Choose $\delta := \frac{\varepsilon}{4(p+2)}$ and fix $y \in BV$ with $\|x - y\|_{BV} \leq \delta$. Then

$$\begin{aligned} \mathrm{Var}(y) &= \|y\|_{BV} - \|y\|_\infty \leq \|x - y\|_{BV} + \mathrm{Var}(x) + \|x\|_\infty - \|y\|_\infty \\ &\leq \delta + \mathrm{Var}(x) + \|x - y\|_\infty \leq \delta + \mathrm{Var}(x) + \|x - y\|_{BV} \\ &\leq 2\delta + \mathrm{Var}(x), \end{aligned} \tag{6.2.2}$$

and for all $s, t \in [0,1]$,

$$\begin{aligned} |x(s) - x(t)| &\leq |x(s) - y(s) - x(t) + y(t)| + |y(s) - y(t)| \\ &\leq \mathrm{Var}(x - y) + |y(s) - y(t)| \leq \delta + |y(s) - y(t)|. \end{aligned} \tag{6.2.3}$$

From (6.2.1) we obtain with the help of (6.2.2) and (6.2.3)

$$\begin{aligned} \sum_{j=1}^{p} |y(s_{j-1}) - y(s_j)| &\overset{(6.2.3)}{\geq} \sum_{j=1}^{p} |x(s_{j-1}) - x(s_j)| - p\delta \overset{(6.2.1)}{\geq} \mathrm{Var}(x) - p\delta - \frac{\varepsilon}{4} \\ &\overset{(6.2.2)}{\geq} \mathrm{Var}(y) - (p+2)\delta - \frac{\varepsilon}{4} = \mathrm{Var}(y) - \frac{\varepsilon}{2}. \end{aligned} \tag{6.2.4}$$

Fix a set $U \subseteq \mathbb{R}$ with $|U|^* \leq \delta$, and let $P : 0 = t_0 < \ldots < t_m = 1$ be a partition of $[0, 1]$. We now consider the partition $T : 0 = \tau_0 < \ldots < \tau_N = 1$ created by putting the s_j and t_j together. For a chain $s_{k-1} \leq t_i < t_{i+1} < \ldots < t_{i+l} \leq s_k$ we have

$$\begin{aligned}&\Big[y(s_{k-1}), y(s_k)\Big]\backslash U \\ &\qquad \subseteq \Big(\Big[y(s_{k-1}), y(t_i)\Big]\backslash U\Big) \cup \bigcup_{j=1}^{l} \Big(\Big[y(t_{i+j-1}), y(t_{i+j})\Big]\backslash U\Big) \cup \Big(\Big[y(t_{i+l}), y(s_k)\Big]\backslash U\Big),\end{aligned}$$

hence

$$\sum_{k=1}^{p} \Big|y(s_{k-1}) - y(s_k)\Big| - p|U|^* \leq \sum_{k=1}^{p} \Big|\Big[y(s_{k-1}), y(s_k)\Big]\backslash U\Big|^* \leq \sum_{j=1}^{N} \Big|\Big[y(\tau_{j-1}), y(t_j)\Big]\backslash U\Big|^*$$

and therefore

$$\sum_{k=1}^{p} \Big|y(s_{k-1}) - y(s_k)\Big| - p|U|^* \leq \sum_{j \notin J(y,U,T)} \Big|y(\tau_{j-1}) - y(\tau_j)\Big|. \tag{6.2.5}$$

Moreover, for $k \in J(y, U, P)$ we have two possibilities. The first is that there exists an index $j \in \{1, \ldots, p\}$ such that $t_{k-1} < s_j < t_k$. In this case, we have

$$\Big|y(t_{k-1}) - y(t_k)\Big| \leq |U|^*.$$

The second case is that there is no such index j. Then $t_{k-1} = \tau_{l-1} < \tau_l = t_k$ for some $l \in \{1, \ldots, N\}$, and in this case, $l \in J(y, U, T)$ and

$$\Big|y(t_{k-1}) - y(t_k)\Big| = \Big|y(\tau_{l-1}) - y(\tau_l)\Big|.$$

But since there are only p points s_j, the first case can occur at most p times. Thus, we obtain

$$\sum_{j \in J(y,U,P)} \Big|y(t_{j-1}) - y(t_j)\Big| \leq \sum_{j \in J(y,U,T)} \Big|y(\tau_{j-1}) - y(\tau_j)\Big| + p|U|^*. \tag{6.2.6}$$

By adding (6.2.5) and (6.2.6) and using (6.2.4) we reach the final estimate

$$\begin{aligned}\sum_{j \in J(y,U,P)} \Big|y(t_{j-1}) - y(t_j)\Big| &\leq \sum_{j \in J(y,U,T)} \Big|y(\tau_{j-1}) - y(\tau_j)\Big| + \sum_{j \notin J(y,U,T)} \Big|y(\tau_{j-1}) - y(\tau_j)\Big| \\ &\qquad - \sum_{k=1}^{p} \Big|y(s_{k-1}) - y(s_k)\Big| + 2p|U|^* \\ &\leq \frac{\varepsilon}{2} + 2p|U|^* \leq \varepsilon\end{aligned}$$

which eventually proves $\mathrm{rVar}(y, U) \leq \varepsilon$ and hence the claim. ■

Since Lipschitz continuity is a fundamental requirement on the generating function $g : \mathbb{R} \to \mathbb{R}$ of the composition operator $C_g : BV \to BV$ to be well-defined, we need to investigate it in a little more detail. Recall that we denote by $\mathrm{lip}(f, [a, b])$ the optimal Lipschitz constant of a function $f : [a, b] \to \mathbb{R}$.

Note that by a result of Dini[4], the Lipschitz constant can also be calculated by the formula

$$\operatorname{lip}(f,[a,b]) = \sup_{y\in[a,b]} \limsup_{x\to y} \frac{|f(x)-f(y)|}{|x-y|}. \tag{6.2.7}$$

If $f \in Lip[a,b]$ and $x : [0,1] \to [a,b]$ belongs to BV, it is easy to see that

$$\operatorname{Var}(f\circ x) \le \operatorname{lip}(f,[a,b]) \operatorname{Var}(x). \tag{6.2.8}$$

The following is a certain converse of this inequality.

Proposition 6.2.5. *Let $R > 0$, $f \in Lip[-R,R]$, $\alpha \in (0,1)$ and $\beta > 0$. Then there exists some $x : [0,1] \to [-R,R]$ with $\operatorname{Var}(x) \le 2\beta$ and*

$$\operatorname{Var}(f\circ x) \ge \alpha\beta \operatorname{lip}(f,[-R,R]).$$

Proof. Due to Dini's formula (6.2.7) we find $u, v \in [-R,R]$ such that $0 < |u-v| \le \beta$ and

$$\left|\frac{f(u)-f(v)}{u-v}\right| \ge \alpha \operatorname{lip}(f,[-R,R]). \tag{6.2.9}$$

Choose $k \in \mathbb{N}$ so that

$$\frac{\beta}{2k} \le |u-v| \le \frac{\beta}{k} \tag{6.2.10}$$

and define $x \in BV$ by

$$x(t) = \begin{cases} u & \text{for } t \in \left\{\frac{1}{2}, \frac{1}{3}, \frac{1}{4}, \dots, \frac{1}{k+1}\right\}, \\ v & \text{otherwise.} \end{cases}$$

Then $x(t) \in \{u,v\} \subseteq [-R,R]$ for all $t \in [0,1]$ and $\operatorname{Var}(x) = 2k|u-v| \le 2\beta$ due to (6.2.10). We obtain with (6.2.9) and (6.2.10)

$$\operatorname{Var}(f\circ x) = 2k|f(u)-f(v)| \ge \alpha \cdot 2k|u-v| \operatorname{lip}(f,[-R,R]) \ge \alpha\beta \operatorname{lip}(f,[-R,R]),$$

as claimed. ■

At this point we recall another link between BV-functions and Lipschitz continuous functions, given by Theorem 1.1.23: A function belongs to BV if and only if it can be written as a composition of a nonexpansive function (i.e. lip ≤ 1) and a monotonically increasing function.

We are now approaching the main result of this section, the proof of the continuity of the operator $C_g : BV \to BV$, defined in (5.0.1), provided that $g \in Lip_{loc}(\mathbb{R})$.
The idea is as follows. Given C_g for some fixed $g \in Lip_{loc}(\mathbb{R})$, approximate C_g by other continuous composition operators C_{g_n} such that the convergence transmits the

[4]The first proof was given by Dini in 1878 in [51], but it can also (and perhaps more elegantly) be deduced from a theorem of Zygmund [138, Chapter VI, §7].

continuity of C_{g_n} to C_g. Recall that in Theorem 5.1.21 we have proven that a composition operator C_f is continuous (even locally Lipschitz continuous) if the generating function f is of class C^1 in $\mathbb{R}$ and has a locally Lipschitz continuous derivative. In particular, if we choose g_n to be of class C^∞ on $\mathbb{R}$, each C_{g_n} would be continuous, and an appropriate continuity preserving type of convergence would make C_g continuous as well. Of course, uniform convergence on balls would do the job. However, we have the following

Theorem 6.2.6. *Let $g_n, g \in Lip_{loc}(\mathbb{R})$ for $n \in \mathbb{N}$ and $R > 0$ be given. The following statements are equivalent.*

(a) (C_{g_n}) converges uniformly to C_g on the set

$$\Big\{x \in BV \mid \|x\|_\infty \leq R \text{ and } \operatorname{Var}(x) \leq 2R\Big\}.$$

(b) The relations $\lim\limits_{n\to\infty} \|g_n - g\|_{[-R,R]} = 0$ and $\lim\limits_{n\to\infty} \operatorname{lip}(g_n - g, [-R, R]) = 0$ hold.

In [31] the authors proved that (b) implies the uniform convergence of (C_{g_n}) to C_g on the closed ball $\mathbb{B}_R(BV) = \{x \in BV \mid \|x\|_{BV} \leq R\}$ for the special case when g_n are certain Bernstein polynomials approximating g. With the same idea one obtains (a) with the general assumptions made in (b):

Proof. For this proof we set

$$B := \Big\{x \in BV \mid \|x\|_\infty \leq R \text{ and } \operatorname{Var}(x) \leq 2R\Big\}.$$

To show "(a)⇒(b)", note that the C_{g_n} converge uniformly to C_g on B. That is why the numbers

$$s_n := \sup_{x\in B} \|C_{g_n}x - C_g x\|_{BV}$$

converge to 0 as $n \to \infty$.
The function $x : [0, 1] \to [-R, R]$, defined by $x(t) = 2Rt - R$, belongs to B, maps $[0, 1]$ homeomorphically to $[-R, R]$ and gives

$$\|g_n - g\|_{[-R,R]} = \|g_n \circ x - g \circ x\|_\infty \leq \|C_{g_n}x - C_g x\|_{BV} \leq s_n$$

which proves the first claim of (b). Proposition 6.2.5, applied with $\alpha = 1/2$ and $\beta = R$ to $f = g_n - g$, yields for each $n \in \mathbb{N}$ a function $x_n \in B$ such that

$$\begin{aligned}\operatorname{lip}(g_n - g, [-R, R]) &\leq 2R^{-1} \operatorname{Var}(g_n \circ x_n - g \circ x_n) \leq 2R^{-1} \|C_{g_n}x_n - C_g x_n\|_{BV}\\ &\leq 2R^{-1}s_n,\end{aligned}$$

and thus proves the second claim of (b). Consequently, (b) is established.
For the reverse implication "(b)⇒(a)" note that for $x \in B$ we have on the one hand

$$\|C_{g_n}x - C_g x\|_\infty = \|g_n \circ x - g \circ x\|_\infty \leq \|g_n - g\|_{[-R,R]}, \tag{6.2.11}$$

and on the other hand

$$\mathrm{Var}(C_{g_n}x - C_g x) \leq \mathrm{lip}(g_n - g, [-R, R])\, \mathrm{Var}(x) \leq 2R\, \mathrm{lip}(g_n - g, [-R, R]) \qquad (6.2.12)$$

by (6.2.8). Since the right hand sides of (6.2.11) and (6.2.12) converge to 0 by (b), and since they do so independently of x, (a) is proven. ■

The condition (b) of Theorem 6.2.6 means that the g_n converge in the topology of the space $Lip_{loc}(\mathbb{R})$ to g, but this is a strong requirement: If g_n is of class C^1 for each $n \in \mathbb{N}$, then g must also be of class C^1, since $C^1(\mathbb{R})$ is a closed subspace of $Lip_{loc}(\mathbb{R})$. This is why the authors in [31] only got that C_g is continuous in BV if $g \in C^1(\mathbb{R})$, because they approximated g by Bernstein polynomials. In order to get continuity of C_g for $g \in Lip_{loc}(\mathbb{R})$, we need another weaker type of convergence which still preserves continuity. It turns out that semi uniform convergence is exactly what we are looking for, and this is the content of our next

Theorem 6.2.7. *Let $g_n, g \in Lip_{loc}(\mathbb{R})$ for $n \in \mathbb{N}$ and $R > 0$ be given. The following statements are equivalent.*

(a) (C_{g_n}) converges semi uniformly to C_g at each $x \in BV$ with $\|x\|_\infty < R$ and $\mathrm{Var}(x) < 2R$.

(b) The relations $\lim\limits_{n\to\infty} \|g_n - g\|_{BV[-r,r]} = 0$ and $\sup\limits_{n\in\mathbb{N}} \mathrm{lip}(g_n - g, [-r, r]) < \infty$ hold for each $r \in (0, R)$.

In particular, if (b) holds for all $r > 0$, then the sequence (C_{g_n}) converges in BV semi uniformly to the operator C_g.

Proof. "(a)⇒(b)": Fix $r \in (0, R)$. We first show that $\|g_n - g\|_{BV[-r,r]} \to 0$ as $n \to \infty$. To this end, define $x(t) := 2rt - r$ on $[0, 1]$. Then x maps $[0, 1]$ strictly increasingly and bijectively onto $[-r, r]$ with $\|x\|_\infty = r < R$ and $\mathrm{Var}(x) = 2r < 2R$. Because of (a), the sequence (C_{g_n}) converges semi uniformly to C_g at x, and this implies that $(C_{g_n}(x))$ converges in BV to $C_g(x)$. This means

$$0 = \lim_{n\to\infty} \|C_{g_n}(x) - C_g(x)\|_{BV} = \lim_{n\to\infty} \|g_n - g\|_{BV[-r,r]} \cdot$$

We now write $f_n := g_n - g$ for $n \in \mathbb{N}$ and show that

$$\sup_{n\in\mathbb{N}} \mathrm{lip}(f_n, [-r, r]) < \infty. \qquad (6.2.13)$$

For each $n \in \mathbb{N}$ pick $u_n, v_n \in [-r, r]$ with $u_n \neq v_n$ and so that

$$\frac{|f_n(u_n) - f_n(v_n)|}{|u_n - v_n|} \geq \frac{1}{2}\, \mathrm{lip}(f_n, [-r, r]) \qquad \text{and} \qquad |u_n - v_n| \leq \frac{1}{n}, \qquad (6.2.14)$$

where (6.2.14) is justified by Dini's formula (6.2.7).
Since $[-r, r]$ is compact, we can assume by passing to suitable subsequences that $u_n \to w$ and $v_n \to w$ for some $w \in [-r, r]$. The constant function $x \equiv w$ for $t \in [0, 1]$ belongs

to BV with $\|x\|_\infty = |w| \le r < R$ and $\mathrm{Var}(x) = 0 < 2R$; hence by (a) we find for $\varepsilon = 1$ some $N_1 \in \mathbb{N}$ and $\delta \in (0, 4R)$ such that

$$\begin{aligned} &\forall n \geq N_1 \ \forall y \in BV: \\ &\qquad \|x-y\|_{BV} \le \delta \ \Rightarrow \ \|C_{g_n}y - C_g y\|_{BV} = \|f_n \circ y\|_{BV} \le 1. \end{aligned} \tag{6.2.15}$$

Since both (u_n) and (v_n) converge to w, there is some $N_2 \in \mathbb{N}$ such that

$$\forall n \geq N_2: \quad \max\Big\{|u_n - w|, |v_n - w|\Big\} \le \frac{\delta}{2}.$$

For each $n \in \mathbb{N}$ let $x_n : [0,1] \to [-r, r]$ be the BV-function constructed in the proof of Proposition 6.2.5 with $\alpha = 1/2$, $\beta = \delta/4$, $u = u_n$, $v = v_n$ and R replaced by r. Then $\|x_n\|_\infty \le r < R$ and $\mathrm{Var}(x_n) \le \delta/2 < 2R$, and thus

$$\|x_n - x\|_{BV} = \|x_n - x\|_\infty + \mathrm{Var}(x_n - x) = \max\{|u_n - w|, |v_n - w|\} + \mathrm{Var}(x_n) \le \delta$$

for $n \geq N_2$, and additionally

$$\mathrm{lip}\Big(f_n, [-r, r]\Big) \le \frac{\mathrm{Var}(f_n \circ x_n)}{\alpha\beta} \le \frac{8\,\|f_n \circ x_n\|_{BV}}{\delta}.$$

From this follows with the help of (6.2.15) $\mathrm{lip}\Big(f_n, [-r, r]\Big) \le 8/\delta$ for $n \geq \max\{N_1, N_2\}$ which proves (6.2.13).

"(b)⇒(a)": Write $f_n := g_n - g$, fix $\varepsilon > 0$ and $x \in BV$ with $\|x\|_\infty < R$ and $\mathrm{Var}(x) < 2R$. Choose $r > 0$ so that $\|x\|_\infty < r < R$. Because of (b),

$$L := \sup_{n \in \mathbb{N}} \mathrm{lip}\Big(f_n, [-r, r]\Big) < \infty.$$

By Lemma 6.2.4 there is some $\delta \in (0, r - \|x\|_\infty]$ such that for all $y \in BV$ and all subsets $U \subseteq \mathbb{R}$,

$$\|x - y\|_{BV} \le \delta \quad \text{and} \quad |U|^* \le \delta \quad \Longrightarrow \quad \mathrm{rVar}(y, U) \le \eta := \frac{\varepsilon}{L + 4R}. \tag{6.2.16}$$

Fix $y \in BV$ with $\|x - y\|_{BV} \le \delta$. Then $\|y\|_\infty \le \|x - y\|_\infty + \|x\|_\infty \le \delta + \|x\|_\infty \le r$ and thus $y(t) \in [-r, r]$ for all $t \in [0, 1]$.

Again by (b), $\|f_n\|_{BV[-r,r]} \to 0$ as $n \to \infty$, and that is why we find some $N \in \mathbb{N}$ such that

$$\|f_n\|_{BV[-r,r]} \le \frac{\eta\delta}{5} \quad \text{for all } n \geq N.$$

Fix such $n \geq N$ and let $\mathcal{I}$ be the system of intervals

$$\mathcal{I} := \Big\{[s, t] \subseteq [-r, r] \mid |f_n(s) - f_n(t)| > \eta|s - t|\Big\}.$$

By Vitali's Covering Lemma we find some countable subsystem $\{I_j \mid j \in J\}$ with index set $J \subseteq \mathbb{N}$ of pairwise disjoint intervals of $\mathcal{I}$ such that

$$U := \bigcup \mathcal{I} \subseteq \bigcup_{j \in J} 5I_j,$$

where for a compact interval $I = [s,t]$ with $s < t$ the symbol $5I$ denotes the compact interval $[s-2(t-s), t+2(t-s)]$. Consequently, U is a set with outer Lebesgue measure

$$\begin{aligned} |U|^* &\leq \sum_{j\in J}^{\infty} |5I_j| = 5\sum_{j\in J}^{\infty} |s_j - t_j| \leq \frac{5}{\eta}\sum_{j\in J}^{\infty} |f_n(s_j) - f_n(t_j)| \\ &\leq \frac{5}{\eta}\|f_n\|_{BV[-r,r]} \leq \delta. \end{aligned} \tag{6.2.17}$$

Now, let $P : 0 = t_0 < \ldots < t_m = 1$ be a partition of $[0,1]$. Therefore, borrowing the notation from Definition 6.2.1,

$$\begin{aligned} &\sum_{j=1}^{m}\Big|f_n\big(y(t_{j-1})\big) - f_n\big(y(t_j)\big)\Big| \\ &\quad= \sum_{j\in J(y,U,P)}\Big|f_n\big(y(t_{j-1})\big) - f_n\big(y(t_j)\big)\Big| + \sum_{j\notin J(y,U,P)}\Big|f_n\big(y(t_{j-1})\big) - f_n\big(y(t_j)\big)\Big| \\ &\quad\leq L\sum_{j\in J(y,U,P)}\Big|y(t_{j-1}) - y(t_j)\Big| + \eta\sum_{j\notin J(y,U,P)}\Big|y(t_{j-1}) - y(t_j)\Big| \\ &\quad\leq L\,\mathrm{rVar}(y,U) + \eta\,\mathrm{Var}(y) \leq \eta(L+4R) \leq \varepsilon. \end{aligned} \tag{6.2.18}$$

Here we have used (6.2.16) and $\mathrm{Var}(y) \leq \|y\|_{BV} \leq \|x-y\|_{BV} + \|x\|_{BV} \leq \delta + 3R \leq 4R$. This shows $\mathrm{Var}(f_n \circ y) \leq \varepsilon$ for all $y \in BV$ with $\|x-y\|_{BV} \leq \delta$.
Finally, for $n \geq N$ and such y,

$$\begin{aligned} \|f_n \circ y\|_{BV} &= \|f_n \circ y\|_\infty + \mathrm{Var}(f_n \circ y) \leq \|f_n\|_{BV[-r,r]} + \varepsilon \leq \varepsilon\left(\frac{\delta}{5(L+4R)} + 1\right) \\ &\leq \varepsilon\left(\frac{R}{5(L+4R)} + 1\right). \end{aligned}$$

This completes the proof. ∎

Let us make two comments. First note that condition (b) of Theorem 6.2.7 is indeed weaker than condition (b) of Theorem 6.2.6, since in Theorem 6.2.7 we only need convergence of (g_n) in BV and boundedness of the Lipschitz constants, whereas in Theorem 6.2.6 we need convergence of (g_n) in $Lip_{loc}(\mathbb{R})$; Example 6.2.9 at the end of this section will illustrate this difference. Second, note that condition (b) of Theorem 6.2.7 gives with the help of Theorem 6.1.10 a criterion under which composition operators are locally equicontinuous in BV.

As a consequence we reach the goal of this section and obtain a new proof for the fact that the composition operator generated by a locally Lipschitz continuous function is continuous in BV.

Theorem 6.2.8. *Let $g \in Lip_{loc}(\mathbb{R})$. Then the operator $C_g : BV \to BV$ is continuous.*

Proof. Fix $R > 0$. Since $g \in Lip_{loc}(\mathbb{R})$ we have $g \in AC[-R,R]$ and $g' \in L_\infty[-R,R]$. The function $h := g'\chi_{[-R,R]}$ then belongs to $L_\infty(\mathbb{R})$. The functions constructed in

Chapter II of [140] are C^∞-functions $h_n : \mathbb{R} \to \mathbb{R}$ and satisfy

$$\|h_n - h\|_{L_1[-R,R]} \le \frac{1}{n} \quad \text{and}$$
$$\|h_n\|_{L_\infty[-R,R]} \le \|h\|_{L_\infty[-R,R]} = \|g'\|_{L_\infty[-R,R]} = \operatorname{lip}\big(g, [-R,R]\big).$$

We now define $g_n : \mathbb{R} \to \mathbb{R}$ by

$$g_n(u) = g(0) + \int_0^u h_n(t)\,\mathrm{d}t.$$

Then each g_n is of class C^∞ with $g_n' = h_n$ on $\mathbb{R}$ and satisfies

$$\begin{aligned}\|g_n - g\|_{BV[-R,R]} &= \|g_n - g\|_{[-R,R]} + \operatorname{Var}(g_n - g, [-R,R]) \le 2\,\|g_n' - g'\|_{L_1[-R,R]}\\ &= 2\,\|h_n - h\|_{L_1[-R,R]} \le \frac{2}{n}.\end{aligned}$$

Moreover,

$$\operatorname{lip}\big(g_n, [-R,R]\big) = \|g_n'\|_{L_\infty[-R,R]} = \|h_n\|_{L_\infty[-R,R]} \le \operatorname{lip}\big(g, [-R,R]\big)$$

for each $n \in \mathbb{N}$. Hence, each g_n generates a composition operator C_{g_n} which maps BV into itself and is continuous (even locally Lipschitz continuous) by Theorem 5.1.21. Finally, the C_{g_n} converge semi uniformly to C_g at each $x \in BV$ with $\|x\|_{BV} < R$ by Theorem 6.2.7, and this leads to the desired continuity of C_g at those x by Theorem 6.1.10. ■

Eventually reaching the end of this section we would like to keep our promise made in Section 6.1 to give a third example of a sequence of functions which converges semi uniformly but not locally uniformly.

Example 6.2.9. Consider the functions

$$g(u) := |u| \quad \text{and} \quad g_n(u) = \sqrt{u^2 + 1/n} \quad \text{for } u \in \mathbb{R}, n \in \mathbb{N}.$$

Then g is Lipschitz continuous on $\mathbb{R}$, but not differentiable at $u = 0$, and each g_n is of class C^∞ on $\mathbb{R}$ with

$$g_n'(u) = \frac{u}{\sqrt{u^2 + 1/n}}.$$

We obtain for fixed $R > 0$ and $n \in \mathbb{N}$,

$$\|g_n - g\|_{[-R,R]} = \sup_{|u| \le R} \left|\sqrt{u^2} - \sqrt{u^2 + 1/n}\right| \le \frac{1}{\sqrt{n}},$$

$$\operatorname{lip}(g_n - g, [-R,R]) = \|g_n' - g'\|_{L_\infty[-R,R]} = \sup_{0<|u|\le R} \left|\frac{|u|}{\sqrt{u^2 + 1/n}} - 1\right| = 1,$$

$$\begin{aligned}\|g_n - g\|_{BV[-R,R]} &= \|g_n - g\|_{[-R,R]} + \|g_n' - g'\|_{L_1[-R,R]}\\ &\le \frac{1}{\sqrt{n}} + 2\int_0^R \left|\frac{u}{\sqrt{u^2 + 1/n}} - 1\right| \mathrm{d}u = \frac{3 + 2\left(\sqrt{nR^2} - \sqrt{nR^2 + 1}\right)}{\sqrt{n}}\\ &\le \frac{3}{\sqrt{n}}.\end{aligned}$$

Thus, condition (b) of Theorem 6.2.6 fails for any $R > 0$ and hence C_{g_n} cannot converge in the space BV locally uniformly to C_g at $x = \mathbb{0} \in BV$. However, condition (b) of Theorem 6.2.7 holds for any $R > 0$ and thus the C_{g_n} converge in the space BV semi uniformly to C_g at $x = \mathbb{0}$. ♢

Note that the function g as well as each function g_n from the previous example are Lipschitz continuous and therefore generate composition operators C_g and C_{g_n} which are continuous in BV by Theorem 6.2.8.

The question is now if the theory developed in this section might be transferred to other BV-spaces. Unfortunately, we do not know the answer to this question, but we conjecture that Theorem 6.2.7 is still true in the spaces WBV_p, YBV_φ and ΛBV, when BV in part (a) is replaced by one of the respective spaces and (b) is kept. If this conjecture is true, then the continuity of C_g in other BV-spaces would follow exactly as in the proof of Theorem 6.2.8. However, the main ingredient for the relevant implication "(b)$\Rightarrow$(a)" in Theorem 6.2.7 is Lemma 6.2.4 and thus the continuity of the restricted variation. It is not at all clear how to prove an analogue of this lemma in other BV-spaces, although its statement seems to be true in such spaces, at least from a (perhaps naive) geometric or graphic point of view.

Chapter 7

Integral Equations

Integral equations often describe specific real world phenomena and are therefore of great interest. For instance, the authors of [102] introduced a second order boundary value problem the solution of which describes the temperature distribution of an adiabatic chemical reactor of length 1. These solutions may be found by rewriting the problem into the nonlinear Hammerstein integral equation

$$u(t) = \frac{1}{\lambda} e^{\lambda(t-1)} Hu + \int_0^1 k(t,s) f\big(u(s)\big)\, \mathrm{d}s,$$

where

$$k(t,s) = \begin{cases} 1 & \text{for } 0 \le s \le t \le 1, \\ e^{\lambda(t-s)} & \text{for } 0 \le t < s \le 1 \end{cases} \qquad \text{and} \qquad f(u) = \begin{cases} \mu(\beta - u)e^u & \text{for } u \le \beta, \\ 0 & \text{for } u > \beta. \end{cases}$$

The constants λ is the Peclet number, μ is the Damkohler number, β is the dimensionless adiabatic temperature rise, and the function u represents the local temperature at a point t of the tube in which the reaction happens. The (possibly nonlinear) functional H models a feedback control system on the reactor that adds or removes heat according to the temperatures detected by some sensors located along the tube, see [41] and references therein for details.
Another example is given by the author of [21] who discusses a model describing the nonlinear age-depending growth of a single population under harvest which generalizes the original model introduced in [70]. The total population $P(t)$ at time t can be described under certain assumptions by a nonlinear Volterra integral equation of the form

$$P(t) = p(t) - h(t) + \int_0^t \exp\left(-\int_0^{t-s} \mu(\tau)\, \mathrm{d}\tau\right) P(s)\beta\big(P(s)\big)\, \mathrm{d}s,$$

where p, μ and h are parameters depending on time and β is a parameter depending on the population size.
Especially the last example motivates the search for BV-solutions to integral equations, because functions of bounded variation may have jumps which then can be interpreted as sudden deaths or births of the population. Moreover, the total increment and

decrement of the population is bounded as there is only a finite amount of biomass in the system under consideration. We will mention two further examples of applications of solutions to integral equations at the beginning of Section 7.1 below.
While there exists a large literature on continuous or integrable solutions of integral equations of the aforementioned type, considerably less is known on BV-solutions.

In this last chapter we apply some of the results from the previous chapters to solve integral equations in the spaces of functions of bounded variation of various types. We will prove existence and sometimes also uniqueness of solutions to either Hammerstein or Volterra integral equations. Our main tool is fixed point theory, so we will impose suitable conditions on the data which make it possible to apply well-known fixed point theorems. Although those conditions have been already considered in the Chapters 4 and 5, we will repeat them here to make the presentation self-contained.
In the first section we will focus on nonlinear Hammerstein integral equations, first only in the space BV and later also in other BV-spaces. The second section is devoted to nonlinear Volterra integral equations where we mainly rewrite our results of the first section. In the third and final section we make some remarks on boundary and initial value problems in the space BV where the boundary and initial conditions are given in a nonclassical coupled setting. Based on Schauder's fixed point theorem we give a simple sufficient condition under which such boundary and initial value problems may be solved and simultaneously generalize the ideas and results developed in the paper [27] that served as a point of departure for many considerations presented here.

7.1 Hammerstein Integral Equations

The first equation that we consider in this chapter is

$$x(t) = h(t) + \lambda \int_0^1 k(t,s)g\big(x(s)\big)\,\mathrm{d}s \quad \text{for } 0 \le t \le 1, \tag{7.1.1}$$

where the functions $h : [0,1] \to \mathbb{R}$, $k : [0,1] \times [0,1] \to \mathbb{R}$ and $g : \mathbb{R} \to \mathbb{R}$ are given and the function $x : [0,1] \to \mathbb{R}$ is unknown. We point out that the role of λ in (7.1.1) is very important. For example, in the problems concerning calculation of either free pulsation of harmonic vibrations of a string or a critical speed of a shaft is reduced to calculating such values of λ for which the corresponding integral equation, being special cases of (7.1.1), have a nontrivial solution.
As mentioned above, in this section we are going to look for solutions x of equation (7.1.1) in the space BV. Borrowing the notation from the Chapters 4 and 5 we may write (7.1.1) equivalently as the operator equation

$$x = h + \lambda(I_k \circ C_g)x, \tag{7.1.2}$$

where I_k denotes the integral operator (4.0.3) generated by $k : [0,1] \times [0,1] \to \mathbb{R}$, and C_g denotes the composition operator (5.0.1) generated by $g : \mathbb{R} \to \mathbb{R}$.
For convenience of the reader we will repeat here some assumptions which appear in previous chapters and will be needed in the study of these equations; we denote them

by (H1), (H2), ... without referring to similarly labeled in the Chapters 4 and 5. The following conditions will be used throughout the sequel; as before, the symbol $\forall' s$ means the indicated property holds only for almost all s.

$$\forall t \in [0,1]: \quad k(t,\cdot) \in L_1, \tag{H1}$$

$$\exists m \in L_1\ \forall' s \in [0,1]: \quad \operatorname{Var}\big(k(\cdot,s)\big) \leq m(s), \tag{H2}$$

$$g \in Lip_{loc}(\mathbb{R}). \tag{H3}$$

Note that condition (H1) is condition (A), and (H2) is (B) for the integral operator I_k of Section 4.3. Condition (H3) is the fundamental acting condition for C_g from Theorem 5.1.19.

The above conditions suffice to obtain the first existence and uniqueness result that has been proven in [29]. Recall that for any of our BV-spaces X we denote by

$$\mathbb{B}_R(X) = \big\{x \in X \mid \|x\|_X \leq R\big\}$$

the closed ball in X with respect to the norm $\|\cdot\|_X$ with radius $R > 0$ and centered at the function $\mathbb{0}$.

Theorem 7.1.1. *Assume (H1), (H2) and (H3), and let $h \in BV$ be fixed. Then for each $R > \|h\|_{BV}$ there is some $\varrho > 0$ such that equation (7.1.1) has for fixed $\lambda \in (-\varrho, \varrho)$ a unique solution in $\mathbb{B}_R(BV)$.*

To be more precise, if $R > \|h\|_{BV}$ is given, the number ϱ can be chosen to be

$$\varrho = \min\left\{\frac{R - \|h\|_{BV}}{\|g\|_{[-R,R]}}, \frac{1}{\operatorname{lip}(g,[-R,R])}\right\} \frac{1}{\|2m + |k(0,\cdot)|\|_{L_1}}. \tag{7.1.3}$$

This also shows that if g grows faster than linear, the number $\|g\|_{[-R,R]}$ grows also faster than linear and hence makes ϱ smaller the larger R is chosen. This means that if we want to make the domain $\mathbb{B}_R(BV)$ of possible solutions large, we have to pay the price that the set of admissible parameters λ for which uniqueness or solvability can be achieved becomes small. We illustrate this in the following two examples. The first shows generally how the minimum in (7.1.3) looks like if the kernel is given in separated kernels. The second is a concrete example of an integral equation with such a kernel and $h = \mathbb{0}$.

Example 7.1.2. If the kernel $k(t,s) = k_1(t)k_2(s)$ is given in separated kernel with $k_1 \in BV$ and $k_2 \in L_1$, then by Proposition 4.3.9 the conditions (H1) and (H2) are fulfilled, and we can simply put $m(s) = \operatorname{Var}(k_1)|k_2(s)|$. Moreover, the norm $\|2m + |k(\cdot,0)|\|_{L_1}$ occurring in (7.1.3) is then

$$\begin{aligned}\|2m + |k(0,\cdot)|\|_{L_1} &= \int_0^1 \Big|2\operatorname{Var}(k_1)|k_2(s)| + |k_1(0)k_2(s)|\Big|\,\mathrm{d}s \\ &= \big(2\operatorname{Var}(k_1) + |k_1(0)|\big)\,\|k_2\|_{L_1}.\end{aligned}$$

Thus, (7.1.3) becomes

$$\varrho = \min\left\{\frac{R - \|h\|_{BV}}{\|g\|_{[-R,R]}}, \frac{1}{\operatorname{lip}(g,[-R,R])}\right\}\frac{1}{\left(2\operatorname{Var}(k_1) + |k_1(0)|\right)\|k_2\|_{L_1}}.$$

Observe that the "larger" k_1, k_2 and g are the smaller becomes ϱ and hence the corresponding set of admissible parameters λ. ♢

As a further example which will frequently serve as a test animal in the sequel we consider for several values of $\alpha > -1$ and $\beta \in \mathbb{R}$ the integral equation

$$x(t) = \lambda t \int_0^1 \left((\alpha+1)|x(s)|^\alpha + 2\beta s\right) \mathrm{d}s \quad \text{for } 0 \le t \le 1 \tag{7.1.4}$$

in the space BV. This equation is of the form (7.1.1) with the functions $h = \mathbb{0}$, $k(t,s) = t$ and $g(u) = (\alpha+1)|u|^\alpha$ if and only if $\beta = 0$. The structure of (7.1.4) dictates that any solution must be of the form $x(t) = ct$ for some $c \in \mathbb{R}$ and is therefore automatically continuous (even analytic). Plugging this into (7.1.4) reveals the *characteristic equation*

$$c = \lambda\left(|c|^\alpha + \beta\right) \tag{7.1.5}$$

which means that a function $x \in BV$ is a solution of (7.1.4) if and only if it has the form $x(t) = ct$ with c satisfying (7.1.5). Theorem 7.1.1 is now applicable if and only if $\alpha \ge 1$ and $\beta = 0$ and guarantees in this case that for $|\lambda|$ smaller than ϱ in (7.1.3) equation (7.1.5) can be solved for c. We illustrate this in the next two examples. We start with $\alpha = 1$ and $\beta = 0$ for which (7.1.4) is a special case of (7.1.1).

Example 7.1.3. For $\alpha = 1$ and $\beta = 0$ equation (7.1.4) reduces to

$$x(t) = 2\lambda t \int_0^1 |x(s)| \,\mathrm{d}s. \tag{7.1.6}$$

In the notation of (7.1.1) we can take $h = \mathbb{0}$, $g(u) = 2|u|$ and $k(t,s) = k_1(t)k_2(s)$ in separated kernels with $k_1(t) = t$ and $k_2(s) = 1$. Obviously, the conditions (H1), (H2) and (H3) are satisfied with $\operatorname{Var}(k_1) = 1$, $k_1(0) = 0$, $\|k_2\|_{L_1} = 1$, $\|g\|_{[-R,R]} = 2R$ and $\operatorname{lip}(g,[-R,R]) = 2$. With the help of Example 7.1.2 we obtain for any $R > 0$ that $\varrho = 1/4$. In particular, Theorem 7.1.1 says that for any $\lambda \in (-1/4, 1/4)$ equation (7.1.6) has only one solution in BV, and clearly, this must be the function $\mathbb{0}$. Indeed, the characteristic equation (7.1.5) reads $c = \lambda|c|$ which has for $|\lambda| < 1/4$ only one solution, namely $c = 0$. ♢

We make two comments on the bound $\varrho = 1/4$ in the previous example. First, this bound and in general the bound given in (7.1.3) is not optimal in the sense that equation (7.1.1) may have a unique solution even for $|\lambda| \ge \varrho$. This is easily seen again in the previous example, because the characteristic equation $c = \lambda|c|$ has for any $|\lambda| \ne 1$ the unique solution $c = 0$. Even worse, for $\lambda = 1$ every $c \ge 0$ and for $\lambda = -1$ every $c \le 0$ is a solution which means that (7.1.6) has uncountably many different solutions in BV. Second, since $\varrho = 1/4$ is independent of R the set of admissible parameters λ is also independent of R which means that for these λ there can be only one solution x to (7.1.6) in the entire space BV. This is of course not always so, and the set of those λ may become small when R is chosen to be large. We illustrate this in the next example.

Example 7.1.4. Consider again equation (7.1.4), but now for $\alpha = 2$ and $\beta = 0$. It then reads

$$x(t) = 3\lambda t \int_0^1 x(s)^2 \,\mathrm{d}s. \tag{7.1.7}$$

In the notation of (7.1.1) we can again take $h = \mathbb{O}$ and $k(t,s) = k_1(t)k_2(s)$ with $k_1(t) = t$ and $k_2(s) = 1$, but this time, $g(u) = 3u^2$. Again, the conditions (H1), (H2) and (H3) are satisfied with $\operatorname{Var}(k_1) = 1$, $k_1(0) = 0$ and $\|k_2\|_{L_1} = 1$. Moreover, $\|g\|_{[-R,R]} = 3R^2$ and $\operatorname{lip}(g, [-R, R]) = 6R$. With the help of Example 7.1.2 we obtain for any $R > 0$,

$$\varrho = \min\left\{\frac{R}{3R^2}, \frac{1}{6R}\right\} \cdot \frac{1}{2} = \frac{1}{12R}.$$

In particular, ϱ gets smaller the larger R is chosen. $\diamondsuit$

As we have seen, the bound ϱ for the admissible parameters λ for which equation (7.1.1) has a BV-solution is in general strongly related to the radius R of the ball in which solutions can be guaranteed. One may enlarge the bound ϱ and therefore also the set of λ by replacing the norm $\|\cdot\|_{BV}$ by a smaller norm. For instance, if we use the norm $\|x\|^*_{BV} = |x(0)| + \operatorname{Var}(x)$ instead of the equivalent norm $\|x\|_{BV} = \|x\|_\infty + \operatorname{Var}(x)$, then it is shown in [29] that ϱ can be chosen to be

$$\varrho = \min\left\{\frac{R - \|h\|^*_{BV}}{\|g\|_{[-R,R]}}, \frac{1}{\operatorname{lip}(g, [-R, R])}\right\} \frac{1}{\|m + |k(0,\cdot)|\|_{L_1}}$$

Therefore, ϱ can be enlarged by a factor of up to 2. For instance, the bound ϱ in Example 7.1.4 would now be $\varrho = 1/(6R)$. However, for the rest of this thesis we stick to the norm $\|\cdot\|_{BV}$, because many estimates are then a little simpler.

Before we turn to a more general equation than (7.1.1), we point out that one might think that Theorem 7.1.1 could also be formulated in the following way.

Theorem 7.1.1*. *Under the assumptions (H1), (H2) and (H3) there exists a number $\varrho > 0$ such that for every λ satisfying $|\lambda| < \varrho$, the equation (7.1.1) has a unique solution in BV.*

Here, ϱ can be any number strictly less than the minimum in (7.1.3). This formulation, however, is not true, because it pretends that equation (7.1.1) has only one unique solution in the entire space BV for $|\lambda| < \varrho$. But the following example shows that under the hypothesis of Theorem 7.1.1 there can be more than one solution in BV.

Example 7.1.5. Consider again the integral equation (7.1.7) from Example 7.1.4. We have seen there that all the conditions (H1), (H2) and (H3) are fulfilled. Theorem 7.1.1* would now say that equation (7.1.7) has only one solution in BV for all $|\lambda| < \varrho$ and some $\varrho > 0$. However, there are in fact two BV-solutions for every $\lambda \neq 0$. To see this, we solve again the characteristic equation (7.1.5) for fixed $\lambda \neq 0$ which now is

$$c = \lambda c^2.$$

The first obvious solution $c = 0$ generates the function $x = \mathbb{0}$ which clearly is a solution to (7.1.7). However, also $c = 1/\lambda$ being the second solution of the characteristic equation generates the solution $x_\lambda(t) := t/\lambda$ with norm $\|x_\lambda\|_{BV} = 2/|\lambda|$ and so Theorem 7.1.1* cannot be true.
From Example 7.1.4 we get $\varrho = 1/(12R)$ for any $R > 0$. Thus, if $0 < |\lambda| < \varrho$, then $\|x_\lambda\|_{BV} = 2/|\lambda| > 24R > R$. This means that if $R > 0$ is chosen arbitrarily, ϱ is given by (7.1.3) and $0 < |\lambda| < \varrho$ is fixed, then equation (7.1.4) has two distinct BV-solutions, but the second one, namely x_λ, does not lie within the ball $\mathbb{B}_R(BV)$ and hence is not covered by Theorem 7.1.1. ♢

As the previous example shows, Theorem 7.1.1 yields uniqueness of the solution x of (7.1.1) indeed only in the ball $\mathbb{B}_R(BV)$. There may be other solutions outside of that ball.

Theorem 7.1.1 has been proved by showing that the operator $T = h + \lambda(I_k \circ C_g)$ has a fixed point in the ball $\mathbb{B}_R(BV)$ that is invariant under T. The conditions given in Theorem 7.1.1 to ensure the existence of such invariant balls have been generalized in the literature. We cite a sample result which uses milder a priori estimates. For example, in [33] the authors impose the following conditions:

- There exists a function $\psi : [0, \infty) \to [0, \infty)$ with $\psi(u) > 0$ for $u > 0$ and

$$|g(u)| \leq \psi\big(|u|\big) \quad \text{for } u \in \mathbb{R}.$$

- For each $R > 0$ there exists a continuous increasing function $\psi_R : [0, \infty) \to [0, \infty)$ with

$$|g(u) - g(v)| \leq \psi_R\big(|u - v|\big) \quad \text{for } |u|, |v| \leq R.$$

It is then shown that under additional appropriate assumptions, for each sufficiently large R there is some $\varrho > 0$ such that equation (7.1.1) has a solution in $\mathbb{B}_R(BV)$ for every $|\lambda| < \varrho$. Clearly, by putting $\psi(r) := \|g\|_{[-r,r]}$ and $\psi_R(r) := \mathrm{lip}(g, [-R, R])r$ in the above conditions, one can recover Theorem 7.1.1.
More general choices for ψ_R, however, are also possible. For instance, one could choose $\psi_R(r) = a(R)\arctan(r)$ or $\psi_R(r) = a(R)\log(1 + r)$ for appropriate functions $a : (0, \infty) \to (0, \infty)$ and therefore enlarge the range of applications of the results presented in [33].

We now consider a slightly more general nonlinear Hammerstein equation by replacing C_g in (7.1.2) by a superposition operator N_g, defined in (5.0.2) and generated by a function $g : [0,1] \times \mathbb{R} \to \mathbb{R}$. Then (7.1.1) reads

$$x(t) = h(t) + \lambda \int_0^1 k(t,s) g\big(s, x(s)\big)\,\mathrm{d}s \quad \text{for } 0 \le t \le 1; \tag{7.1.8}$$

note that our sample equation (7.1.4) has this form. The operator equation (7.1.2) becomes

$$x = h + \lambda (I_k \circ N_g) x. \tag{7.1.9}$$

Again, we impose three further conditions.

$$\sup_{t\in[0,1]} \|k(t,\cdot)\|_{L_1} < \infty, \tag{H4}$$

$$\sup_{\tau\in[0,1]} \operatorname{Var}\left(\int_0^\tau k(\cdot,s)\,\mathrm{d}s\right) < \infty, \tag{H5}$$

$$g \in Lip_{loc}([0,1]\times\mathbb{R}). \tag{H3*}$$

Note that (H3*) is a natural generalization of (H3) to two dimensions. Also note that (H4) implies condition (A) while (H5) coincides with condition (C) of Section 4.3. Moreover, (H1) and (H2) together imply (H4). This is, because if

$$|k(t,s)| \le \big|k(0,s) - k(t,s)\big| + |k(0,s)| \le \operatorname{Var}\big(k(\cdot,s)\big) + |k(0,s)| \le m(s) + |k(0,s)|$$

holds for all $t \in [0,1]$ and almost all $s \in [0,1]$ and some function $m \in L_1$, then

$$\|k(t,\cdot)\|_{L_1} \le \|m\|_{L_1} + \|k(0,\cdot)\|_{L_1},$$

and so the supremum in (H4) becomes at most $\|m\|_{L_1} + \|k(0,\cdot)\|_{L_1}$. Also, observe that (H3*) implies condition (G) of Section 5.2 which plays a crucial role in Theorem 5.2.12, because for $R > 0$, any partition $0 = t_0 < \ldots < t_n = 1$ of $[0,1]$ and points $u_0, \ldots, u_n \in [-R, R]$ with

$$\sum_{j=1}^n |u_{j-1} - u_j| \le R$$

we obtain

$$\sum_{j=1}^n |g(t_{j-1}, u_j) - g(t_j, u_j)| \le L_R \sum_{j=1}^n |t_{j-1} - t_j| = L_R,$$

$$\sum_{j=1}^n |g(t_{j-1}, u_{j-1}) - g(t_{j-1}, u_j)| \le L_R \sum_{j=1}^n |u_{j-1} - u_j| \le RL_R,$$

where

$$\begin{aligned} L_R &:= \operatorname{lip}(g, [0,1]\times[-R,R]) \\ &= \sup\left\{ \frac{\big|g(s,u) - g(t,v)\big|}{|s-t| + |u-v|} \mid s,t \in [0,1], u,v \in [-R,R], (s,u) \ne (t,v) \right\}. \end{aligned} \tag{7.1.10}$$

Thus, (G) follows with $M_R := \max\{L_R, RL_R\}$.

With these new conditions at hand the following result was proven in [32].

Theorem 7.1.6. *Assume (H3*), (H4) and (H5), and let $h \in BV$ be fixed. Then for each $R > \|h\|_{BV}$ there is some $\varrho > 0$ such that equation (7.1.8) has for fixed $\lambda \in (-\varrho, \varrho)$ a unique solution in $\mathbb{B}_R(BV)$.*

To be more precise, if $R > \|h\|_{BV}$, the number ϱ can be chosen so that

$$\varrho = \min\left\{\frac{R - \|h\|_{BV}}{\Big(2S_1 + \|k(0,\cdot)\|_{L_1}\Big)\Big(L_R + RL_R + |g(0,0)|\Big)}, \frac{1}{L_R S_2}\right\}, \tag{7.1.11}$$

where S_1 is the supremum in (H5), S_2 is the supremum in (H4) and $L_R = \mathrm{lip}(g, [0,1] \times [-R, R])$ is defined as in (7.1.10).

A few remarks are in order. First, an analogue to Theorem 7.1.6 in the sense of Theorem 7.1.1* is also wrong, as we have seen in Example 7.1.5. Nevertheless, Theorem 7.1.6 as stated here has a wider range of applications than Theorem 7.1.1, for two reasons. First, superposition operators are of course more general than composition operators. Second, there are kernel functions which satisfy the condition (H4) and (H5) but not condition (H2).

Example 7.1.7. Let $k(t,s) = \chi_{\mathbb{Q}}(t-s)$ be the kernel function that we have already studied in the Examples 4.3.3 and 4.3.8. We have seen there that k satisfies (H1) and (H5) but not (H2). Since, for fixed $t \in [0,1]$, we have

$$\|k(t,\cdot)\|_{L_1} = \int_0^1 \chi_{\mathbb{Q}}(t-s)\,\mathrm{d}s = 0,$$

the kernel function k also satisfies (H4). In particular, Theorem 7.1.6 is applicable provided that $g \in Lip_{loc}([0,1] \times \mathbb{R})$, but Theorem 7.1.1 is not. However, for this particular kernel, the integral equations (7.1.1) and (7.1.8) have only the solution $x = h$, since the integral vanishes in both equations. Note that in this case, Theorem 7.1.6 yields indeed uniqueness of solutions in the entire space BV, because $\varrho = \infty$ in (7.1.11) for any $R > \|h\|_{BV}$. ◇

It is clear that Theorem 7.1.6 is applicable to our test equation (7.1.4) with $k(t,s) = t$ and $g(t,u) = (\alpha+1)|u|^\alpha + 2\beta t$ if and only if $\alpha \geq 1$ and β is arbitrary.
As we have seen in Example 7.1.3 the integral equation (7.1.6) can have infinitely many solutions. We now consider a similar example which has for $\beta > 0$ at most two solutions and illustrates the bound (7.1.11).

Example 7.1.8. This time, we have a look at (7.1.4) for $\alpha = 1$ and arbitrary β which now reads

$$x(t) = 2\lambda t \int_0^1 \Big(|x(s)| + \beta s\Big)\,\mathrm{d}s. \tag{7.1.12}$$

In the notation of (7.1.8) we have $h = \mathbb{0}$, $k(t,s) = t$ and $g(t,u) = 2|u| + 2\beta t$. Thus,

$\|k(0,\cdot)\|_{L_1} = 0$, $|g(0,0)| = 0$ and

$$S_1 = \sup_{\tau\in[0,1]} \operatorname{Var}\left(\int_0^\tau k(\cdot,s)\,\mathrm{d}s\right) = 1,$$
$$S_2 = \sup_{t\in[0,1]} \|k(t,\cdot)\|_{L_1} = 1,$$
$$L_R = \operatorname{lip}\left(g,[0,1]\times[-R,R]\right) = 2\max\{1,|\beta|\}.$$

We obtain from (7.1.11),

$$\rho(R) = \min\left\{\frac{R}{4(1+R)\max\{1,|\beta|\}}, \frac{1}{2\max\{1,|\beta|\}}\right\} = \frac{R}{4(1+R)\max\{1,|\beta|\}} < \frac{1}{4}$$

for $R > 0$. Since $\rho(R) \to 1/(4\max\{1,|\beta|\}) =: \varrho_0 > 0$ as $R \to \infty$, Theorem 7.1.6 now even states that equation (7.1.12) has for every $\lambda \in (-\varrho_0,\varrho_0)$ a unique solution in the entire space BV.
Indeed, the characteristic equation of (7.1.12) is

$$c = \lambda(|c|+\beta). \tag{7.1.13}$$

For $\beta > 0$ it has no solution if $|\lambda| \geq 1$ and a unique solution if $|\lambda| < 1$, namely $c = \lambda\beta/(1-|\lambda|)$. For $\beta = 0$ is has only $c = 0$ as a solution if $|\lambda| \neq 1$, but any $c \leq 0$ if $\lambda = -1$ and any $c \geq 0$ if $\lambda = 1$, as we have seen in the discussion after Example 7.1.3. In this case, (7.1.13) and therefore (7.1.12) have infinitely many solutions. For $\beta < 0$, equation (7.1.13) has one or two solutions, namely $c = \lambda\beta/(1-\lambda)$ for $\lambda \notin (0,1]$ and $c = \lambda\beta/(1+\lambda)$ for $\lambda \notin [-1,0]$.
As before, the bound $\varrho_0 \leq 1/4$ is not optimal. ◇

In contrast to equation (7.1.6) from Example 7.1.3 which has a unique solution for every $|\lambda| \neq 1$ and infinitely many solutions for $|\lambda| = 1$, equation (7.1.12) has no solutions whatsoever for $|\lambda| \geq 1$ and $\beta > 0$.

We have now solved (7.1.4) for $\alpha = 1$ and arbitrary β. In the next example we will solve it also for $\alpha = 2$ and arbitrary β. Since then the characteristic equation (7.1.5) reduces to a simple quadratic equation, we do not need so many case distinctions as in Example 7.1.8.

Example 7.1.9. Consider (7.1.4) for $\alpha = 2$ but arbitrary β, that is,

$$x(t) = \lambda t \int_0^1 \left(3x(s)^2 + 2\beta s\right)\mathrm{d}s \tag{7.1.14}$$

with characteristic equation $c = \lambda(c^2+\beta)$ which is equivalent to the quadratic equation

$$\lambda c^2 - c + \lambda\beta = 0. \tag{7.1.15}$$

For $\lambda = 0$ there is only the solution $c = 0$. For $\lambda \neq 0$, this quadratic equation has two solutions if $4\lambda^2\beta < 1$, namely

$$c = \frac{1 \pm \sqrt{1-4\lambda^2\beta}}{2\lambda},$$

only one if $4\lambda^2\beta = 1$, namely $c = 1/(2\lambda)$, and no (real) solution if $4\lambda^2\beta > 1$. In particular, for $\beta = 0$ we regain the two solutions $c = 0$ and $c = 1/\lambda$ found in Example 7.1.5. ♢

Let us come back for a second to condition (H3*) which states that $g : [0,1] \times \mathbb{R} \to \mathbb{R}$ is locally Lipschitz continuous with respect to both arguments. We will see later that for our purposes this condition is too restrictive. Thus, we will rather use the following weaker condition

$$\forall R > 0\ \exists L_R > 0\ \forall t \in [0,1]\ \forall u, v \in [-R, R] : \quad \left|g(t,u) - g(t,v)\right| \le L_R|u - v| \quad \text{(H6)}$$

which is precisely condition (B) of Section 5.2 and may be reformulated equivalently as

$$\forall R > 0\ \exists L_R > 0\ \forall u, v \in [-R, R] : \quad \|g(\cdot,u) - g(\cdot,v)\|_\infty \le L_R|u - v|.$$

It imposes a local Lipschitz condition on the second variable, uniformly in the first, but does not impose a Lipschitz condition on the first variable. Clearly, (H6) is indeed weaker than (H3*), and here is an explicit

Example 7.1.10. The function $g(t,u) = \sqrt{t}\,u$ for $t \in [0,1]$ and $u \in \mathbb{R}$ satisfies (H6) with $L_R = 1$ for all $R > 0$, but it does not satisfy (H3*), because $t \mapsto \sqrt{t}$ is not Lipschitz continuous near $t = 0$. ♢

We will see later in our main Theorem 7.1.16 that it suffices to require the inequality in (H6) to be true for only almost all $t \in [0,1]$.

Since the interest in finding solutions to equations like (7.1.1) and (7.1.8) is motivated by problems from physics, biology, economics and other sciences, sometimes it is necessary to consider continuous BV-solutions.
Here we impose the following two conditions

$$\begin{aligned} &\forall \varepsilon > 0\ \exists \delta > 0\ \forall t_1, t_2 \in [0,1]\ \forall' s \in [0,1] : \\ &\qquad |t_1 - t_2| \le \delta \ \Rightarrow\ \left|k(t_1,s) - k(t_2,s)\right| \le \varepsilon, \end{aligned} \quad \text{(H7)}$$

$$\begin{aligned} &\forall \varepsilon > 0\ \exists \delta > 0\ \forall t_1, t_2, \tau \in [0,1] : \\ &\qquad |t_1 - t_2| \le \delta \ \Rightarrow\ \left|\int_0^\tau k(t_1,s) - k(t_2,s)\,\mathrm{d}s\right| \le \varepsilon. \end{aligned} \quad \text{(H8)}$$

Note that (H8) is precisely condition (E) from Section 4.3 and is crucial in Theorem 4.3.14 which provides a necessary and sufficient condition for the integral operator I_k to be bounded on $BV \cap C$. Moreover, it is clear that (H7) implies (H8). The reverse implication, however, does not hold in general.

Example 7.1.11. Consider a kernel in separated kernels with $k(t,s) = k_1(t)k_2(s)$, where k_1 is injective and continuous and $k_2 \in L_1 \backslash L_\infty$. Then

$$\left|k(t_1,s) - k(t_2,s)\right| = |k_1(t_1) - k_1(t_2)||k_2(s)|$$

gets unbounded for fixed $t_1 \neq t_2$, because k_2 is essentially unbounded. Thus, (H7) fails. However, (H8) is satisfied. To see this fix $\varepsilon > 0$ and pick $\delta > 0$ so small that $|k_1(t_1) - k_1(t_2)| \leq \varepsilon / \|k_2\|_{L_1}$ for $|t_1 - t_2| \leq \delta$; note that k_1 is also uniformly continuous and $k_2 \notin L_\infty$ implies $\|k_2\|_{L_1} > 0$. Then

$$\left| \int_0^T k(t_1, s) - k(t_2, s)\,\mathrm{d}s \right| = |k_1(t_1) - k_2(t_2)| \left| \int_0^T k_2(s)\,\mathrm{d}s \right| \leq |k_1(t_1) - k_2(t_2)|\, \|k_2\|_{L_1} \leq \varepsilon,$$

and (H8) is established.

For a sturdy example take, for instance, $k_1(t) = t$ and $k_2(s) = 1/\sqrt{s}$ for $0 < s \leq 1$ and $k_2(0) = 0$. ◇

The stronger condition (H7) is now essential for the following result that has been proven in the paper [29].

Theorem 7.1.12. *Assume (H1), (H2), (H3) and (H7), and let $h \in BV \cap C$ be fixed. Then for each $R > \|h\|_{BV}$ there is some $\varrho > 0$ such that equation (7.1.1) has for fixed $\lambda \in (-\varrho, \varrho)$ a unique solution in $\mathbb{B}_R(BV \cap C)$, where $BV \cap C$ is equipped with the norm $\|\cdot\|_{BV}$.*

Since the kernel $k(t, s) = t$ clearly satisfies (H7), our test equation (7.1.4) can be solved using Theorem 7.1.12 if and only if $\alpha \geq 1$ and $\beta = 0$. As we have seen, any solution to (7.1.4) must be of the form $x(t) = ct$ and hence is automatically continuous.

The proofs of all the preceding existence results heavily base on the Banach-Caccioppoli Contraction Principle, which explains why we also obtain uniqueness of solutions. We now present a result that has been proven with the fixed point theorem of Schauder. The advantage is that we may impose weaker conditions on g and that we get existence of solutions for *every* $\lambda \in \mathbb{R}$. But of course, we have to pay a price for this: We loose uniqueness of solutions in balls.

The first result in this direction deals with equation (7.1.1) for $h = \mathbb{0}$, that is, with fixed points of the operator $\lambda(I_k \circ C_g)$. For this we need to impose the following conditions on $g : \mathbb{R} \to \mathbb{R}$.

$$\exists q > 1\ \forall R > 0 : \quad g \in RBV_q[-R, R], \tag{H9}$$

$$\lim_{|u| \to \infty} \frac{|g(u)|}{|u|} = 0. \tag{H10}$$

Here, RBV_q denotes the Riesz space introduced in Definition 1.2.24. Condition (H10) means that g obeys a strictly sublinear growth for large values of the argument. The following result was proven in [31].

Theorem 7.1.13. *Under the assumptions (H1), (H2), (H9) and (H10) the equation (7.1.1) has for $h = \mathbb{0}$ and every $\lambda \in \mathbb{R}$ a solution $x \in BV$.*

For our test equation (7.1.4) with $\beta = 0$ the condition (H9) is equivalent to $\alpha \geq 0$ whereas condition (H10) restricts α to be less than 1. Thus, in total, Theorem (7.1.13) can be applied to (7.1.4) if and only if $0 \leq \alpha < 1$ and $\beta = 0$. We give an explicit example for the case $\alpha = 1/2$.

Example 7.1.14. Equation (7.1.4) reads for $\alpha = 1/2$ and $\beta = 0$,

$$x(t) = \frac{3}{2}\lambda t \int_0^1 \sqrt{|x(s)|}\,\mathrm{d}s \tag{7.1.16}$$

with characteristic equation (7.1.5)

$$c = \lambda\sqrt{|c|}.$$

This equation has for every $\lambda \in \mathbb{R}$ at most two solutions, namely the unique solution $c = 0$ for $\lambda = 0$, the two solutions $c = 0$ and $c = \lambda^2$ for $\lambda > 0$, and the two solutions $c = 0$ and $c = -\lambda^2$ for $\lambda < 0$. ◊

Thus, Example 7.1.14 shows that we cannot expect uniqueness of the solutions given by Theorem 7.1.13.

In fact, the authors of [32] proved a more general result, namely Theorem 7.1.13 for arbitrary $h \in BV$ and the nonautonomous case $g : [0,1] \times \mathbb{R} \to \mathbb{R}$. More precisely, they showed that

$$H(\tau, x)(t) := \tau h(t) + \tau\lambda \int_0^1 k(t,s) g\big(s, x(s)\big)\,\mathrm{d}s \quad \text{for } 0 \le t, \tau \le 1$$

defines (under some additional hypothesis) an admissible compact homotopy joining the operator $H(1,x) = h + \lambda(I_k \circ N_g)x$ and the operator $H(0,x) = \mathbb{0}$ on a suitable ball in BV, and then applied the Leray-Schauder degree on that ball. The growth condition (H10) has then to be replaced by its nonautonomous version

$$\lim_{|u|\to\infty} \frac{\|g(\cdot,u)\|_{L_\infty}}{|u|} = 0. \tag{7.1.17}$$

The following example which generalizes Example 7.1.14 shows that even under the hypothesis (7.1.17) solutions to the integral equation (7.1.8) may not be unique.

Example 7.1.15. Consider again (7.1.4) for $\alpha = 1/2$, but this time for arbitrary β. It reads

$$x(t) = \lambda t \int_0^1 \left(\frac{3}{2}\sqrt{|x(s)|} + 2\beta s\right)\mathrm{d}s \tag{7.1.18}$$

with corresponding characteristic equation

$$c = \lambda\left(\sqrt{|c|} + \beta\right). \tag{7.1.19}$$

To solve this equation we consider for $\beta \in \mathbb{R}$ the function φ_β, defined by

$$\varphi_\beta : D_\beta \to \mathbb{R},\ t \mapsto \frac{t}{\sqrt{|t|} + \beta} \quad \text{and} \quad D_\beta := \begin{cases} \mathbb{R} & \text{for } \beta > 0, \\ \mathbb{R}\backslash\{-\beta^2, \beta^2\} & \text{for } \beta \le 0. \end{cases}$$

For $\beta > 0$ the function $\varphi_\beta : \mathbb{R} \to \mathbb{R}$ is bijective justifying why the equation $\varphi_\beta(c) = \lambda$ and hence also (7.1.19) has for given $\lambda \in \mathbb{R}$ exactly one solution c. For $\beta = 0$ equation

(7.1.19) has at most two solutions, namely one for $\lambda = 0$ and two for $\lambda \neq 0$, as we have seen in Example 7.1.14. We therefore focus on $\beta < 0$; Figure 7.1.1 shows φ_β for some fixed $\beta < 0$.

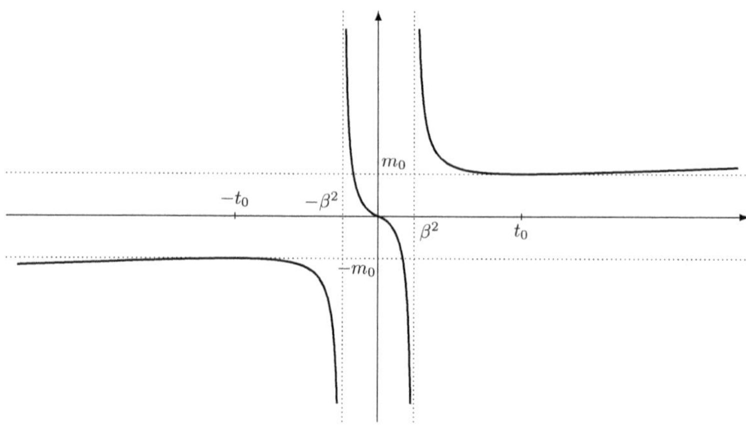

Figure 7.1.1: The function φ_β for $\beta < 0$.

Since the function φ_β is symmetric with respect to $(0,0)$, we only consider it for $t > 0$. As

$$\lim_{t \to \beta^2+} \varphi_\beta(t) = \infty = \lim_{t \to \infty} \varphi_\beta(t),$$

the function φ_β must have a minimum in (β^2, ∞). Since

$$\varphi_\beta'(t) = \frac{2\beta + \sqrt{t}}{2\left(\beta + \sqrt{t}\right)^2} \quad \text{for } t > 0,$$

this minimum must be located at $t_0 := 4\beta^2$ at which φ_β is

$$m_0 := \varphi_\beta(t_0) = 4\beta^2/(\beta + 2|\beta|) = -4\beta > 0,$$

and φ_β is strictly decreasing in (β^2, t_0) and strictly increasing in (t_0, ∞). Consequently, for $\lambda < m_0$ equation (7.1.19) has no solution $c > \beta^2$, for $\lambda = m_0$ it has exactly one solution $c > \beta^2$, namely $c = t_0$, and for $\lambda > m_0$ it has two solutions $c > \beta^2$. Since

$$\varphi_\beta(0) = 0 \qquad \text{and} \qquad \lim_{t \to \beta^2-} \varphi_\beta(t) = -\infty \qquad \text{and} \qquad \varphi_\beta'(t) < 0 \quad \text{for } t \in \left(0, \beta^2\right),$$

φ_β is strictly decreasing from 0 to $-\infty$ on $[0, \beta^2)$. Thus, for $\lambda \leq 0$ equation (7.1.19) has one further solution in $[0, \beta^2)$.

Due to the symmetry we conclude that for $\lambda > -m_0$, equation (7.1.19) has no solution $c < -\beta^2$, for $\lambda = -m_0$ it has exactly one solution $c < -\beta^2$, namely $c = -t_0$, and for $\lambda < -m_0$ it has two solutions $c < -\beta^2$. Moreover, for $\lambda > 0$ equation (7.1.19) has one further solution in $(-\beta^2, 0)$. ◇

Since we now have fully solved equation (7.1.4) for the three cases $\alpha \in \{1/2, 1, 2\}$ we organize in Table 7.1.1 the number of solutions for all combinations of the parameters α, β and λ.

Table 7.1.1: Number of solutions of equation (7.1.4) for $\alpha \in \{1/2, 1, 2\}$.

parameters			number of solutions
$\alpha = 2$	$\beta = 0$	$\lambda \neq 0$	2
		$\lambda = 0$	1
	$\beta \neq 0$	$\lambda = 0$	1
		$0 < 4\beta\lambda^2 < 1$	2
		$4\beta\lambda^2 = 1$	1
		$4\beta\lambda^2 > 1$	0
$\alpha = 1$	$\beta > 0$	$\lvert\lambda\rvert < 1$	1
		$\lvert\lambda\rvert \geq 1$	0
	$\beta = 0$	$\lvert\lambda\rvert = 1$	∞
		$\lvert\lambda\rvert \neq 1$	1
	$\beta < 0$	$\lvert\lambda\rvert > 1$	2
		$\lvert\lambda\rvert \leq 1$	1
$\alpha = 1/2$	$\beta > 0$	$\lambda \in \mathbb{R}$	1
	$\beta = 0$	$\lambda \neq 0$	2
		$\lambda = 0$	1
	$\beta < 0$	$\lvert\lambda\rvert > -4\beta$	3
		$\lvert\lambda\rvert = -4\beta$	2
		$\lvert\lambda\rvert < -4\beta$	1

We now generalize the previously discussed equations (7.1.1) and (7.1.8) and consider the equation

$$x(t) = h\big(t, x(t)\big) + \lambda f\big(t, x(t)\big) \int_0^1 k(t,s) g\big(s, x(s)\big)\,\mathrm{d}s \quad \text{for } 0 \leq t \leq 1 \tag{7.1.20}$$

which can be written as the operator equation

$$x = N_h x + \lambda N_f x (I_k \circ N_g) x. \tag{7.1.21}$$

Clearly, one can recover (7.1.1) by setting $h(t,u) = \tilde{h}(t)$, $f(t,u) = 1$ and $g(t,u) = \tilde{g}(u)$ and (7.1.8) by letting $h(t,u) = \tilde{h}(t)$ and $f(t,u) = 1$.

In order to solve equation (7.1.20) in its full generality we use the fixed point theorem of Banach-Caccioppoli. Since we do not know any nontrivial condition for compactness of the superposition operator we also solve (7.1.20) - but only special cases - with a variant of Darbo's fixed point theorem which is also a generalization of Krasnoselskii's fixed point theorem. The former will get us a unique solution but only for small λ while the latter lets us find solutions for any λ which may not be unique. However, in any case we need due to the generality of the equation (7.1.20) a lot of conditions on the data involved. Although all these conditions have been presented in previous chapters we will repeat them here for the convenience of the reader.

For a function $g : [0,1] \times D \to \mathbb{R}$ with $D \in \{[0,1], \mathbb{R}\}$ and a space X of real-valued functions on $[0,1]$ and a space Y of real-valued functions on D we impose

$$g(\cdot, 0) \in X, \tag{H11(X)}$$

$$\forall u \in \mathbb{R} : \quad g(\cdot, u) \in X, \tag{H11*(X)}$$

$$\forall t \in [0,1] : \quad g(t, \cdot) \in Y, \tag{H12(Y)}$$

as well as

$$\forall R > 0 \; \exists L_R > 0 \; \exists z_R \in X \; \forall s,t \in [0,1] \; \forall u,v \in [-R,R] : \\ |g(s,u) - g(s,v) - g(t,u) + g(t,v)| \leq L_R |z(s) - z(t)||u - v|, \tag{H13(X)}$$

where the last condition has been recovered from (v) of Theorem 5.2.31. Observe that (H12(L_1)) is precisely condition (H1). Moreover, we define

$$\forall R > 0 \; \exists a_R \in L_\infty \; \forall t \in [0,1] \; \forall u,v \in [-R,R] : \\ |g(t,u) - g(t,v)| \leq a_R(t)|u - v|, \tag{H6*}$$

and that is a light generalization of condition (H6) which was defined as

$$\forall R > 0 \; \exists L_R > 0 \; \forall u,v \in [-R,R] : \quad \|g(\cdot,u) - g(\cdot,v)\|_\infty \leq L_R|u - v|. \tag{H6}$$

So, for measurable g clearly (H6) implies (H6*) with $a_R \equiv L_R$, but not vice versa, because in (H6*) the function a_R may be unbounded. Finally, for a function $k : [0,1] \times [0,1] \to \mathbb{R}$ and a BV-space X we recall the (B)-type conditions of Section 4.3 which we have summarized as follows (also see Table 4.3.2).

$$\exists \theta > 0 \; \exists m \in L_1 \; \forall' s \in [0,1] : \quad \mathrm{Var}_X\big(\theta k(\cdot, s)\big) \leq m(s), \tag{H14}$$

where Var_X stands for the variation of the space X, that is, $\mathrm{Var}_{BV} = \mathrm{Var}$, $\mathrm{Var}_{WBV_p} = \mathrm{Var}_p$, $\mathrm{Var}_{YBV_\varphi} = \mathrm{Var}_\varphi$, $\mathrm{Var}_{\Lambda BV} = \mathrm{Var}_\Lambda$ and $\mathrm{Var}_{RBV_p} = \mathrm{RVar}_p$. As we have seen in Section 4.3 the scaling factor θ is relevant only in the space YBV_φ when φ is arbitrary, because (H14) is precisely condition (B_X) from that section. In all other BV-space BV, WBV_p, ΛBV and RBV_p we can always assume $\theta = 1$; nevertheless, Table 4.3.2 lists (H14) in all BV-spaces separately. Armed with this arsenal of conditions we are now in position to formulate and proof our main result of this section.

Theorem 7.1.16. *Let X be any of the spaces BV, WBV_p, YBV_φ, ΛBV or RBV_p, and let $f, g, h : [0,1] \times \mathbb{R} \to \mathbb{R}$ and $k : [0,1] \times [0,1] \to \mathbb{R}$ be functions. Assume that*

(i) k satisfies (H12(L_1)) and (H14),

(ii) g satisfies (H11(L_∞)) and (H6*),*

(iii) f satisfies (H11(X)), (H12($C^1(\mathbb{R})$)) and (H13(X)),

(iv) $\partial_2 f$ *satisfies (H11(B)) and (H6),*

(v) h *satisfies (H11(X)), (H12(*$C^1(\mathbb{R})$*)) and (H13(X)) with*

$$|h(s,u) - h(s,v) - h(t,u) + h(t,v)| \le E_R|w_R(s) - w_R(t)||u - v|$$

and $w_R \in X$*,*

(vi) $\partial_2 h$ *satisfies (H11(B)) and (H6) with*

$$\|\partial_2 h(\cdot,u) - \partial_2 h(\cdot,v)\|_\infty \le C_R|u - v|.$$

Then for each $R > 0$ *satisfying*

$$\sup_{\|x\|_X \le R} \left\|h\Big(\cdot, x(\cdot)\Big)\right\|_X < R \quad \text{and} \quad 7RC_R + E_R\Phi_X(w_R) + \|\partial_2 h(\cdot,0)\|_\infty < 1 \tag{7.1.22}$$

there is some $\varrho > 0$ *such that equation (7.1.20) has for fixed* $\lambda \in (-\varrho, \varrho)$ *a unique solution in* $\mathbb{B}_R(X)$*, where* Φ_X *is as in Table 1.2.1.*

Before we give the proof, let us summarize and comment the requirements (i)–(vi) in Theorem 7.1.16 even at the risk of being redundant in a more accessible and compressed way. The (H11)-type conditions in (ii)–(vi) mean nothing but

$$g(\cdot,u) \in L_\infty \ \forall u \in \mathbb{R}, \quad f(\cdot,0), h(\cdot,0) \in X, \quad \partial_2 f(\cdot,0), \partial_2 h(\cdot,0) \in B. \tag{7.1.23}$$

The (H12)-type conditions in (i), (iii) and (v) read

$$\forall t \in [0,1]: \ k(t,\cdot) \in L_1, \quad f(t,\cdot), h(t,\cdot) \in C^1(\mathbb{R}) \tag{7.1.24}$$

and justify simultaneously the existence of $\partial_2 h$ and $\partial_2 f$ in (7.1.23). The condition (H14) in (i) is again

$$\exists \theta > 0 \ \exists m \in L_1 \ \forall' s \in [0,1]: \ \mathrm{Var}_X\Big(\theta k(\cdot,s)\Big) \le m(s). \tag{7.1.25}$$

The (H6)-type and (H13)-type conditions in (ii)–(vi) can be summarized as follows: For each $R > 0$ there are constants $A_R, B_R, C_R, D_R, E_R > 0$ and functions $z_R, w_R \in X$ such that

$$\forall' t \in [0,1] \ \forall u,v \in [-R,R]: \quad \Big|g(t,u) - g(t,v)\Big| \le A_R|u - v|, \tag{7.1.26}$$

as well as for all $u, v \in [-R, R]$,

$$\|\partial_2 f(\cdot,u) - \partial_2 f(\cdot,v)\|_\infty \le B_R|u - v|, \tag{7.1.27}$$

$$\|\partial_2 h(\cdot,u) - \partial_2 h(\cdot,v)\|_\infty \le C_R|u - v|, \tag{7.1.28}$$

$$\begin{aligned} &\forall s,t \in [0,1]: \\ &\qquad |f(s,u) - f(s,v) - f(t,u) + f(t,v)| \le D_R|z_R(s) - z_R(t)||u - v|, \end{aligned} \tag{7.1.29}$$

$$\begin{aligned} &\forall s,t \in [0,1]: \\ &\qquad |h(s,u) - h(s,v) - h(t,u) + h(t,v)| \le E_R|w_R(s) - w_R(t)||u - v|. \end{aligned} \tag{7.1.30}$$

These conditions together with the Theorems 4.3.21, 5.2.31 and 5.2.34 will then guarantee that the single components N_h, N_f and $I_k \circ N_g$ in equation (7.1.21) satisfy a Lipschitz condition in such a way that the entire operator $N_h + \lambda N_f \cdot (I_k \circ N_g)$ maps the ball $\mathbb{B}_R(X)$ into itself and is a contraction provided that λ is sufficiently small and R is sufficiently large.
Let us now take a deep breath for the

Proof of Theorem 7.1.16. Fix $R > 0$ so that (7.1.22) holds and let the quantities $A_R, B_R, C_R, D_R, E_R > 0$ and $z_R, w_R \in X$ be as in (7.1.26)–(7.1.30). We start with $I_k \circ N_g$ and define $\gamma_X = \gamma_X(k, m, \theta)$ to be the bound on the norm of I_k as in (4.3.6), that is,

$$\gamma_X := \|k(0,\cdot)\|_{L_1} + \theta^{-1} \begin{cases} 2\,\|m\|_{L_1} & \text{for } X = BV, \\ 2\,\|m\|_{L_1}^{1/p} & \text{for } X = WBV_p, \\ \left(\varphi^{-1}(1) + 1\right) \max\{1, \|m\|_{L_1}\} & \text{for } X = YBV_\varphi, \\ \left(1 + \lambda_1^{-1}\right) \|m\|_{L_1} & \text{for } X = \Lambda BV, \\ 2\,\|m\|_{L_1}^{1/p} & \text{for } X = RBV_p, \end{cases}$$

where θ and m are as in (H14). Then γ_X is well-defined and finite by (i). By Theorem 4.3.21 the integral operator I_k maps L_∞ into X and satisfies

$$\|I_k x\|_X \le \gamma_X \|x\|_{L_\infty} \quad \text{for } x \in L_\infty. \tag{7.1.31}$$

By (ii) the function g satisfies all conditions of Theorem 5.2.34 with the Lipschitz constant A_R given in (7.1.26). Accordingly, the operator N_g maps X into L_∞ with

$$\|N_g x - N_g y\|_{L_\infty} \le A_R \|x - y\|_X \quad \text{for } x, y \in \mathbb{B}_R(X).$$

This, the linearity of I_k and (7.1.31) imply for such x and y,

$$\|(I_k \circ N_g)x - (I_k \circ N_g)y\|_X \le \gamma_X A_R \|x - y\|_X, \tag{7.1.32}$$

as well as

$$\begin{aligned} \|(I_k \circ N_g)x\|_X &\le \|(I_k \circ N_g)x - (I_k \circ N_g)\mathbb{0}\|_X + \|I_k N_g \mathbb{0}\|_X \\ &\le \gamma_X A_R \|x\|_X + \gamma_X \|g(\cdot, 0)\|_{L_\infty} \\ &\le \gamma_X \left(R A_R + \|g(\cdot, 0)\|_{L_\infty}\right). \end{aligned} \tag{7.1.33}$$

We now turn to the operators N_f and N_h. By (iii)–(vi) the functions f and h satisfy all hypotheses of Theorem 5.2.31 with the necessary constants given in (7.1.27) and (7.1.29) for f and in (7.1.28) and (7.1.30) for h. Accordingly, the operators N_f and N_h map the space X into itself, and by (5.2.23) they satisfy the estimates

$$\|N_f x - N_f y\|_X \le \left(7 R B_R + D_R \Phi_X(z_R) + \|\partial_2 f(\cdot, 0)\|_\infty\right) \|x - y\|_X, \tag{7.1.34}$$

$$\|N_h x - N_h y\|_X \le \left(7 R C_R + E_R \Phi_X(w_R) + \|\partial_2 h(\cdot, 0)\|_\infty\right) \|x - y\|_X \tag{7.1.35}$$

for all $x, y \in \mathbb{B}_R(X)$. Moreover,

$$\begin{aligned}\|N_f x\|_X &\leq \|N_f x - N_f \mathbb{0}\|_X + \|N_f \mathbb{0}\|_X \\ &\leq 7R^2 B_R + R D_R \Phi_X(z_R) + R \|\partial_2 f(\cdot, 0)\|_\infty + \|f(\cdot, 0)\|_X \end{aligned} \tag{7.1.36}$$

for these x. Setting

$$\begin{aligned} L_1 &:= 7RC_R + E_R \Phi_X(w_R) + \|\partial_2 h(\cdot, 0)\|_\infty, \\ M_1 &:= \sup_{\|x\|_X \leq R} \left\| h\Big(\cdot, x(\cdot)\Big) \right\|_X, \\ L_2 &:= 7RB_R + D_R \Phi_X(z_R) + \|\partial_2 f(\cdot, 0)\|_\infty, \\ M_2 &:= 7R^2 B_R + R D_R \Phi_X(z_R) + R \|\partial_2 f(\cdot, 0)\|_\infty + \|f(\cdot, 0)\|_X, \\ L_3 &:= \gamma_X A_R, \\ M_3 &:= \gamma_X \left(R A_R + \|g(\cdot, 0)\|_{L_\infty} \right), \end{aligned}$$

as well as $T_1 := N_h$, $T_2 := N_f$ and $T_3 := I_k \circ N_g$, we thus have in total

$$\|T_j x - T_j y\|_X \leq L_j \|x - y\|_X \quad \text{and} \quad \|T_j x\|_X \leq M_j \quad \text{for } x, y \in \mathbb{B}_R(X), j \in \{1, 2, 3\}.$$

This follows for $j = 1$ from (7.1.35) and the definition of M_1, for $j = 2$ from (7.1.34) and (7.1.36), and for $j = 3$ from (7.1.32) and (7.1.33).
By (7.1.22), we have $M_1 < R$ and $L_1 < 1$, and we can therefore pick $\varrho > 0$ so that

$$M_1 + \varrho M_2 M_3 \leq R \quad \text{and} \quad L_1 + \varrho(M_2 L_3 + M_3 L_2) \leq 1. \tag{7.1.37}$$

Any fixed $\lambda \in (-\varrho, \varrho)$ then satisfies

$$M_1 + |\lambda| M_2 M_3 < R \quad \text{and} \quad L_1 + |\lambda|(M_2 L_3 + M_3 L_2) < 1. \tag{7.1.38}$$

It remains to show that $T := T_1 + \lambda T_2 \cdot T_3$ maps $\mathbb{B}_R(X)$ into itself and is a contraction for $\lambda \in (-\varrho, \varrho)$. On the one hand, for fixed $x \in \mathbb{B}_R(X)$ we obtain from the left part of the estimates in (7.1.38),

$$\begin{aligned}\|Tx\|_X = \|T_1 x + \lambda(T_2 x)(T_3 x)\|_X &\leq \|T_1 x\|_X + |\lambda| \, \|T_2 x\|_X \, \|T_3 x\|_X \\ &\leq M_1 + |\lambda| M_2 M_3 \leq R; \end{aligned}$$

in particular, T maps $\mathbb{B}_R(X)$ into itself. Note that we have here used the fact that X is a normalized algebra. On the other hand, for $x, y \in \mathbb{B}_R(X)$ we get

$$\begin{aligned} &\|Tx - Ty\|_X \\ &\quad = \|T_1 x - T_1 y + \lambda(T_2 x)(T_3 x) - \lambda(T_2 y)(T_3 y)\|_X \\ &\quad \leq \|T_1 x - T_1 y\|_X + |\lambda| \, \|(T_2 x)(T_3 x) - (T_2 x)(T_3 y)\|_X \\ &\qquad\qquad + |\lambda| \, \|(T_2 x)(T_3 y) - (T_2 y)(T_3 y)\|_X \\ &\quad \leq L_1 \|x - y\|_X + |\lambda| \, \|T_2 x\|_X \, \|T_3 x - T_3 y\|_X + |\lambda| \, \|T_2 x - T_2 y\|_X \, \|T_3 y\|_X \\ &\quad \leq L_1 \|x - y\|_X + |\lambda| M_2 L_3 \|x - y\|_X + |\lambda| M_3 L_2 \|x - y\|_X \\ &\quad = \Big(L_1 + |\lambda| \big(M_2 L_3 + M_3 L_2 \big) \Big) \|x - y\|_X ; \end{aligned} \tag{7.1.39}$$

in particular, T is a contraction because of the right estimate in (7.1.38). Thus, the Banach-Caccioppoli Fixed Point Theorem guarantees the existence of a unique solution to (7.1.20) in $\mathbb{B}_R(X)$. The proof is complete. ■

The crucial condition in Theorem 7.1.16 is of course (7.1.22) and we make some remarks on that condition. It guarantees that there exists at least one radius R such that the operator N_h maps the ball $\mathbb{B}_R(X)$ into itself and is a contraction on that ball. This is a mandatory requirement, because Theorem 7.1.16 shall also be applicable for $f \equiv 0$. But in this case, equation (7.1.20) reduces to the fixed point problem $N_h x = x$ which is solvable with the Banach-Caccioppoli Fixed Point Theorem only if N_h itself meets all its requirements.
Furthermore, the interplay between the radius R and the bound ϱ for the admissible parameters λ is given by (7.1.37). To find such ϱ depending on R, one has to calculate the parameters $L_1, L_2, L_3, M_1, M_2, M_3$ and then pick ϱ so that (7.1.37) is satisfied. We illustrate this on our test equation (7.1.4) in the following

Example 7.1.17. Consider again the equation (7.1.4) for the parameters $\alpha > -1$ and $\beta \in \mathbb{R}$ in the space BV. In the notation of (7.1.20) we can put $h(t,u) = 0$, $f(t,u) = 1$, $k(t,s) = t$ and $g(t,u) = (\alpha+1)|u|^\alpha + 2\beta t$. This implies for the quantities in (7.1.26)–(7.1.30) that $A_R = (\alpha+1)|\alpha|R^{\alpha-1}$, $B_R = C_R = D_R = E_R = 0$ and $z_R = w_R = \mathbb{0}$; in particular, Theorem 7.1.16 is applicable only if $\alpha \geq 1$. Due to $\gamma_{BV} = \gamma_{BV}(k, m, \theta) = \gamma_{BV}(\mathbb{1}, 1) = 2$, the numbers $L_1, L_2, L_3, M_1, M_2, M_3$ then become

$$L_1 = 0, \qquad L_2 = 0, \qquad L_3 = 2(\alpha+1)|\alpha|R^{\alpha-1},$$
$$M_1 = 0, \qquad M_2 = 1, \qquad M_3 = 2\Big((\alpha+1)|\alpha|R^\alpha + 2|\beta|\Big).$$

Therefore, condition (7.1.37) now reads

$$2\varrho\Big((\alpha+1)|\alpha|R^\alpha + 2|\beta|\Big) \leq R, \tag{7.1.40}$$
$$2\varrho(\alpha+1)|\alpha|R^{\alpha-1} \leq 1. \tag{7.1.41}$$

This explains that for $\alpha = 1$ the bound ϱ may be chosen independently of R as we have seen in Example 7.1.3 and the discussion thereafter, because in this case (7.1.40) grows linearly while (7.1.41) remains constant as $R \to \infty$. For $\alpha > 1$, however, (7.1.40) grows faster than linearly while (7.1.41) becomes unbounded as $R \to \infty$. To compensate this, ϱ has to be picked smaller the larger R is chosen. ◊

In case that N_h is a constant operator, that is, $h(t,u) = \tilde{h}(t)$ for some $\tilde{h} \in X$, condition (7.1.22) reduces to $\left\|\tilde{h}\right\|_X < R$ which is the condition $R > \|h\|_{BV}$ in the Theorems 7.1.1 and 7.1.6. Moreover, if g does not depend on t, that is, $g(t,u) = \tilde{g}(u)$, then the condition (H6*) posed in Theorem 7.1.16 reduces to $g \in Lip_{loc}(\mathbb{R})$ which is exactly (H3) of Theorem 7.1.1. In this sense, our very general Theorem 7.1.16 covers both Theorem 7.1.1 and Theorem 7.1.6. However, this is not entirely true, as in Theorem 7.1.6 the conditions (H4) and (H5) required for the kernel function k are milder[1] than

[1]See Example 7.1.7.

(H2) used in Theorem 7.1.1. Nevertheless, Theorem 7.1.16 has a much wider range of applications due to its generality.

We now look for (possibly unique) *continuous* solutions to (7.1.20) in our BV-spaces and impose the following additional condition on the kernel function $k : [0,1] \times [0,1] \to \mathbb{R}$ of the integral operator I_k.

$$\forall \varepsilon > 0 \ \exists \delta > 0 \ \forall t_1, t_2 \in [0,1] : \quad |t_1 - t_2| \leq \delta \ \Rightarrow \ \|k(t_1, \cdot) - k(t_2, \cdot)\|_{L_1} \leq \varepsilon. \tag{H15}$$

This is precisely condition (F) from Section 4.3. We then have the following variant of Theorem 7.1.16 which generalizes Theorem 7.1.12.

Theorem 7.1.18. *Let X be any of the spaces BV, WBV_p, YBV_φ, ΛBV or RBV_p. Assume that all hypotheses of Theorem 7.1.16 are met. Moreover, assume in addition that the functions f and h therein are continuous and that the kernel k satisfies the additional assumption (H15). Then for each $R > 0$ satisfying (7.1.22) there is some $\varrho > 0$ such that equation (7.1.20) has for fixed $\lambda \in (-\varrho, \varrho)$ a unique solution in $\mathbb{B}_R(X \cap C)$. Here, the space $X \cap C$ is equipped with the norm $\|\cdot\|_X$.*

Proof. As we have seen in the proof of Theorem 7.1.16 for any $R > 0$ satisfying (7.1.22) there is some $\varrho > 0$ such that the operator T_λ, defined by

$$T_\lambda x := N_h x + \lambda N_f x (I_k \circ N_g) x,$$

maps the ball $\mathbb{B}_R(X)$ into itself and is a contraction on that ball for every $\lambda \in (-\varrho, \varrho)$. Now, since we assume in addition that f and h are continuous, the superposition operators N_h and N_f map C into itself. Moreover, since N_g maps X into L_∞ by Theorem 5.2.34, it clearly also maps $X \cap C$ into L_∞. The additional assumption (H15) on the kernel k now guarantees that I_k maps L_∞ into C by Theorem 4.3.16. Thus, T_λ maps even the ball $\mathbb{B}_R(X \cap C)$ into itself and is a contraction on that ball for every $\lambda \in (-\varrho, \varrho)$. The claim follows now from the Banach-Caccioppoli Fixed Point Theorem. ∎

Note that Theorem 7.1.18 can be applied to our test equation (7.1.4) if and only if $\alpha \geq 1$.

We illustrate again the interplay between the different quantities A_R, B_R, C_R, D_R and E_R, but now on a more abstract level by investigating the following special case of equation (7.1.20) which is still a slight generalization of (7.1.8).

Example 7.1.19. Consider the equation

$$x(t) = a(t)x(t) + b(t) + \lambda \int_0^1 k(t,s) g\big(s, x(s)\big) \, ds \tag{7.1.42}$$

with the given data $a, b : [0,1] \to \mathbb{R}$, $k : [0,1] \times [0,1] \to \mathbb{R}$ and $g : [0,1] \times \mathbb{R} \to \mathbb{R}$. This equation is indeed a special case of (7.1.20) with $f(t,u) = 1$ and $h(t,u) = a(t)u + b(t)$. In particular, $\partial_2 f(t,u) = 0$ and $\partial_2 h(t,u) = a(t)$, as well as $h(\cdot, 0) = b$ and $f(\cdot, 0) = \mathbb{1}$.

We now check the hypothesis of Theorem 7.1.16 and work ourselves through the list (7.1.23)–(7.1.30) of required conditions. Line (7.1.23) is satisfied if $g(\cdot, u) \in L_\infty$ for each $u \in \mathbb{R}$, $a \in B$ and $b \in X$. (7.1.24) is satisfied if $k(t, \cdot) \in L_1$ for each $t \in [0,1]$. Condition (7.1.25) holds if $\mathrm{Var}_X(\theta k(\cdot, s)) \leq m(s)$ for some $m \in L_1$, some $\theta > 0$ and almost all $s \in [0,1]$. Additionally, we need

$$\left|g(t,u) - g(t,v)\right| \leq A_R|u - v| \quad \text{for almost all } t \in [0,1] \text{ and all } u, v \in [-R, R]$$

in order to fulfill (7.1.26). The lines (7.1.27), (7.1.28) and (7.1.29) are satisfied with $B_R = C_R = D_R = 0$ and $z_R = \mathbb{O}$. Finally, (7.1.30) holds with $E_R = 1$ and $w_R = a$, and so $a \in B$ needs to be replaced by the stronger requirement $a \in X$.
Since $h(t, x(t)) = a(t)x(t) + b(t)$ we have $\|h(\cdot, x(\cdot))\|_X \leq \|a\|_X R + \|b\|_X$ for $\|x\|_X \leq R$. For (7.1.22) we also need $7RC_R + E_R\Phi_X(w_R) + \|\partial_2 h(\cdot, 0)\|_\infty = \Phi_X(a) + \|a\|_\infty = \|a\|_X$. If now $\|a\|_X < 1$, then (7.1.22) is satisfied for all sufficiently large R, and Theorem 7.1.16 says that for each such R there is some $\varrho > 0$ such that (7.1.42) has for fixed $\lambda \in (-\varrho, \varrho)$ a unique solution in $\mathbb{B}_R(X)$. ◇

As we have seen in Example 7.1.3 and the comments thereafter an integral equation of the form (7.1.20) may have more than one solution. The question is now if we can get uniqueness of solutions in the entire space with the help of the fixed point theorem of Banach-Caccioppoli. Of course, this is possible only if the operator $N_h + \lambda N_f(I_k \circ N_g)$ is a contraction on the entire space; in particular, N_h and N_f should be contractions, and N_f should map the entire space into a ball of fixed radius if we want to impose as less restrictions to k and g as possible. According to the Theorems 5.2.28 and 5.2.29 this is doable only if the generating functions f and h degenerate to "almost" affine functions; we therefore consider only the equation (7.1.42) for this purpose.

Theorem 7.1.20. *Let X be any of the spaces BV, WBV_p, YBV_φ, ΛBV or RBV_p. Assume that the functions $g : [0,1] \times \mathbb{R} \to \mathbb{R}$ and $k : [0,1] \times [0,1] \to \mathbb{R}$ satisfy (i) and (ii) of Theorem 7.1.16 and the additional condition*

$$\lim_{R \to \infty} A_R < \infty, \tag{7.1.43}$$

where A_R is as in (7.1.26). Finally, assume that the functions $a, b : [0,1] \to \mathbb{R}$ belong to X and satisfy $\|a\|_X < 1$. Then there is some $\varrho > 0$ such that equation (7.1.42) has for fixed $\lambda \in (-\varrho, \varrho)$ a unique solution in the entire space X.

Note that A_R is increasing with respect to R, and this justifies that the limit in (7.1.43) exists at least in the extended sense.

By Example 7.1.19, all conditions on the data in Theorem 7.1.16 are satisfied. Accordingly, there is some $\varrho > 0$ such that equation (7.1.42) has for every $\lambda \in (-\varrho, \varrho)$ a solution in X. However, ϱ may depend on the radius R of the ball in which solutions exist. In order to show that there is for each such λ only one solution in the entire space X we need to prove that ϱ can be chosen to be the same for infinitely many arbitrarily large values of R, and this is what we are going to do now.

Proof of Theorem 7.1.20. As we have discussed in Example 7.1.19, we have $h(t,u) = a(t)u + b(t)$ and $f(t,u) = 1$. From (7.1.43) we see that there is a number $A > 0$ such that $\gamma_X A_R \le A$ for all $R > 0$, where γ_X is as in the proof of Theorem 7.1.16. Because of $\|a\|_X < 1$ we can pick $\varrho > 0$ so that

$$\|a\|_X + \varrho A \le 1, \tag{7.1.44}$$

In the proof of Theorem 7.1.16, we also have considered the numbers $M_j(R) = M_j$ and $L_j(R) = L_j$ for $j \in \{1,2,3\}$, and we do so here again. For these we have with the results found in Example 7.1.19,

$$\begin{aligned}
L_1(R) &= \Phi_X(a) + \|\partial_2 h(\cdot,0)\|_\infty = \|a\|_X\,, \\
M_1(R) &= \sup_{\|x\|_X \le R} \Big\|h\big(\cdot, x(\cdot)\big)\Big\|_X \le \|a\|_X\, R + \|b\|_X\,, \\
L_2(R) &= 0, \\
M_2(R) &= \|f(\cdot,0)\|_X = 1, \\
L_3(R) &= \gamma_X A_R, \\
M_3(R) &= \gamma_X\Big(RA_R + \|g(\cdot,0)\|_{L_\infty}\Big).
\end{aligned}$$

We now fix $\lambda \in (-\varrho, \varrho)$ and get

$$\begin{aligned}
M_1(R) + \lambda M_2(R) M_3(R) &\le \|a\|_X\, R + \|b\|_X + \lambda\gamma_X\Big(RA_R + \|g(\cdot,0)\|_{L_\infty}\Big) \\
&\le R\Big(\|a\|_X + \lambda A\Big) + \lambda\gamma_X\, \|g(\cdot,0)\|_{L_\infty} + \|b\|_X\,.
\end{aligned}$$

Because of (7.1.44) we have $\|a\|_X + \lambda A < 1$, and this is why we can find an $R_0 > 0$ such that

$$M_1(R) + \lambda M_2(R) M_3(R) \le R \quad \text{for all } R \ge R_0. \tag{7.1.45}$$

Moreover, for all $R > 0$ we also have

$$\begin{aligned}
L_1(R) + \lambda\Big(M_2(R)L_3(R) + M_3(R)L_2(R)\Big) &\le \|a\|_X + \lambda\gamma_X A_R \le \|a\|_X + \lambda A \\
&< \|a\|_X + \varrho A \le 1.
\end{aligned} \tag{7.1.46}$$

From (7.1.45) and (7.1.46) we conclude similarly as in the proof of Theorem 7.1.16 that the operator $T : X \to X$, defined by

$$Tx(t) = a(t)x(t) + b(t) + \lambda \int_0^1 k(t,s) g\big(s, x(s)\big)\, \mathrm{d}s,$$

has a unique fixed point in the ball $\mathbb{B}_R(X)$ for every $R \ge R_0$; in particular, equation (7.1.42) has exactly one solution in the entire space X. ■

For our test equation (7.1.4) we have seen in Example 7.1.17 that $A_R = (\alpha+1)|\alpha|R^{\alpha-1}$. In particular, we can apply Theorem 7.1.20 to this example if and only if $\alpha = 1$, because (7.1.43) forces A_R to stay bounded as $R \to \infty$.

Applying the Banach-Caccioppoli Contraction Principle to equation (7.1.20) is also interesting from a computational point of view, because it provides an explicit method on how to actually find the solution. Indeed, if the existence of a solution is guaranteed on a ball $\mathbb{B}_R(X)$ one may calculate it explicitly by an iterative process: Starting with $x_0 := \mathbb{0}$ the recursively defined sequence (x_n) of functions, defined by

$$x_{n+1} = N_h x_n + \lambda N_f x_n (I_k \circ N_g) x_n,$$

converges in the BV-type norm $\|\cdot\|_X$ to the unique solution $x \in \mathbb{B}_R(X)$. Moreover, the speed of convergence is given by the *a priori* estimate

$$\|x_n - x\|_X \leq \frac{q^n}{1-q} \|x_1\|_X ,$$

where, as we have seen in (7.1.39), the number $q = L_1 + |\lambda|(M_2L_3 + M_3L_2)$ is a contraction constant of the operator $N_h + \lambda N_f \cdot (I_k \circ N_g)$ and the quantities M_2, M_3, L_1, L_2, L_3 are as in (7.1.38). Let us illustrate this in the following example in the space BV.

Example 7.1.21. Let

$$h(t) = \begin{cases} -\dfrac{1}{240} & \text{for } 0 \leq t \leq \frac{1}{2}, \\ \dfrac{1}{4} - \dfrac{1}{240e^{1/16}} & \text{for } \frac{1}{2} < t \leq 1, \end{cases}$$

and consider the integral equation

$$x(t) = h(t) + \frac{1}{30} e^{-x(t)^2} \int_0^1 x(s)\,\mathrm{d}s. \tag{7.1.47}$$

In the notation of (7.1.20) we have $f(t,u) = e^{-u^2}$, $g(t,u) = u$ and $k(t,s) = 1$. Therefore, for $R = 1$ the quantities in (7.1.25)–(7.1.30) become

$$m = \mathbb{0}, \quad \theta = 1, \quad A_R = 1, \quad B_R = 2, \quad C_R = D_R = E_R = 0, \quad z_R = w_R = \mathbb{0}.$$

Since $\gamma_{BV} = \gamma_{BV}(k, m, \theta) = 1$ in this case, the numbers in (7.1.37) are

$$L_1 = 0, \quad M_1 = \|h\|_{BV} = \frac{121}{240} - \frac{1}{120e^{1/16}}, \quad L_2 = 14, \quad M_2 = 15, \quad L_3 = 1, \quad M_3 = 1.$$

Thus, for $\lambda = 1/30$, we have

$$\begin{aligned} M_1 + \lambda M_2 M_3 &= \frac{121}{240} - \frac{1}{120e^{1/16}} + \frac{1}{2} < 1 = R, \\ L_1 + \lambda(M_2L_3 + M_3L_2) &= \frac{29}{30} < 1. \end{aligned} \tag{7.1.48}$$

Consequently, the requirements in (7.1.38) are satisfied which means that (7.1.47) has a unique solution $x \in \mathbb{B}_1(BV)$ by Theorem 7.1.16.
In order to find the solution x explicitly, one may now compute the iterates

$$x_{n+1}(t) = h(t) + \frac{1}{30} e^{-x_n(t)^2} \int_0^1 x_n(s)\,\mathrm{d}s \quad \text{for } n \in \mathbb{N}_0,$$

where $x_0 := \mathbb{0}$. Note that we already know from the Banach-Caccioppoli Fixed Point Theorem that (x_n) converges in BV to x. Since h is of the form

$$h(t) = \begin{cases} a_1 & \text{for } 0 \le t \le \frac{1}{2}, \\ b_1 & \text{for } \frac{1}{2} < t \le 1, \end{cases} \quad a_1 = -\frac{1}{240}, \quad b_1 := \frac{1}{4} - \frac{1}{240e^{1/16}},$$

each iterate x_n is also of this form, that is,

$$x_n(t) = \begin{cases} a_n & \text{for } 0 \le t \le \frac{1}{2}, \\ b_n & \text{for } \frac{1}{2} < t \le 1, \end{cases} \quad \text{for } n \in \mathbb{N}_0,$$

where the sequences (a_n) and (b_n) satisfy the coupled recurrence relation

$$a_{n+1} = a_1 + \frac{1}{60}e^{-a_n^2}(a_n + b_n) \quad \text{and} \quad b_{n+1} = b_1 + \frac{1}{60}e^{-b_n^2}(a_n + b_n)$$

and the initial states $a_0 = b_0 = 0$. Computing the first iterates of (a_n) and (b_n) gives the numbers listed in Table 7.1.2.

Table 7.1.2: Approximate values of (a_n) and (b_n) for $n \in \{0, \dots, 5\}$.

n	0	1	2	3	4	5
a_n	0.000000	−0.004167	−0.000135	−0.000004	0.000000	0.000000
b_n	0.000000	0.246086	0.249881	0.249996	0.250000	0.250000

This suggests that the solution x is given by

$$x(t) = \begin{cases} 0 & \text{for } 0 \le t \le \frac{1}{2}, \\ \frac{1}{4} & \text{for } \frac{1}{2} < t \le 1, \end{cases}$$

and it is easy to check that x is indeed the (unique) solution in $\mathbb{B}_1(BV)$.
Note that a contraction constant q of the operator defining the right hand side of (7.1.47) is given by (7.1.48), that is, $q = 29/30$. In particular, the speed of convergence of the iterates x_n to the solution x may be estimated by

$$\|x_n - x\|_{BV} \le 30 \left(\frac{29}{30}\right)^n \|h\|_{BV}.$$

As Table 7.1.2 shows, in this example the convergence seems to be much faster. ◇

In order to solve equation (7.1.20) we have used the Contraction Principle of Banach-Caccioppoli. As we have seen in the discussion around Theorem 7.1.13 one is temped to try other fixed point theorems to solve (7.1.20), and this is what we are going to do in the following.
The operator $T = N_h + \lambda N_f \cdot (I_k \circ N_g)$ in (7.1.21) is the sum of the two operators $T_1 = N_h$ and $T_2 = \lambda N_f \cdot (I_k \circ N_g)$. If we can manage to arrange that T_1 is a contraction and T_2 is continuous and compact such that its sum maps a closed ball into itself, then a variant of Darbo's fixed point theorem would deliver a fixed point and hence a solution

to (7.1.20) in that ball. The advantage of this *ansatz* is that the operator T_2 may now only be continuous and not necessarily a contraction which allows us to impose milder conditions on g. The price we pay is then again that we cannot expect to get uniqueness of solutions. Unfortunately, there is another problem we have to overcome: We do not have a (nontrivial) sufficient condition on the superposition operator N_f to be compact, except those letting N_f degenerate to either a multiplication or a constant operator. This way, we can only solve special cases of (7.1.20), and we will focus on one in particular which reads

$$x(t) = h\big(t, x(t)\big) + \lambda \int_0^1 k(t,s) g\big(s, x(s)\big)\,\mathrm{d}s. \tag{7.1.49}$$

Here is the formulation of the Darbo like fixed point theorem we were talking about.

Theorem 7.1.22. *Let E be a nonempty, bounded, closed and convex subset of a Banach space X. Assume that the operators $T_1, T_2 : E \to X$ satisfy the following conditions.*

(i) T_1 is a contraction,

(ii) T_2 is continuous and compact,

(iii) $T_1x + T_2x \in E$ for all $x \in E$.

Then the operator $T = T_1 + T_2$ has a fixed point in E.

Observe that Theorem 7.1.22 indeed follows from the classical fixed point theorem of Darbo, because if T_1 is a contraction with contraction constant $L \in [0,1)$, and if T_2 is compact, then $\mu((T_1 + T_2)(M)) \leq L\mu(M)$ for all $M \subseteq E$ and μ being either the Kuratowski or the Hausdorff measure of non-compactness [47, 83]. Also note that Theorem 7.1.22 is a stronger version of a fixed point theorem of Krasnoselskii [82]. Therein, condition (iii) is replaced by the much more restrictive condition $T_1x+T_2y \in E$ for all $x, y \in E$.

We also remark that the continuity requirement in (ii) cannot be dropped.

Example 7.1.23. Set $X = E = [0,1]$ and $T_1 := \mathbb{0}$ and $T_2 := \chi_{\{0\}}$. Then T_1 is a contraction, T_2 is compact but not continuous on E, and $T_1x + T_2x = x \in \{0,1\} \subseteq E$ for all $x \in E$. However, $T = T_1 + T_2$ has no fixed point in E, because $Tx = x$ is equivalent to $\chi_{\{0\}}(x) = x$ which has no solution in E. ♢

To use Theorem 7.1.22 for equation (7.1.49) we impose the following new conditions on $g : [0,1] \times \mathbb{R} \to \mathbb{R}$.

$$\forall R > 0\ \exists M_R > 0\ \forall' t \in [0,1]\ \forall u \in [-R,R] : \quad \big|g(t,u)\big| \leq M_R, \tag{H16}$$

$$\forall' t \in [0,1] : \quad g(t,\cdot) \in C(\mathbb{R}). \tag{H17}$$

In addition, we need the following condition on the kernel function $k : [0,1] \times [0,1] \to \mathbb{R}$.

$$\forall \theta > 0\ \exists m_\theta \in L_1\ \forall' s \in [0,1] : \quad \mathrm{Var}_X\big(\theta k(\cdot, s)\big) \leq m_\theta(s), \tag{H18}$$

where Var_X denotes again the respective variation in the space X as in (H14). This condition is precisely condition (B_X^*) in Section 4.3 and equivalent to (H14) for any of our BV-spaces except for $X = YBV_\varphi$, as we have seen in Example 4.3.18. For a precise formulation of (H18) for each individual BV-space we refer the reader to Table 4.3.2. With these new conditions at hand, we obtain

Theorem 7.1.24. *Let X be any of the spaces BV, WBV_p, YBV_φ, ΛBV or RBV_p, and let $g, h : [0,1] \times \mathbb{R} \to \mathbb{R}$ and $k : [0,1] \times [0,1] \to \mathbb{R}$ be functions. Assume that*

(i) k satisfies (H12(L_1)) and (H18),

(ii) g satisfies (H11(L_∞)), (H16) and (H17),*

(iii) h satisfies (H11(X)), (H12($C^1(\mathbb{R})$)) and (H13(X)) with

$$|h(s,u) - h(s,v) - h(t,u) + h(t,v)| \leq E_R|w_R(s) - w_R(t)||u - v|$$

and $w_R \in X$,

(iv) $\partial_2 h$ satisfies (H11(B)) and (H6) with

$$\|\partial_2 h(\cdot,u) - \partial_2 h(\cdot,v)\|_\infty \leq C_R|u - v|.$$

Then for each $R > 0$ satisfying

$$\sup_{\|x\|_X \leq R} \left\| h\big(\cdot, x(\cdot)\big) \right\|_X < R \quad \text{and} \quad 7RC_R + E_R\Phi_X(w_R) + \|\partial_2 h(\cdot,0)\|_\infty < 1 \qquad (7.1.50)$$

there is some $\varrho > 0$ such that equation (7.1.49) has for fixed $\lambda \in (-\varrho, \varrho)$ a solution in $\mathbb{B}_R(X)$.

Proof. For $R > 0$ let

$$S_R := \sup_{\|x\|_X \leq R} \left\| h\big(\cdot, x(\cdot)\big) \right\|_X .$$

Condition (7.1.50) together with (iii), (iv) and Theorem 5.2.31 guarantees that the operator $T_1 := N_h$ maps the ball $\mathbb{B}_R(X)$ into itself with

$$\|T_1 x\|_X = \|N_h x\|_X \leq S_R \quad \text{for } x \in \mathbb{B}_R(X). \qquad (7.1.51)$$

Moreover, T_1 is a contraction on $\mathbb{B}_R(X)$ by (5.2.23) and (7.1.50).

We now deal with the operator $T_2 := I_k \circ N_g$ and show first that T_2 maps the space X into itself. If $\gamma_X = \gamma(k,m,\theta)$ denotes the same constant as in the proof of Theorem 7.1.16, we have by (i) and Theorem 4.3.21 that I_k maps the space L_∞ into X and is bounded with

$$\|I_k x\|_X \leq \gamma_X \|x\|_{L_\infty} \quad \text{for } x \in L_\infty.$$

Moreover, (ii) implies that for each $R > 0$ there is some $M_R > 0$ such that $|g(t,u)| \leq M_R$ for almost all $t \in [0,1]$ and all $u \in [-R,R]$. In particular, N_g maps the space X into L_∞ with

$$\|N_g x\|_{L_\infty} \leq M_R \quad \text{for } x \in \mathbb{B}_R(X). \tag{7.1.52}$$

This means that $T_2 = I_k \circ N_g$ maps the space X into itself and is bounded with

$$\|T_2 x\|_X = \|(I_k \circ N_g)x\|_X \leq \gamma_X M_R \quad \text{for } x \in \mathbb{B}_R(X). \tag{7.1.53}$$

We now show that $T_2 : X \to X$ is continuous. To this end, fix $x \in X$ and take a sequence (x_n) of functions in X that converges in X to the function x. In particular, the sequence (x_n) lies in a ball $\mathbb{B}_R(X)$ for some $R > 0$ and converges pointwise to x. By (ii), $g(t,\cdot)$ is continuous for almost every fixed $t \in [0,1]$; in particular, the functions $N_g x_n = g(\cdot, x_n(\cdot))$ converge to $N_g x = g(\cdot, x(\cdot))$ almost everywhere on $[0,1]$, and because of (7.1.52) they form a bounded sequence in L_∞. By Proposition 4.3.22 we conclude that the functions $T_2 x_n = (I_k \circ N_g)x_n$ converge in X to $T_2 x = (I_k \circ N_g)x$. This shows that T_2 is continuous.

A similar argument yields that T_2 is compact. Indeed, if (x_n) is a bounded sequence in X we find by Helly's Theorem 1.2.28 a subsequence $(x_{n_j})_j$ which converges pointwise to some function $x \in X$. Exactly as above we get that the functions $T_2 x_{n_j} = I_k \circ N_g x_{n_j}$ converge in X to $T_2 x = I_k \circ N_g x$ as $j \to \infty$.

We now fix $R > 0$ with (7.1.50). Then $S_R < R$, and this is why we can pick $\varrho > 0$ so that

$$S_R + \varrho \gamma_X M_R \leq R. \tag{7.1.54}$$

Using the estimates (7.1.51) and (7.1.53) this implies for $T_\lambda := T_1 + \lambda T_2$ with $\lambda \in (-\varrho, \varrho)$ fixed,

$$\begin{aligned} \|T_\lambda x\|_X &\leq \|T_1 x\|_X + |\lambda| \, \|T_2 x\|_X \leq S_R + |\lambda| \gamma_X M_R \\ &\leq S_R + \varrho \gamma_X M_R \leq R \qquad \text{for } x \in \mathbb{B}_R(X) \end{aligned}$$

which means that T_λ maps the ball $\mathbb{B}_R(X)$ into itself. By Theorem 7.1.22, applied with $E = \mathbb{B}_R(X)$, we obtain that T_λ has a fixed point which is a solution to (7.1.49) in the ball $\mathbb{B}_R(X)$. ∎

Of course, if we impose additionally that h is continuous and that k satisfies (H15) we get an analogue of Theorem 7.1.18. The proof is exactly the same as before. The additional hypothesis on h and k guarantee that the operators N_h and $I_k \circ N_g$ map the space C into itself. We therefore obtain

Theorem 7.1.25. *Let X be any of the spaces BV, WBV_p, YBV_φ, ΛBV or RBV_p. Assume that all hypotheses of Theorem 7.1.24 are met. Moreover, assume in addition that the functions h therein is continuous and that the kernel k satisfies the additional assumption (H15). Then for each $R > 0$ satisfying (7.1.50) there is some $\varrho > 0$ such that equation (7.1.49) has for fixed $\lambda \in (-\varrho, \varrho)$ a solution in $\mathbb{B}_R(X \cap C)$. Here, the space $X \cap C$ is equipped with the norm $\|\cdot\|_X$.*

The two differences between the Theorems 7.1.16 and 7.1.24 and the Theorems 7.1.18 and 7.1.25 is that in the latter two the conditions on g are milder for the price that we loose uniqueness of solutions. In particular, the latter two theorems are applicable to our test equation (7.1.4) if and only if $\alpha \geq 0$. As we have seen in Example 7.1.15, unique solutions cannot be guaranteed.

As a last existence result concerning equation (7.1.49) we give criteria under which it has at least one solution in X for *every* $\lambda \in \mathbb{R}$.

Theorem 7.1.26. *Let X be any of the spaces BV, WBV_p, YBV_φ, ΛBV or RBV_p. Assume that the functions $g, h : [0,1] \times \mathbb{R} \to \mathbb{R}$ and $k : [0,1] \times [0,1] \to \mathbb{R}$ satisfy all conditions of Theorem 7.1.24 and the additional assumptions*

$$\forall R > 0 : \; L_R < 1 \quad \textit{and} \quad \limsup_{R\to\infty} \frac{R - S_R}{M_R + 1/R} = \infty, \tag{7.1.55}$$

where

$$L_R := 7RC_R + E_R\Phi_X(w_R) + \|\partial_2 h(\cdot, 0)\|_\infty ,$$

$$M_R := \sup_{|u|\leq R} \|g(\cdot, u)\|_{L_\infty} \qquad \textit{and} \qquad S_R := \sup_{\|x\|_X \leq R} \|h(\cdot, x(\cdot))\|_X .$$

Then equation (7.1.49) has for every $\lambda \in \mathbb{R}$ a solution in X.

Proof. Fix $\lambda \in \mathbb{R}$ and let γ_X be as in the proof of Theorem 7.1.24. Due to (7.1.55) there is some $R > 0$ such that $L_R < 1$ and $R - S_R \geq \gamma_X |\lambda| (M_R + 1/R)$ which implies

$$S_R + |\lambda|\gamma_X M_R \leq R.$$

As a consequence, the operator $T_\lambda := T_1 + \lambda T_2$, where $T_1 := N_h$ and $T_2 := I_k \circ N_g$, maps the ball $\mathbb{B}_R(X)$ into itself, T_1 is a contraction on that ball and λT_2 is continuous and compact, as we have shown in the proof of Theorem 7.1.24. Theorem 7.1.26 is now an immediate consequence of Theorem 7.1.22. ∎

Theorem 7.1.26 can be understood as a generalization of Theorem 7.1.13. Let us now check how the condition (7.1.55) looks like for our test equation (7.1.4).

Example 7.1.27. Let $\alpha \geq 0$. As we have seen in Example 7.1.17, $L_R = S_R = 0$, where L_R and S_R are now as in Theorem 7.1.26. Moreover, since $g(t,u) = (\alpha+1)|u|^\alpha + 2\beta t$, we get

$$(\alpha+1)R^\alpha - 2|\beta| \leq M_R \leq (\alpha+1)R^\alpha + 2|\beta|,$$

where M_R is also as in Theorem 7.1.26. We obtain

$$\frac{R}{(\alpha+1)R^\alpha + 2|\beta| + 1/R} \leq \frac{R - S_R}{M_R + 1/R} \leq \frac{R}{(\alpha+1)R^\alpha - 2|\beta| + 1/R}$$

and hence

$$\lim_{R\to\infty} \frac{R - S_R}{M_R + 1/R} = \begin{cases} \infty & \text{for } 0 \leq \alpha < 1, \\ 1/2 & \text{for } \alpha = 1, \\ 0 & \text{for } \alpha > 1 \end{cases}$$

which means that (7.1.55) is satisfied if and only if $0 \leq \alpha < 1$. ♢

Thus, Theorem 7.1.26 is applicable to equation (7.1.4) if and only if $0 \leq \alpha < 1$ and β is arbitrary. This explains why the equations (7.1.12) and (7.1.14) do not have solutions for every λ as we have seen in the Examples 7.1.8 and 7.1.9, respectively, whereas equation (7.1.18) has solutions for every λ (for $\beta < 0$ and $|\lambda| \geq -4\beta$ even more than one), as we have found out in Example 7.1.15.

The following Table 7.1.3 shows the combinations of parameters α and β for which the theorems discussed so far may be applied to solve our test equation (7.1.4).

Table 7.1.3: Parameters in (7.1.4) for which the preceding theorems may be applied.

Theorem	parameters	
7.1.1	$\alpha \geq 1$	$\beta = 0$
7.1.6	$\alpha \geq 1$	$\beta \in \mathbb{R}$
7.1.12	$\alpha \geq 1$	$\beta = 0$
7.1.13	$0 \leq \alpha < 1$	$\beta = 0$
7.1.16	$\alpha \geq 1$	$\beta \in \mathbb{R}$
7.1.18	$\alpha \geq 1$	$\beta \in \mathbb{R}$
7.1.20	$\alpha = 1$	$\beta \in \mathbb{R}$
7.1.24	$\alpha \geq 0$	$\beta \in \mathbb{R}$
7.1.25	$\alpha \geq 0$	$\beta \in \mathbb{R}$
7.1.26	$0 \leq \alpha < 1$	$\beta \in \mathbb{R}$

Especially for boundary and initial value problems which we will discuss briefly in Section 7.3, the following more abstract equation will turn out to be very handy. Let us consider

$$x(t) = Ax(t) + \lambda \int_0^1 k(t,s)g\big(s, x(s)\big)\,\mathrm{d}s \quad \text{for } 0 \leq t \leq 1 \tag{7.1.56}$$

in a BV-space X, where $A : X \to X$ is a *linear* operator. In order to solve this equation, we need a variant of Darbo's Theorem 7.1.22 for balls. The idea for the proof is taken from [49] and relies on the fixed point theorem of Schauder.

Theorem 7.1.28. *Let $(X, \|\cdot\|)$ be a Banach algebra. Assume that the operators $A, S : X \to X$ have the following properties.*

(i) A is linear and bounded with $\|A^n\|_{X\to X} < 1$ for some $n \in \mathbb{N}$.

(ii) S is continuous and compact with $S\big(\mathbb{B}_R(X)\big) \subseteq \mathbb{B}_{R'}(X)$.

If the radii R and R' are related by

$$\frac{R'}{1 - \|A^n\|_{X\to X}} \sum_{j=0}^{n-1} \left\|A^j\right\|_{X\to X} \leq R, \tag{7.1.57}$$

then the operator $A + S$ has a fixed point in $\mathbb{B}_R(X)$.

Proof. Because of (i) the Neumann series of A converges, and $I - A$ is invertible with

$$(I-A)^{-1} = \left(\sum_{j=0}^{\infty} A^{jn}\right) \sum_{j=0}^{n-1} A^j.$$

In particular, $(I - A)^{-1}$ is bounded with

$$\left\|(I-A)^{-1}\right\|_{X\to X} \leq \frac{1}{1-\|A^n\|_{X\to X}} \sum_{j=0}^{n-1} \left\|A^j\right\|_{X\to X}.$$

Since S maps the ball $\mathbb{B}_R(X)$ into the ball $\mathbb{B}_{R'}(X)$ by (ii), the composition $T := (I-A)^{-1} \circ S$ maps the ball $\mathbb{B}_R(X)$ into the ball $\mathbb{B}_{R''}(X)$, where R'' is given by

$$R'' = \frac{R'}{1-\|A^n\|_{X\to X}} \sum_{j=0}^{n-1} \left\|A^j\right\|_{X\to X}.$$

Consequently, if $R'' \leq R$, and this is just the relation (7.1.57), the operator T maps the ball $\mathbb{B}_R(X)$ into itself. Moreover, since S is continuous and compact and $(I-A)^{-1}$ is linear and bounded, the operator T is also continuous and compact. By Schauder's Fixed Point Theorem, T has a fixed point $x \in \mathbb{B}_R(X)$ which means $(I-A)^{-1}Sx = x$. Equivalently, $Sx = (I-A)x = x - Ax$ and hence $Ax + Sx = x$. This completes the proof. ∎

Using Gelfand's Formula

$$\mathfrak{R}(A) = \lim_{n\to\infty} \|A^n\|_{X\to X}^{1/n}$$

for the spectral radius $\mathfrak{R}(A)$ of a linear operator $A : X \to X$ on a Banach space X (see [85]) one may replace condition (i) of Theorem 7.1.28 by $\mathfrak{R}(A) < 1$. While $\mathfrak{R}(A) < 1$ is equivalent to $\|A^n\|_{X\to X} < 1$ for some $n \in \mathbb{N}$, it is *not* equivalent to $\|A\|_{X\to X} < 1$. Furthermore, having spectral radius less than one does not mean that the operator norm is bounded somehow in terms of the spectral radius. We show both in the following example.

Example 7.1.29. For $c \in \mathbb{R}$ consider the linear operator $A_c : BV \to BV$, defined by $A_c x(t) = ctx(0)$. Then $\|A_c\|_{BV\to BV} = 2|c|$, but since

$$A_c^2 x(t) = A_c(A_c x)(t) = ct(A_c x)(0) = 0 \quad \text{for all } x \in BV, t \in [0,1],$$

we have $\|A_c^2\|_{BV\to BV} = 0$. By picking c arbitrarily large we see that the norms of the operators A_c may become arbitrarily large although the operators A_c themselves still have spectral radius 0. ◇

Note that if $\|A\|_{X\to X} < 1$ which means that A itself is a contraction, then the relation (7.1.57) reduces to $R' \leq R(1 - \|A\|_{X\to X})$. We can now apply Theorem 7.1.28 to equation (7.1.56).

Theorem 7.1.30. *Let X be any of the spaces BV, WBV_p, YBV_φ, ΛBV or RBV_p. Assume that the functions $g : [0,1] \times \mathbb{R} \to \mathbb{R}$ and $k : [0,1] \times [0,1] \to \mathbb{R}$ satisfy (i) and (ii) of Theorem 7.1.24. Moreover, assume that $A : X \to X$ is a bounded linear operator with $\|A^n\|_{X\to X} < 1$ for some $n \in \mathbb{N}$. Then for each $R > 0$ there is some $\varrho > 0$ such that equation (7.1.56) has for fixed $\lambda \in (-\varrho, \varrho)$ a solution in $\mathbb{B}_R(X)$.*

Proof. As we have seen in the proof of Theorem 7.1.24 the operator $\tilde{S} := I_k \circ N_g$ maps the space X into itself, is continuous and compact and satisfies the estimate (7.1.53), that is,

$$\left\|\tilde{S}x\right\|_X \leq \gamma_X M_R \quad \text{for } x \in \mathbb{B}_R(X), R > 0, \tag{7.1.58}$$

where γ_X is as in the proof of Theorem 7.1.16 and $M_R := \sup_{|u|\leq R} \|g(\cdot,u)\|_{L_\infty}$.
The inequality (7.1.58) means that $\tilde{S}$ maps the ball $\mathbb{B}_R(X)$ into the ball $\mathbb{B}_{\gamma_X M_R}(X)$ for any $R > 0$. Note that M_R is finite for each $R > 0$ as g satisfies (H16). We now fix $R > 0$. Since $\|A^n\|_{X\to X} < 1$ we may pick $\varrho > 0$ so that

$$\varrho \frac{\gamma_X M_R}{1 - \|A^n\|_{X\to X}} \sum_{j=0}^{n-1} \left\|A^j\right\|_{X\to X} \leq R. \tag{7.1.59}$$

Then, for each $\lambda \in (-\varrho, \varrho)$ the operator $S := \lambda\tilde{S}$ maps the ball $\mathbb{B}_R(X)$ into the ball $\mathbb{B}_{R'}(X)$, where $R' = |\lambda|\gamma_X M_R$. Because of (7.1.59), the two radii R and R' satisfy (7.1.57). Thus, $A + S = A + \lambda\tilde{S}$ has a fixed point in $\mathbb{B}_R(X)$ by Theorem 7.1.28. ∎

As an application we look at the following

Example 7.1.31. Consider the integral equation

$$x(t) = \frac{\omega}{2} t \int_0^1 x\Big(\sin^2(\omega s)\Big)\,\mathrm{d}s + \lambda \int_0^1 \sin(ts)\cos\Big(\omega x(s) + s\Big)\,\mathrm{d}s \tag{7.1.60}$$

in the space BV for a constant $\omega > 0$. In the notation of (7.1.56) we have $k(t,s) = \sin(ts)$ which clearly satisfies (H12(L_1)) and (H18), $g(t,u) = \cos(\omega u + t) \leq 1$ for all $t \in [0,1]$ and $u \in \mathbb{R}$ which obviously fulfills (H11*(L_∞)), (H16) and (H17), as well as

$$Ax(t) = \frac{\omega}{2} t \int_0^1 x\Big(\sin^2(\omega s)\Big)\,\mathrm{d}s$$

which defines a linear operator $A : BV \to BV$ with

$$\|Ax\|_{BV} = \|Ax\|_\infty + \mathrm{Var}(Ax) = \omega \left|\int_0^1 x\Big(\sin^2(\omega s)\Big)\,\mathrm{d}s\right| \leq \omega \|x\|_{BV}.$$

Consequently, $\|A\|_{BV\to BV} \leq \omega$, and the function $x = \mathbb{1}$ shows even $\|A\|_{BV\to BV} = \omega$. Since, by Theorem 1.1.20,

$$\mathrm{Var}\Big(\theta k(\cdot,s)\Big) = \theta \int_0^1 \Big|\partial_1 k(t,s)\Big|\,\mathrm{d}t = \theta \int_0^1 s\Big|\cos(ts)\Big|\,\mathrm{d}t \leq \theta s$$

we can take $\theta = 1$ and $m(s) = s$ in (H14) and get

$$\gamma_{BV} = \gamma_{BV}(k,m,\theta) = \|k(0,\cdot)\|_{L_1} + 2\theta^{-1}\|m\|_{L_1} = 1,$$

as well as $M_R = 1$, where γ_{BV} and M_R are as in the proof of Theorem 7.1.30. Moreover, the estimate (7.1.59) is certainly satisfied if $\omega \in (0,1)$ and $0 < \varrho \leq (1-\omega)R$, because then A is a contraction. Thus, for $\omega \in (0,1)$ Theorem 7.1.30 says that for any $R > 0$ and $|\lambda| < (1-\omega)R$ equation (7.1.60) has a solution $x \in \mathbb{B}_R(BV)$. ◇

Note that equation (7.1.60) cannot be written in the form (7.1.49) and hence cannot be solved using Theorem 7.1.24 or one of its successive results. Later in Example 7.3.4 we will see that integral equations of type (7.1.56) may indeed be solved with Theorem 7.1.30 even when A is not a contraction.

If one replaces X in Theorem 7.1.30 by $X \cap C$ and adjusts the requirements on the data, we obtain the following result which is very similar to Theorem 7.1.25.

Theorem 7.1.32. *Let X be any of the spaces BV, WBV_p, YBV_φ, ΛBV or RBV_p. Assume that the functions $g : [0,1] \times \mathbb{R} \to \mathbb{R}$ and $k : [0,1] \times [0,1] \to \mathbb{R}$ satisfy (i) and (ii) of Theorem 7.1.24 and that k meets in addition (H15). Moreover, assume that $A : X \cap C \to X \cap C$ is a linear operator with $\|A^n\|_{X\cap C \to X \cap C} < 1$ for some $n \in \mathbb{N}$. Then for each $R > 0$ there is some $\varrho > 0$ such that equation (7.1.56) has for fixed $\lambda \in (-\varrho, \varrho)$ a solution in $\mathbb{B}_R(X \cap C)$. Here, $X \cap C$ is equipped with the norm $\|\cdot\|_X$.*

The proof rests on the same ideas as presented in the proofs of the Theorems 7.1.25 and 7.1.30 and will be omitted. Observe that since k and A in Example 7.1.31 even satisfy the hypotheses of Theorem 7.1.32, any solution to (7.1.60) must be continuous. Let us summarize in Table 7.1.4 which of the preceding theorems may be applied to the various Hammerstein integral equations considered in this section. As always, X serves as a placeholder for one of the spaces BV, WBV_p, YBV_φ, ΛBV and RBV_p.

Table 7.1.4: Theorems solving Hammerstein integral equations of several types.

By Theorem	equation	has	solution in	for
7.1.1	(7.1.1)	exactly one	$\mathbb{B}_R(BV)$	small $\|\lambda\|$
7.1.6	(7.1.8)	exactly one	$\mathbb{B}_R(BV)$	small $\|\lambda\|$
7.1.12	(7.1.1)	exactly one	$\mathbb{B}_R(BV \cap C)$	small $\|\lambda\|$
7.1.13	(7.1.1)	at least one	BV	all λ and $h = \mathbb{0}$
7.1.16	(7.1.20)	exactly one	$\mathbb{B}_R(X)$	small $\|\lambda\|$
7.1.18	(7.1.20)	exactly one	$\mathbb{B}_R(X \cap C)$	small $\|\lambda\|$
7.1.20	(7.1.42)	exactly one	X	small $\|\lambda\|$
7.1.24	(7.1.49)	at least one	$\mathbb{B}_R(X)$	small $\|\lambda\|$
7.1.25	(7.1.49)	at least one	$\mathbb{B}_R(X \cap C)$	small $\|\lambda\|$
7.1.26	(7.1.49)	at least one	X	all λ
7.1.30	(7.1.56)	at least one	$\mathbb{B}_R(X)$	small $\|\lambda\|$
7.1.32	(7.1.56)	at least one	$\mathbb{B}_R(X \cap C)$	small $\|\lambda\|$

7.2 Volterra Integral Equations

In this short section we translate the theorems discussed in the previous section onto Volterra integral equations, that is, equations of the form

$$x(t) = h\big(t, x(t)\big) + \lambda f\big(t, x(t)\big) \int_0^t \nu(t, s) g\big(s, x(s)\big) \,\mathrm{d}s \quad \text{for } 0 \le t \le 1, \tag{7.2.1}$$

for given data $f, g, h : [0,1] \times \mathbb{R} \to \mathbb{R}$ and a Volterra kernel $\nu : [0,1] \times [0,1] \to \mathbb{R}$, that is, $\nu(t, s) = 0$ for $0 \le t < s \le 1$. Note that any Volterra integral equation (7.2.1) can be understood as a Hammerstein integral equation (7.1.49) where the kernel k is replaced by the Volterra kernel ν. Therefore, each result of the previous section remains true for the equation (7.2.1) and its various special cases. However, due to the special structure of the Volterra kernel, conditions imposed on the kernel k in Section 7.1 may now be rephrased in a more relaxed form, and we already did so for most of the conditions considered so far in Section 4.3. For convenience of the reader we again repeat the needed conditions here to make the presentation self-contained.

Let us begin with the Volterra version of equation (7.1.1) which is

$$x(t) = h(t) + \lambda \int_0^t \nu(t, s) g\big(x(s)\big) \,\mathrm{d}s \quad \text{for } 0 \le t \le 1 \tag{7.2.2}$$

and the two corresponding conditions (H1) and (H2). Since these two conditions are precisely (A) and (B) of Section 4.3, their Volterra equivalent is given by

$$\forall t \in [0,1] : \quad \nu(t, \cdot) \in L_1[0, t], \tag{V1}$$

$$\exists m \in L_1 \ \forall' s \in [0,1] : \quad |\nu(s, s)| + \operatorname{Var}\big(\nu(\cdot, s), [s, 1]\big) \le m(s), \tag{V2}$$

and these are the conditions (VA) and (VB) of Section 4.3.

With (H1) and (H2) replaced by (V1) and (V2) we obtain from Theorem 7.1.1,

Theorem 7.2.1. *Assume (V1), (V2) and (H3), and let $h \in BV$ be fixed. Then for each $R > \|h\|_{BV}$ there is some $\varrho > 0$ such that equation (7.2.2) has for fixed $\lambda \in (-\varrho, \varrho)$ a unique solution in $\mathbb{B}_R(BV)$.*

The bound ϱ for the admissible parameters λ given in (7.1.3) is now

$$\varrho = \min\left\{ \frac{R - \|h\|_{BV}}{\|g\|_{[-R,R]}}, \frac{1}{\operatorname{lip}(g, [-R, R])} \right\} \frac{1}{2\,\|m\|_{L_1}}, \tag{7.2.3}$$

because $\nu(0, s) = 0$ for all $s \in (0, 1]$. This may enlarge the set of the λ, as the following example demonstrates.

Example 7.2.2. We consider the Hammerstein and Volterra equation

$$x(t) = \lambda \int_0^1 \kappa(t,s)x(s)\,\mathrm{d}s \qquad \text{and} \qquad x(t) = \lambda \int_0^t \kappa(t,s)x(s)\,\mathrm{d}s \tag{7.2.4}$$

in the space BV, where

$$\kappa(t,s) = \begin{cases} \max\left\{\dfrac{s-t}{s}, \dfrac{t-s}{1-s}\right\} & \text{for } 0 < s < 1, \\ 1 & \text{for } s \in \{0,1\} \end{cases}$$

which is shown in the picture below as a function of t only.

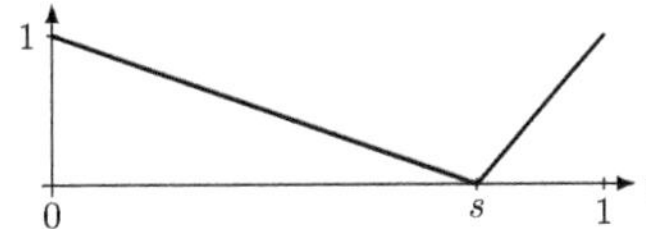

Figure 7.2.1: $\kappa(\cdot, s)$ for some $s \in (0,1)$.

In the notation of (7.1.1) and (7.2.2) we can put $h(t) = 0$ and $g(u) = u$. Then $\|g\|_{[-R,R]} = R$ and $\mathrm{lip}(g, [-R,R]) = 1$. For fixed $s \in (0,1)$ we have $\mathrm{Var}\big(\kappa(\cdot,s)\big) = 2$. For $s \le t \le 1$ we have $\kappa(t,s) = \frac{t-s}{1-s}$ and thus $|\kappa(s,s)| + \mathrm{Var}(\kappa(\cdot,s),[s,1]) = 1$. Consequently, we take $m_1 \equiv 2$ for m in (H2) and $m_2 \equiv 1$ for m in (V2). Due to $\kappa(0,\cdot) = \mathbb{1}$ the corresponding ϱ become $\varrho_1 = 1/5$ in (7.1.3) and $\varrho_2 = 1/2$ in (7.2.3). $\diamondsuit$

Since the assumptions on the kernel in Theorem 7.1.13 are the same as in Theorem 7.1.1, the Volterra equivalent of the former becomes

Theorem 7.2.3. *Under the assumptions (V1), (V2), (H9) and (H10) equation (7.2.2) has for $h = \mathbb{0}$ and every $\lambda \in \mathbb{R}$ a solution $x \in BV$.*

Observe that we could not expect uniqueness of solutions in Theorem 7.1.13 and neither can we here.

Example 7.2.4. Consider the Volterra equation

$$x(t) = \frac{3}{2}\lambda t \int_0^t \sqrt{|x(s)|}\,\mathrm{d}s \tag{7.2.5}$$

which is a Volterra equivalent to (7.1.16). Obviously, $x = \mathbb{0}$ is a solution. However, the *ansatz* $x(t) = ct^p$ for $p > 0$ yields that $p = 4$ and $2c = \lambda\sqrt{|c|}$. This equation has for any $\lambda \in \mathbb{R}$ a solution, namely $c = \lambda^2/4$ for $\lambda \ge 0$ and $c = -\lambda^2/4$ for $\lambda < 0$. In particular, (7.2.5) has for any $\lambda \neq 0$ at least two distinct solutions. $\diamondsuit$

The slightly more general equation (7.1.8) reads in the Volterra version

$$x(t) = h(t) + \lambda \int_0^t \nu(t,s) g\big(s, x(s)\big)\,\mathrm{d}s \quad \text{for } 0 \le t \le 1. \tag{7.2.6}$$

The kernel conditions (H4) and (H5) turn into

$$\sup_{t \in [0,1]} \|\nu(t,\cdot)\|_{L_1[0,t]} < \infty, \tag{V3}$$

$$\sup_{\tau \in [0,1]} \mathrm{Var}\left(\int_0^{\min\{\tau,\cdot\}} \nu(\cdot,s)\,\mathrm{d}s\right) < \infty. \tag{V4}$$

Here, (V4) is precisely (VC) of Section 4.3 and means that the Jordan variation of the function $t \mapsto \int_0^{\min\{\tau,t\}} \nu(t,s)\,\mathrm{d}s$ stays bounded as τ runs through $[0,1]$. As we have seen in Example 4.3.28 an arbitrary kernel satisfying (H1) and (V4) does not have to satisfy (H5). In the next example we show that an arbitrary kernel k may satisfy (H1), (V3) and (V4) but not (H5).

Example 7.2.5. Consider the kernel $k : [0,1] \times [0,1] \to \mathbb{R}$, defined by

$$k(t,s) = \begin{cases} 1/\sqrt{st} & \text{for } 0 < st \leq 1, \\ 0 & \text{for } st = 0. \end{cases}$$

For each $t \in [0,1]$ we have $k(t,s) = 0$ if $t = 0$ and $k(t,s) = t^{-1/2}/\sqrt{s}$ for $0 < s \leq 1$. Thus, $k(t,\cdot) \in L_1$ for each $t \in [0,1]$ and so k fulfills (H1). Furthermore,

$$\sup_{t\in[0,1]} \|k(t,\cdot)\|_{L_1[0,t]} = \sup_{t\in(0,1]} \frac{1}{\sqrt{t}} \int_0^t \frac{1}{\sqrt{s}}\,\mathrm{d}s = 2$$

showing that k meets (V3).
Letting

$$h(\tau,t) := \int_0^{\min\{\tau,t\}} k(t,s)\,\mathrm{d}s$$

we have for any $\tau \in [0,1]$ and $t \in (0,1]$,

$$h(\tau,t) = \int_0^{\min\{\tau,t\}} \frac{1}{\sqrt{st}}\,\mathrm{d}s = 2\sqrt{\min\{\tau/t, 1\}},$$

and for $t = 0$ we get $h(\tau,0) = 0$. Consequently,

$$\sup_{\tau\in[0,1]} \operatorname{Var}\Big(h(\tau,\cdot)\Big) = \sup_{\tau\in[0,1]} \Big(4 - 2\sqrt{\tau}\Big) = 4$$

showing that k satisfies (V4). However, for $t, \tau \in (0,1]$,

$$\int_0^{\tau} k(t,s)\,\mathrm{d}s = \frac{2\sqrt{\tau}}{\sqrt{t}}$$

which is unbounded with respect to t. Consequently, (H5) cannot hold. ◇

With the respective Volterra versions (V3) and (V4) for (H4) and (H5) at hand, Theorem 7.1.6 reads as follows.

Theorem 7.2.6. *Assume (H3*), (V3) and (V4), and let $h \in BV$ be fixed. Then for each $R > \|h\|_{BV}$ there is some $\varrho > 0$ such that equation (7.2.6) has for fixed $\lambda \in (-\varrho, \varrho)$ a unique solution in $\mathbb{B}_R(BV)$.*

The number ϱ from (7.1.11) bounding the set of parameters λ for which Theorem 7.2.6 may be applied is then

$$\varrho = \min\left\{ \frac{R - \|h\|_{BV}}{2S_1\Big(L_R + RL_R + |g(0,0)|\Big)}, \frac{1}{L_R S_2} \right\}, \tag{7.2.7}$$

where S_1 is the supremum in (V4), S_2 is the supremum in (V3) and $L_R = \text{lip}(g, [0,1] \times [-R, R])$ is defined as in (7.1.10).

For searching continuous BV-solutions to (7.2.6) we have to adapt condition (H7) to Volterra kernels. Let us assume for a moment that ν is a Volterra kernel satisfying (H7), and let $\varepsilon > 0$ be given. Then there is some $\delta > 0$ such that for each $t_1, t_2 \in [0,1]$ with $|t_1 - t_2| \leq \delta$ there is a measurable set $I \subseteq [0,1]$ of measure 1 so that $|\nu(t_1, s) - \nu(t_2, s)| \leq \varepsilon$ for all $s \in I$. For $t_1 < s \leq t_2$ this implies $|\nu(t_2, s)| \leq \varepsilon$, and for $t_2 < s \leq t_1$ it gives $|\nu(t_1, s)| \leq \varepsilon$. Therefore, (H7) implies the following two conditions.

$$\forall \varepsilon > 0 \; \exists \delta > 0 \; \forall t \in [0,1] \; \forall' s \in \left[0,1\right] : \qquad 0 \leq t - s \leq \delta \;\Rightarrow\; |\nu(t,s)| \leq \varepsilon \tag{V5a}$$

$$\forall \varepsilon > 0 \; \exists \delta > 0 \; \forall t_1, t_2 \in [0,1] \; \forall' s \in \left[0, \min\{t_1, t_2\}\right] : \qquad |t_1 - t_2| \leq \delta \;\Rightarrow\; |\nu(t_1, s) - \nu(t_2, s)| \leq \varepsilon. \tag{V5b}$$

For a Volterra kernel ν these two conditions together imply in turn (H7). To see this pick for a given $\varepsilon > 0$ a number $\delta > 0$ such that (V5a) and (V5b) hold simultaneously. Let $t_1, t_2 \in [0,1]$ be so that $|t_1 - t_2| \leq \delta$. Then by (V5a) there are measurable sets $I_{a,1}, I_{a,2} \subseteq [0,1]$ of measure 1 such that $|\nu(t_1, s)| \leq \varepsilon$ for all $s \in I_{a,1}$ and $|\nu(t_2, s)| \leq \varepsilon$ for all $s \in I_{a,2}$. By (V5b) we find another measurable set $I_b \subseteq [0, \min\{t_1, t_2\}]$ of full measure $|I_b| = \min\{t_1, t_2\}$ such that $|\nu(t_1, s) - \nu(t_2, s)| \leq \varepsilon$ for all $s \in I_b$. Then $J := I_{a,1} \cap I_{a_2} \cap \left(I_b \cup (\min\{t_1, t_2\}, 1]\right)$ has measure 1. Now, fix $s \in J$. If $s \leq t_1 \leq t_2$ we have $s \in I_b$ and thus $|\nu(t_1, s) - \nu(t_2, s)| \leq \varepsilon$. For $t_1 < s \leq t_2$ we have $s \in I_{a,2}$ and hence $|\nu(t_1, s) - \nu(t_2, s)| = |\nu(t_2, s)| \leq \varepsilon$. Similarly, $|\nu(t_1, s) - \nu(t_2, s)| = |\nu(t_1, s)| \leq \varepsilon$ for $t_2 < s \leq t_1$, because then $s \in I_{a,1}$. Finally, for $t_1, t_2 < s$ we have $|\nu(t_1, s) - \nu(t_2, s)| = 0 \leq \varepsilon$. Consequently, (H7) holds.

An arbitrary kernel k satisfying (V5a) and (V5b) does not have to satisfy (H7). Here is an example of such a kernel.

Example 7.2.7. Define $k : [0,1] \times [0,1] \to \mathbb{R}$ by $k(t,s) = \chi_{\{0\}}(t)$. For $t \in (0,1]$ we have $|k(t,s) - k(0,s)| = 1$ for all $s \in [0,1]$ and hence (H7) is violated.
However, (V5a) and (V5b) hold. To see this fix $t \in [0,1]$. If $t > 0$ then $k(t,s) = 0$ for all $s \in [0,1]$. If $t = 0$, the set of all s with $0 \leq s \leq t$ contains only $s = 0$ and hence is a null set. Thus, (V5a) is satisfied. For (V5b) fix $t_1, t_2 \in [0,1]$. If $m := \min\{t_1, t_2\} > 0$, then $t_1, t_2 \geq m > 0$, and for $s \in [0, m]$ we have $k(t_1, s) = k(t_2, s) = 0$. If $m = 0$, the set $[0, m]$ contains only $s = 0$ and hence is a null set again. Consequently, also (V5b) is satisfied. ◇

With the condition (H7) replaced by (V5a) and (V5b) and the kernel k replaced by a Volterra kernel ν, Theorem 7.1.12 can now be reformulated as follows.

Theorem 7.2.8. *Assume (V1), (V2), (H3), (V5a) and (V5b), and let $h \in BV \cap C$ be fixed. Then for each $R > \|h\|_{BV}$ there is some $\varrho > 0$ such that equation (7.2.2) has for fixed $\lambda \in (-\varrho, \varrho)$ a unique solution in $\mathbb{B}_R(BV \cap C)$, where $BV \cap C$ is equipped with the norm $\|\cdot\|_{BV}$.*

We now come to the main result, namely Theorem 7.1.16 which solves the most general Hammerstein integral equation (7.1.20). In order to translate this theorem onto the equation (7.2.1) we have to look only at the conditions made in part (i) of Theorem 7.1.16. Besides (H12(L_1)) which is equivalent to (H1) we also imposed (H14) which is equivalent to (B_X) from Section 4.3. For Volterra kernels, (H1) is precisely (V1), and the Volterra version of (B_X) is just (VB_X) from Section 4.3 again which may be found for each BV-space individually in Table 4.3.4. Therefore, we say that a Volterra kernel ν satisfies (V6) if and only if it satisfies condition (VB_X) given in that table. We then obtain from Theorem 7.1.16 the following Volterra version.

Theorem 7.2.9. *Let X be any of the spaces BV, WBV_p, YBV_φ, ΛBV or RBV_p, and let $f, g, h : [0,1] \times \mathbb{R} \to \mathbb{R}$ be functions and $\nu : [0,1] \times [0,1] \to \mathbb{R}$ be a Volterra kernel. Assume that f, g and h satisfy (ii)–(vi) of Theorem 7.1.16. Moreover, assume that ν satisfies (V1) and (V6). Then for each $R > 0$ satisfying (7.1.22) there is some $\varrho > 0$ such that equation (7.2.1) has for fixed $\lambda \in (-\varrho, \varrho)$ a unique solution in $\mathbb{B}_R(X)$.*

For the Volterra equivalent of equation (7.1.42) which is given by

$$x(t) = a(t)x(t) + b(t) + \lambda \int_0^t \nu(t,s) g\big(s, x(s)\big)\, \mathrm{d}s \quad \text{for } 0 \le t \le 1 \tag{7.2.8}$$

we get the following version of Theorem 7.1.20.

Theorem 7.2.10. *Let X be any of the spaces BV, WBV_p, YBV_φ, ΛBV or RBV_p. Assume that the Volterra kernel $\nu : [0,1] \times [0,1] \to \mathbb{R}$ satisfies (V1) and (V6). Moreover, assume that the function $g : [0,1] \times \mathbb{R} \to \mathbb{R}$ satisfies (ii) of Theorem 7.1.16 and the additional condition*

$$\lim_{R \to \infty} A_R < \infty,$$

where A_R is as in (7.1.26). Finally, assume that the functions $a, b : [0,1] \to \mathbb{R}$ belong to X and satisfy $\|a\|_X < 1$. Then there is some $\varrho > 0$ such that equation (7.2.8) has for fixed $\lambda \in (-\varrho, \varrho)$ a unique solution in the entire space X.

In order to translate Theorem 7.1.18 we need a Volterra equivalent of condition (H15) which is precisely condition (F) of Section 4.3. In the Volterra setting it is equivalent to condition (VF) of the same section and reads

$$\begin{aligned} &\forall \varepsilon > 0 \ \exists \delta > 0 \ \forall t_1, t_2 \in [0,1] : \quad |t_1 - t_2| \le \delta \\ &\quad \Rightarrow \int_0^{\min\{t_1,t_2\}} \Big| g(t_1,s) - g(t_2,s) \Big|\, \mathrm{d}s + \int_{\min\{t_1,t_2\}}^{\max\{t_1,t_2\}} \Big| g\big(\max\{t_1,t_2\}, s\big) \Big|\, \mathrm{d}s \le \varepsilon. \end{aligned} \tag{V7}$$

As we have already seen in Example 4.3.30 and the remark prior to Theorem 4.3.32 an arbitrary kernel satisfying (V7) does not have to satisfy (H15). With this condition at hand we obtain the following Volterra version of Theorem 7.1.18.

Theorem 7.2.11. *Let X be any of the spaces BV, WBV_p, YBV_φ, ΛBV or RBV_p, and let $f, g, h : [0,1] \times \mathbb{R} \to \mathbb{R}$ be functions and $\nu : [0,1] \times [0,1] \to \mathbb{R}$ be a Volterra kernel. Assume that f, g and h satisfy (ii)–(vi) of Theorem 7.1.16. Moreover, assume that f and h therein are continuous and that the kernel ν satisfies (V1), (V6) and (V7). Then for each $R > 0$ satisfying (7.1.22) there is some $\varrho > 0$ such that equation (7.2.1) has for fixed $\lambda \in (-\varrho, \varrho)$ a unique solution in $\mathbb{B}_R(X \cap C)$. Here, the space $X \cap C$ is equipped with the norm $\|\cdot\|_X$.*

We pass to equation (7.1.49) which in Volterra style is given by

$$x(t) = h\big(t, x(t)\big) + \lambda \int_0^t \nu(t,s) g\big(s, x(s)\big)\,\mathrm{d}s \quad \text{for } 0 \le t \le 1. \tag{7.2.9}$$

Since the Theorems 7.1.24, 7.1.25 and 7.1.26 concerning this equation have been proven with Darbo's Fixed Point Theorem 7.1.22 which needed condition (H18) instead of (H14) we now replace (H18) by its Volterra equivalent. As we have seen in Section 4.3 this is precisely condition (VB_X^*) given in Table 4.3.4, and so we say that a Volterra kernel ν satisfies (V8) if and only if it satisfies condition (VB_X^*) given in that table. If we now replace (H18) in the Theorems 7.1.24, 7.1.25 and 7.1.26 by (V8) we obtain their Volterra versions.

Theorem 7.2.12. *Let X be any of the spaces BV, WBV_p, YBV_φ, ΛBV or RBV_p. Assume that the functions $g, h : [0,1] \times \mathbb{R} \to \mathbb{R}$ satisfy (ii)–(iv) of Theorem 7.1.24. Moreover, assume that the Volterra kernel $\nu : [0,1] \times [0,1] \to \mathbb{R}$ satisfies (V1) and (V8). Then for each $R > 0$ satisfying (7.1.50) there is some $\varrho > 0$ such that equation (7.2.9) has for fixed $\lambda \in (-\varrho, \varrho)$ a solution in $\mathbb{B}_R(X)$.*

Theorem 7.2.13. *Let X be any of the spaces BV, WBV_p, YBV_φ or ΛBV, and let $g, h : [0,1] \times \mathbb{R} \to \mathbb{R}$ be functions and $\nu : [0,1] \times [0,1] \to \mathbb{R}$ be a Volterra kernel. Assume that that g and h satisfy (ii)–(iv) of Theorem 7.1.24. Moreover, assume in addition that the function h therein is continuous and that the kernel ν satisfies (V1), (V7) and (V8). Then for each $R > 0$ satisfying (7.1.50) there is some $\varrho > 0$ such that equation (7.2.9) has for fixed $\lambda \in (-\varrho, \varrho)$ a solution in $\mathbb{B}_R(X \cap C)$. Here, the space $X \cap C$ is equipped with the norm $\|\cdot\|_X$.*

Theorem 7.2.14. *Let X be any of the spaces BV, WBV_p, YBV_φ, ΛBV or RBV_p, and let $g, h : [0,1] \times \mathbb{R} \to \mathbb{R}$ be functions and $\nu : [0,1] \times [0,1] \to \mathbb{R}$ be a Volterra kernel. Assume that the functions $g, h : [0,1] \times \mathbb{R} \to \mathbb{R}$ satisfy (ii)–(iv) of Theorem 7.1.24 with (7.1.55). Moreover, assume that the kernel ν satisfies (V1) and (V8). Then equation (7.2.9) has for every $\lambda \in \mathbb{R}$ a solution in X.*

Finally, we consider equation (7.1.56). Its Volterra version is

$$x(t) = Ax(t) + \lambda \int_0^t \nu(t,s) g\big(s, x(s)\big)\,\mathrm{d}s \quad \text{for } 0 \le t \le 1. \tag{7.2.10}$$

We obtain the following Volterra variant of Theorem 7.1.30.

Theorem 7.2.15. *Let X be any of the spaces BV, WBV_p, YBV_φ, ΛBV or RBV_p, and let $g : [0,1] \times \mathbb{R} \to \mathbb{R}$ be a function, $A : X \to X$ be a linear and bounded operator and $\nu : [0,1] \times [0,1] \to \mathbb{R}$ be a Volterra kernel. Assume that g satisfies (ii) of Theorem 7.1.24, that $\|A^n\|_{X \to X} < 1$ for some $n \in \mathbb{N}$ and that ν satisfies (V1) and (V8). Then for each $R > 0$ there is some $\varrho > 0$ such that equation (7.2.10) has for fixed $\lambda \in (-\varrho, \varrho)$ a solution in $\mathbb{B}_R(X)$.*

Similarly, one may get a Volterra version of Theorem 7.1.32 by replacing (H1) by (V1), (H15) by (V7) and (H18) by (V8).

We end this section with two further comments. If the solution of a nonlinear equation, like those considered in this chapter, is not unique, it is of some interest to have information on the topological structure of the solution set. One prominent example is the R_δ-property which means that the set of solutions is homeomorphic to the intersection of a decreasing sequence of absolute retracts.
Below we cite a sample result of this type for solutions $x \in \Lambda BV \cap C$ of the Volterra integral equation (7.2.6). Since this equation can be reformulated as a fixed point problem (7.1.9) where the integral operator I_k induced by a kernel k has to be replaced by a Volterra operator V_ν induced by a Volterra kernel ν, our discussion will rely upon the following structural result on fixed point sets of continuous operators in C which was proven in [145].

Proposition 7.2.16. *Let $T : C \to C$ be a continuous operator which satisfies the following four conditions.*

(i) The set $T(C) \subseteq C$ is equicontinuous.

(ii) There exist $t_0 \in [0,1]$ and $y_0 \in \mathbb{R}$ such that $Tx(t_0) = y_0$ for all $x \in C$.

(iii) The operator T is locally defined[2]*.*

(iv) Every sequence (x_n) in C satisfying

$$\lim_{n\to\infty} \|x_n - Tx_n\|_\infty = 0$$

has an accumulation point in C.

Then the fixed point set of T is a compact R_δ-set.

Condition (i) in Proposition 7.2.16 suggests to use some Arzelà-Ascoli type result, while condition (iv) is usually called a *Palais-Smale condition*; this is an important ingredient of topological and variational methods in nonlinear analysis.
In order to apply Proposition 7.2.16 to the solution set of equation (7.2.6) recall that a function $g : [0,1] \times \mathbb{R} \to \mathbb{R}$ is said to satisfy an *L_p-Carathéodory condition* if $t \mapsto g(t,u)$ is Lebesgue measurable for each $u \in \mathbb{R}$, $u \mapsto f(t,u)$ is continuous for almost each

[2]See Definition 5.2.47.

$t \in [0,1]$, and $|f(t,u)| \le m_p(t)$ for almost all $t \in [0,1]$ and some function $m_p \in L_p$. Moreover, we need the following technical hypothesis

$$\forall \varepsilon > 0\ \exists \delta > 0\ \forall t_1, t_2 \in [0,1]:$$
$$0 \le t_2 - t_1 \le \delta \quad \Longrightarrow \quad \int_0^{t_1} \Big| k(t_1, s) - k(t_2, s) \Big| m_p(s)\, \mathrm{d}s \le \varepsilon \qquad \text{(V9)}$$

which is some modification of (V7) and involves the function m_p from the Carathéodory condition. The following result was proven in [30].

Theorem 7.2.17. *Suppose that $g : [0,1] \times \mathbb{R} \to \mathbb{R}$ satisfies an L_p-Carathéodory condition for some $p \in (1, \infty]$, and let $h \in BV \cap C$ and $\lambda \in \mathbb{R}$. Assume that the Volterra kernel ν satisfies (V1), (V6) with $m \in L_q$ and $1/p + 1/q = 1$ and (V9). Then the set of all $x \in BV \cap C$ solving (7.2.6) is a compact R_δ-set.*

The proof basically rests on showing that the operator $T = h + I_\nu \circ N_g$ satisfies all the conditions imposed in Proposition 7.2.16. Note that (ii) is satisfied for $t_0 = 0$ and $y_0 = h(0)$ and (iii) is clearly true as we have seen at the end of Section 5.2. The most restrictive condition is of course condition (i), and this is the reason why Theorem 7.2.17 cannot be generalized so easily to equation (7.2.1). This is because if $h : [0,1] \times \mathbb{R} \to \mathbb{R}$ generates a superposition operator N_h from C into itself such that the set $N_h(C)$ is an equicontinuous subset of C, then N_h actually degenerates to a constant operator. To see this, first note that the acting condition $N_h(C) \subseteq C$ implies that h is continuous with respect to its first argument. Second, the equicontinuity of $N_h(C)$ implies

$$\forall t \in [0,1]\ \forall \varepsilon > 0\ \exists \delta > 0\ \forall x \in C\ \forall \tau \in [0,1]:$$
$$|t - \tau| \le \delta \quad \Longrightarrow \quad \Big| N_h x(t) - N_h x(\tau) \Big| \le \varepsilon. \qquad (7.2.11)$$

We now fix $u, v \in \mathbb{R}$, $t \in [0,1]$ and $\varepsilon > 0$. Then we pick $\delta > 0$ according to (7.2.11) and choose $\tau \in [0,1]$ so that $0 < |t - \tau| \le \min\{\varepsilon, \delta\}$. The function $x : [0,1] \to \mathbb{R}$, defined by

$$x(s) = \frac{t - s}{t - \tau} v + \frac{\tau - s}{\tau - t} u,$$

belongs to C and satisfies $x(t) = u$ and $x(\tau) = v$. Consequently, from (7.2.11) we now obtain $|h(t,u) - h(\tau,v)| \le \varepsilon$. Since $h(\cdot, v)$ is continuous at t, it follows by letting $\varepsilon \to 0$ that $\tau \to t$ and hence $h(t,u) = h(t,v)$. But this means that h is actually independent of its second argument and hence N_h degenerates to a constant operator, as claimed.

As a final remark we point out that every result in this section guarantees a solution of a Volterra integral equation that lives on the entire interval $[0,1]$. Due to the special structure of the Volterra operator it is sometimes possible to achieve solutions only on a subinterval $[0,T]$ of $[0,1]$ with the benefit that the underlying Volterra integral equation may have solutions on that subinterval for a larger set of parameters λ. One sample result concerning equation (7.2.2) was proven in [29]: Indeed, the author has shown that under the assumptions (V1), (V2) and (H3) for each $R > \|h\|_{BV}$ there is some $T \in (0,1]$ and $\varrho > 0$ such that the Volterra integral equation (7.2.2) has a unique solution $x \in \mathbb{B}_R(BV[0,T])$ for any $|\lambda| < \varrho$. The interplay between ϱ and T is given by the relation

$$\varrho = \min\left\{\frac{R - \|h\|_{BV}}{\|g\|_{[-R,R]}}, \frac{1}{\operatorname{lip}(g,[-R,R])}\right\}\frac{1}{2\|m\|_{L_1[0,T]}}.$$

Note that the only difference to formula (7.2.3) is that here the L_1-norm of m is taken only on $[0,T]$. This shows that since $\|m\|_{L_1[0,T]} \to 0$ as $T \to 0+$ one can make ϱ large by taking T small. In particular, equation (7.2.2) now has solutions for every λ, but the larger $|\lambda|$ is chosen the smaller may be the domain on which the solutions live.

7.3 Boundary and Initial Value Problems

It is well known that boundary value problems are closely related to Hammerstein integral equations like (7.1.1) and initial value problems are closely related to Volterra integral equations like (7.2.2). In this section we give examples illustrating these relations in more detail. We discuss some problems which may be solved by means of our results obtained in the Theorems 7.1.30 and 7.2.15. While there is a vast literature on continuous solutions of such problems, considerably less is known on BV-solutions. We will be interested in solutions primarily in the spaces BV equipped with $\|\cdot\|_{BV}$.

Boundary Value Problems

We start with a boundary value problem in a nonclassical setting. Consider the second order equation

$$x''(t) = -\lambda g\big(t, x(t)\big) \quad \text{for } 0 \le t \le 1, \tag{7.3.1}$$

subject to the coupled *boundary conditions*

$$x(0) = A_0 x \quad \text{and} \quad x(1) = A_1 x, \tag{7.3.2}$$

where $A_0, A_1 : BV \to \mathbb{R}$ are given linear functionals. In the following we refer to the boundary value problem (7.3.1) together with the boundary conditions (7.3.2) by the symbol (BVP). In order to solve (BVP) with Theorem 7.1.30, we consider the Hammerstein integral equation

$$x(t) = Ax(t) + \lambda \int_0^1 \kappa(t,s) g\big(s, x(s)\big)\,\mathrm{d}s \quad \text{for } 0 \le t \le 1, \tag{7.3.3}$$

where

$$\kappa(t,s) = \begin{cases} t(1-s) & \text{for } 0 \le t < s \le 1, \\ (1-t)s & \text{for } 0 \le s \le t \le 1, \end{cases} \tag{7.3.4}$$

is the usual Green's function of the second order derivative and A is a linear operator from BV into itself. The bridge between (7.3.3) and our (BVP) is now built by our next result. Denoting by

$$AC^1 := \{x \in C^1 \mid x' \in AC\}$$

the space of all continuously differentiable functions with derivative in AC, it says that any $x \in BV$ solving (7.3.3) automatically belongs to AC^1, provided that the linear operator $A : BV \to BV$ is defined by

$$Ax(t) = (1-t)A_0x + tA_1x \quad \text{for } 0 \leq t \leq 1, \tag{7.3.5}$$

where A_0 and A_1 are the linear functionals used in (7.3.2).

Proposition 7.3.1. *Let $g : [0,1] \times \mathbb{R} \to \mathbb{R}$ satisfy (H11*(L_∞)), (H16) and (H17), and let $A : BV \to BV$ be defined by (7.3.5). Then any function $x \in BV$ solving (7.3.3) is differentiable in $[0,1]$ and has an absolutely continuous derivative. Moreover, it satisfies (7.3.2) and solves (7.3.1) almost everywhere.*
If, in addition, g is continuous in $[0,1] \times \mathbb{R}$, then x is of class C^2 and solves (BVP) everywhere on $[0,1]$.

Proof. Assume that (7.3.3) is satisfied for some $x \in BV$ and some $\lambda \in \mathbb{R}$. First observe that $h(s) := g(s, x(s))$ belongs to L_∞ because of (H11*(L_∞)), (H16) and (H17). Moreover, we set

$$\begin{aligned}\varphi(t) &:= \int_0^1 \kappa(t,s) g\big(s, x(s)\big)\, ds \\ &= \int_0^t (1-t)sh(s)\, ds - \int_1^t t(1-s)h(s)\, ds \qquad \text{for } 0 \leq t \leq 1.\end{aligned}$$

By [150], the function φ belongs to AC with

$$\varphi'(t) = -\int_0^t sh(s)\, ds - \int_1^t (1-s)h(s)\, ds \quad \text{for almost all } t \in [0,1],$$

but since the right hand side is again in AC we conclude that $\varphi \in C^1$ with $\varphi' \in AC$. Moreover, we obtain

$$\varphi''(t) = -th(t) - (1-t)h(t) = -h(t) \quad \text{for almost all } t \in [0,1].$$

In addition, by definition of A the function Ax is affine and hence of class C^2 with $(Ax)'' = \mathbb{0}$. From (7.3.3) follows

$$x(t) = Ax(t) + \lambda\varphi(t) \quad \text{for } 0 \leq t \leq 1;$$

in particular, this shows that (7.3.1) holds indeed almost everywhere in $[0,1]$. Moreover, since $\varphi(0) = \varphi(1) = 0$, we obtain $x(0) = Ax(0) = A_0x$ and $x(1) = Ax(1) = A_1x$. Consequently, the first part of the proof is complete.
If, in addition, g is continuous, then so must be x'' which means that x is of class C^2 and solves (BVP) everywhere on $[0,1]$. ∎

According to Proposition 7.3.1, in order to find a solution $x \in AC^1$ of (BVP) - and by this we mean a function $x \in AC^1$ that satisfies the boundary conditions (7.3.2) and the equation (7.3.1) almost everywhere on $[0,1]$ - all we have to do is to make sure that the norms of the two linear functionals A_0 and A_1 behave in such a way that the norm of the iterate operator A^n shrinks below 1 for some $n \in \mathbb{N}$. We give two sufficient conditions for this in the following theorem the ideas of which come from [27].

Theorem 7.3.2. *Assume that the function $g : [0,1] \times \mathbb{R} \to \mathbb{R}$ satisfies (H11*(L_∞)), (H16) and (H17). Moreover, assume that the linear functionals $A_0, A_1 : BV \to \mathbb{R}$ are bounded and satisfy one of the following two conditions.*

(a) $\|A_0\|_{BV\to\mathbb{R}} + 2\|A_0 - A_1\|_{BV\to\mathbb{R}} < 1.$

(b) $A_0\mathbb{1} = A_1\mathbb{1} = 0$ *and* $|A_0 w - A_1 w| < 1$, *where* $w(t) := t$ *for* $t \in [0,1]$.

Then for each $R > 0$ there is some $\varrho > 0$ such that (BVP) has for fixed $\lambda \in (-\varrho, \varrho)$ a solution $x \in \mathbb{B}_R(AC^1)$. Here, the space AC^1 is equipped with the norm $\|\cdot\|_{BV}$.
If, in addition, g is continuous in $[0,1] \times \mathbb{R}$, then every such solution is of class C^2.

Proof. Define A as in Proposition 7.3.1, that is, $Ax = (\mathbb{1} - w)A_0x + wA_1x$. Since A_0 and A_1 are supposed to be bounded and linear, then so is A. We show for either of the two options (a) and (b) that there is some $n \in \mathbb{N}$ such that $\|A^n\|_{BV\to BV} < 1$. Once this is done, Theorem 7.1.30 tells us that for each $R > 0$ there is some $\varrho > 0$ such that the integral equation (7.3.3) has for fixed $\lambda \in (-\varrho, \varrho)$ a solution $x \in \mathbb{B}_R(BV)$. By Proposition 7.3.1, the solution x belongs to AC^1, has the correct boundary values according to (7.3.2) and satisfies (7.3.1) almost everywhere. Note that κ satisfies (H12(L_1)) and (H18), because $\mathrm{Var}(\kappa(\cdot, s)) = 2s(1-s)$ for all $s \in [0,1]$, and hence Theorem 7.1.30 is applicable. If g is continuous, then x is twice continuously differentiable.

It remains to show $\|A^n\|_{BV\to BV} < 1$ for some $n \in \mathbb{N}$ provided that A_0 and A_1 satisfy (a) or (b). Let us start with (a). We have for any $x \in BV$,

$$\begin{aligned}\|Ax\|_{BV} &= \|\mathbb{1}A_0x + w(A_1 - A_0)x\|_{BV} \leq \|\mathbb{1}\|_{BV}|A_0x| + \|w\|_{BV}\left|(A_1 - A_0)x\right| \\ &\leq \|A_0\|_{BV\to\mathbb{R}}\|x\|_{BV} + 2\|A_0 - A_1\|_{BV\to\mathbb{R}}\|x\|_{BV}.\end{aligned}$$

Consequently,

$$\|A\|_{BV\to BV} \leq \|A_0\|_{BV\to\mathbb{R}} + 2\|A_0 - A_1\|_{BV\to\mathbb{R}} < 1$$

by (a), showing that A is a contraction. In this case, we may take $n = 1$.

We now assume that A_0 and A_1 satisfy option (b). Note that in this case, $A\mathbb{1} = \mathbb{0}$. By induction, we first prove that the iterates of A are given by

$$A^{n+2}x = \big((\mathbb{1} - w)A_0w + wA_1w\big)\big(A_1w - A_0w\big)^n\big(A_1x - A_0x\big) \tag{7.3.6}$$

for $x \in BV, n \in \mathbb{N}_0$, where we set $0^0 := 1$. First, we have

$$\begin{aligned} A(Ax) &= A\Big((\mathbb{1} - w)A_0x + wA_1x\Big) = A(\mathbb{1} - w)A_0x + AwA_1x = Aw(A_1x - A_0x) \\ &= \Big((\mathbb{1} - w)A_0w + wA_1w\Big)(A_1x - A_0x), \end{aligned}$$

and this is (7.3.6) for $n = 0$. Moreover,

$$\begin{aligned} A_1(Ax)-&A_0(Ax) \\ &= (A_1 - A_0)\Big((\mathbb{1} - w)A_0x + wA_1x\Big) \\ &= (A_1 - A_0)(\mathbb{1} - w)A_0x + (A_1 - A_0)wA_1x = (A_1 - A_0)w(A_1x - A_0x) \\ &= (A_1w - A_0w)(A_1x - A_0x). \end{aligned}$$

From this we deduce that if (7.3.6) has been proven for some $n \in \mathbb{N}_0$, then

$$\begin{aligned} A^{n+3}x = A^{n+2}(Ax) &= \Big((\mathbb{1} - w)A_0w + wA_1w\Big)\Big(A_1w - A_0w\Big)^n\Big(A_1(Ax) - A_0(Ax)\Big) \\ &= \Big((\mathbb{1} - w)A_0w + wA_1w\Big)\Big(A_1w - A_0w\Big)^{n+1}(A_1x - A_0x). \end{aligned}$$

By induction, (7.3.6) is established. As a consequence we get for $n \geq 2$,

$$\|A^n\|_{BV\to BV} \leq \|(\mathbb{1} - w)A_0w + wA_1w\|_{BV}\Big|A_1w - A_0w\Big|^{n-2}\Big(\|A_0\|_{BV\to\mathbb{R}} + \|A_1\|_{BV\to\mathbb{R}}\Big),$$

and since $|A_1w - A_0w| < 1$ by (b) we find some $n \in \mathbb{N}$ such that $\|A^n\|_{BV\to BV} < 1$. ■

Let us now pass to some example showing how to apply Theorem 7.3.2. The following is similar to an example from [27] and builds on option (a) of Theorem 7.3.2.

Example 7.3.3. Consider the boundary value problem

$$\left.\begin{aligned} x''(t) &= -\lambda g\big(t, x(t)\big) \quad \text{for } 0 \leq t \leq 1, \\ x(0) &= \tfrac{1}{7}x(1/2) + \tfrac{1}{6}x(2/3), \\ x(1) &= \tfrac{1}{7}x(1/4) + \tfrac{1}{6}x(4/5) \end{aligned}\right\} \tag{7.3.7}$$

with g satisfying (H11$^*(L_\infty)$), (H16) and (H17). We are interested in finding a solution $x \in AC^1$. In the notation used in (7.3.2) we define our functionals A_0 and A_1 by

$$A_0x := \tfrac{1}{7}x(1/2) + \tfrac{1}{6}x(2/3) \quad \text{and} \quad A_1x := \tfrac{1}{7}x(1/4) + \tfrac{1}{6}x(4/5).$$

Then A_0 and A_1 are bounded linear functionals on BV with

$$|A_0x| \leq \tfrac{1}{7}\|x\|_\infty + \tfrac{1}{6}\|x\|_\infty \leq \tfrac{13}{42}\|x\|_{BV}$$

and

$$\begin{aligned} |A_0x - A_1x| &= \Big|\tfrac{1}{7}\Big(x(1/2) - x(1/4)\Big) + \tfrac{1}{6}\Big(x(2/3) - x(4/5)\Big)\Big| \\ &\leq \tfrac{1}{7}\operatorname{Var}(x) + \tfrac{1}{6}\operatorname{Var}(x) \leq \tfrac{13}{42}\|x\|_{BV}. \end{aligned}$$

Thus,

$$\|A_0\|_{BV\to\mathbb{R}} + 2\,\|A_0 - A_1\|_{BV\to\mathbb{R}} \le \frac{39}{42} < 1$$

which means that A_0 and A_1 satisfy option (a) of Theorem 7.3.2. Accordingly, (7.3.7) has for small $|\lambda|$ an AC^1-solution. Observe that A_0 and A_1 do not satisfy option (b), as $A_0\mathbb{1} = A_1\mathbb{1} = 13/42 \neq 0$. ◊

The next example shows that in some cases only option (b) in Theorem 7.3.2 can be used.

Example 7.3.4. Consider the boundary value problem

$$\left.\begin{aligned} x''(t) &= -\lambda g\big(t, x(t)\big) \quad \text{for } 0 \le t \le 1,\\ x(0) &= 3x(1/2) - 3x(2/3),\\ x(1) &= 2x(1/4) - 2x(4/5) \end{aligned}\right\} \tag{7.3.8}$$

with g satisfying (H11*(L_∞)), (H16) and (H17). We are interested in finding a solution $x \in AC^1$. In the notation used in (7.3.2) we define our functionals A_0 and A_1 by

$$A_0x := 3x(1/2) - 3x(2/3) \quad \text{and} \quad A_1x := 2x(1/4) - 2x(4/5).$$

Then A_0 and A_1 are bounded linear functionals on BV with $A_0\mathbb{1} = A_1\mathbb{1} = 0$ and, writing $w(t) = t$ as in Theorem 7.3.2,

$$|A_0w - A_1w| = \left|\frac{3}{2} - 2 - \frac{1}{2} + \frac{8}{5}\right| = \frac{3}{5} < 1$$

which means that A_0 and A_1 satisfy option (b) of Theorem 7.3.2. Accordingly, (7.3.8) has for small $|\lambda|$ an AC^1-solution. Observe that A_0 and A_1 do not satisfy option (a), because

$$\frac{A_0\chi_{[0,1/2]}}{\left\|\chi_{[0,1/2]}\right\|_{BV}} = \frac{3}{2}$$

and so $\|A_0\|_{BV\to\mathbb{R}} \ge 3/2 > 1$. ◊

Unfortunately, in some cases, neither option (a) nor option (b) of Theorem 7.3.2 can be used. We give a third example.

Example 7.3.5. Consider the boundary value problem

$$\left.\begin{aligned} x''(t) &= -\lambda g\big(t, x(t)\big) \quad \text{for } 0 \le t \le 1,\\ x(0) &= x(1/3) + x(2/3),\\ x(1) &= -\tfrac{1}{2}x(1/3) - \tfrac{1}{2}x(2/3), \end{aligned}\right\} \tag{7.3.9}$$

with g satisfying (H11*(L_∞)), (H16) and (H17). In the notation used in (7.3.2) we define our functionals A_0 and A_1 by

$$A_0x := x(1/3) + x(2/3) \quad \text{and} \quad A_1x := -\tfrac{1}{2}x(1/3) - \tfrac{1}{2}x(2/3).$$

Then A_0 and A_1 are bounded linear functionals on BV. However, $A_0\mathbb{1} = 2 \neq 0$, and so option (b) of Theorem 7.3.2 cannot be used. But (a) cannot be used either, because the same equality also shows $\|A_0\|_{BV\to\mathbb{R}} \ge 2$. ◊

We therefore generalize the ideas of Theorem 7.3.2. Due to the special structure of the linear operator A defined in (7.3.5) it is possible to give an exact formula for its spectral radius. For this purpose we prove first an abstract result about the spectral radius of linear operators of a slightly more general form than (7.3.5) which might be of its own interest.

Proposition 7.3.6. *Let $(X, \|\cdot\|)$ be a real Banach space, let $A_0, A_1 : X \to \mathbb{R}$ be bounded linear functionals and let $v, w \in X$ be fixed. Then for the spectral radius $\mathfrak{R}$ of the operator $A : X \to X$, defined by*

$$Ax = vA_0x + wA_1x \quad \text{for } x \in X, \tag{7.3.10}$$

we have the relationship

$$\mathfrak{R}(A) = \mathfrak{R}\begin{pmatrix} A_0v & A_1v \\ A_0w & A_1w \end{pmatrix}. \tag{7.3.11}$$

Proof. For this proof we set

$$C := \begin{pmatrix} A_0v & A_1v \\ A_0w & A_1w \end{pmatrix}.$$

We first prove the "$\leq$"-part in (7.3.11) and begin by showing that the iterates of A can be written in the form

$$A^n x = va_nx + wb_nx \quad \text{for } x \in X, \tag{7.3.12}$$

where $a_n, b_n : X \to \mathbb{R}$ are linear functionals satisfying for all $x \in X$ the linear recursions

$$a_{n+1}x = a_nvA_0x + a_nwA_1x \qquad \text{and} \qquad a_1x := A_0x, \tag{7.3.13}$$

$$b_{n+1}x = b_nvA_0x + b_nwA_1x \qquad \text{and} \qquad b_1x := A_1x. \tag{7.3.14}$$

Indeed, once the formula (7.3.12) for A^n has been established for some $n \in \mathbb{N}$, we get

$$\begin{aligned} A^{n+1}x &= A^n(Ax) = va_n(Ax) + wb_n(Ax) \\ &= v \cdot \Big(a_nvA_0x + a_nwA_1x\Big) + w \cdot \Big(b_nvA_0x + b_nwA_1x\Big) \\ &= va_{n+1}x + wb_{n+1}x. \end{aligned} \tag{7.3.15}$$

Plugging v and w for x into the recursion formulas (7.3.13) and (7.3.14) we see that the four numbers a_nv, a_nw, b_nv and b_nw put into the matrix

$$B_n := \begin{pmatrix} a_nv & b_nv \\ a_nw & b_nw \end{pmatrix}$$

in turn satisfy the matrix recursion $B_{n+1} = CB_n$ for all $n \in \mathbb{N}$ with $B_1 = C$. Thus, $B_{n+1} = CB_n$ and hence

$$B_n = C^n \quad \text{for all } n \in \mathbb{N}. \tag{7.3.16}$$

Setting

$$M := \max\left\{ \|v\|\,\|A_0\|_{X\to\mathbb{R}},\ \|v\|\,\|A_1\|_{X\to\mathbb{R}},\ \|w\|\,\|A_0\|_{X\to\mathbb{R}},\ \|w\|\,\|A_1\|_{X\to\mathbb{R}} \right\}$$

we obtain from (7.3.15)

$$\begin{aligned}\left\|A^{n+1}\right\|_{X\to X} &\le \|v\| \left(|a_n v|\,\|A_0\|_{X\to\mathbb{R}} + |a_n w|\,\|A_1\|_{X\to\mathbb{R}}\right)\\ &\quad + \|w\| \left(|b_n v|\,\|A_0\|_{X\to\mathbb{R}} + |b_n w|\,\|A_1\|_{X\to\mathbb{R}}\right)\\ &\le M\Big(|a_n v| + |a_n w| + |b_n v| + |b_n w|\Big) \le 2M\,\|B_n\|_\infty = 2M\,\|C^n\|_\infty,\end{aligned}$$

where $\|\cdot\|_\infty$ denotes the row sum norm of a matrix. Consequently, by Gelfand's formula,

$$\mathfrak{R}(A) = \lim_{n\to\infty} \left\|A^{n+1}\right\|_{X\to X}^{1/n} \le \lim_{n\to\infty} \Big(2M\,\|C^n\|_\infty\Big)^{1/n} = \lim_{n\to\infty} \|C^n\|_\infty^{1/n} = \mathfrak{R}(C),$$

and this is the inequality "$\le$" in (7.3.11).

We now prove the remaining inequality "$\ge$" in (7.3.11) and distinguish between the cases when v and w are linearly dependent respectively independent in X.

Case 1: Assume $w = \lambda v$ for some $\lambda \in \mathbb{R}$. In this case, we have

$$C = \begin{pmatrix} A_0 v & A_1 v \\ \lambda A_0 v & \lambda A_1 v \end{pmatrix}$$

with $\mathfrak{R}(C) = |A_0 v + \lambda A_1 v|$. Moreover, we get for the linear functional $L := A_0 + \lambda A_1$,

$$\begin{aligned} Ax &= v(A_0 + \lambda A_1)x = vLx,\\ A^2x &= A(Ax) = vL(vLx) = vLvLx,\\ A^3x &= A^2(Ax) = vLvL(vLx) = v(Lv)^2Lx \end{aligned}$$

and inductively

$$A^n x = v(Lv)^{n-1}Lx \quad \text{for all } n \in \mathbb{N}, x \in X, \tag{7.3.17}$$

where we set $0^0 := 1$ again. If $v = 0$, we also have $w = 0$ and hence $A = 0$ and $C = 0$ which implies $\mathfrak{R}(A) = \mathfrak{R}(C) = 0$. We therefore assume $v \ne 0$. If $Lx = 0$ for all $x \in X$, we have $A_0 = -\lambda A_1$ which implies on the one hand $Ax = 0$ for all $x \in X$ and hence $\mathfrak{R}(A) = 0$, and on the other hand

$$C = \begin{pmatrix} -\lambda A_1 v & A_1 v \\ -\lambda^2 A_1 v & \lambda A_1 v \end{pmatrix}$$

with $\mathfrak{R}(C) = 0 = \mathfrak{R}(A)$. We therefore suppose that there is some $y \in X$ with $Ly \ne 0$ and $\|y\| = 1$. Consequently, by (7.3.17),

$$\|A^n\|_{X\to X} \ge \|A^n y\| = \|v\|\,|A_0 v + \lambda A_1 v|^{n-1} |Ly|$$

and hence by Gelfand's formula

$$\mathfrak{R}(A) = \lim_{n\to\infty} \|A^n\|_{X\to X}^{1/n} \geq \left|A_0 v + \lambda A_1 v\right| = \mathfrak{R}(C).$$

This proves the "$\geq$"-part in (7.3.11) for Case 1.

Case 2: Assume $w \neq \mu v$ for all $\mu \in \mathbb{R}$. By definition, the spectral radius of the operator A on the real space X is the spectral radius of its complexification $A : X_{\mathbb{C}} \to X_{\mathbb{C}}$, defined by $A(x+iy) := A(x) + iA(y)$, where $X_{\mathbb{C}} := \{x + iy \mid x, y \in X\}$ denotes the complexification of X equipped with the norm

$$\|x+iy\|_{X_{\mathbb{C}}} = \max_{t\in[0,2\pi]} \|\cos(t)x + \sin(t)y\|$$

and the common addition and multiplication with complex scalars. In addition, we have

$$\|x\| = \|x\|_{X_{\mathbb{C}}} = \|ix\|_{X_{\mathbb{C}}} \text{ for all } x \in X \qquad \text{and} \qquad \|A\|_{X\to X} = \|A\|_{X_{\mathbb{C}}\to X_{\mathbb{C}}}.$$

Similarly, we complexify the functionals A_0 and A_1 by extending them to the functionals $A_0, A_1 : X_{\mathbb{C}} \to \mathbb{C}$ via

$$A_0(x+iy) = A_0x + iA_0y \qquad \text{and} \qquad A_1(x+iy) = A_1x + iA_1y.$$

By splitting into real and imaginary parts we see that then

$$Az = vA_0z + wA_1z \quad \text{for all } z \in X_{\mathbb{C}}.$$

Let now $\lambda \in \mathbb{C}$ be an eigenvalue of C^T with eigenvector $u = (u_1, u_2) \in \mathbb{C}^2 \backslash \{(0,0)\}$, that is,

$$u^TC = \lambda u^T$$

which means

$$u_1A_0v + u_2A_0w = \lambda u_1 \qquad \text{and} \qquad u_1A_1v + u_2A_1w = \lambda u_2. \tag{7.3.18}$$

Since v and w are linearly independent in X, there are $x, y \in X$ with

$$Ax = \mathrm{Re}(u_1)\, v + \mathrm{Re}(u_2)\, w \qquad \text{and} \qquad Ay = \mathrm{Im}(u_1)\, v + \mathrm{Im}(u_2)\, w.$$

Setting $z := x + ix \in X_{\mathbb{C}}$, this leads to

$$Az = A(x+iy) = Ax + iAy = vu_1 + wu_2,$$

and since v and w are linearly independent in X and $u = (u_1, u_2) \neq (0,0)$, we conclude $Az \neq 0$. We obtain from (7.3.18)

$$\begin{aligned} A(Az) &= vA_0(Az) + wA_1(Az) = v(u_1A_0v + u_2A_0w) + w(u_1A_1v + u_2A_1w) \\ &= \lambda(u_1v + u_2w) = \lambda Az. \end{aligned}$$

Since $Az \neq 0$ we conclude that Az is an eigenvector and λ is an eigenvalue of (the complexification of) A. Consequently, $\mathfrak{R}(A) \geq \mathfrak{R}(C^T) = \mathfrak{R}(C)$. ∎

Let us look at an example in the very simple case that $X = \mathbb{R}$.

Example 7.3.7. For $X = \mathbb{R}$ consider the functional $Ax = vA_0x + wA_1x$, where $A_0x := ax$, $A_1x := bx$ and $a, b, v, w \in \mathbb{R}$ are constants. Then $Ax = (av + bw)x$, and the iterates of A are given by $A^nx = (av + bw)^nx$. Thus, $\mathfrak{R}(A) = |av + bw|$. On the other hand it is easy to see that the matrix

$$C = \begin{pmatrix} A_0v & A_1v \\ A_0w & A_1w \end{pmatrix} = \begin{pmatrix} av & bv \\ aw & bw \end{pmatrix}$$

has the two eigenvalues 0 and $av + bw$. Again, $\mathfrak{R}(C) = |av + bw|$, in accordance with Proposition 7.3.6. ◇

The following refinement of Theorem 7.3.2 is now immediate.

Theorem 7.3.8. *Assume that the function $g : [0,1] \times \mathbb{R} \to \mathbb{R}$ satisfies (H11*(L_∞)), (H16) and (H17). Moreover, assume that the linear functionals $A_0, A_1 : BV \to \mathbb{R}$ are bounded with*

$$\mathfrak{R}\begin{pmatrix} A_0v & A_1v \\ A_0w & A_1w \end{pmatrix} < 1, \tag{7.3.19}$$

where the functions $v, w \in BV$ are defined by $v(t) = 1 - t$ and $w(t) = t$ for $t \in [0,1]$. Then for each $R > 0$ there is some $\varrho > 0$ such that (BVP) has for fixed $\lambda \in (-\varrho, \varrho)$ a solution $x \in \mathbb{B}_R(AC^1)$. Here, the space AC^1 is equipped with the norm $\|\cdot\|_{BV}$.
If, in addition, g is continuous on $[0,1] \times \mathbb{R}$, then every such solution is of class C^2.

Proof. The argument is similar as in the proof of Theorem 7.3.2. Accordingly, we only need to show that the operator $Ax = vA_0x + wA_1x$ satisfies $\|A^n\|_{BV\to BV} < 1$ for some $n \in \mathbb{N}$. But this is clear, because (7.3.19) in combination with Proposition 7.3.6 yields that $\mathfrak{R}(A) < 1$. ■

Before we turn back to Example 7.3.5, let us mention that Theorem 7.3.2 is completely covered by Theorem 7.3.8. Indeed, in the proof of Theorem 7.3.2 we have shown that under the option (a) or (b) we have $\mathfrak{R}(A) < 1$ and thus

$$\mathfrak{R}\begin{pmatrix} A_0v & A_1v \\ A_0w & A_1w \end{pmatrix} < 1$$

by Proposition 7.3.6. Moreover, Theorem 7.3.8 has three further advantages: First, since it uses the spectral radius of the operator A and not its norm, Theorem 7.3.8 is independent of the norm used in BV. Indeed, one may show that Theorem 7.3.2 still holds if BV is equipped with the smaller norm $\|x\|_{BV}^* = |x(0)| + \mathrm{Var}(x)$. In this case, option (a) can be replaced by

$$\|A_0\|_{BV\to\mathbb{R}}^* + \|A_0 - A_1\|_{BV\to\mathbb{R}}^* < 1, \tag{a*}$$

where $\|\cdot\|_{BV\to\mathbb{R}}^*$ is the operator norm for linear functionals from $(BV, \|\cdot\|_{BV}^*)$ into $(\mathbb{R}, |\cdot|)$; for instance, this was done in [27] in the subspace $BV \cap C$ of BV, equipped with the norm $\|\cdot\|_{BV}^*$ for functionals A_0 and A_1, defined via Riemann-Stieltjes integrals.

Second, (7.3.19) is easier to varify than the options given in Theorem 7.3.2. And third, in the following example, in which we come back to Example 7.3.5, we show that neither Theorem 7.3.2 nor its modification with (a*) may be applied to (BVP) while Theorem 7.3.8 may be, indeed. So Theorem 7.3.8 is stronger than Theorem 7.3.2.

Example 7.3.9. Consider again the boundary value problem from Example 7.3.5 with the functionals

$$A_0x := x(1/3) + x(2/3) \quad \text{and} \quad A_1x := -\tfrac{1}{2}x(1/3) - \tfrac{1}{2}x(2/3).$$

There we have seen that $A_0\mathbb{1} = 2 \neq 0$ holds and hence that Theorem 7.3.2 cannot be applied. But neither can its modification with (a*), because the same equality shows $\|A_0\|^*_{BV\to\mathbb{R}} \geq 2$, as $\|\mathbb{1}\|_{BV} = \|\mathbb{1}\|^*_{BV} = 1$.
However, the matrix in (7.3.8) becomes

$$\begin{pmatrix} A_0v & A_1v \\ A_0w & A_1w \end{pmatrix} = \begin{pmatrix} 1 & -1/2 \\ 1 & -1/2 \end{pmatrix}$$

and has the eigenvalues 0 and 1/2. Thus, Theorem 7.3.8 may be applied. ◇

Summarizing the previous examples, we always considered boundary conditions of the form

$$x(0) = ax(\tau_1) + bx(\tau_2) \quad \text{and} \quad x(1) = cx(\sigma_1) + dx(\sigma_2), \tag{7.3.20}$$

where $a, b, c, d \in \mathbb{R}$ and $\tau_1, \tau_2, \sigma_1, \sigma_2 \in [0, 1]$ are given. The estimate (7.3.19) now says that Theorem 7.3.8 may be applied if and only if

$$\mathfrak{R}\begin{pmatrix} a(1-\tau_1) + b(1-\tau_2) & c(1-\sigma_1) + d(1-\sigma_2) \\ a\tau_1 + b\tau_2 & c\sigma_1 + d\sigma_2 \end{pmatrix} < 1. \tag{7.3.21}$$

Instead of considering "local" boundary conditions of type (7.3.20), we may - of course - also look at "global" boundary conditions like, for instance,

$$x(0) = \int_0^1 k_0(s)x(s)\,\mathrm{d}s \quad \text{and} \quad x(1) = \int_0^1 k_1(s)x(s)\,\mathrm{d}s,$$

where $k_0, k_1 \in L_1$. Then A_0 and A_1 are defined by

$$A_0x = \int_0^1 k_0(s)x(s)\,\mathrm{d}s \quad \text{and} \quad A_1x = \int_0^1 k_1(s)x(s)\,\mathrm{d}s$$

and thus Theorem 7.3.8 may be applied if and only if

$$\mathfrak{R}\begin{pmatrix} \displaystyle\int_0^1 k_0(s)(1-s)\,\mathrm{d}s & \displaystyle\int_0^1 k_1(s)(1-s)\,\mathrm{d}s \\ \displaystyle\int_0^1 k_0(s)s\,\mathrm{d}s & \displaystyle\int_0^1 k_1(s)s\,\mathrm{d}s \end{pmatrix} < 1.$$

Example 7.3.10. Consider the boundary value problem

$$\left.\begin{aligned} x''(t) &= -\lambda g\big(t, x(t)\big) \quad \text{for } 0 \le t \le 1, \\ x(0) &= x(1) = \frac{1}{2}\int_0^1 x(s)\,\mathrm{d}s. \end{aligned}\right\} \tag{7.3.22}$$

We then have $k_0 = k_1 \equiv 1/2$ and

$$\mathfrak{R}\begin{pmatrix} \displaystyle\int_0^1 k_0(s)(1-s)\,\mathrm{d}s & \displaystyle\int_0^1 k_1(s)(1-s)\,\mathrm{d}s \\ \displaystyle\int_0^1 k_0(s)s\,\mathrm{d}s & \displaystyle\int_0^1 k_1(s)s\,\mathrm{d}s \end{pmatrix} = \mathfrak{R}\begin{pmatrix} 1/4 & 1/4 \\ 1/4 & 1/4 \end{pmatrix} = \frac{1}{2} < 1.$$

Thus, under the assumptions of Theorem 7.3.8 we find for small $|\lambda|$ a solution $x \in AC^1$. ◇

Before we turn to initial value problems we point out that condition (7.3.19) is *not* necessary for (BVP) to have a solution in AC^1.

Example 7.3.11. Consider the boundary value problem

$$\left.\begin{aligned} x''(t) &= \lambda x(t)\big(2+4t^2\big) \quad \text{for } 0 \le t \le 1, \\ x(0) &= e^{-1/4}x(1/2), \\ x(1) &= e^{3/4}x(1/2). \end{aligned}\right\} \tag{7.3.23}$$

In the notation of (7.3.20) we have $a = e^{-1/4}$, $b = d = 0$, $c = e^{3/4}$ and $\tau_1 = \sigma_1 = 1/2$. Thus, the spectral radius of the matrix in (7.3.21) becomes

$$\mathfrak{R}\left[\frac{1}{2}\begin{pmatrix} e^{-1/4} & e^{3/4} \\ e^{-1/4} & e^{3/4} \end{pmatrix}\right] = \frac{1+e}{2e^{1/4}} > 1.$$

Nevertheless, it is easy to check that the function $x(t) = \exp(t^2)$ is an (even analytic) solution to the boundary value problem (7.3.23) for $\lambda = 1$. ◇

The theory developed in this section may be applied to other similar boundary value problems than those we have considered in the examples so far. Instead of inundating the reader in too much technicalities, we skip the details and will be brief, because the arguments are similar as those used before.
Consider the third order boundary value problem

$$\left.\begin{aligned} x'''(t) &= -\lambda g\big(t, x(t)\big) \quad \text{for } 0 \le t \le 1, \\ x'(0) &= A_0 x, \\ x'(1) &= A_1 x, \end{aligned}\right\} \tag{7.3.24}$$

where $A_0, A_1 : BV \to \mathbb{R}$ are bounded linear functionals. We are looking for solutions x in the space

$$AC^2 := \{x \in C^2 \mid x'' \in AC\}$$

by which we mean a function $x \in AC^2$ that satisfies the differential equation

$$x'''(t) = -\lambda g\big(t, x(t)\big)$$

almost everywhere in $[0,1]$ and has the correct boundary values $x'(0) = A_0 x$ and $x'(1) = A_1 x$. In order to find such a solution we solve the integral equation

$$x(t) = Ax(t) + \lambda \int_0^t \int_0^1 \kappa(\tau, s) g\big(s, x(s)\big)\, \mathrm{d}s\, \mathrm{d}\tau \quad \text{for } 0 \le t \le 1 \tag{7.3.25}$$

instead of (7.3.3) in the space BV, where κ is again the Green's function (7.3.4), and the linear operator $A : BV \to BV$ is now given by

$$Ax := -\frac{1}{2}(1-t)^2 A_0 x + \frac{1}{2} t^2 A_1 x = v A_0 x + w A_1 x,$$

where

$$v(t) = -\frac{1}{2}(1-t)^2 \quad \text{and} \quad w(t) = \frac{1}{2} t^2 \quad \text{for } 0 \le t \le 1.$$

For $x \in AC^2$ the outer integral in (7.3.25) defines a differentiable function. Similarly as in Proposition 7.3.1 one may show that any function $x \in BV$ satisfying (7.3.25) is a solution in AC^2 to the boundary value problem (7.3.24). Note that for the first derivative we have

$$x'(t) = (1-t) A_0 x + t A_1 x + \lambda \int_0^1 \kappa(t, s) g\big(s, x(s)\big)\, \mathrm{d}s \quad \text{for } 0 \le t \le 1 \tag{7.3.26}$$

and so indeed $x'(0) = A_0 x$ and $x'(1) = A_1 x$.
Now, in order to solve (7.3.25) we can use Fubini's Theorem to reduce the double integral to a single one and transform the integral equation into

$$x(t) = Ax(t) + \lambda \int_0^1 \hat{\kappa}(t, s) g\big(s, x(s)\big)\, \mathrm{d}s \quad \text{for } 0 \le t \le 1, \tag{7.3.27}$$

where

$$\hat{\kappa}(t, s) = \int_0^t \kappa(\tau, s)\, \mathrm{d}\tau = \frac{1}{2} \begin{cases} s\big((2-t)t - s\big) & \text{for } 0 \le s \le t \le 1, \\ t^2(1-s) & \text{for } 0 \le t < s \le 1. \end{cases}$$

Consequently, under the hypotheses of Theorem 7.3.8 (with v and w as above), we may solve (7.3.27) and therefor also (7.3.25) exactly as we solved (BVP), namely with the help of Theorem 7.1.30.

Initial Value Problems

The theory developed so far in this section can also be used to solve initial value problems with coupled nonclassical initial conditions. Such problems will be treated now. Consider again the second order equation (7.3.1), that is,

$$x''(t) = -\lambda g\big(t, x(t)\big) \quad \text{for } 0 \le t \le 1, \tag{7.3.28}$$

but this time subject to the coupled *initial conditions*

$$x(0) = A_0 x \quad \text{and} \quad x'(0) = A_1 x, \tag{7.3.29}$$

where $A_0, A_1 : BV \to \mathbb{R}$ are given linear functionals. In the following we refer to the initial value problem (7.3.28) together with the initial conditions (7.3.29) by the symbol (IVP). In order to solve (IVP) we proceed similarly as we did to solve (BVP), but this time we use Theorem 7.2.15 instead of Theorem 7.1.30. To this purpose we consider the Volterra integral equation

$$x(t) = Ax(t) + \lambda \int_0^t \nu(t,s) g\big(s, x(s)\big)\, ds \quad \text{for } 0 \le t \le 1, \tag{7.3.30}$$

where the Volterra kernel ν is given by

$$\nu(t,s) = \begin{cases} s - t & \text{for } 0 \le s \le t \le 1, \\ 0 & \text{for } 0 \le t < s \le 1, \end{cases} \tag{7.3.31}$$

and A is a linear operator from BV into itself. The bridge between (7.3.30) and our (IVP) is now built by the following proposition which is a perfect analogue to Proposition 7.3.1.

Proposition 7.3.12. *Let $g : [0,1] \times \mathbb{R} \to \mathbb{R}$ satisfy (H11*(L_∞)), (H16) and (H17), and let $A : BV \to BV$ be defined by*

$$Ax(t) = A_0 x + tA_1 x \quad \textit{for } 0 \le t \le 1. \tag{7.3.32}$$

Then any function $x \in BV$ solving (7.3.30) is differentiable in $[0,1]$ and has an absolutely continuous derivative. Moreover, it satisfies (7.3.29) and solves (7.3.28) almost everywhere.
If, in addition, g is continuous in $[0,1] \times \mathbb{R}$, then x is of class C^2 and solves (IVP) everywhere on $[0,1]$.

Proof. Assume that (7.3.30) is satisfied for some $x \in BV$ and some $\lambda \in \mathbb{R}$. First observe that $h(s) := g(s, x(s))$ belongs to L_∞ because of (H11*(L_∞)), (H16) and (H17). Moreover, we set

$$\varphi(t) := \int_0^t \nu(t,s) g\big(s, x(s)\big)\, ds = \int_0^t (s-t) h(s)\, ds \quad \text{for } 0 \le t \le 1.$$

By [150], the function φ belongs to AC with

$$\varphi'(t) = -\int_0^t h(s)\,\mathrm{d}s \quad \text{for almost all } t \in [0,1],$$

but since the right hand side is again in AC we conclude that $\varphi \in C^1$ with $\varphi' \in AC$. Moreover, we obtain

$$\varphi''(t) = -h(t) \quad \text{for almost all } t \in [0,1].$$

In addition, by definition of A the function Ax is linear and hence of class C^2 with $(Ax)'' = \mathbb{0}$. From (7.3.30) follows

$$x(t) = Ax(t) + \lambda\varphi(t) \quad \text{for } 0 \leq t \leq 1;$$

in particular, this shows that (7.3.28) holds indeed almost everywhere in $[0,1]$. Moreover, since $\varphi(0) = \varphi'(0) = 0$, we obtain $x(0) = Ax(0) = A_0x$ and $x'(0) = (Ax)'(0) = A_1x$. Consequently, the first part of the proof is complete.
If, in addition, g is continuous, then so must be x'' which means that x is of class C^2 and solves (IVP) everywhere on $[0,1]$. ∎

Proposition 7.3.12 says that any $x \in BV$ solving (7.3.30) automatically belongs to AC^1. Consequently, in order to find a solution $x \in AC^1$ of (IVP) - and by this we mean a function $x \in AC^1$ that satisfies the initial conditions (7.3.29) and the equation (7.3.28) almost everywhere on $[0,1]$ - all we have to do is to make sure that the norms of the two linear functionals A_0 and A_1 behave in such a way that the norm of the iterate operator A^n, where A is now given by (7.3.32), shrinks below 1 for some $n \in \mathbb{N}$. In Theorem 7.3.2 (a) we have given such a criterion which was so strong that the operator A in (7.3.5) was a contraction. A similar criterion for the operator A in (7.3.32) would now read

$$\max\Big\{\,\|A_0\|_{BV\to\mathbb{R}}\,,\,\|A_0 + A_1\|_{BV\to\mathbb{R}}\,\Big\} + \|A_1\|_{BV\to\mathbb{R}} < 1. \tag{7.3.33}$$

Indeed, since $t \mapsto Ax(t)$ for fixed $x \in BV$ parameterizes a straight line through the points $(0, A_0x)$ and $(1, A_0x + A_1x)$ we can calculate its BV-norm explicitly and obtain

$$\begin{aligned}\|Ax\|_{BV} &= \|Ax\|_\infty + \mathrm{Var}(Ax) = \max\Big\{|A_0x|, |A_0x + A_1x|\Big\} + |A_1x| \\ &\leq \left(\max\Big\{\,\|A_0\|_{BV\to\mathbb{R}}\,,\,\|A_0 + A_1\|_{BV\to\mathbb{R}}\,\Big\} + \|A_1\|_{BV\to\mathbb{R}}\right)\|x\|_{BV}\,.\end{aligned}$$

Thus, the estimate (7.3.33) guarantees $\|A\|_{BV\to BV} < 1$, that is, A is a contraction.

Example 7.3.13. Consider the initial value problem

$$\left.\begin{aligned} x''(t) &= -\lambda g\big(t, x(t)\big) \quad \text{for } 0 \leq t \leq 1, \\ x(0) &= \tfrac{1}{7}x(1/2) + \tfrac{1}{6}x(2/3), \\ x'(0) &= \tfrac{1}{7}x(1/4) + \tfrac{1}{6}x(4/5), \end{aligned}\right\} \tag{7.3.34}$$

with g satisfying (H11*(L_∞)), (H16) and (H17). Note that the "right hand sides" of (7.3.34) are the same as in (7.3.7). Exactly as in Example 7.3.3 we define

$$A_0x := \tfrac{1}{7}x(1/2) + \tfrac{1}{6}x(2/3) \quad \text{and} \quad A_1x := \tfrac{1}{7}x(1/4) + \tfrac{1}{6}x(4/5).$$

Then A_0 and A_1 are bounded linear functionals on BV with

$$|A_0x|, |A_1x| \leq \tfrac{1}{7}\|x\|_\infty + \tfrac{1}{6}\|x\|_\infty \leq \tfrac{13}{42}\|x\|_{BV},$$

$$\begin{aligned} |A_0x + A_1x| &= \tfrac{1}{7}\big|x(1/2) + x(1/4)\big| + \tfrac{1}{6}\big|x(2/3) + x(4/5)\big| \\ &\leq \tfrac{2}{7}\|x\|_\infty + \tfrac{1}{3}\|x\|_\infty \leq \tfrac{13}{21}\|x\|_{BV}. \end{aligned}$$

Thus,

$$\max\Big\{\|A_0\|_{BV\to\mathbb{R}}, \|A_0 + A_1\|_{BV\to\mathbb{R}}\Big\} + \|A_1\|_{BV\to\mathbb{R}} \leq \frac{13}{14} < 1$$

which means that A_0 and A_1 satisfy (7.3.33). Accordingly, (7.3.34) has for small $|\lambda|$ an AC^1-solution, analogously to the boundary value problem (7.3.7). ◊

One may now think that option (a) of Theorem 7.3.2 always implies (7.3.33) or vice versa. But no such implication is true, as the following example shows.

Example 7.3.14. Define the two functionals $A_0, A_1 : BV \to \mathbb{R}$ by

$$A_0x := A_1x := \frac{1}{2}x(\tau)$$

for some $\tau \in [0, 1]$. Then $\|A_0\|_{BV\to\mathbb{R}} = \|A_1\|_{BV\to\mathbb{R}} = 1/2$, $\|A_0 - A_1\|_{BV\to\mathbb{R}} = 0$ and $\|A_0 + A_1\|_{BV\to\mathbb{R}} = 1$. Thus, the estimate (7.3.33) is violated, while (a) of Theorem 7.3.2 is fulfilled.

On the other hand, if we instead define

$$A_0x := \frac{1}{3}x(\tau) \quad \text{and} \quad A_1x := -\frac{1}{3}x(\tau)$$

again for some fixed $\tau \in [0, 1]$, then $\|A_0\|_{BV\to\mathbb{R}} = \|A_1\|_{BV\to\mathbb{R}} = 1/3$, $\|A_0 - A_1\|_{BV\to\mathbb{R}} = 2/3$ and $\|A_0 + A_1\|_{BV\to\mathbb{R}} = 0$. In this case, (a) of Theorem 7.3.2 is violated while (7.3.33) is fulfilled. ◊

We now jump to Theorem 7.3.8 and see how it looks like in the setting of (IVP). Since the structure of the linear operator A in (7.3.32) is covered by Proposition 7.3.6 we have a general method to calculate the spectral radius of A. Accordingly, we have the following analogue to Theorem 7.3.8.

Theorem 7.3.15. *Assume that the function $g : [0,1] \times \mathbb{R} \to \mathbb{R}$ satisfies (H11*(L_∞)*), (H16) and (H17). Moreover, assume that the linear functionals $A_0, A_1 : BV \to \mathbb{R}$ are bounded with*

$$\mathfrak{R}\begin{pmatrix} A_0\mathbb{1} & A_1\mathbb{1} \\ A_0 w & A_1 w \end{pmatrix} < 1, \tag{7.3.35}$$

where the function $w \in BV$ is defined by $w(t) = t$ for $t \in [0,1]$.
Then for each $R > 0$ there is some $\varrho > 0$ such that (IVP) has for fixed $\lambda \in (-\varrho, \varrho)$ a solution $x \in \mathbb{B}_R(AC^1)$. Here, the space AC^1 is equipped with the norm $\|\cdot\|_{BV}$.
If, in addition, g is continuous on $[0,1] \times \mathbb{R}$, then every such solution is of class C^2.

Proof. The argument is similar as in the proof of Theorem 7.3.8; we just have to change four things. The first is that we now consider the linear operator $A : BV \to BV$, defined by

$$Ax = \mathbb{1}A_0 x + wA_1 x \quad \text{for } x \in BV,$$

where $w(t) = t$. Second, we now use (7.3.35) together with Proposition 7.3.6 to guarantee that $\mathfrak{R}(A) < 1$ and that we therefore find some $n \in \mathbb{N}$ such that $\|A^n\|_{BV \to BV} < 1$. Third, Theorem 7.2.15 yields a solution $x \in BV$ of the Volterra equation (7.3.30). Fourth and finally, Proposition 7.3.12 tells us that x in fact belongs to AC^1 and solves (IVP). Moreover, it also tells us that if g is continuous, then x is of class C^2 and therefore a classical solution to (IVP). ■

Let us have a look back at Example 7.3.14.

Example 7.3.16. Consider the initial value problem

$$\left.\begin{aligned} x''(t) &= -\lambda g\big(t, x(t)\big) \quad \text{for } 0 \le t \le 1, \\ x(0) &= \tfrac{1}{2}x(1/2), \\ x'(0) &= \tfrac{1}{2}x(1/2), \end{aligned}\right\} \tag{7.3.36}$$

with g satisfying (H11*(L_∞)), (H16) and (H17). Using the functionals $A_0 x := \frac{1}{2}x(1/2)$ and $A_1 x := \frac{1}{2}x(1/2)$ which are precisely the functionals used in Example 7.3.14 for $\tau = 1/2$ we have seen there that (7.3.33) is violated; even worse, the operator $Ax = \mathbb{1}A_0 x + wA_1 x$ with $w(t) = t$ is no contraction in BV, as $\|A\mathbb{1}\|_{BV} = 1$. However,

$$\mathfrak{R}\begin{pmatrix} A_0\mathbb{1} & A_1\mathbb{1} \\ A_0 w & A_1 w \end{pmatrix} = \mathfrak{R}\begin{pmatrix} 1/2 & 1/2 \\ 1/4 & 1/4 \end{pmatrix} = \frac{3}{4} < 1$$

and so Theorem 7.3.15 tells us that (7.3.36) has for small $|\lambda|$ an AC^1-solution. ◇

Let us again look at the general kind of initial value problems we have considered so far in the previous examples. Similar to (7.3.20) they have the form

$$x(0) = ax(\tau_1) + bx(\tau_2) \quad \text{and} \quad x'(0) = cx(\sigma_1) + dx(\sigma_2), \tag{7.3.37}$$

where $a, b, c, d \in \mathbb{R}$ and $\tau_1, \tau_2, \sigma_1, \sigma_2 \in [0, 1]$ are given. The estimate (7.3.35) now says that Theorem 7.3.15 may be applied if and only if

$$\mathfrak{R}\begin{pmatrix} a+b & c+d \\ a\tau_1 + b\tau_2 & c\sigma_1 + d\sigma_2 \end{pmatrix} < 1. \tag{7.3.38}$$

From this it is easily seen that there is also no inclusion between the two estimates (7.3.21) and (7.3.38).

Example 7.3.17. Let $\tau_1 = \sigma_1 = 1/3$, $\tau_2 = \sigma_2 = 2/3$ and $c = d = 0$. Then the matrix in (7.3.21) becomes

$$\begin{pmatrix} 2a/3 + b/3 & 0 \\ a/3 + 2b/3 & 0 \end{pmatrix}$$

with spectral radius $|2a + b|/3$, whereas the matrix in (7.3.38) turns into

$$\begin{pmatrix} a+b & 0 \\ a/3 + 2b/3 & 0 \end{pmatrix}$$

with spectral radius $|a + b|$. Consequently, for $a = -1/2$ and $b = 5/2$ the spectral radius of the first matrix is $1/2$ while that of the second matrix is 2. For $a = 11/2$ and $b = -5$, however, it is exactly the other way round. ♢

Similarly as we did for boundary value problems instead of considering "local" initial conditions of type (7.3.37), we may also look at "global" initial conditions like, for instance,

$$x(0) = \int_0^1 k_0(s)x(s)\,\mathrm{d}s \quad \text{and} \quad x'(0) = \int_0^1 k_1(s)x(s)\,\mathrm{d}s,$$

where $k_0, k_1 \in L_1$. Then A_0 and A_1 are defined by

$$A_0x = \int_0^1 k_0(s)x(s)\,\mathrm{d}s \quad \text{and} \quad A_1x = \int_0^1 k_1(s)x(s)\,\mathrm{d}s$$

and thus Theorem 7.3.15 may be applied if and only if

$$\mathfrak{R}\begin{pmatrix} \int_0^1 k_0(s)\,\mathrm{d}s & \int_0^1 k_1(s)\,\mathrm{d}s \\ \int_0^1 k_0(s)s\,\mathrm{d}s & \int_0^1 k_1(s)s\,\mathrm{d}s \end{pmatrix} < 1.$$

Example 7.3.18. Consider the initial value problem

$$\left.\begin{aligned} x''(t) &= -\lambda g\big(t, x(t)\big) \quad \text{for } 0 \le t \le 1, \\ x(0) &= x'(0) = \int_0^1 sx(s)\,\mathrm{d}s. \end{aligned}\right\} \tag{7.3.39}$$

We then have $k_0(s) = k_1(s) = s$ and

$$\mathfrak{R}\begin{pmatrix} \int_0^1 k_0(s)\,\mathrm{d}s & \int_0^1 k_1(s)\,\mathrm{d}s \\ \int_0^1 k_0(s)s\,\mathrm{d}s & \int_0^1 k_1(s)s\,\mathrm{d}s \end{pmatrix} = \mathfrak{R}\begin{pmatrix} 1/2 & 1/2 \\ 1/3 & 1/3 \end{pmatrix} = \frac{5}{6} < 1.$$

Thus, under the assumptions of Theorem 7.3.15 we find for small $|\lambda|$ a solution $x \in AC^1$. ♢

As a last remark we point out that condition (7.3.35) is *not* necessary for (IVP) to have a solution in AC^1.

Example 7.3.19. Consider the initial value problem

$$\left.\begin{aligned} x''(t) &= -4\lambda\left(4t^2x(t)+\sqrt{1-x(t)^2}\right) \quad \text{for } 0 \le t \le 1, \\ x(0) &= \sqrt{2}\,x\left(\sqrt{\pi/8}\right), \\ x'(0) &= 0. \end{aligned}\right\} \tag{7.3.40}$$

In the notation of (7.3.37) we have $a = \sqrt{2}$, $b = c = d = 0$ and $\tau_1 = \sqrt{\pi/8}$. Thus, the spectral radius of the matrix in (7.3.38) becomes

$$\mathfrak{R}\begin{pmatrix} \sqrt{2} & 0 \\ \sqrt{\pi}/2 & 0 \end{pmatrix} = \sqrt{2} > 1.$$

Nevertheless, it is easy to check that the function $x(t) = \cos\left(2t^2\right)$ is an (even analytic) solution to the initial value problem (7.3.40) for $\lambda = 1$. ◇

To conclude, the situation here is basically the same as for (BVP).

References

[1] A. Acosta, W. Aziz, J. Matkowski, N. Merentes: *Uniformly continuous composition operator in the space of φ-variation functions in the sense of Riesz.* Fasc. Math. **43** (2010), pp. 5–11.

[2] P. S. Aleksandrov: *Introduction to Set Theory and Theory of Functions.* Russian: Nauka, Moscow (1948). Engl. transl.: Verlag H. Deutsch, Frankfurt (1994).

[3] T. Andô: *On products of Orlicz spaces.* Math. Ann. **140**, no. 3 (1960), pp. 174–186.

[4] J. Appell: *Analysis in Beispielen und Gegenbeispielen.* Springer, Heidelberg (2009).

[5] J. Appell: *Higher Analysis in Examples and Counterexamples* (to appear).

[6] J. Appell, J. Banaś, N. Merentes: *Bounded variation and around.* De Gruyter, Berlin (2013).

[7] J. Appell, T. D. Benavides: *Nonlinear Hammerstein equations and functions of bounded Riesz-Medvedev variation.* Topol. Methods Nonlinear Anal. **47** (2016), pp. 319–332.

[8] J. Appell, N. Guanda, N. Merentes, J. L. Sánchez: *Boundedness and continuity properties of nonlinear composition operators: A survey.* Comm. Appl. Anal. **15** (2011), pp. 153–182.

[9] J. Appell, N. Merentes, J. L. S. Hernández: *Locally Lipschitz composition operators in spaces of functions of bounded variation.* Ann. Mat. Pura Appl. **190** (2011), pp. 33–43.

[10] J. Appell, S. Reinwand: *Funktionen mit Stammfunktion I: Charakterisierungen und Eigenschaften.* Math. Semesterber. **65**, no. 2 (2018), pp. 195–210.

[11] J. Appell, A.-K. Roos: *Stetigkeit, Monotonie, Oszillation, Variation: Naheliegendes und Überraschendes.* Math. Semesterber. **61**, no. 2 (2014), pp. 233–248.

[12] J. Appell, P. P. Zabrejko: *Nonlinear Superposition Operators.* Cambridge Tracts in Math. **95** (1990).

[13] J. Appell, P. P. Zabrejko: *Remarks on the composition operator problem in various function spaces.* Complex Var. Elliptic Equ. **55**, no. 8 (2010), pp. 727–737.

[14] C. Arzelà: *Intorno alla continuità della somma di infinità di funzioni continue.* Rend. Sess. Accad. Sci. Ist. Bologna (1883/1884), pp. 79–84.

[15] C. Arzelà: *Sulle serie di funzioni.* Mem. R. Accad. Sci. Ist. Bologna **5**, no. 8 (1900), pp. 131–186, 701–744.

[16] F. R. Astudillo-Villaba, R. E. Castillo, J. C. Ramos-Fernández: *Multiplication Operators on the Spaces of Functions of Bounded p-Variation in Wiener's Sense.* Real Anal. Exchange **42**, no. 2 (2017), pp. 329–44.

[17] F. Astudillo-Villalba, J. Ramos-Fernández: *Multiplication operators on the space of functions of bounded variation.* Demonstratio Math. **50**, no. 1 (2017), pp. 105–115.

[18] W. Aziz, J. A. Guerrero, K. Maldonado, N. Merentes: *Locally Defined Operators in the Space of Functions of Bounded Riesz-Variation.* J. Math. (Hindawi) **2015** (2015), pp. 1–4.

[19] A. Baernstein: *On the Fourier series of functions of bounded Φ-variation.* Studia Math. **42** (1972), pp. 243–248.

[20] J. Borsik: *Maximal classes for some families of Darboux-like and quasicontinuous-like functions.* In: Monograph on the occasion of the 100th birthday anniversary of Zygmunt Zahorski, Wydawnictwo Politechniki Slaskiej, Gliwice (2015).

[21] F. Brauer: *Nonlinear age-dependent population growth under harvesting.* Comput. Math. Appl. **9**, no. 3 (1883), pp. 345–352.

[22] A. M. Bruckner: *Differentiation of Real Functions.* Springer, Berlin, Heidelberg, New York (1978).

[23] A. M. Bruckner, J. G. Ceder: *Darboux Continuity.* Jahresber. Deutsch. Math.-Verein. **67** (1965), pp. 93–117.

[24] A. M. Bruckner, J. Mařík, C. E. Weil: *Some Aspects of Products of Derivatives.* Amer. Math. Monthly **99**, no. 2 (1992), pp. 134–145.

[25] D. Bugajewska: *On the superposition operator in the space of functions of bounded variation, revisited.* Math. Comput. Modelling **52** (2010), pp. 791–796.

[26] D. Bugajewska, D. Bugajewski, P. Kasprzak, P. Maćkowiak: *Nonautonomous superposition operators in the space of functions of bounded variation.* Topol. Methods Nonlinear Anal. **48** (2016), pp. 637–660.

[27] D. Bugajewska, G. Infante, P. Kasprzak: *Solvability of Hammerstein Integral Equations with Applications to Boundary Value Problems.* Z. Anal. Anwend. **36** (2017), pp. 393–417.

[28] D. Bugajewska, S. Reinwand: *Some Remarks on Multiplier Spaces II: BV-Type Spaces.* Z. Anal. Anwend. **38**, no. 3 (2019), pp. 309–327.

[29] D. Bugajewski: *On BV-solutions of some nonlinear integral equations.* Integral Equations Operator Theory **46** (2003), pp. 387–398.

[30] D. Bugajewski, K. Czudek, J. Gulgowski, J. Sadowski: *On some nonlinear operators in ΛBV-spaces.* J. Fixed Point Theory Appl. **19** (2017), pp. 2785–2818.

[31] D. Bugajewski, J. Gulgowski, P. Kasprzak: *On continuity and compactness of some nonlinear operators in the spaces of functions of bounded variation.* Ann. Mat. Pura Appl. **195** (2016), pp. 1513–1530.

[32] D. Bugajewski, J. Gulgowski, P. Kasprzak: *On integral operators and nonlinear integral equations in the spaces of functions of bounded variation.* J. Math. Anal. Appl. **444**, no. 1 (2016), pp. 230–250.

[33] D. Bugajewski, D. O'Regan: *Existence results for BV-solutions of nonlinear integral equations.* J. Integral Equations Appl. **15**, no. 4 (2003), pp. 343–357.

[34] R. Caccioppoli: *Sulle quadratura delle superficie piane e curve.* Atti Accad. Naz. Lincei Rend. Lincei Mat. Appl. **6**, no. 6 (1927), pp. 142–146.

[35] R. Caccioppoli: *Sulle coppie di funzioni a variazione limitata.* Rc. Accad. Sci. fis. mat. Napoli **34** (1928), pp. 83–88.

[36] P. Calabrese: *A note on alternating series.* Amer. Math. Monthly **69** (1962), pp. 215–217.

[37] C. Carathéodory: *Vorlesungen über reelle Funktionen.* De Gruyter, Leipzig, Berlin (1918).

[38] A. Caserta, G. Maio, L. Holá: *Arzelà's Theorem and strong uniform convergence on bornologies.* J. Math. Anal. Appl. **371** (2010), pp. 384–392.

[39] V. G. Celidze, A. G. Dzvaršeišvili: *The Theory of the Denjoy Integral and Some Applications.* Russian: Tbilisi Univ. Press, Tbilisi (1978). Engl. Translation: World Scientific, Singapore, 1989.

[40] H. C. Chaparro: *On multipliers between bounded variation spaces.* Ann. Funct. Anal. **9**, no. 3 (2018), pp. 376–383.

[41] F. Cianciaruso, G. Infante, P. Pietramala: *Solutions to perturbed Hammerstein integral equations with applications.* Nonlinear Anal. Real World Appl. **33** (2017), pp. 317–347.

[42] J. Ciemnoczołowski, W. Orlicz: *Inclusion theorems for classes of functions of generalized bounded variation.* Ann. Soc. Math. Pol. Series I: Comment. Math. **24** (1984), pp. 181–194.

[43] K. Ciesielski: *Characterizing Derivatives by Preimages of Sets.* Real Anal. Exchange **23** (1999).

[44] E. Conway, J. Smoller: *Global solutions of the cauchy problem for quasilinear first-order equations in several space variables.* Comm. Pure Appl. Math. **19**, no. 1 (1966), pp. 95–105.

[45] H. T. Croft: *A Note on a Darboux Continuous Function.* J. Lond. Math. Soc. **s1-38**, no. 1 (1963), pp. 9–10.

[46] H. H. D. Bugajewska, D. Bugajewski: *BV_φ-solutions of nonlinear integral equations.* J. Math. Anal. Appl. **287** (2003), pp. 265–278.

[47] G. Darbo: *Punti uniti in transformazioni a condominio non compacto.* Rend. Semin. Mat. Univ. Padova **24** (1955), pp. 84–92.

[48] E. De Giorgi: *Su una teoria generale della misura $(r-1)$-dimensionale in uno spazio adr dimensioni.* Ann. Mat. Pura Appl. **36** (1954), pp. 191–213.

[49] B. Dhage: *On a fixed point theorem of Krasnoselkii-Schaefer type.* Electron. J. Qual. Theory Differ. Equ. **6** (2002), pp. 1–9.

[50] U. Dini: *Sulle serie a termini positivi.* Ann. Univ. Toscane **9** (1867), pp. 29–69.

[51] U. Dini: *Fondamenti per la teorica delle funzioni di variabili reali.* T. Nistri e C., Pisa (1878).

[52] J. Dirichlet: *Sur la convergence des séries trigonométriques qui servent à représenter une fonction arbitraire entre des limites données.* J. Reine Angew. Math. **4** (1829), pp. 157–169.

[53] R. Dougherty, A. S. Kechris: *The complexity of antidifferentiation.* Adv. Math. **88**, no. 2 (1991), pp. 145 – 169.

[54] R. Drozdowski, J. Jędrzejewski, A. Sochaczweska: *On the Almost Uniform Convergence.* Sci. Iss. Math. Jan Długosz University Częstochowa **18** (2013), pp. 11–17.

[55] R. M. Dudley, R. Norvaiša: *Concrete Functional Calculus.* Springer, New York (2010).

[56] L. C. Evans: *Partial differential equations.* American Mathematical Society, Providence, R.I. (2010).

[57] H. Fast: *Une remarque sur la propriété de Weierstrass.* Colloq. Math. **7**, no. 1 (1959), pp. 75–77.

[58] H. Federer: *Geometric Measure Theory.* Springer Berlin Heidelberg, first edition (1996).

[59] R. Fleissner: *Distant bounded variation and products of derivatives.* Fund. Math. **94** (1977), pp. 65–73.

[60] R. Fleissner: *A note on Baire 1 Darboux functions.* Real Anal. Exchange **3** (1978), pp. 104–106.

[61] A. Fonda: *The Kurzweil-Henstock Integral for Undergraduates: A Promenade Along the Marvelous Theory of Integration.* Birkhäuser, Basel (2018).

[62] J. Fourier: *Théorie analytique de la chaleur.* F. Didot, Paris (1881).

[63] D. Fraňková: *Regulated functions.* Math. Bohem. **116**, no. 1 (1991), pp. 20–59.

[64] C. Freiling: *On the Problem of Characterizing Derivatives.* Real Anal. Exchange **23**, no. 2 (1999), pp. 805–812.

[65] O. E. Galkin: *Sequences of composition operators in spaces of functions of bounded Φ-variation.* Math. Notes **85**, no. 3 (2009), pp. 328–339.

[66] R. G. Gibson, T. Natkaniec: *Darboux like functions.* Real Anal. Exchange **22**, no. 2 (1996), pp. 492–533.

[67] M. L. Goldman, P. P. Zabrejko: *Theorems about the integrability of products of functions for the Kurzweil-Henstock integral.* Dokl. Nats. Akad. Nauk Belarusi **59**, no. 6 (2015), pp. 1–9. (Russian).

[68] R. A. Gordon: *The integrals of Lebesgue, Denjoy, Perron and Henstock.* Grad. Stud. Math. Vol. 4 (1994).

[69] J. A. Guerrero, O. Mejía, N. Merentes: *Locally Defined Operators and Locally Lipschitz Composition Operators in the Space $WBV_{p(\cdot)}([a,b])$.* Adv. Pure Math. **6** (2016), pp. 727–744.

[70] M. E. Gurtin, R. C. MacCamy: *Nonlinear Age-dependent Population Dynamics.* Arch. Ration. Mech. Anal. **54** (1974), pp. 281–300.

[71] E. Helly: *Über lineare Funktionaloperationen.* Österreich. Akad. Wiss. Math.-Natur. Kl. Sitzungsber. II **121** (1912), pp. 265–297.

[72] H. Heuser: *Lehrbuch der Analysis, Teil 1.* Teubner, Stuttgart (1993).

[73] L. Holá, T. Šalát: *Graph convergence, uniform, quasi-uniform and continuous convergence and some characterizations of compactness.* Acta Math. Univ. Comenian **54–55** (1988), pp. 121–131.

[74] C. Jordan: *Sur la série de Fourier.* C. R. Math. Acad. Sci. Paris **92** (1881), pp. 228–230.

[75] M. Josephy: *Composing Functions of Bounded Variation.* Proc. Amer. Math. Soc. **83**, no. 2 (1981), pp. 354–356.

[76] R. Kannan, C. K. Krueger: *Advanced Analysis on the Real Line.* Springer, Berlin (1996).

[77] K. Knopp: *Theorie und Anwendung der unendlichen Reihen,* volume 2. Springer, Berlin (1922).

[78] G. Köhler: *Analysis.* Heldermann, Lemgo (2006).

[79] P. Kolwicz, K. Leśnik, L. Maligranda: *Pointwise multipliers of Calderón-Lozanovskii spaces.* Math. Nachr. **286**, no. 8–9 (2013), pp. 876–907.

[80] P. Kolwicz, K. Leśnik, L. Maligranda: *Pointwise products of some Banach function spaces and factorization.* J. Funct. Anal. **266**, no. 2 (2014), pp. 616–659.

[81] P. Kolwicz, K. Leśnik, L. Maligranda: *Symmetrization, factorization and arithmetic of quasi-Banach function spaces.* J. Math. Anal. Appl. **470** (2019), pp. 1136–1166.

[82] M. A. Krasnosel'skij: *On a fixed point principle.* Tr. Semin. Funk. Anal. Voron. Gos. Univ. **6** (1958), pp. 87–90. (Russian).

[83] C. Kuratowski: *Sur les espaces complets.* Fund. Math. **15** (1930), pp. 301–309.

[84] D. S. Kurtz, C. W. Swartz: *Theories Of Integration: The Integrals Of Riemann, Lebesgue, Henstock-Kurzweil, and McShane.* World Scientific (2004).

[85] P. D. Lax: *Functional Analysis.* Wiley-Interscience, New York (2002).

[86] H. Lebesgue: *Sur l'intégration des fonctions discontinues.* Ann. Sci. Éc. Norm. Supér. **3**, no. 27 (1910), pp. 361–450.

[87] P.-Y. Lee: *Lanzhou Lectures on Henstock Integration.* World Scientific, Singapore (1989).

[88] T. Y. Lee: *Henstock-Kurzweil Integration on Euclidean Spaces.* World Scientific, Singapore (2011).

[89] K. Lichawski, J. Matkowski, J. Miś: *Locally defined operators in the space of differentiable functions.* Bull. Pol. Acad. Sci. Math. **37** (1989), pp. 315–325.

[90] A. LINDENBAUM: *Sur quelques propriétés des fonctions de variable réelle.* Ann. Soc. Pol. Math. **6** (1927), pp. 129–130.

[91] S. ŁOJASIEWICZ, A. V. FERREIRA: *An Introduction to the Theory of Real Functions.* Wiley, Chichester (West Sussex), New York (1988).

[92] E. R. LOVE, L. C. YOUNG: *Sur une classe de fonctionnelles linéaires.* Fund. Math. **28** (1937), pp. 243–257.

[93] M. LUPA: *Form of Lipschitzian operator of substitution in some class of functions.* Sci. Bull. Łodzka Technical Univ. **21** (1989), pp. 87–96.

[94] A. G. LYAMIN: *On the acting problem for the Nemytskij operator in the space of functions of bounded variation.* 11th School Theory Oper. Function Spaces, Chel'yabinsk (1986), pp. 63–64. (Russian).

[95] P. MAĆKOWIAK: *A counterexample to Lyamin's theorem.* Proc. Amer. Math. Soc. **142**, no. 5 (2014), pp. 1773–1776.

[96] P. MAĆKOWIAK: *On the continuity of superposition operators in the space of functions of bounded variation.* Aequationes Math. **91**, no. 4 (2017), pp. 759–777.

[97] L. MALIGRANDA: *Orlicz Spaces and Interpolation.* Sem. Mat. Univ. Campinas **5** (1989), pp. 1–206.

[98] L. MALIGRANDA, E. NAKAI: *Pointwise multipliers of Orlicz spaces.* Arch. Math. **95**, no. 3 (2010), pp. 251–256.

[99] L. MALIGRANDA, L. E. PERSSON: *Generalized duality of some Banach function spaces.* Indag. Math. **51**, no. 3 (1989), pp. 323–338.

[100] L. MALIGRANDA, W. WNUK: *Landau type theorem for variable Lebesgue spaces.* Comment. Math. **55**, no. 2 (2015), pp. 119–126.

[101] M. MARCUS, V. J. MIZEL: *Complete characterization of functions which act, via superposition, on Sobolev spaces.* Trans. Amer. Math. Soc. **251** (1979), pp. 187–218.

[102] L. MARKUS, N. R. AMUNDSON: *Nonlinear Boundary-Value Problems Arising in Chemical Reactor Theory.* J. Differential Equations **4** (1968), pp. 102–113.

[103] A. MATKOWSKA: *On characterization of Lipschitzian operators of substitution in the class of Hölder continuous functions.* Sci. Bull. Łodzka Tech. Univ. **17** (1984), pp. 81–85.

[104] A. MATKOWSKA, J. MATKOWSKI, N. MERENTES: *Remark on globally Lipschitz composition operators.* Demonstratio Math. **28**, no. 1 (1995), pp. 171–175.

[105] J. Matkowski: *Form of Lipschitz operators of substitution in Banach spaces of differentiable functions.* Sci. Bull. Łodzka Tech. Univ. **17** (1985), pp. 5–10.

[106] J. Matkowski: *Characterization of globally Lipschitzian composition operators in the Sobolev space* $W_p^n[a,b]$. Sci. Bull. Łodzka Tech. Univ. **24** (1993), pp. 90–99.

[107] J. Matkowski, N. Merentes: *Characterization of globally Lipschitzian composition operators in the Banach space* $BV_p^1[a,b]$. Arch. Math. (Basel) **28** (1992), pp. 181–186.

[108] J. Matkowski, J. Miś: *On a Characterization of Lipschitzian Operators of Substitution in the Space* $BV[a,b]$. Math. Nachr. **117**, no. 1 (1984), pp. 155–159.

[109] J. Matkowski, M. Wróbel: *Locally defined operators in the space of Whitney differentiable functions.* Nonlinear Anal. **68** (2008), pp. 2873–3232.

[110] J. Matkowski, M. Wróbel: *Represetation theorem for locally defined operators in the space of Whitney differentiable functions.* Manuscripta Math. **129** (2009), pp. 437–448.

[111] J. Mařík: *Multipliers of summable derivatives.* Real Anal. Exchange **8**, no. 2 (1982), pp. 486–493.

[112] J. Mařík: *Transformation and multiplication of derivatives.* Contemp. Math. **42** (1985), pp. 119–134.

[113] J. Mařík, C. E. Weil: *Multipliers of spaces of derivatives.* Math. Bohem. **129**, no. 2 (2004), pp. 181–217.

[114] I. Maximoff: *On continuous transformation of some functions into an ordinary derivative.* Ann. Sc. Norm. Super. Pisa Cl. Sci. **Ser. 2, 12**, no. 3–4 (1947), pp. 147–160.

[115] S. Mazurkiewicz: *Über die Menge der differenzierbaren Funktionen.* Fund. Math. **27**, no. 1 (1936), pp. 244–249.

[116] R. Meise, D. Vogt: *Einführung in die Funktionalanalysis.* Vieweg (1992).

[117] O. Mejía, N. Merentes, B. Rzepka: *Locally Lipschitz Composition Operators in Space of the Functions of Bounded* $\kappa\Phi$*-Variation.* J. Funct. Spaces (Hindawi) **2014** (2014), pp. 1–8.

[118] A. P. Morse: *Convergence in variation and related topics.* Trans. Amer. Math. Soc. **41**, no. 1 (1937), pp. 48–83.

[119] M. Muresan: *A Concrete Approach to Classical Analysis.* Springer New York (2009).

[120] J. Musielak: *Orlicz Spaces and Modular Spaces.* Springer Berlin, Heidelberg (1983).

[121] J. Musielak, W. Orlicz: *On generalized variations (I).* Studia Math. **18** (1959), pp. 11–41.

[122] T. Natkaniec: *On quasi-continuous functions having Darboux property.* Math. Pannon. **3** (1992), pp. 81–96.

[123] T. Natkaniec, W. Orwat: *Variations on Products and Quotients of Darboux Functions.* Real Anal. Exchange **15**, no. 1 (1989), pp. 193–202.

[124] S. M. Nicu Boboc: *Sur la détermination d'une fonction par les valeurs prises sur un certain ensemble.* Ann. Sci. Éc. Norm. Supér. **3e série, 76**, no. 2 (1959), pp. 151–159.

[125] O. A. Oleinik: *Discontinuous solutions of non-linear differential equations.* Uspekhi Mat. Nauk **12**, no. 3 (1957), pp. 3–73.

[126] O. A. Oleinik: *Construction of a generalized solution of the Cauchy problem for a quasi-linear equation of first order by the introduction of "vanishing viscosity".* Uspekhi Mat. Nauk **14**, no. 2 (1959), pp. 159–164.

[127] R. O'Neil: *Fractional Integration in Orlicz Spaces I.* Trans. Amer. Math. Soc. **115** (1965), pp. 300–328.

[128] J. C. Oxtoby: *Measure and Category.* Springer, Berlin (1980).

[129] S. Perlman, D. Waterman: *Some remark on functions of Λ-bounded variation.* Proc. Amer. Math. Soc. **47**, no. 1 (1979), pp. 113–118.

[130] P. B. Pierce, D. Waterman: *On the invariance of classes ΦBV, ΛBV under composition.* Proc. Amer. Math. Soc. **132**, no. 3 (2003), pp. 755–760.

[131] D. Preiss: *Algebra generated by derivatives.* Real Anal. Exchange **8**, no. 1 (1982), pp. 208–216.

[132] D. Preiss, M. Tartaglia: *On Characterizing Derivatives.* Proc. Amer. Math. Soc. **123**, no. 8 (1995), pp. 2417–2420.

[133] T. Radakovič: *Über Darbouxsche und stetige Funktionen.* Monatsh. Math. Phys. **38** (1931), pp. 117–122.

[134] S. Reinwand: *Types of Convergence Which Preserve Continuity.* Real Anal. Exchange **45**, no. 1 (to appear).

[135] F. Riesz: *Untersuchungen über Systeme integrierbarer Funktionen.* Math. Ann. **69** (1910), pp. 449–497.

[136] F. Riesz: *Sur certains systèmes singuliers d'équations intégrales.* Ann. Sci. Éc. Norm. Supér. **28** (1911), pp. 33–62.

[137] H. E. Robbins: *A Note on the Riemann Integral.* Amer. Math. Monthly **50**, no. 10 (1943), pp. 617–618.

[138] S. Saks: *Theory of the Integral.* Dover Publications Inc., New York, 2nd revised edition (1964).

[139] R. Salem: *Essais sur les séries trigonométriques.* Thèses de l'entre-deux-guerres 230, Hermann & cie, Paris (1940).

[140] R. Showalter: *Hilbert Space Methods in Partial Differential Equations.* Monograph in: Electron. J. Differential Equations (1998).

[141] W. Sierpiński: *Sur une propriété de fonctions réelles quelconques définies dans les espaces métriques.* Matematiche (Catania) **8**, no. 2 (1953), pp. 73–78.

[142] G. F. Simmons, J. K. Hammitt: *Introduction to topology and modern analysis.* McGraw-Hill, New York (1963).

[143] R. M. Solovay: *A model of set-theory in which every set of reals is Lebesgue measurable.* Ann. Mat. Pura Appl. **2**, no. 92 (1970), pp. 1–56.

[144] P. Szczuka: *Sums and Products of Darboux Internally Quasi-Continuous Functions.* An. Ştiinţ. Univ. Al. I. Cuza Iaşi Mat. (N.S.) **2** (2016), pp. 493–500.

[145] S. Szufla: *On the structure of solution sets of nonlinear equations.* Proc. Conf. Diff. Equ. Optim. Contr., Zielona Góra (1989), pp. 33–39.

[146] B. S. Thomson: *On Riemann Sums.* Real Anal. Exchange **37**, no. 1 (2011), pp. 221–242.

[147] A. C. M. van Rooij, W. H. Schikhof: *A Second Course on Real Functions.* Cambridge University Press (1982).

[148] G. Vitali: *Sul problema della misura dei gruppi di punti di una retta.* Tip. Gamberini e Parmeggiani, Bologna (1905).

[149] V. Volterra: *Sui principii del Calcolo integrale.* Giornale Mat. Battaglini **14** (1881), pp. 333–372.

[150] J. Šremr: *On Differentiation of a Lebesgue Integral with respect to a Parameter.* Math. Appl. **1** (2012), pp. 91–116.

[151] D. Waterman: *On convergence of Fourier series of functions of generalized bounded variation.* Studia Math. **44** (1972), pp. 107–117.

[152] D. Waterman: *On Λ-bounded variation.* Studia Math. **57** (1976), pp. 33–45.

[153] D. Werner: *Funktionalanalysis.* Springer Spektrum, eighth edition (2018).

[154] N. Wiener: *The Quadratic Variation of a Function and its Fourier Coefficients.* J. Math. Phys. **3**, no. 2 (1924), pp. 72–94.

[155] J. Wolff: *On the quasi-uniform convergence.* Proc. R. Neth. Acad. Arts Sci. **22** (1920), pp. 650–655.

[156] M. Wróbel: *Locally defined operators and a partial solution of a conjecture.* Nonlinear Anal. **72** (2010), pp. 495–506.

[157] M. Wróbel: *Representation theorem for local operators in the space of continuous and monotone functions.* J. Math. Anal. Appl. **372** (2010), pp. 45–54.

[158] M. Wróbel: *Locally Defined Operators in the Space of Functions of Bounded φ-Variation.* Real Anal. Exchange **38**, no. 1 (2012/2013), pp. 79–94.

[159] L. C. Young: *An Inequality of the Hölder-type, connected with Stieltjes Integration.* Acta Math. **67** (1936), pp. 251–282.

[160] L. C. Young: *Sur une généralisation de la notion de variation de puissance p-ième bornée au sens de M. Wiener, et sur la convergence des séries de Fourier.* C. R. Math. Acad. Sci. Paris **204** (1937), pp. 470–472.

[161] L. C. Young: *General inequalities for Stieltjes integrals and the convergence of Fourier series.* Math. Ann. **115** (1938), pp. 581–612.

[162] W. H. Young: *A theorem in the theory of functions of a real variable.* Rend. Circ. Mat. Palermo **24** (1907), pp. 187–192.

[163] W. H. Young: *On the distinction of right and left at points of discontinuity.* Quart. J. Pure Appl. Math. **39** (1908), pp. 67–83.

www.ingramcontent.com/pod-product-compliance
Ingram Content Group UK Ltd.
Pitfield, Milton Keynes, MK11 3LW, UK
UKHW021652190726
13853UKWH00001B/208

9 783736 974036